DATE DUE

VOLUME SEVENTY SEVEN

VITAMINS AND HORMONES
Ghrelin

VITAMINS AND HORMONES

Editorial Board

TADHG P. BEGLEY
ANTHONY R. MEANS
BERT W. O'MALLEY
LYNN RIDDIFORD
ARMEN H. TASHJIAN, JR.

VOLUME SEVENTY SEVEN

VITAMINS AND HORMONES
Ghrelin

Editor-in-Chief

GERALD LITWACK
*Former Professor and Chair
Department of Biochemistry and Molecular Pharmacology
Thomas Jefferson University Medical College
Philadelphia, Pennsylvania*

*Former Visiting Scholar
Department of Biological Chemistry
David Geffen School of Medicine at UCLA
Los Angeles, California*

AMSTERDAM • BOSTON • HEIDELBERG • LONDON
NEW YORK • OXFORD • PARIS • SAN DIEGO
SAN FRANCISCO • SINGAPORE • SYDNEY • TOKYO
Academic Press is an imprint of Elsevier

Cover photo credit: PDB ID: 1p7x ("Theoretical Model of Human Ghrelin Precursor")
V. Umashankar, Vels College of Science, affiliated to
University of Madras, India
Ab Initio Structure of Human Ghrelin Precursor

Academic Press is an imprint of Elsevier
84 Theobalad's Road, London WC1X 8RR, UK
Radarweg 29, PO Box 211, 1000 AE Amsterdam, The Netherlands
Linacre House, Jordan Hill, Oxford OX2 8DP, UK
30 Corporate Drive, Suite 400, Burlington, MA 01803, USA
525 B Street, Suite 1900, San Diego, CA 92101-4495, USA

First edition 2008

Copyright © 2008 Elsevier Inc. All rights reserved.

No part of this publication may be reproduced, stored in a retrieval system or transmitted in any form or by any means electronic, mechanical, photocopying, recording or otherwise without the prior written permission of the publisher

Permissions may be sought directly from Elsevier's Science & Technology Rights Department in Oxford, UK: phone (+44) (0) 1865 843830; fax (+44) (0) 1865 853333; email: permissions@elsevier.com. Alternatively you can submit your request online by visiting the Elsevier web site at http://elsevier.com/locate/permissions, and selecting *Obtaining permission to use Elsevier material*

Notice

No responsibility is assumed by the publisher for any injury and/or damage to persons or property as a matter of products liability, negligence or otherwise, or from any use or operation of any methods, products, instructions or ideas contained in the material herein. Because of rapid advances in the medical sciences, in particular, independent verification of diagnoses and drug dosages should be made

ISBN: 978-0-12-373685-7
ISSN: 0083-6729

For information on all Academic Press publications
visit our website at books.elsevier.com

Printed and bound in USA
07 08 09 10 11   10 9 8 7 6 5 4 3 2 1

Working together to grow
libraries in developing countries

www.elsevier.com | www.bookaid.org | www.sabre.org

ELSEVIER   BOOK AID International   Sabre Foundation

Former Editors

ROBERT S. HARRIS
Newton, Massachusetts

JOHN A. LORRAINE
University of Edinburgh
Edinburgh, Scotland

PAUL L. MUNSON
University of North Carolina
Chapel Hill, North Carolina

JOHN GLOVER
University of Liverpool
Liverpool, England

GERALD D. AURBACH
Metabolic Diseases Branch
National Institute of
Diabetes and Digestive and
Kidney Diseases
National Institutes of Health
Bethesda, Maryland

KENNETH V. THIMANN
University of California
Santa Cruz, California

IRA G. WOOL
University of Chicago
Chicago, Illinois

EGON DICZFALUSY
Karolinska Sjukhuset
Stockholm, Sweden

ROBERT OLSEN
School of Medicine
State University of New York
at Stony Brook
Stony Brook, New York

DONALD B. MCCORMICK
Department of Biochemistry
Emory University School of
Medicine, Atlanta, Georgia

Contents

Contributors	*xiii*
Preface	*xvii*

1. The Structure of Ghrelin — 1
Andreas Kukol

 I. Introduction — 2
 II. Background — 2
 III. Results of Structural Studies — 5
 IV. Discussion — 9
 V. Conclusions — 11
 References — 11

2. Biochemistry of Ghrelin Precursor Peptides — 13
Chris J. Pemberton and A. Mark Richards

 I. Introduction — 14
 II. Ghrelin Gene Precursors — 15
 III. Posttranslational Products of Proghrelin — 17
 IV. Distribution of Proghrelin Peptides — 22
 V. Proghrelin Peptides in Lower Vertebrates — 24
 VI. Summary — 25
 References — 25

3. Structure of Mammalian and Nonmammalian Ghrelins — 31
Masayasu Kojima, Takanori Ida, and Takahiro Sato

 I. Introduction — 32
 II. Mammalian Ghrelin — 33
 III. Nonmammalian Ghrelin — 36
 IV. Ghrelin and Motilin Family — 40
 V. Activity Change by Fatty Acid Chain Length of Ghrelin — 41
 VI. Putative Ghrelin Acyl-Modifying Enzyme — 41
 VII. Conclusion — 43
 References — 43

4. The Growth Hormone Secretagogue Receptor — 47
Conrad Russell Young Cruz and Roy G. Smith

- I. Introduction: A Case of Reverse Pharmacology — 48
- II. Genetics and Molecular Biology — 49
- III. Regulation of Gene Expression — 55
- IV. Signal Transduction — 58
- V. GHSR1a: Tissue Distribution and Functions in Different Organ Systems — 62
- VI. Other Ligands — 74
- VII. Evidence for Other Receptors — 77
- VIII. Future Directions — 78
- References — 79

5. Basic Aspects of Ghrelin Action — 89
Yolanda Pazos, Felipe F. Casanueva, and Jesus P. Camiña

- I. Introduction — 90
- II. Structure of GHSR1a: A G-Protein–Coupled Receptor — 91
- III. How to Define the Role of the System Ghrelin/GHSR1a? — 96
- IV. Are There Alternative Ligands for the GHSR1a? — 97
- V. Endocytosis of GHSR1a — 98
- VI. Homo- or Heteromeric Complexes for GHSR1a — 99
- VII. GHSR1a: G-Protein-Signaling Pathways — 101
- VIII. A Brief Commentary: "New" Receptors for Ghrelin and Desacyl Ghrelin — 107
- IX. Concluding Remarks — 108
- References — 109

6. Appetite and Metabolic Effects of Ghrelin and Cannabinoids: Involvement of AMP-Activated Protein Kinase — 121
Hinke van Thuijl, Blerina Kola, and Márta Korbonits

- I. Introduction — 122
- II. Ghrelin — 122
- III. Cannabinoids — 127
- IV. Adenosine Monophosphate-Activated Protein Kinase — 131
- V. The Effects of Ghrelin and Cannabinoids on AMPK — 135
- References — 138

7. Ghrelin and Feedback Systems — 149
Katsunori Nonogaki

- I. Introduction — 150
- II. Regulation of Ghrelin Secretion — 151

III. Afferent Pathways of Ghrelin from the Stomach to the Hypothalamus	158
IV. Hyperphagia, Obesity, and Des-acyl Ghrelin	159
V. Hypothalamic Gene Expression and Plasma Des-acyl Ghrelin	161
VI. Conclusion	161
References	162

8. Ghrelin Gene-Related Peptides Modulate Rat White Adiposity — 171

Andrés Giovambattista, Rolf C. Gaillard, and Eduardo Spinedi

I. Introduction	172
II. Effects of Ghrelin on Rat Retroperitoneal Adipocyte Endocrine Functions	176
III. Desacyl Ghrelin as a Potential Physiological Modulator of Adiposity	183
IV. Discussion and Remarks	193
References	199

9. Cardiac, Skeletal, and Smooth Muscle Regulation by Ghrelin — 207

Adelino F. Leite-Moreira, Amândio Rocha-Sousa, and Tiago Henriques-Coelho

I. Introduction	208
II. Structure and Distribution of Ghrelin	210
III. Contractile Effects of Ghrelin	217
References	230

10. Ghrelin and Bone — 239

Martijn van der Velde, Patric Delhanty, Bram van der Eerden, Aart Jan van der Lely, and Johannes van Leeuwen

I. Bone Balance: Resorption and Formation	239
II. Interplay Between the Gastrointestinal System and Bone: The Effect of Gastrectomy/Fundectomy on Bone	240
III. Effects of GH and GHS on Bone Metabolism	242
IV. Correlation Between Ghrelin and Bone Parameters in Clinical Studies	244
V. Effects of Ghrelin on Osteoblastic Cells *In Vitro*	245
VI. Conclusions	249
References	249

11. Ghrelin in Pregnancy and Lactation — 259

Jens Fuglsang

I. Introduction	260
II. GHSRs in Pregnancy	260

III.	Ghrelin in the Fallopian Tubes, Uterus, and Placenta	262
IV.	Ghrelin Secretion in Pregnancy	262
V.	Placenta as a Source of Ghrelin?	265
VI.	Physiological Actions of Ghrelin in Pregnancy	266
VII.	Lactation	273
VIII.	Fetal Ghrelin	274
IX.	Future Directions for Research	276
X.	Summary	276
	References	277

12. Ghrelin and Reproduction: Ghrelin as Novel Regulator of the Gonadotropic Axis — 285

Manuel Tena-Sempere

I.	Introduction: Ghrelin is a Multifunctional Regulator with Key Roles in Energy Balance	286
II.	Neuroendocrine Control of Reproduction: The Gonadotropic Axis	287
III.	Reproduction and the Energy Status are Functionally Linked	288
IV.	Ghrelin as Putative Regulator of the Gonadotropic Axis	289
V.	Role of Ghrelin in the Control of Gonadotropin Secretion	289
VI.	Putative Roles of Ghrelin in Puberty Onset	291
VII.	Molecular Diversity of Ghrelin: Reproductive Effects of UAG and Obestatin	292
VIII.	Expression and Direct Actions of Ghrelin in the Gonads	293
IX.	Futures Perspectives and Conclusions	295
	References	298

13. Ghrelin and Prostate Cancer — 301

Fabio Lanfranco, Matteo Baldi, Paola Cassoni, Martino Bosco, Corrado Ghé, and Giampiero Muccioli

I.	Prostate Cancer	302
II.	Influences of Hormones on Prostate Cancer Progression	303
III.	The Peptide Hormone Ghrelin and Prostate Cancer	307
IV.	Conclusions	315
	References	316

14. Novel Connections Between the Neuroendocrine and Immune Systems: The Ghrelin Immunoregulatory Network — 325

Dennis D. Taub

I.	Neuroendocrine–Immune Interactions	326
II.	The Growth Hormone Secretagogue Receptor	327

III.	Ghrelin: Hormone or Cytokine?	329
IV.	Conclusions	341
	References	342

Index   *347*

Contributors

Matteo Baldi
Department of Internal Medicine, Division of Endocrinology and Metabolism, University of Turin, 10126 Turin, Italy

Martino Bosco
Department of Biomedical Science and Oncology, Division of Pathology, University of Turin, 10126 Turin, Italy

Jesus P. Camiña
Laboratory of Molecular Endocrinology, Research Area, Complejo Hospitalario Universitario de Santiago (CHUS), Santiago de Compostela, Spain
CIBER de Fisiopatología, Obesidad y Nutricion (CB06/03) Instituto de Salud Carlos III, Spain

Felipe F. Casanueva
Laboratory of Molecular Endocrinology, Research Area, Complejo Hospitalario Universitario de Santiago (CHUS), Department of Medicine, University of Santiago de Compostela Santiago de Compostela, Spain and CIBER de Fisiopatología, Obesidad y Nutricion (CB06/03) Instituto de Salud Carlos III, Spain

Paola Cassoni
Department of Biomedical Science and Oncology, Division of Pathology, University of Turin, 10126 Turin, Italy

Conrad Russell Young Cruz
Translational Biology and Molecular Medicine Program, Baylor College of Medicine, One Baylor Plaza, Houston, Texas 77030

Patric Delhanty
Department of Internal Medicine, Erasmus MC, Dr. Molewaterplein 50, 3015 CE Rotterdam, The Netherlands

Bram van der Eerden
Department of Internal Medicine, Erasmus MC, Dr. Molewaterplein 50, 3015 CE Rotterdam, The Netherlands

Jens Fuglsang
Gynaecological/Obstetrical Research Laboratory, Aarhus University Hospital, Skejby Hospital, DK-8200 Aarhus N, Denmark; and Gynaecological/Obstetrical Department, Aarhus University Hospital, Skejby Hospital, DK-8200 Aarhus N, Denmark

Rolf C. Gaillard
Division of Endocrinology, Diabetology and Metabolism, University Hospital (CHUV), CH 1011 Lausanne, Switzerland

Corrado Ghé
Department of Anatomy, Pharmacology and Forensic Medicine, Division of Pharmacology, University of Turin, 10125 Turin, Italy

Andrés Giovambattista
Neuroendocrine Unit, Multidisciplinary Institute on Cell Biology (CONICET-CICPBA), 1900 La Plata, Argentina

Tiago Henriques-Coelho
Department of Physiology, Faculty of Medicine, University of Porto, Alameda Professor Hernâni Monteiro, 4200-319 Porto, Portugal

Takanori Ida
Molecular Genetics, Institute of Life Science, Kurume University, Kurume, Fukuoka, Japan

Masayasu Kojima
Molecular Genetics, Institute of Life Science, Kurume University, Kurume, Fukuoka, Japan

Blerina Kola
Department of Endocrinology, Barts and The London School of Medicine, Queen Mary University of London, London EC1M 6BQ, United Kingdom

Márta Korbonits
Department of Endocrinology, Barts and The London School of Medicine, Queen Mary University of London, London EC1M 6BQ, United Kingdom

Andreas Kukol
School of Life Sciences, University of Hertfordshire Hatfield AL10 9AB, United Kingdom

Fabio Lanfranco
Department of Internal Medicine, Division of Endocrinology and Metabolism, University of Turin, 10126 Turin, Italy

Johannes van Leeuwen
Department of Internal Medicine, Erasmus MC, Dr. Molewaterplein 50, 3015 CE Rotterdam, The Netherlands

Adelino F. Leite-Moreira
Department of Physiology, Faculty of Medicine, University of Porto, Alameda Professor Hernâni Monteiro, 4200-319 Porto, Portugal

Aart Jan van der Lely
Department of Internal Medicine, Erasmus MC, Dr. Molewaterplein 50, 3015 CE Rotterdam, The Netherlands

Giampiero Muccioli
Department of Anatomy, Pharmacology and Forensic Medicine, Division of Pharmacology, University of Turin, 10125, Turin, Italy

Katsunori Nonogaki
Center of Excellence, Division of Molecular Metabolism and Diabetes, Tohoku University Graduate School of Medicine, Miyagi 980-8575, Japan

Yolanda Pazos
Laboratory of Molecular Endocrinology, Research Area, Complejo Hospitalario Universitario de Santiago (CHUS), Santiago de Compostela, Spain
CIBER de Fisiopatología, Obesidad y Nutricion (CB06/03) Instituto de Salud Carlos III, Spain

Chris J. Pemberton
Christchurch Cardioendocrine Research Group, Department of Medicine, Christchurch School of Medicine and Health Sciences, University of Otago, PO Box 4345, Christchurch, New Zealand

A. Mark Richards
Christchurch Cardioendocrine Research Group, Department of Medicine, Christchurch School of Medicine and Health Sciences, University of Otago, PO Box 4345, Christchurch, New Zealand

Amândio Rocha-Sousa
Department of Physiology, Faculty of Medicine, University of Porto, Alameda Professor Hernâni Monteiro, 4200-319 Porto, Portugal

Takahiro Sato
Molecular Genetics, Institute of Life Science, Kurume University, Kurume, Fukuoka, Japan

Roy G. Smith
Translational Biology and Molecular Medicine Program, Huffington Center on Aging, Department of Molecular and Cellular Biology and Department of Medicine, Baylor College of Medicine, One Baylor Plaza, Houston, Texas 77030

Eduardo Spinedi
Neuroendocrine Unit, Multidisciplinary Institute on Cell Biology (CONICET-CICPBA), 1900 La Plata, Argentina
Division of Endocrinology, Diabetology and Metabolism, University Hospital (CHUV), CH 1011 Lausanne, Switzerland

Dennis D. Taub
Laboratory of Immunology, National Institute on Aging (NIH), Baltimore, Maryland 21224

Manuel Tena-Sempere
Physiology Section, Department of Cell Biology, Physiology and Immunology, University of Córdoba, 14004 Córdoba, Spain

Hinke van Thuijl
Department of Endocrinology, Barts and The London School of Medicine, Queen Mary University of London, London EC1M 6BQ, United Kingdom

Martijn van der Velde
Department of Internal Medicine, Erasmus MC, Dr. Molewaterplein 50, 3015 CE Rotterdam, The Netherlands

Preface

Ghrelin, a relatively newly discovered neuroendocrine hormone, is active in the growth hormone and insulin-like growth factor axis and affects many tissues. This collection represents a timely review of the state-of-the-art of our newer knowledge about this hormone.

The book begins with a chapter entitled: "The Structure of Ghrelin" by A. Kukol followed by "Biochemistry of Ghrelin Precursor Peptides" by C. J. Pemberton and A. M. Richards. Structural aspects are reviewed by M. Kojima, T. Ida, and T. Sato in a chapter: "Structure of Mammalian and Nonmammalian Ghrelins." The ghrelin receptor is the subject of a chapter by C. R. Y. Cruz and R. G. Smith entitled: "The Growth Hormone Secretagogue Receptor." Y. Pazos, F. F. Casanueva, and J. P. Camiña report on the "Basic Aspects of Ghrelin Action." "Appetite and Metabolic Effects of Ghrelin and Cannabinoids: Involvement of AMP-Activated Protein Kinase (AMPK)" is a contribution of H. van Thuijl, B. Kola, and M. Korbonits. K. Nonogaki writes on "Ghrelin and Feedback Systems."

The focus of this collection then shifts to effects on organ systems and disease, as represented by: "Ghrelin Gene-Related Peptides Modulate Rat White Adiposity" by A. Giovambattista, R. C. Gaillard, and E. Spinedi. Along similar lines, A. F. Leite-Moreira, A. Rocha-Sousa, and T. Henriques-Coelho offer: "Cardiac, Skeletal and Smooth Muscle Regulation by Ghrelin." "Ghrelin and Bone" is covered by M. van der Velde, P. Delhanty, B. van der Eerden, A. J. van der Lely, and J. van Leeuwen. "Ghrelin in Pregnancy and Lactation" is the topic of J. Fuglsang while M. Tena-Sempere reviews: "Ghrelin and Reproduction: Ghrelin as a Novel Regulator of the Gonadotropic Axis." F. Lanfranco, M. Baldi, P. Cassoni, M. Bosco, C. Ghe and G. Muccioli offer a review entitled: "Ghrelin and Prostate Cancer". Finally, the connection between ghrelin and the neuroendocrine and immune systems is reviewed by D. D. Taub in a work: "Novel Connections Between the Neuroendocrine and Immune Systems: The Ghrelin Immunoregulatory Network".

The picture on the cover is a structure of a ghrelin precursor peptide from the Protein Data Bank (ID number 1P7X).

This volume will continue to cover topics of interest involving hormones and vitamins. The excellent assistance of Renske van Dijk and Tari Broderick of Academic Press/Elsevier is highly appreciated.

Gerald Litwack
Toluca Lake, California
April 2007

CHAPTER ONE

The Structure of Ghrelin

Andreas Kukol*

Contents

I. Introduction	2
II. Background	2
A. Experimental methods	2
B. Computational methods	4
III. Results of Structural Studies	5
A. Nuclear magnetic resonance	5
B. MD simulations	6
IV. Discussion	9
V. Conclusions	11
Acknowledgments	11
References	11

Abstract

The structure of ghrelin, a 28-residue octanoylated peptide hormone, is only known up to the level of primary structure identifying an active core of residues 1–5 or 1–4 including octanoyl-Ser3 as necessary to elicit receptor response. This chapter reviews the results and limitations of experimental and computer modeling studies, which have appeared in the literature. The ^{1}H NMR spectroscopy experimental studies revealed an unstructured and/or fast interconverting peptide at acidic pH, while molecular dynamics (MD) simulation studies at neutral pH pointed to a stable conformation over a time period of 25 ns in water and in the presence of a lipid bilayer. The significance of these findings is discussed with regards to the pH difference, the timescales accessible to simulation and NMR spectroscopy, and the limitations of computational modeling. MD simulations of ghrelin in the presence of a lipid membrane revealed that the octanoyl side chain did not insert into the lipid bilayer, but instead the peptide bound to the lipid headgroups with residues Arg15, Lys16, Glu17, and Ser18, which are located in a hairpin-like bend in the structure. The implications of these findings with regards to a recently obtained homology model of the ghrelin receptor are discussed. © 2008 Elsevier Inc.

---

* School of Life Sciences, University of Hertfordshire, Hatfield AL10 9AB, United Kingdom

I. Introduction

Ghrelin is a peptide hormone secreted from endocrine cells in the stomach acting as the natural agonist of the growth hormone secretagogue receptor (GHSR), a G-protein–coupled receptor. It plays an important role in the control of growth hormone secretion and also in appetite regulation and food intake. Since its discovery by Kojima *et al.* (1999), many reviews have appeared in the literature (Kojima and Kangawa, 2005; Korbonits *et al.*, 2004) and most recent functional aspects are extensively discussed in this volume. Structural studies of ghrelin have concentrated on the primary structure, that is, the sequence of amino acid residues, which consists of 28 residues with an octanoyl group connected to Ser3 (Fig. 1). The active core required for agonist potency of the human GHSR is the N-terminal tetrapeptide Gly-Ser-Ser (*n*-octanoyl)-Phe-COOH (Bednarek *et al.*, 2000). The GHSR is a seven transmembrane helix G-protein–coupled receptor, which has been discovered long before its natural ligand ghrelin was known (Guillemin *et al.*, 1982).

An important goal in structural biology is to obtain the three-dimensional (3D) tertiary structure of a biomolecule, which is the focus of this chapter. In order to develop synthetic agonists and antagonists for the GHSR, which may be used as drugs for the treatment of pathophysiological conditions related to ghrelin and its receptor, it is advantageous to know the 3D structure of the natural agonist ghrelin. The structure may be helpful to understand pathophysiological conditions, binding to other proteins or lipid bilayers, or it may be used as a pharmacophore model that defines essential structural features a synthetic agonist should possess, which is useful even in the absence of a structure for the ghrelin receptor. This chapter critically reviews experimental and theoretical studies relating to the 3D structure of full-length human ghrelin with emphasis on molecular dynamics (MD) simulations. The structure of the GHSR is not known, although a homology model has been presented (Pedretti *et al.*, 2006), which will be discussed briefly in relation to ghrelin structure.

II. Background
A. Experimental methods

The main experimental methods to yield high-resolution structures of proteins are x-ray crystallography and nuclear magnetic resonance (NMR) spectroscopy. While x-ray crystallography is based on the analysis of diffraction patterns of crystals, NMR spectroscopy is the method of choice for structural analysis of smaller proteins (up to 25 kDa) in solution (Wuthrich, 1986).

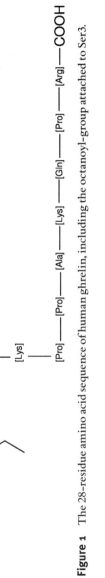

Figure 1 The 28-residue amino acid sequence of human ghrelin, including the octanoyl-group attached to Ser3.

NMR spectroscopy relies on the absorption of radiowaves by nuclei (^{1}H, ^{13}C, ^{15}N, and so on) in a magnetic field. If the bonding structure of a molecule is known, as it is the case for proteins, 2D- and 3D-NMR spectroscopy is used to obtain torsion angles from measured coupling constants and distances from the nuclear overhauser effect (NOE). These data constrain the conformational space a molecule can adopt and with a sizeable number of such constraints, five or more per residue, an atomic 3D structure is calculated with the help of computers, usually involving *in vacuo* MD calculations (Section II.B) and energy minimization (EM). While there is no lower size limit for NMR spectroscopy, the analysis of spectra becomes difficult if the molecule exists in two or more conformations, in particular if these conformations interconvert between each other on a timescale, which is fast compared to the millisecond time resolution in analyzing chemical shifts of ^{1}H NMR spectroscopy. The possibility of temperature reduction in order to decrease the rate of interconversion is limited for biological samples by the freezing point of water. Thus, an NMR structural analysis of small flexible proteins/peptides is often not possible.

B. Computational methods

Computational methods rely usually on a quantum mechanical or classical mechanical representation of a molecule. Given the size of biomolecules, quantum mechanical calculations are too time consuming; thus, a classical mechanical representation is often chosen in which all molecules in the system are represented by ball-and-spring models. The force field describes the equations and parameters for the potential energy of the system in dependence of the atom coordinates. The force field takes into account bond lengths between atoms, bond angles, torsion angles, electrostatic and van der Waals interactions. The parameters of the force field are obtained from quantum mechanical calculations, vibration frequencies, or chosen to reproduce thermodynamic properties using simulation; for example, for the GROMOS96 force field used in the ghrelin study, parameters have been chosen to reproduce the experimental heat of vaporization and density of aliphatic hydrocarbons (van Gunsteren *et al.*, 1999). In order to speed up calculations, GROMOS96 uses the united atom approach, which subsumes nonpolar hydrogen atoms into their adjacent carbon atom.

In order to obtain stable, physically realistic conformations of molecules, the procedures of EM and MD simulation are applied. EM is always used after initial construction of a molecule in order to prepare the system for a subsequent MD simulation. EM tries to minimize the potential energy of the system as calculated from the force field by incrementally changing the atom coordinates according to a minimization algorithm. EM is only able to find local minima and cannot overcome energy barriers. MD simulations calculate the "real" thermal fluctuations of the system at a given temperature and pressure over a defined period of time usually in the range of picoseconds or nanoseconds limited by the computational power or the simulation time available. A landmark MD simulation was the 1-μs simulation of a 36-residue protein in explicit solvent starting from a fully extended structure, which folded into a marginally stable state that resembled the native experimentally known conformation (Duan and Kollman, 1998). In case the folding pathway is not of interest, but only the global energy minimum, simulated annealing (SA) MD is a powerful technique to find the most stable conformation. Rather than trying to mimic a natural process, SAMD is a computational technique to find the best solution to a problem, which has a large number of possible solutions (Kirkpatrick *et al.*, 1983). SAMD simulations are started at a very high temperature and the system is progressively cooled down. At high temperature, the system is able to effectively explore the conformational space overcoming high-energy barriers, while as the temperature is reduced, lower energy states become more probable and eventually at zero temperature the system should reach the global energy minimum.

Inherent limitations of MD and SAMD simulations are the accuracy of the force field, the length of the simulation, for example, picoseconds to nanoseconds, which may be short compared to the timescale of conformational changes of proteins (microseconds to hours), and the execution of only one or a few simulations compared to the statistical average over 6×10^{14} molecules in an experiment using 1 ml of a 1-μM solution. Thus, the statistical significance of one MD simulation even carried out over several microseconds may be questioned. In the search for an energy minimum, it is important to start the MD simulation from several different conformations as well as repeating the simulation of the same starting structure at different random initial atom velocities as exemplified in the conformational search protocol for α-helical bundles in a lipid bilayer (Beevers and Kukol, 2006b,c).

III. Results of Structural Studies

As mentioned in Section I, structural studies of ghrelin have concentrated on the primary structure, that is, the sequence of amino acid residues and octanoylation of Ser3. So far only two studies relating to the 3D structure have appeared in the literature, one NMR study (Silva Elipe *et al.*, 2001) and MD simulation studies in explicit water and in the presence of an explicit lipid bilayer (Beevers and Kukol, 2006a).

A. Nuclear magnetic resonance

Using ^{1}H NMR spectroscopy, a structural analysis of full-length human ghrelin and various truncated analogues including the octanoylated segment of the first five residues, which has shown binding to the human ghrelin receptor (GHSR), has been performed (Silva Elipe *et al.*, 2001). NMR spectra were recorded at a temperature of 298 K in an H_2O/D_2O mixture

(90:10) with a small amount of trifluoroacetic acid, which was necessary to increase solubility. This resulted in an acidic pH between 1.1 and 3.1 for the various samples. The low dispersion of chemical shifts indicated an undefined random coil structure in all cases. Furthermore, NOE connectivities were only detected between hydrogen atoms bonded to neighboring atoms, for example, HC_α–NH and within each individual amino acid residue. This indicates the absence of secondary structure elements like α-helices or β-sheets as well as a high mobility of the random coil structures. An exception was the five-residue long truncated ghrelin peptide, which showed two NOEs between octanoyl-Ser3 and Phe4 indicating some rigidity between the β-CH_2 group of Ser and the NH group of Phe4; unfortunately, the derived distance has not been reported.

It can be concluded that under the experimental conditions chosen for the NMR experiment, human ghrelin shows a random coil structure within the millisecond time frame accessible to standard ^{1}H NMR experiments. A truncated ghrelin composed of residues 1–5 shows some rigidity between octanoyl-Ser3 and Phe4 but overall adopts a random coil conformation as well.

B. MD simulations

1. Structure in aqueous solution

The full-length human ghrelin peptide in water was subjected to SAMD followed by 10 ns constant temperature MD at pH 7 in order to investigate, if ghrelin is able to fold into a unique structure stable in the time frame accessible to MD simulations (Beevers and Kukol, 2006a). Twenty random starting structures of an extended conformation were subjected to SAMD, reducing the temperature from 450 to 273 K over a period of 2 ns. The resulting structures were analyzed for clusters of similar structures because the independent folding of different random starting structures into a similar structure is indicative of a unique energy minimum, which may be prevalent in solution. One cluster of 13 structures was found, the members of which share a similar fold. The representative structure of this cluster (Fig. 2A), which was most similar to all other members of the cluster, has no clearly defined fold apart from a bulge at residues His9 to Gln13 indicating the onset of α-helix formation and a short loop consisting of residues Ser18 to Lys20. A subsequent 10-ns MD simulation at constant temperature of 298 K showed that the SAMD structure had not achieved its equilibrium fold, but underwent further structural changes. The endpoint structure of the 10-ns simulation after EM (Fig. 2B) showed a clear formation of a short α-helix from Pro7 to Glu13 and the formation of a hairpin structure with Glu17 to Lys20 in the bending region. These structural elements formed after 6 ns in the constant temperature simulation and remained stable throughout the last 4 ns.

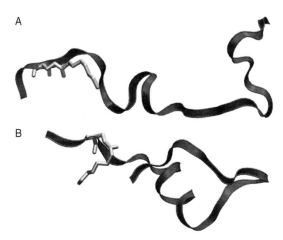

Figure 2 Peptide backbone structure of ghrelin with the octanoyl chain highlighted. (A) Representative structure of a cluster of similar structures obtained through 2-ns SAMD simulation reducing the temperature from 450 to 273 K. (B) Structure after 10-ns MD simulation in water at 298 K followed by EM.

2. Lipid-binding properties of ghrelin

Since the GHSR is a transmembrane protein, ghrelin would naturally approach the vicinity of the cellular lipid membrane when binding to the GHSR. Furthermore, the hydrophobic octanoyl chain might either function as a lipid anchor increasing the local ghrelin concentration at the membrane surface or alternatively participate in direct binding to the GHSR. In order to investigate these questions and the stability of the structure obtained, a 15-ns MD simulation of ghrelin in the presence of a lipid bilayer in water was performed (Beevers and Kukol, 2006a). The system for simulation consisted of 128 dimyristoyl phosphatidylcholine (DMPC) molecules, 7184 water molecules, and 5 chloride ions neutralizing the +5 positive charge of the peptide. The structure in Fig. 2B was placed in the lipid bilayer system in such a position that the octanoyl chain points to the lipid bilayer in order to facilitate potential insertion in the lipid bilayer (Fig. 3). The progress of the simulation is shown in Fig. 4 in terms of the root mean square deviation (RMSD) of the backbone coordinates with respect to the starting structure. The RMSD is a measure of overall structural difference to a reference structure (chosen here at $t = 0$) with values up to 0.5-nm backbone RMSD being indicative of normal structural fluctuations, while values above 1 nm would indicate a conformational change. The characteristic fold of the structure did not change significantly throughout the simulation with an RMSD of around 0.25 nm after 15 ns. This is also apparent from the visual appearance of the snapshots shown at various times in Fig. 4. However, the orientation of the peptide with respect to the lipid bilayer changed during the simulation. During the first 3 ns, the

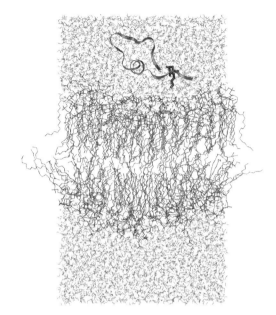

Figure 3 The system used for MD simulation of ghrelin in the vicinity of a lipid bilayer. The system shown consists of ghrelin, 5 chloride ions, 128 DMPC molecules, and 7184 water molecules.

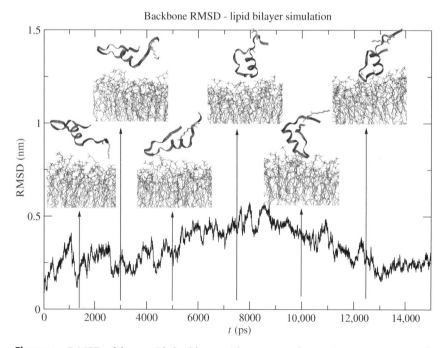

Figure 4 RMSD of the peptide backbone with respect to the starting structure at $t = 0$. The arrows indicate the time points of snaphots displayed above the curve. Water molecules are not shown.

peptide approached the lipid bilayer and remained in its initial orientation. After about 5 ns, the peptide started to reorient such that the N- and C-terminus pointed away from the lipid bilayer toward the water phase, while the loop formed a close contact with the lipid bilayer headgroups. In particular, the residues Arg14, Lys16, Glu17, and Ser18 were in close contact with the lipid headgroups. Most notably, the octanoyl side chain of Ser3 pointed into the water phase, although it was able to make contact with the lipid headgroups during the first 2 ns of the simulation.

In summary, the MD simulation studies revealed a stable fold of the ghrelin peptide in solution, which remained unchanged during a further 15-ns simulation in a lipid bilayer/water system. The octanoyl chain did not insert into the lipid bilayer, but a positively charged loop structure interacted with the zwitterionic lipid headgroups.

IV. Discussion

Contrary to the NMR experiments, the MD simulations revealed a stable fold of the 28-residue ghrelin peptide, a loop structure with a short stretch of α-helix. Possible reasons for these differences are that the MD simulation has revealed only one of several conformations, which is stable over a time range of 25 ns but is able to adopt other conformations, which are, however, short-lived compared to the millisecond time resolution of standard ^{1}H NMR spectroscopy. One should also take into account the pH difference between the NMR experiment at acidic pH and the MD simulation at neutral pH. At acidic pH, the number of charges of ghrelin increases from +5 to +9, resulting in significant electrostatic repulsion between residues considering the small size of the peptide. The increased positive charge could very well prevent ghrelin from adopting a defined conformation. It is perhaps no surprise that some rigidity has been found in a truncated peptide composed of residues 1–5, which would only acquire a +1 charge at acidic conditions due to its protonated N-terminus. While ghrelin is secreted from the stomach under acidic conditions, physiologically more relevant is the conformation at neutral pH, when it approaches the lipid membrane-bound receptors, which are expressed mainly in the hypothalamus and the pituitary (Howard *et al.*, 1996).

The structural features of the octanoyl-serine residue in ghrelin are reminiscent of a detergent molecule composed of a hydrophilic headgroup and a hydrophobic tail. One might expect the hydrophobic octanoyl group to act as a lipid anchor attaching ghrelin to the lipid membrane, thus facilitating binding to the receptor. However, the results of the MD simulation in the presence of a lipid bilayer lead to the hypothesis that the GHSR-binding site consists of a hydrophobic pocket of a size, which can accommodate the octanoyl chain (Beevers and Kukol, 2006a). This postulate is in line with the minimal structural requirements of ghrelin necessary to elicit GHSR response, which consist of the first five residues including the octanoyl group (Bednarek et al., 2000). This hypothesis has been strengthened by the construction of a human GHSR model based on local homology modeling (Pedretti et al., 2006). This model has been subjected to computational docking studies using a tetrapeptide consisting of the first four residues of ghrelin including the octanoyl chain. The docking studies revealed that the octanoyl chain is accommodated in a hydrophobic pocket lined by extracellular loop 2 and shows interaction with residues Pro192, Trp193, Pro200, Ala204, and Val205 (Pedretti et al., 2006). For size comparison, the ghrelin model shown in Fig. 2B has been placed tentatively on a surface model of the GHSR coordinates obtained from Vistoli (Pedretti et al., 2006) in Fig. 5. Although no docking or MD simulation has been performed, the possibility of the interaction between the C-terminus of ghrelin and the receptor exists, facilitated by the hairpin-like structure.

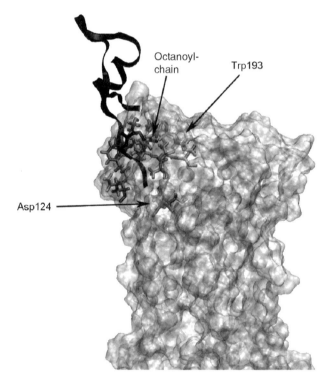

Figure 5 Superposition of the ghrelin model on the GHSR homology model. Residues of the GHSR, which have been proposed to interact with the octanoyl-chain are highlighted in light gray, while residues which have been proposed to interact with residues 1–5 of ghrelin are shown in dark gray. The interacting residues are based on a docking study with truncated ghrelin 1–4 (Pedretti et al., 2006).

V. Conclusions

The subject of the 3D structure of ghrelin is hampered by a lack of experimental and computational investigations. Based on computational modeling, it may be concluded that ghrelin has a defined structure at neutral pH in aqueous solution and furthermore that the octanoyl side chain does make direct contact to the GHSR on binding rather than function as a mere lipid anchor. The emergence of GHSR homology models, which can be used for computational docking, is clearly an advantage for understanding of the receptor-bound ghrelin structure. However, homology models need to be based on experimentally known structures, which are scarce in the area of G-protein–coupled receptors and transmembrane proteins in general. Most homology models of G-protein–coupled receptors are based on the crystal structure of bovine rhodopsin (Palczewski *et al.*, 2000), although current approaches predict the structure of fragments based on local homology to various proteins (Ginalski *et al.*, 2005). A docking model of a truncated ghrelin to GHSR has confirmed the postulated role of the octanoyl side chain (Pedretti *et al.*, 2006). Further studies using all atom MD simulations of the GHSR/ghrelin in a lipid bilayer could expand the existing docking model and provide further insight into the GHSR–ghrelin interaction.

However, modeling results need to be backed up by experiments; NMR investigations taking into account the pH dependence of the charge state of ghrelin and measurements at lower temperatures in order to reduce structural fluctuations are clearly necessary in order to advance the field and ultimately contribute to development of novel drugs for the benefit of human health.

ACKNOWLEDGMENTS

Figures of molecular structures have been prepared with VMD (Humphrey *et al.*, 1996) using the molecular surface plug-in (Varshney *et al.*, 1994).

The author is grateful to Prof. G. Vistoli for providing the coordinates of the GHSR model, to Dr. A. Beevers for carefully reading the chapter, and to the Biotechnology and Biological Sciences Research Council, United Kingdom (grant no. 88/B19450) for ongoing support.

REFERENCES

Bednarek, M. A., Feighner, S. D., Pong, S. S., McKee, K. K., Hreniuk, D. L., Silva, M. V., Warren, V. A., Howard, A. D., Van Der Ploeg, L. H., and Heck, J. V. (2000). Structure-function studies on the new growth hormone-releasing peptide, ghrelin: Minimal sequence of ghrelin necessary for activation of growth hormone secretagogue receptor 1a. *J. Med. Chem.* **43**(23), 4370–4376.

Beevers, A. J., and Kukol, A. (2006a). Conformational flexibility of the peptide hormone ghrelin in solution and lipid membrane bound: A molecular dynamics study. *J. Biomol. Struct. Dyn.* **23**(4), 357–364.

Beevers, A. J., and Kukol, A. (2006b). Systematic molecular dynamics searching in a lipid bilayer: Application to the glycophorin A and oncogenic ErbB-2 transmembrane domains. *J. Mol. Graph. Model.* **25**, 226–233.

Beevers, A. J., and Kukol, A. (2006c). The transmembrane domain of the oncogenic mutant ErbB-2 receptor: A structure obtained from site-specific infrared dichroism and molecular dynamics. *J. Mol. Biol.* **361**, 945–953.

Duan, Y., and Kollman, P. A. (1998). Pathways to a protein folding intermediate observed in a 1-microsecond simulation in aqueous solution. *Science* **282**(5389), 740–744.

Ginalski, K., Grishin, N. V., Godzik, A., and Rychlewski, L. (2005). Practical lessons from protein structure prediction. *Nucleic Acids Res.* **33**(6), 1874–1891.

Guillemin, R., Brazean, P., Bohlen, P., Esch, F., Ling, N., and Wehrenberg, W. B. (1982). Growth hormone-releasing factor from a human pancreatic tumor that caused acromegaly. *Science* **218**, 585–587.

Howard, A. D., Feighner, S. D., Cully, D. F., Arena, J. P., Liberator, P. A., Rosenblum, C. I., Hamelin, M., Hreniuk, D. L., Palyha, O. C., Anderson, J., Paress, P. S., Diaz, C., et al. (1996). A receptor in pituitary and hypothalamus that functions in growth hormone release. *Science* **273**(5277), 974–977.

Humphrey, W., Dalke, A., and Schulten, K. (1996). VMD: Visual molecular dynamics. *J. Mol. Graph.* **14**(1), 33–38, 27–38.

Kirkpatrick, S., Gelatt, C. D., and Vecchi, M. P. (1983). Optimization by simulated annealing. *Science* **220**, 671–680.

Kojima, M., and Kangawa, K. (2005). Ghrelin: Structure and function. *Physiol. Rev.* **85**(2), 495–522.

Kojima, M., Hosoda, H., Date, Y., Nakazato, M., Matsuo, H., and Kangawa, K. (1999). Ghrelin is a growth-hormone-releasing acylated peptide from stomach. *Nature* **402**(6762), 656–660.

Korbonits, M., Goldstone, A. P., Gueorguiev, M., and Grossman, A. B. (2004). Ghrelin—a hormone with multiple functions. *Front. Neuroendocrinol.* **25**(1), 27–68.

Palczewski, K., Kumasaka, T., Hori, T., Behnke, C. A., Motoshima, H., Fox, B. A., Le Trong, I., Teller, D. C., Okada, T., Stenkamp, R. E., Yamamoto, M., and Miyano, M. (2000). Crystal structure of rhodopsin: A G protein-coupled receptor. *Science* **289**(5480), 739–745.

Pedretti, A., Villa, M., Pallavicini, M., Valoti, E., and Vistoli, G. (2006). Construction of human ghrelin receptor (hGHS-R1a) model using a fragmental prediction approach and validation through docking analysis. *J. Med. Chem.* **49**(11), 3077–3085.

Silva Elipe, M. V., Bednarek, M. A., and Gao, Y. D. (2001). 1H NMR structural analysis of human ghrelin and its six truncated analogs. *Biopolymers* **59**(7), 489–501.

van Gunsteren, W. F., Daura, X., and Mark, A. E. (1999). GROMOS force field. *Encyclopedia Comput. Chem.* **2**, 1211–1216.

Varshney, A., Brooks, F. P., and Wright, M. V. (1994). Linearly scalable computation of smooth molecular surfaces. *IEEE Comput. Graph. Appl.* **14**, 19–25.

Wuthrich, K. (1986). "NMR of Proteins and Nucleic Acids." Wiley, New York.

CHAPTER TWO

BIOCHEMISTRY OF GHRELIN PRECURSOR PEPTIDES

Chris J. Pemberton* *and* A. Mark Richards*

Contents

I. Introduction	14
II. Ghrelin Gene Precursors	15
A. Organization of the ghrelin gene	15
B. Promoter activity	15
III. Posttranslational Products of Proghrelin	17
A. Ghrelin and desacyl ghrelin	17
B. C-ghrelin peptides	20
IV. Distribution of Proghrelin Peptides	22
V. Proghrelin Peptides in Lower Vertebrates	24
VI. Summary	25
References	25

Abstract

Since its discovery in 1999, the stomach-derived hormone ghrelin has been studied intensively. Proghrelin is 94 amino acids long in mammals and this undergoes proteolytic processing to produce ghrelin [residues 1–28 of proghrelin(1–94)] and the C-terminal peptide C-ghrelin, which likely contains the entire 66 amino acids of the prohormone C-terminus. The accumulating data identifies ghrelin as having important roles in growth hormone (GH) release, appetite, metabolism, energy balance, cardiovascular function, reproduction, and bone growth. The most striking feature of ghrelin is that it can be acylated at its third amino acid residue (usually Ser), usually in the form of *n*-octanoyl group (C8:0). Approximately 10–20% of circulating ghrelin is acylated and this feature confers its GH releasing ability, mediated by the GH secretagogue receptor (GHSR). In contrast, the remaining 80–90% of circulating ghrelin is desacylated. Desacyl ghrelin was initially thought to be inactive, but recent *in vivo* and *in vitro* evidences have identified biological actions for this peptide, independent of GHSR. Whether C-ghrelin has bioactivity remains to be

---

* Christchurch Cardioendocrine Research Group, Department of Medicine, Christchurch School of Medicine and Health Sciences, University of Otago, PO Box 4345, Christchurch, New Zealand

determined, but it is known that plasma concentrations of this peptide respond to endocrine and metabolic manipulations in the same fashion as ghrelin itself. A third putative proghrelin peptide, termed "obestatin" has been mooted, but confirmatory biochemical and functional evidences supporting the existence of this peptide have not been forthcoming, suggesting it to be a biochemical miscalculation. This chapter will address biochemical aspects of proghrelin peptides and point to potential avenues for future work. © 2008 Elsevier Inc.

I. INTRODUCTION

The discovery of ghrelin as an endogenous ligand regulating growth hormone (GH) secretion occurred against the backdrop of an intense research effort focused around the GH secretagogue receptor (GHSR) and has been described in detail elsewhere (Kojima and Kangawa, 2005; Korbonits *et al.*, 2004; van der Lely *et al.*, 2004). That ghrelin is primarily synthesized in and secreted from stomach tissue (Kojima *et al.*, 1999) was an unexpected, but upon reflection, not surprising discovery because, in its role as the prime recipient of energy intake, the stomach/gastrointestinal tract is in a unique position to signal to the pituitary/hypothalamus specific information regarding nutrient content and volume. Such a pattern of discovery is not without parallel as the identification of the cardiac natriuretic peptides resulted from careful testing of atrial tissue extracts for their ability to induce the natriuretic and diuretic activity of the kidney (De Bold, 1979).

The single most important feature of the discovery of ghrelin was the painstaking attention to detail observed regarding the hormone's biochemistry. Thus, ghrelin was purified from the rat stomach through four steps of chromatography: gel filtration, two ion-exchange HPLC steps, and a final reverse-phase high performance liquid chromatography (RP-HPLC) procedure in combination with its ability to stimulate GHSR-mediated Ca^{2+} influx in a stable cell line (Kojima *et al.*, 1999). However, after the 28-amino acid sequence was obtained from Edman degradation, verification that synthetic ghrelin eluted consistent with endogenous peptide on RP-HPLC was not forthcoming and synthetic peptide did not stimulate GHSR-mediated GH release. Thus, analysis of the endogenous material on mass spectrometry (MS) revealed a previously unappreciated posttranslational modification consisting of an acylate addition to Ser3. When C8:0 modified synthetic ghrelin was rerun on RP-HPLC, it eluted consistent with endogenous purified peptide and it stimulated GH release. Without such a careful biochemical approach, it is unlikely that ghrelin's physiological roles would have been identified so rapidly, with a remarkable 2200-plus publications appearing in the 8 years since its discovery.

II. Ghrelin Gene Precursors

A. Organization of the ghrelin gene

The gene encoding ghrelin is present in mammalian and nonmammalian species including birds, fish, and amphibians. The gene for human proghrelin is located on chromosome 3 at position p25–26 (Kishimoto et al., 2003), whereas the gene for GHSR is at position q26–27 on the same chromosome (Smith et al., 2001). In humans, rats, and mice, the ghrelin gene comprises five exons (Kanamoto et al., 2004; Kojima et al., 1999; Tanaka et al., 2001a). The first exon is only 20 bp and appears to be a noncoding region (Fig. 1). Putative transcriptional initiation sites in the ghrelin promoter have been identified at proximal positions −32 and −30 in the human and rat, with distal sites also at −555 and −474 bp, respectively (taking the translation start site as +1). Mature ghrelin (i.e., the 28-amino acid GH releasing peptide) is encoded by exons 2 and 3 with the remaining proghrelin peptide sequence contained in exons 4 and 5 (Jeffrey et al., 2005a; Kojima et al., 1999; Tanaka et al., 2001a). Variants of rat and mouse proghrelin have been identified in which Gln14 (coded by CAG) provides an alternative splicing signal to generate two different ghrelin mRNAs (Hosoda et al., 2000b). This alternative splicing results in a des-Gln14-ghrelin precursor which is otherwise identical to proghrelin and exists in significant amounts (~20% in rat stomach) (Hosoda et al., 2000b). This des-Gln14 variant mRNA also exists in human stomach but only in minor amounts (Hosoda et al., 2003). There also exist exon-deleted variants of proghrelin in human and mouse tissues. Thus, in human breast cancer tissue, an exon 3 deleted form has been identified (Jeffrey et al., 2005b) with an exon 4 deleted mouse homolog present in multiple tissues (Jeffrey et al., 2005a). Both of these variants results in full length mature ghrelin, but a much reduced C-terminal region that potentially encodes a unique C-terminal tail of 16 and 11 amino acids in the human and mouse, respectively.

B. Promoter activity

Basal transcription activity of the human ghrelin promoter appears to require an essential sequence downstream from the distal initiation site (the noncoding exon 1 region and proximal region of intron 1), whereas in the rat the noncoding exon 1 is not required (Wei et al., 2005). In the rat and mouse, a TATATAA element appears to be important for defining the transcriptional start point (Tanaka et al., 2001a; Wei et al., 2005), whereas in humans (Kishimoto et al., 2003) the role of the TATATAA element does not appear to be as important. Multiple putative binding sites for transcription factors have been identified including AP2, basic helix-loop-helix

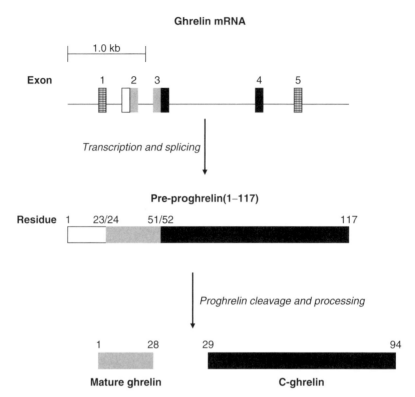

Figure 1 The human ghrelin gene is encoded by five exons. Exon 1 is short and encodes the 5′-untranslated region (hatched bar). A splicing product generated from exon 2 through to exon 4 is the predominant form of human ghrelin mRNA *in vivo* which is translated into a 117-amino acid ghrelin precursor (preproghrelin). Cellular ER processing to remove the signal peptide and subsequent protease cleavage of the proghrelin (1–94) precursor results in the production of the 28-amino acid mature ghrelin peptide (present in acyl and desacyl forms) along with the 66-amino acid C-terminal fragment, C-ghrelin. In rodents, pig, and ruminants, another splicing variant encoding des-Gln14-ghrelin is produced by alternative splicing at the end of intron 2.

(bHLH), PEA-3, Myb, NF-IL-6, hepatocyte nuclear factor-5, and NF-B, and half-sites for estrogen and glucocorticoid response elements (Kanamoto *et al.*, 2004; Kishimoto *et al.*, 2003; Tanaka *et al.*, 2001a). Interestingly, a typical GC or CAAT box element has not been identified in the ghrelin promoter and several reports have indicated that ghrelin gene expression, and its control, may be specific to cell type (Kanamoto *et al.*, 2004; Kishimoto *et al.*, 2003; Nakai *et al.* 2004). Ghrelin promoter activity can also be stimulated by glucagon and its second messenger cAMP (Kishimoto *et al.*, 2003) but it is unclear how this is achieved as there are no AP1 or CRE responsive sites in the promoter region, at least up to -2000 bp.

III. Posttranslational Products of Proghrelin

A. Ghrelin and desacyl ghrelin

The amino acid sequences of mammalian proghrelin(1–94) precursors are well conserved (Fig. 2). In humans and rodents, preproghrelin is a 117-amino acid peptide. After translocation through the endoplasmic reticulum membrane, the 23-amino acid signal peptide is cleaved, resulting in proghrelin(1–94). Mature ghrelin is generated from proghrelin(1–94) by cleavage between residues Arg28-Ala29, indicating Pro27-Arg28 is used as a proteolytic cleavage recognition site (Kojima *et al.*, 1999). Such a recognition site is present in other hormones [e.g., atrial natriuretic peptide (ANP)] which is cleaved by the cardiac tissue enzyme Corin (Yan *et al.*, 2000). It is unknown if a Corin-like enzyme exists in the stomach, but recent Western blotting, MS, and immunocytochemical evidences suggest that proghrelin(1–94) can be cleaved by prohormone convertase 1/3 (PC1/3) in mouse stomach tissue to generate *bona fide* ghrelin *in vivo* (Zhu *et al.*, 2006) and that PC1/3-deficient mice have an excess expression of proghrelin mRNA.

The region of highest homology between mature mammalian ghrelin peptides exists in the N-terminal 10 residues which are 100% identical. This conservation presumably provides an enzymatic "footprint" for acylation of the third Ser residue. The remaining C-terminal 18 amino acids of mature ghrelin (i.e., residues 11–28) are not 100% identical across species. Furthermore, bovine and ovine mature ghrelins are 27-amino acid peptides that lack Gln14 (like rat des-Gln14-ghrelin). This is because there is only one AG splice acceptor site between exons 2 and 3 in these species, yielding only one mRNA that gives rise to the 27-residue ghrelin. However, bovine and ovine mature ghrelins are still produced from a Pro-Arg recognition site (Fig. 2).

The most striking feature of mature ghrelin is the addition of an acyl group to Ser3. The biological actions of acylated ghrelin are numerous and have been reviewed in detail (Kojima and Kangawa, 2005; Korbonits *et al.*, 2004; van der Lely *et al.*, 2004). It is notable that the acyl addition is present

```
             1                          28 29                                                      94
Human    GSSFLSPEHQRVQQRKESKKPPAKLQPR  ALAGWLRPEDGGQAEGAEDELEVRFNAPFDVGIKLSGVQYQQHSQALGKFLQDILWEEAKEAPADK
Rat      GSSFLSPEHQKAQQRKESKKPPAKLQPR  ALEGWLHPEDRGQAEEAEEELEIRFNAPFDVGIKLSGAQYQQHGRALGKFLQDILWEEVKEAPANK
Mouse    GSSFLSPEHQKAQQRKESKKPPAKLQPR  ALEGWLHPEDRGQAEETEEELEIRFNAPFDVGIKLSGAQYQQHGRALGKFLQDILWEEVKEAPADK
Bovine   GSSFLSPEHQKLQ-RKEAKKPSGRLKPR  TLEGQFDPEVGSQAEGAEDELEIRFNAPFNIGIKLAGAQSLQHGQTLGKFLQDILWEEAEETLANE
Porcine  GSSFLSPEHQKVQQRKESKKPAAKLKPR  ALEGWLGPEDSGEVEGTEDKLEIRFNAPCDVGIKLSGAQSDQHGQPLGKFLQDILWEEVTEAPADK
Canine   GSSFLSPEHQKLQQRKESKKPPAKLQPR  ALEGSLGPEDTSQVEEAEDELEIRFNAPFDVGIKLSGPQYHQHGQALGKFLQEVLWEDTNEALADE
Ovine    GSSFLSPEHQKLQ-RKEPKKPSGRLKPR  ALEGQFDPDVGSQEEGAEDELEIRFNAPFNIGIKLSGAQSLQHGQTLGKFLQDILWEEAEETLADE
```

Figure 2 Sequence alignment of mammalian proghrelin(1–94) precursors (single letter notation for amino acids is used). The N-terminal 10 amino acids of all mature ghrelin peptides (bound by the box) is identical, and contains the acyl-modified Ser3 (bold and underlined). The 66-amino acid C-ghrelin peptide shows much greater variation in sequence across species.

in all species studied to date, including mammals, birds, amphibians, and fish. The nature of this acyl addition is complex, comprising octanoylated (C8:0), decanoylated (C10:0), and possibly decenoylated (C10:1) forms in the human stomach at least (Hosoda et al., 2003). All of these forms stimulate GH release, bind to the GHSR and presumably, induce adiposity and positive effects upon feeding (Hosoda et al., 2003). The enzyme(s) responsible for the addition of acyl groups to mature ghrelin have not been identified, but emerging evidence suggests that acylation is at least partially dependent upon nutrient intake. Thus, increasing the dietary proportion of medium chain fatty acid or medium chain triacylglycerol in rats results in significant increases in stomach proportions of acylated ghrelin forms, without altering total peptide production (Nishi et al., 2005). Furthermore, introduction of n-heptanoic acid or glycerol triheptanoate into the diet of rats resulted in production of nonendogenous forms of acylated ghrelin (e.g., C7:0), indicating that the enzyme(s) responsible for acylation may utilize dietary fat sources for such modifications. This effect may not be limited to mammals as excess dietary octanoic acid administration, given intraperitoneal or oral, also results in increases in octanoyl modified ghrelin in the proventriculus (glandular portion of the stomach) of neonatal chicks (Yamato et al., 2005).

In terms of immunoreactivity, desacyl ghrelin, in which the acyl addition at Ser3 is removed, constitutes the predominant circulating form of ghrelin. Thus, in humans (Ariyasu et al., 2001; Bang et al., 2006; Patterson et al., 2005) and rats (Hosoda et al., 2000a), desacyl ghrelin constitutes $\geq 80\%$ of the circulating peptide. In contrast, rat (Hosoda et al., 2000a) and human stomach tissues (Date et al., 2000; Hosoda et al., 2003) contain an $\sim 2:1$ ratio of desacyl to acylated forms. This indicates that just prior to or upon entry to the circulation, acylated ghrelin is further converted to the nonacylated form.

The precise *in vivo* mechanism of conversion of ghrelin from acylated to desacylated forms remains elusive. It is known that plasma octanoyl ghrelin binds to high-density lipoprotein (HDL) forms that contain esterase, paraoxonase, and clusterin activities (Beaumont et al., 2003). Lysophospholipase I has been identified as a putative ghrelin desacylation enzyme in rat stomach tissue (Shanado et al., 2004) but the activity of this enzyme is not found in rat plasma. In rat serum however, desoctanoylation of ghrelin is inhibited by bis-*p*-nitrophenylphosphate (a carboxylesterase inhibitor); furthermore, purified carboxylesterase can remove octanoyl groups from ghrelin (De Vriese et al., 2004). In human serum, octanoyl ghrelin is converted to desoctanoyl forms by enzymatic activity that is phenylmethylsulfonyl fluoride (PMSF), eserine salicylate, and sodium fluoride sensitive, indicating dependence on Ser protease/esterase and butyrylcholinesterase activities (De Vriese et al., 2004). Furthermore, after desoctanoylation, ghrelin was further degraded by N-terminal proteolysis at multiple sites between Ser2 and Glu8 in a variety of tissues. Thus, the desoctanoylation and subsequent proteolysis of ghrelin appears to involve multiple complex steps and may exhibit species variations in the precise enzymes responsible.

Circulating desacylated ghrelin was initially thought to represent nothing more than an inactive form of the hormone as it does not bind to GHSR1a receptor, which mediates acylated ghrelin bioactivity (Kojima *et al.*, 1999). Emerging evidence strongly suggests that this is not the case. For example, ghrelin and desacyl ghrelin both inhibit doxorubicin-induced apoptosis in adult and H9c2 cardiomyocytes and endothelial cells via activation of the extracellular signal-related kinase (ERK)-1/2 and Akt Ser kinase pathways (Baldanzi *et al.*, 2002). Interestingly, in the H9c2 cell line, both acyl and desacyl ghrelin bound to high-affinity sites, despite this cell line not expressing GHSR1a. Further *in vivo* evidence for a metabolic role of desacyl ghrelin has come from experimental and clinical data. Thus, by using rats deficient in GH in order to exclude its compounding effects, Thompson *et al.* (2004) demonstrated direct peripheral adipogenic actions of acyl and desacyl ghrelin in tibial bone marrow. This action was not replicated by the use of L-163255 (a potent GHSR1a agonist) suggesting that the adipogenic actions of both acyl and desacyl ghrelin in this tissue were independent of GHSR1a activation. In humans, co-infusion of desacyl ghrelin with acyl ghrelin abolished increases in plasma glucose and decreases in insulin that were observed with acyl ghrelin infusion alone (Broglio *et al.*, 2004). However, desacyl ghrelin did not alter the GH, prolactin, adrenocorticotrophin, or cortisol responses to acyl ghrelin, suggesting that the counteractions of desacyl ghrelin were limited to metabolic actions. Experimental bioactivity evidence suggests that, (1) desacyl ghrelin can antagonize acyl-ghrelin-induced increases in glucose output in porcine hepatic cell cultures *in vitro* (Gauna *et al.*, 2005); (2) central, but not peripheral, administration of desacyl ghrelin can induce increases in feeding in rats and mice *in vivo*, via Orexin expressing neurons in the lateral hypothalamus (Toshinai *et al.*, 2006); (3) desacyl ghrelin is an endothelium-independent vasodilator with potency comparable with acyl ghrelin to antagonize endothelin-induced vasoconstriction in human arteries *in vitro* (Kleinz *et al.*, 2006); and (4) desacyl ghrelin can stimulate proliferation of human osteoblast via a mechanism involving mitogen-activated protein kinase (MAPK) and phosphoinositide-3 kinase (PI-3K) signaling pathways (Delhanty *et al.*, 2006). Furthermore, mice that overexpress desacyl ghrelin show a small phenotype, with a 20% reduction in longitudinal growth, but no alterations in body mass index (BMI), feeding behavior, nutritional condition, and body fat mass compared with wild-type controls (Ariyasu *et al.*, 2005). Furthermore, these transgenic desacyl animals had significantly lower serum insulin-like growth factor 1 (IGF-1) concentrations and a markedly reduced GH response to exogenous acyl ghrelin administration.

A consistent feature of the emerging bioactivity of desacyl ghrelin is that it appears to be independent of the GHSR1a receptor. In all of the studies outlined earlier, acyl and desacyl ghrelin bioactivities were apparent despite the absence of GHS receptor types. Of interest, Delhanty *et al.* (2006) note that in their human bone biopsies and osteoblast cell cultures, GHSR1a mRNA was not detected, but that of its related receptor GHSR1b was. GHSR1b was originally suggested to be an inactive counterpart to GHSR1a based on its inability to stimulate GH secretion (Howard *et al.*, 1996). However, there are few reports directly assessing the ability of GHSR1b to be stimulated by acyl or desacyl ghrelin forms and what influence this may have on GHSR1a function. Furthermore, the existence of functional, alternate receptor types that can mediate desacyl (and acyl) ghrelin bioactivity cannot be discounted. For example, peripherally administered desacyl ghrelin can pass through the blood–brain barrier (BBB) in rats—as has been demonstrated for acyl ghrelin (Banks *et al.*, 2002)—and induce reductions in food intake via pathways that do not involve vagal afferent activity (Chen *et al.*, 2005). These results stand in marked contrast to those of Toshinai *et al.* (2006) who found the reverse was true, that is, desacyl ghrelin *promotes* food intake. In the study of Chen *et al.* (2005) however, the pattern of central neuronal *c-fos* activation in desacyl ghrelin stimulated animals differed from that induced by acyl ghrelin and remarkably, the actions of desacyl ghrelin could be inhibited by corticotrophin-releasing factor receptor type 2 (CRF-R2) blockade, suggesting that this may be an alternative pathway of ghrelin action. Given that Urocortin-1, an endogenous ligand for CRF-R2, can induce decreases in circulating total ghrelin levels in man (Davis *et al.*, 2004), it is conceivable that there may be links between these two hormone systems. However, care will need to be taken in experiments addressing this issue as the contrasting results of Chen *et al.* (2005) and Toshinai *et al.* (2006) reveal.

B. C-ghrelin peptides

In contrast with ghrelin, there is a paucity of reports describing potential peptides derived from the C-terminal region of proghrelin (C-ghrelin). An initial report, using a single antibody assay directed to the C-terminus of proghrelin(1–94), described putative peptides derived from the C-terminus in the human circulation, ranging in weight between M_r 3500 and 7000 (Pemberton *et al.*, 2003). In 2005, it was reported that a second biologically active peptide was potentially secreted from the C-terminal domain of proghrelin(1–94). This peptide was proposed to comprise the sequence proghrelin (53–75) and to be amidated at its C-terminus, a common posttranslational feature of biologically active peptides with Gly residues located at, or near, their C-terminus. Because it appeared to have anti-orexigenic effects and it inhibited jejunal contraction, the peptide was termed "obestatin" indicating its ability to "suppress" obesity and act in opposition to ghrelin (Zhang *et al.*, 2005). Furthermore, obestatin was reported to bind to and stimulate the activity of the G-protein orphan receptor GPR39 (McKee et al., 1997), transcripts for which were found in the important target regions of pituitary and hypothalamus. Such a finding suggested that the proghrelin gene coded for hormones with opposing actions in metabolism and energy balance and could have partially explained the intriguing observation that deletion of the ghrelin gene affects neither overall growth nor overall appetite (Sun et al., 2003; Wortley et al., 2004).

However, it is increasingly unlikely that this is in fact true. Recent reports have been unable to replicate: (1) the anti-orexigenic findings of Zhang et al. (2005), (2) the actions of "obestatin" upon the GPR39 receptor, (3) the finding that hypothalamic GPR39 receptor populations exist in rodents, and (4) that "obestatin" itself exists as an endogenous peptide in rodents or humans. Thus, several studies in rodents from multiple groups have not been able to find any effect of "obestatin" upon food intake, body weight, body composition, energy expenditure, or hypothalamic peptides involved in energy balance regulation (Gourcerol et al., 2006; Nogueiras et al., 2006; Samson et al., 2006). Clearance and distribution studies do not support "obestatin" as a stable peptide involved in energy balance as it diffused across the BBB *in vivo* in mice in a random, spurious fashion (Pan et al., 2006), suggesting it does not utilize a specific, saturable transport system like that described for ghrelin (Banks et al., 2002). Furthermore, "obestatin" was rapidly degraded by RBE4 cerebral microvessel endothelial cells and that ^{125}I-obestatin had a plasma half-life of ≥ 2 min, whereas ghrelin exhibited saturable binding characteristics and has a plasma half-life of ~ 10 min (Nagaya et al., 2001; Pan et al., 2006). Studies addressing the ability of "obestatin" to stimulate the GPR39 receptor have failed to reproduce its putative bioactivity (Holst et al., 2006; Lauwers et al., 2006; Nogueiras et al., 2006) and mRNA transcripts for GPR39 have not been found in mouse or rat hypothalamic tissues (Jackson et al., 2006; Nogueiras et al., 2006). Studies on the tissue distribution and concentrations of "obestatin" in rats and humans have reported either no detectable (Bang et al., 2006) or very little detectable "obestatin" in rat stomach tissues (Chanoine et al., 2006).

Further to this, the biochemical identification of "obestatin" as a distinct, endogenous peptide in stomach tissue extracts and the circulation of rats and humans has not been confirmed (Bang et al., 2006). In this study, specific immunoassays directed to the N-terminus, C-terminus, and mid-region of proghrelin(29–94) all detected the same molecular species of $M_r \sim 7000$, suggesting that the only forms present were close to the full length 66-amino acid C-ghrelin peptide (Bang et al., 2006). Furthermore, natural congener peptides logically formed from the putative generation of "obestatin" (i.e., proghrelin(29–52) and proghrelin(76–94)) were not found. Interestingly, this report revealed that C-ghrelin peptides can interact with solid chromatography matrices, retarding their elution on size exclusion chromatography and giving inaccurate estimates of molecular weight unless sufficient organic or acidic solvent is present in the buffer system. This is consistent with the fact that all known mammalian C-ghrelin sequences contain ~20% Glu and Asp residues rendering it an acidic protein; indeed, acidic proteins and peptides are well known to interact with chromatography matrices.

Thus, it appears unlikely that "obestatin" is an endogenous peptide with antagonistic actions to ghrelin. Rather, it may be that the majority of tissue stored and circulating C-ghrelin peptides in mammals are consistent in length with the complete C-terminal sequence of proghrelin, that is, C-ghrelin (29–94) (Fig. 2). Furthermore, it may also be unlikely that C-ghrelin acts in opposition to ghrelin, as it appears to respond to endocrine- and feeding-induced manipulations in the same manner as ghrelin (Shiiya *et al.*, 2002) at least in humans (Bang *et al.*, 2006). Thus, plasma C-ghrelin concentrations were significantly reduced by feeding and oral glucose ingestion and tended to be inhibited by exogenous i.p. injection of glucagon, all phenomenon shared by ghrelin itself. However, it cannot be completely discounted that C-ghrelin peptides could antagonize elements of ghrelin bioactivity, confirmation of which will require precise biochemical characterization of such peptides and carefully controlled *in vivo* infusions.

Also of interest will be the future identification of circulating forms and biological activity (if any), of the unique C-ghrelin sequences encoded by the exons 3 and 4 deleted variants identified by Jeffrey *et al.* (2005a,b) in human and mouse tissues, respectively. Of particular interest will be the verification of these potential peptides as *bona fide* circulating entities capable of influencing metabolism and energy balance.

IV. Distribution of Proghrelin Peptides

In mammals, the site of maximum gene expression of ghrelin is the stomach, followed closely by the gastrointestinal tract (Kojima *et al.*, 1999). In agreement with this, the tissue with the highest enrichment of proghrelin peptides (i.e., ghrelin and C-ghrelin) is the stomach (Ariyasu *et al.*, 2001; Bang *et al.*, 2006; Hosoda *et al.*, 2000a) with the next highest amounts found in the duodenum, jejunum, ileum, cecum, and colon (Hosoda *et al.*, 2000a). Other sites of notable proghrelin peptide immunoreactivity are the submaxillary gland, kidney, pancreas, and pituitary. Small amounts are also found in the cardiovascular (Iglesias *et al.* 2004; Pemberton *et al.*, 2004), reproductive (Tena-Sempere *et al.*, 2002), (Tanaka *et al.*, 2001b) and adrenal (Hosoda *et al.*, 2000a) organs, suggesting they may possess local paracrine or autocrine ghrelin systems as the GHSR is also found in these tissues (Gnanapavan *et al.*, 2002).

In stomach, the cell type synthesizing and secreting ghrelin has been identified as the X/A-cell, which are most abundant in the fundus (Sakata *et al.*, 2002; Tanaka-Shintani and Watanabe, 2005). The X/A-cell is one of the four distinct forms found in the mucosal layer of the stomach and makes up ~20% of the endocrine cell population (Date et al., 2000; Rindi et al., 2002). Ghrelin within the X/A-cells is stored in compact, electron-dense granules (Date et al., 2000; Yabuki et al., 2004) but it is unknown whether C-ghrelin peptides are stored in similar granules, although presumably they would be present in the same cell type. In agreement with this tissue data, it has been shown that *ex vivo* isolated stomach preparations secrete ghrelin (Kamegai et al., 2004; Shimada et al., 2003) and that the number of X/A-like cells in the fetal stomach and fetal stomach ghrelin concentrations have a positive correlation; thus, stomach populations of X/A-cells increase after birth in concert with stomach concentrations of ghrelin (Hayashida et al., 2002). As described earlier, the desacyl form comprises the large majority of stored and circulating ghrelin. Thus, in X/A-cell secretory granules, desacyl ghrelin constitutes ~80% of total tissue ghrelin concentrations (Hosoda et al., 2000a). C-ghrelin levels in stomach are approximately the same as those of total ghrelin (Bang et al., 2006) although only one form was identified on chromatography.

The concentration of ghrelin in the mammalian brain is very low compared with stomach levels. However, central ghrelin distribution is found in important appetite controlling regions such as the hypothalamic arcuate nucleus (Kojima et al., 1999; Lu et al., 2002) and hypothalamic neurons adjacent to the third ventricle between the dorsal, ventral, paraventricular, and arcuate hypothalamic nuclei (Cowley et al., 2003). These neurons send efferent fibers to central locations that contain neuropeptide Y (NPY) and agouti-related protein (AgRP) to possible potentiate release of these orexigenic peptides. Importantly, ghrelin mRNA and peptide is present in the pituitary in significant amounts (Korbonits et al., 2004) where it influences the release of GH.

Fetal pancreatic ghrelin mRNA and peptide levels are higher than stomach levels and still remain high into the neonatal period (Chanoine and Wong, 2004; Date et al., 2002). However, the pancreas cell type responsible for ghrelin production is unclear. Given that pancreatic ghrelin expression is highest in the prenatal and neonatal periods, it has been suggested that ghrelin has a close link with insulin which has been proposed to continue into adulthood (Prado et al., 2004), but this remains controversial.

The kidney contains a local ghrelin/GHSR system with the predominant site of expression being the glomerulus (Gnanapavan et al., 2002; Mori et al., 2000). Kidney tissue extracts have both acylated and desacyl ghrelin indicating that the enzyme responsible for ghrelin acylation is present in multiple tissues as well. Given that plasma concentrations of ghrelin correlate inversely with the degree of renal dysfunction (Yoshimoto et al., 2002) and that kidney membranes contain significant amounts of ghrelin degrading enzymes (De Vriese et al., 2004), it is likely that the kidney is a predominant organ for ghrelin clearance and degradation.

V. Proghrelin Peptides in Lower Vertebrates

Ghrelin appears to be well conserved in lower vertebrates as fish (Kaiya et al., 2003a), birds (Kaiya et al., 2002), and amphibians (Galas et al., 2002; Kaiya et al., 2001); all express the gene and produce functional ghrelin peptide that releases GH in all species (Fig. 3). Their homology to human ghrelin is low, ranging from 29% for bullfrog through to 44% for fish and 54% for bird forms of the hormone, respectively (Kojima and Kangawa, 2005). A feature of interest is that fish encode for short versions of ghrelin (ranging in length from 19 to 21 amino acids), amphibians, and birds contain 26- to 28-amino acid length forms whereas mammalian forms are predominately 27/28 amino acids long. The length of putative C-ghrelin peptides among lower vertebrates is highly variable, but like their higher vertebrate counterparts, there does not appear to be a consensus recognition site which could be proteolytically cleaved to generate C-ghrelin fragments of smaller lengths. A notable secondary feature of ghrelin in fish is the addition of an amidated Val or Ile residue to its C-terminus but it does not appear that this

Mammal
Human	GSSFLSPEHQRVQQRKESKKPPAKLQPR
Rat	GSSFLSPEHQKAQQRKESKKPPAKLQPR
Mouse	GSSFLSPEHQKAQQRKESKKPPAKLQPR
Porcine	GSSFLSPEHQKVQQRKESKKPAAKLKPR
Canine	GSSFLSPEHQKLQQRKESKKPPAKLQPR
Ovine	GSSFLSPEHQKLQRKEPKKPSGRLKPR
Bovine	GSSFLSPEHQKLQRKEAKKPSGRLKPR

Amphibian
Bullfrog	GLTFLSPADMQKIAERQSQNKLRHGNMN

Birds
Chicken	GSSFLSPTYKNIQQQKDTRKPTARLH
Turkey	GSSFLSPAYKNIQQQKDTRKPTARLH
Emu	GSSFLSPDYKKIQQRKDPRKPTTKLH

Fish
Goldfish	GTSFLSPAQKPQGRRPPRM
Trout	GSSFLSPSQKPQVRQGKGKPPRV
Seabream	GSSFLSPSQKPQNRGKSSRV
Eel	GSSFLSPSQRPQGKDKKPPRV

Figure 3 Mature ghrelin forms derived from their proghrelin precursors in higher vertebrates (mammals) through amphibians to birds and fish. Note that the highest region of sequence conservation resides in the first seven amino acids, possibly indicating a common, conserved enzyme (or enzymes) is responsible for ghrelin acylation across all these species. Bullfrog ghrelin has a unique substitution in which Ser3 is replaced by Thr as the acylation site.

amidation confers additional bioactivity (Kaiya *et al.*, 2003a,b,c; Unniappan *et al.*, 2002). Bullfrog ghrelin contains Thr instead of Ser at the acylation site in position 3, but this residue still contains a hydroxyl group in its side chain allowing for efficient acylation (Kaiya *et al.*, 2001).

VI. Summary

Ghrelin presents as a unique hormone in all vertebrates, and it is given that status by a posttranslational modification that had not been previously observed. A huge number of studies have been carried out addressing its biological activity in multiple disciplines which has resulted new and important insights in these disciplines. However, several key pieces of information concerning ghrelin are still unknown: (1) What is the enzyme responsible for acylating ghrelin? (2) Why does genetic deletion of ghrelin not result in major changes in phenotype? (3) Does C-ghrelin, the congener peptide resulting from cleavage of proghrelin, have biological actions as well? (4) Are there as yet unidentified receptors that can mediate biological actions of proghrelin peptides? Answers to these questions will prove pivotal to our continuing understanding of the fascinating biology and biochemistry of ghrelin.

REFERENCES

Ariyasu, H., Takaya, K., Tagami, T., Ogawa, Y., Hosoda, K., Akamizu, T., Suda, M., Koh, T., Natsui, K., Toyooka, S., Shirakami, G., Usui, T., *et al.* (2001). Stomach is a major source of circulating ghrelin, and feeding state determines plasma ghrelin-like immunoreactivity levels in humans. *J. Clin. Endocrinol. Metab.* **86,** 4753–4758.

Ariyasu, H., Takaya, K., Iwakura, H., Hosoda, H., Akamizu, T., Arai, Y., Kangawa, K., and Nakao, K. (2005). Transgenic mice over expressing des-acyl ghrelin show small phenotype. *Endocrinology* **146,** 355–364.

Baldanzi, G., Filigheddu, N., Cutrupi, S., Catapano, F., Bonissoni, S., Fubini, A., Malan, D., Baj, G., Granata, R., Broglio, F., Papotti, M., Surico, N., *et al.* (2002). Ghrelin and desacyl ghrelin inhibit cell death in cardiomyocytes and endothelial cells through ERK1/2 and PI 3-kinase/AKT. *J. Cell Biol.* **159,** 1029–1037.

Bang, A. S., Soule, S. G., Yandle, T. G., Richards, A. M., and Pemberton, C. J. (2007). Characterisation of proGhrelin peptides in mammalian tissue and plasma. *J. Endocrinol.* **192,** 313–323.

Banks, W. A., Tschop, M., Robinson, S. M., and Heiman, M. L. (2002). Extent and direction of ghrelin transport across the blood-brain barrier is determined by its unique primary structure. *J. Pharmacol. Exp. Ther.* **302,** 822–827.

Beaumont, N. J., Skinner, V. O., Tan, T. M., Ramesh, B. S., Byrne, D. J., MacColl, G. S., Keen, J. N., Bouloux, P. M., Mikhailidis, D. P., Bruckdorfer, K. R., Vanderpump, M. P., and Srai, K. S. (2003). Ghrelin can bind to a species of high density lipoprotein associated with paraoxonase. *J. Biol. Chem.* **278,** 8877–8880.

Broglio, F., Gottero, C., Prodam, F., Gauna, C., Muccioli, G., Papotti, M., Abribat, T., Van Der Lely, A. J., and Ghigo, E. (2004). Non-acylated ghrelin counteracts the metabolic but not the neuroendocrine response to acylated ghrelin in humans. *J. Clin. Endocrinol. Metab.* **89**, 3062–3065.

Chanoine, J. P., and Wong, A. C. (2004). Ghrelin gene expression is markedly higher in fetal pancreas compared with fetal stomach: Effect of maternal fasting. *Endocrinology* **145**, 3813–3820.

Chanoine, J. P., Wong, A. C., and Barrios, V. (2006). Obestatin, acylated and total ghrelin concentrations in the perinatal rat pancreas. *Horm. Res.* **66**, 81–88.

Chen, C. Y., Inui, A., Asakawa, A., Fujino, K., Kato, I., Chen, C. C., Ueno, N., and Fujimiya, M. (2005). Des-acyl ghrelin acts by CRF type 2 receptors to disrupt fasted stomach motility in conscious rats. *Gastroenterology* **129**, 8–25.

Cowley, M. A., Smith, R. G., Diano, S., Tschop, M., Pronchuk, N., Grove, K. L., Strasburger, C. J., Bidlingmaier, M., Esterman, M., Heiman, M. L., Garcia-Segura, L. M., Nillni, E. A., *et al.* (2003). The distribution and mechanism of action of ghrelin in the CNS demonstrates a novel hypothalamic circuit regulating energy homeostasis. *Neuron* **37**, 649–661.

Date, Y., Kojima, M., Hosoda, H., Sawaguchi, A., Mondal, M. S., Suganuma, T., Matsukura, S., Kangawa, K., and Nakazato, M. (2000). Ghrelin, a novel growth hormone-releasing acylated peptide, is synthesized in a distinct endocrine cell type in the gastrointestinal tracts of rats and humans. *Endocrinology* **141**, 4255–4261.

Date, Y., Nakazato, M., Hashiguchi, S., Dezaki, K., Mondal, M. S., Hosoda, H., Kojima, M., Kangawa, K., Arima, T., Matsuo, H., Yada, T., and Matsukura, S. (2002). Ghrelin is present in pancreatic alpha-cells of humans and rats and stimulates insulin secretion. *Diabetes* **51**, 124–129.

Davis, M. E., Pemberton, C. J., Yandle, T. G., Lainchbury, J. G., Rademaker, M. T., Nicholls, M. G., Frampton, C. M., and Richards, A. M. (2004). Urocortin-1 infusion in normal humans. *J. Clin. Endocrinol. Metab.* **89**, 1402–1409.

De Bold, A. J., Borenstein, H. B., Veress, A. T., and Sonnenberg, H. (1979). A rapid and potent natriuretic response to intravenous injection of atrial myocyte extract in rats. *Life Sci.* **28**, 89–94.

De Vriese, C., Gregoire, F., Lema-Kisoka, R., Waelbroeck, M., Robberecht, P., and Delporte, C. (2004). Ghrelin degradation by serum and tissue homogenates: Identification of the cleavage sites. *Endocrinology* **145**, 4997–5005.

Delhanty, P. J., van der Eerden, B. C., van der Velde, M., Gauna, C., Pols, H. A., Jahr, H., Chiba, H., van der Lely, A. J., and van Leeuwen, J. P. (2006). Ghrelin and unacylated ghrelin stimulate human osteoblast growth via mitogen-activated protein kinase (MAPK)/phosphoinositide 3-kinase (PI3K) pathways in the absence of GHS-R1a. *J. Endocrinol.* **188**, 37–47.

Galas, L., Chartrel, N., Kojima, M., Kangawa, K., and Vaudry, H. (2002). Immunohistochemical localization and biochemical characterization of ghrelin in the brain and stomach of the frog *Rana esculenta*. *J. Comp. Neurol.* **450**, 34–44.

Gauna, C., Delhanty, P. J., Hofland, L. J., Janssen, J. A., Broglio, F., Ross, R. J., Ghigo, E., and van der Lely, A. J. (2005). Ghrelin stimulates, whereas des-octanoyl ghrelin inhibits, glucose output by primary hepatocytes. *J. Clin. Endocrinol. Metab.* **90**, 1055–1060.

Gnanapavan, S., Kola, B., Bustin, S. A., Morris, D. G., McGee, P., Fairclough, P., Bhattacharya, S., Carpenter, R., Grossman, A. B., and Korbonits, M. (2002). The tissue distribution of the mRNA of ghrelin and subtypes of its receptor, GHS-R, in humans. *J. Clin. Endocrinol. Metab.* **87**, 2988–2991.

Gourcerol, G., Million, M., Adelson, D. W., Wang, Y., Wang, L., Rivier, J., St-Pierre, D. H., and Tache, Y. (2006). Lack of interaction between peripheral injection of CCK and obestatin in the regulation of gastric satiety signalling in rodents. *Peptides* **27**, 2811–2819.

Hayashida, T., Nakahara, K., Mondal, M. S., Date, Y., Nakazato, M., Kojima, M., Kangawa, K., and Murakami, N. (2002). Ghrelin in neonatal rats: Distribution in stomach and its possible role. *J. Endocrinol.* **173,** 239–245.

Holst, B., Egerod, K. L., Schild, E., Vickers, S. P., Cheetham, S., Gerlach, L. O., Storjohann, L., Stidsen, C. E., Jones, R., Beck-Sickinger, A. G., and Schwartz, T. W. (2007). GPR39 signalling is stimulated by zinc ions but not by obestatin. *Endocrinology*, **148,** 13–20.

Hosoda, H., Kojima, M., Matsuo, H., and Kangawa, K. (2000a). Ghrelin and des-acyl ghrelin: Two major forms of rat ghrelin peptide in gastrointestinal tissue. *Biochem. Biophys. Res. Commun.* **279,** 909–913.

Hosoda, H., Kojima, M., Matsuo, H., and Kangawa, K. (2000b). Purification and characterization of rat des-Gln14-Ghrelin, a second endogenous ligand for the growth hormone secretagogue receptor. *J. Biol. Chem.* **275,** 21995–22000.

Hosoda, H., Kojima, M., Mizushima, T., Shimizu, S., and Kangawa, K. (2003). Structural divergence of human ghrelin. Identification of multiple ghrelin-derived molecules produced by posttranslational processing. *J. Biol. Chem.* **278,** 64–70.

Howard, A. D., Feighner, S. D., Cully, D. F., Arena, J. P., Liberator, P. A., Rosenblum, C. I., Hamelin, M., Hreniuk, D. L., Palyha, O. C., Anderson, J., Paress, P. S., Diaz, C., et al. (1996). A receptor in pituitary and hypothalamus that functions in growth hormone release. *Science* **273,** 974–977.

Iglesias, M. J., Pineiro, R., Blanco, M., Gallego, R., Dieguez, C., Gualillo, O., Gonzalez-Juanatey, J. R., and Lago, F. (2004). Growth hormone releasing peptide (ghrelin) is synthesized and secreted by cardiomyocytes. *Cardiovasc. Res.* **62,** 481–488.

Jackson, V. R., Hothacker, H. P., and Civelli, O. (2006). GPR39 receptor expression in the mouse brain. *Neuroreport* **17,** 813–816.

Jeffrey, P. L., Duncan, R. P., Yeh, A. H., Jaskolski, R. A., Hammond, D. S., Herington, A. C., and Chopin, L. K. (2005a). Expression of the Ghrelin axis in the mouse; an exon 4-delected mouse proghrelin variant encodes a novel C-terminal peptide. *Endocrinology* **146,** 432–440.

Jeffrey, P. L., Murray, R. E., Yeh, A. H., McNamara, J. F., Duncan, R. P., Francis, G. D., Herington, A. C., and Chopin, L. K. (2005b). Expression and function of the ghrelin axis, including a novel preproGhrelin isoform, in human breast cancer tissues and cell lines. *Endocr. Relat. Cancer* **12,** 839–850.

Kaiya, H., Kojima, M., Hosoda, H., Koda, A., Yamamoto, K., Kitajima, Y., Matsumoto, M., Minamitake, Y., Kikuyama, S., and Kangawa, K. (2001). Bullfrog ghrelin is modified by *n*-octanoic acid at its third n residue. *J. Biol. Chem.* **276,** 40441–40448.

Kaiya, H., Van Der Geyten, S., Kojima, M., Hosoda, H., Kitajima, Y., Matsumoto, M., Geelissen, S., Darras, V. M., and Kangawa, K. (2002). Chicken ghrelin: Purification, cDNA cloning, and biological activity. *Endocrinology* **143,** 3454–3463.

Kaiya, H., Kojima, M., Hosoda, H., Moriyama, S., Takahashi, A., Kawauchi, H., and Kangawa, K. (2003a). Peptide purification, complementary deoxyribonucleic acid (DNA) and genomic DNA cloning, and functional characterization of ghrelin in rainbow trout. *Endocrinology* **144,** 5215–5226.

Kaiya, H., Kojima, M., Hosoda, H., Riley, L. G., Hirano, T., Grau, E. G., and Kangawa, K. (2003b). Amidated fish ghrelin: Purification, cDNA cloning in the Japanese eel and its biological activity. *J. Endocrinol.* **176,** 415–423.

Kaiya, H., Kojima, M., Hosoda, H., Riley, L. G., Hirano, T., Grau, E. G., and Kangawa, K. (2003c). Identification of tilapia ghrelin and its effects on growth hormone and prolactin release in the tilapia, *Oreochromis mossambicus. Comp. Biochem. Physiol. B Biochem. Mol. Biol.* **135,** 421–429.

Kamegai, J., Tamura, H., Shimizu, T., Ishii, S., Sugihara, H., and Oikawa, S. (2004). Effects of insulin, leptin, and glucagon on ghrelin secretion from isolated perfused rat stomach. *Regul. Pept.* **119**, 77–81.

Kanamoto, N., Akamizu, T., Tagami, T., Hataya, Y., Moriyama, K., Takaya, K., Hosoda, H., Kojima, M., Kangawa, K., and Nakao, K. (2004). Genomic structure and characterisation of the 5′-flanking region of the human ghrelin gene. *Endocrinology* **145**, 4144–4153.

Kishimoto, M., Okimura, Y., Nakata, H., Kudo, T., Iguchi, G., Takahashi, Y., Kaji, H., and Chihara, K. (2003). Cloning and characterization of the 5(′)-flanking region of the human ghrelin gene. *Biochem. Biophys. Res. Commun.* **305**, 186–192.

Kleinz, M. J., Maguire, J. J., Skepper, J. N., and Davenport, A. P. (2006). Functional and immunocytochemical evidence for a role of ghrelin and des-octanoyl ghrelin in the regulation of vascular tone in man. *Cardiovasc. Res.* **69**, 227–235.

Kojima, M., and Kangawa, K. (2005). Ghrelin: Structure and function. *Physiol. Rev.* **85**, 495–522.

Kojima, M., Hosoda, H., Date, Y., Nakazato, M., Matsuo, H., and Kangawa, K. (1999). Ghrelin is a growth-hormone-releasing acylated peptide from stomach. *Nature* **402**, 656–660.

Korbonits, M., Goldstone, A. P., Gueorguiev, M., and Grossman, A. B. (2004). Ghrelin—a hormone with multiple functions. *Front. Neuroendocrinol.* **25**, 27–68.

Lauwers, E., Landuyt, B., Arckens, L., Schoofs, L., and Luyten, W. (2006). Obestatin does not activate orphan G protein-coupled receptor GPR39. *Biochem. Biophys. Res. Commun.* **351**(1), 21–25.

Lu, S., Guan, J. L., Wang, Q. P., Uehara, K., Yamada, S., Goto, N., Date, Y., Nakazato, M., Kojima, M., Kangawa, K., and Shioda, S. (2002). Immunocytochemical observation of ghrelin-containing neurons in the rat arcuate nucleus. *Neurosci. Lett.* **321**, 157–160.

McKee, K. K., Tan, C. P., Palyha, O. C., Liu, J., Feighner, S. D., Hreniuk, D. L., Smith, R. G., Howard, A. D., and Van der Ploeg, L. H. (1997). Cloning and characterization of two human G protein-coupled receptor genes (GPR38 and GPR39) related to the growth hormone secretagogue and neurotensin receptors. *Genomics* **46**, 426–434.

Mori, K., Yoshimoto, A., Takaya, K., Hosoda, K., Ariyasu, H., Yahata, K., Mukoyama, M., Sugawara, A., Hosoda, H., Kojima, M., Kangawa, K., and Nakao, K. (2000). Kidney produces a novel acylated peptide, ghrelin. *FEBS Lett.* **486**, 213–216.

Nagaya, N., Kojima, M., Uematsu, M., Yamagishi, M., Hosoda, H., Oya, H., Hayashi, Y., and Kangawa, K. (2001). Hemodynamic and hormonal effects of human ghrelin in healthy volunteers. *Am. J. Physiol. Regul. Integr. Comp. Physiol.* **280**, R1483–R1487.

Nakai, N., Kaneko, M., Nakao, N., Fujikawa, T., Nakashima, K., Ogata, M., and Tanaka, M. (2004). Identification of promoter region of ghrelin gene in human medullary thyroid carcinoma cell line. *Life Sci.* **75**, 2193–2201.

Nishi, Y., Hiejima, H., Hosoda, H., Kaiya, H., Mori, K., Fukue, Y., Yanase, T., Nawata, H., Kangawa, K., and Kojima, M. (2005). Ingested medium-chain fatty acids are directly utilized for the acyl modification of ghrelin. *Endocrinology* **146**, 2255–2264.

Nogueiras, R., Pfluger, P., Tovar, S., Myrtha, A., Mitchell, S., Morris, A., Perez-Tilve, D., Vazquez, M. J., Wiedmer, P., Castaneda, T. R., Dimarchi, R., Tschop, M., *et al.* (2007). Effects of obestatin on energy balance and growth hormone secretion in rodents. *Endocrinology*, **148**, 21–26.

Pan, W., Tu, H., and Kastin, A. J. (2006). Differential BBB interactions of three ingestive peptides: Obestatin, ghrelin, and adiponectin. *Peptides* **27**, 911–916.

Patterson, M., Murphy, K. G., le Roux, C. W., Ghatei, M. A., and Bloom, S. R. (2005). Characterization of ghrelin-like immunoreactivity in human plasma. *J. Clin. Endocrinol. Metab.* **90**, 2205–2211.

Pemberton, C., Wimalasena, P., Yandle, T., Soule, S., and Richards, M. (2003). C-terminal proGhrelin peptides are present in the human circulation. *Biochem. Biophys. Res. Commun.* **310**, 567–573.

Pemberton, C. J., Tokola, H., Bagi, Z., Koller, A., Pontinen, J., Ola, A., Vuolteenaho, O., Szokodi, I., and Ruskoaho, H. (2004). Ghrelin induces vasoconstriction in the rat coronary vasculature with altering cardiac peptide secretion. *Am. J. Physiol.* **287**, H1522–H1529.

Prado, C. L., Pugh-Bernard, A. E., Elghazi, L., Sosa-Pineda, B., and Sussel, L. (2004). Ghrelin cells replace insulin-producing cells in two mouse models of pancreas development. *Proc. Natl. Acad. Sci. USA* **101**, 2924–2929.

Rindi, G., Necchi, V., Savio, A., Torsello, A., Zoli, M., Locatelli, V., Raimondo, F., Cocchi, D., and Solcia, E. (2002). Characterisation of gastric ghrelin cells in man and other mammals: Studies in adult and foetal tissues. *Histochem. Cell Biol.* **117**, 511–519.

Sakata, I., Nakamura, K., Yamazaki, M., Matsubara, M., Hayashi, Y., Kangawa, K., and Sakai, T. (2002). Ghrelin-producing cells exist as two types of cells, closed- and opened-type cells, in the rat gastrointestinal tract. *Peptides* **23**, 531–536.

Samson, W. K., White, M. M., Price, C., and Ferguson, A. V. (2007). Obestatin acts in the brain to inhibit thirst. *Am. J. Physiol. Regul. Integr. Comp. Physiol.* **292**, R637–R643.

Shanado, Y., Kometani, M., Uchiyama, H., Koizumi, S., and Teno, N. (2004). Lysophospholipase I identified as a ghrelin deacylation enzyme in rat stomach. *Biochem. Biophys. Res. Commun.* **325**, 1487–1494.

Shiiya, T., Nakazato, M., Mizuta, M., Date, Y., Mondal, M. S., Tanaka, M., Nozoe, S., Hosoda, H., Kangawa, K., and Matsukura, S. (2002). Plasma ghrelin levels in lean and obese humans and the effect of glucose on ghrelin secretion. *J. Clin. Endocrinol. Metab.* **87**, 240–244.

Shimada, M., Date, Y., Mondal, M. S., Toshinai, K., Shimbara, T., Fukunaga, K., Murakami, N., Miyazato, M., Kangawa, K., Yoshimatsu, H., Matsuo, H., and Nakazato, M. (2003). Somatostatin suppresses ghrelin secretion from the rat stomach. *Biochem. Biophys. Res. Commun.* **302**, 520–525.

Smith, R. G., Leonard, R., Bailey, A. R., Palyha, O., Feighner, S., Tan, C., McKee, K. K., Pong, S. S., Griffin, P., and Howard, A. (2001). Growth hormone secretagogue receptor family members and ligands. *Endocrine* **14**, 9–14.

Sun, Y., Ahmed, S., and Smith, R. G. (2003). Deletion of ghrelin impairs neither growth nor appetite. *Mol. Cell. Biol.* **23**, 7973–7981.

Tanaka, M., Hayashida, Y., Iguchi, T., Nakao, N., Nakai, N., and Nakashima, K. (2001a). Organization of the mouse ghrelin gene and promoter: Occurrence of a short noncoding first exon. *Endocrinology* **142**, 3697–3700.

Tanaka, M., Hayashida, Y., Nakao, N., Nakai, N., and Nakashima, K. (2001b). Testis-specific and developmentally induced expression of a ghrelin gene-derived transcript that encodes a novel polypeptide in the mouse. *Biochim. Biophys. Acta* **1522**, 62–65.

Tanaka-Shintani, M., and Watanabe, M. (2005). Distribution of ghrelin-immunoreactive cells in human gastric mucosa: Comparison with that of parietal cells. *J. Gastroenterol.* **40**, 345–349.

Tena-Sempere, M., Barreiro, M. L., Gonzalez, L. C., Gaytan, F., Zhang, F. P., Caminos, J. E., Pinilla, L., Casanueva, F. F., Dieguez, C., and Aguilar, E. (2002). Novel expression and functional role of ghrelin in rat testis. *Endocrinology* **143**, 717–725.

Thompson, N. M., Gill, D. A., Davies, R., Loveridge, N., Houston, P. A., Robinson, I. C., and Wells, T. (2004). Ghrelin and des-octanoyl ghrelin promote adipogenesis directly *in vivo* by a mechanism independent of the type 1a growth hormone secretagogue receptor. *Endocrinology* **145**, 234–242.

Toshinai, K., Yamaguchi, H., Sun, Y., Smith, R. G., Yamanaka, A., Sakurai, T., Date, Y., Mondal, M. S., Shimbara, T., Kawagoe, T., Murakami, N., Miyazato, M., *et al.* (2006).

Des-acyl ghrelin induces food intake by a mechanism independent of the growth hormone secretatgogue receptor. *Endocrinology* **147,** 2306–2314.

Unniappan, S., Lin, X., Cervini, L., Rivier, J., Kaiya, H., Kangawa, K., and Peter, R. E. (2002). Goldfish ghrelin: Molecular characterization of the complementary deoxyribonucleic acid, partial gene structure and evidence for its stimulatory role in food intake. *Endocrinology* **143,** 4143–4146.

van der Lely, A., Tschop, M., Heiman, M. L., and Ghigo, E. (2004). Biological, physiological, pathophysiological and pharmacological aspects of ghrelin. *Endocr. Rev.* **25,** 426–457.

Wei, W., Wang, G., Qi, X., Englander, E. W., and Greeley, G. H., Jr. (2005). Characterization and regulation of the rat and human ghrelin promoters. *Endocrinology* **146,** 1611–1625.

Wortley, K. E., Anderson, K. D., Garcia, K., Murray, J. D., Malinova, L., Liu, R., Moncrieffe, M., Thabet, K., Cox, H. J., Yancopoulos, G. D., Wiegand, S. J., and Sleeman, M. W. (2004). Genetic deletion of ghrelin does not decrease food intake but influences metabolic fuel preference. *Proc. Natl. Acad. Sci. USA* **101,** 8227–8232.

Yabuki, A., Ojima, T., Kojima, M., Nishi, Y., Mifune, H., Matsumoto, M., Kamimura, R., Masuyama, T., and Suzuki, S. (2004). Characterization and species differences in gastric ghrelin cells from mice, rats and hamsters. *J. Anat.* **205,** 239–246.

Yamato, M., Sakata, I., Wada, R., Kaiya, H., and Sakai, T. (2005). Exogenous administration of octanoic acid accelerates octanoylated ghrelin production in the proventriculus of neonatal chicks. *Biochem. Biophys. Res. Commun.* **333,** 583–589.

Yan, W., Wu, F., Morser, J., and Wu, Q. (2000). Corin, a transmembrane cardiac serine protease, acts as a pro-atrial natriuretic peptide-converting enzyme. *Proc. Natl. Acad. Sci. USA* **97,** 8525–8529.

Yoshimoto, A., Mori, K., Sugawara, A., Mukoyama, M., Yahata, K., Suganami, T., Takaya, K., Hosoda, H., Kojima, M., Kangawa, K., and Nakao, K. (2002). Plasma ghrelin and desacyl ghrelin concentrations in renal failure. *J. Am. Soc. Nephrol.* **13,** 2748–2752.

Zhang, J. V., Ren, E. G., Avsian-Kretchmer, O., Luo, C.-W., Rauch, R., Klein, C., and Hsueh, A. J. (2005). Obestatin, a peptide encoded the ghrelin gene, opposes ghrelin's effects on food intake. *Science* **310,** 996–999.

Zhu, X., Cao, Y., Voodg, K., and Steiner, D. F. (2006). On the processing of proghrelin to ghrelin. *J. Biol. Chem.* **281**(50), 38867–38870.

CHAPTER THREE

Structure of Mammalian and Nonmammalian Ghrelins

Masayasu Kojima,* Takanori Ida,* and Takahiro Sato*

Contents

I. Introduction	32
II. Mammalian Ghrelin	33
A. Molecular forms of ghrelin in rat stomach	33
B. Multiple forms of ghrelin in human stomach	34
C. Molecular forms of ghrelin in several tissues	35
D. Other mammalian ghrelin	35
III. Nonmammalian Ghrelin	36
A. Bird ghrelin	36
B. Fish ghrelins	37
C. Frog ghrelin	39
D. Reptile ghrelin	40
IV. Ghrelin and Motilin Family	40
V. Activity Change by Fatty Acid Chain Length of Ghrelin	41
VI. Putative Ghrelin Acyl-Modifying Enzyme	41
VII. Conclusion	43
References	43

Abstract

The discovery of ghrelin has elucidated the role of the stomach as an important organ in the regulation of growth hormone (GH) release and energy homeostasis. Ghrelin is a peptide hormone in which Ser3 Thr3 in frogs) is modified by an n-octanoic acid; this modification is essential for ghrelin's activity. Ghrelin and motilin, another gastric peptide, structurally and functionally define a peptide superfamily; these two factors may have evolved from a common ancestral peptide. Ghrelin is found in both mammalian species as well as nonmammalian species, such as frogs, birds, and fish. Moreover, ghrelin structure, particularly that of the acyl-modification regions, is highly conserved throughout vertebrate species. All of the ghrelin peptides that have been identified are modified by a fatty acid, primarily n-octanoic acid. These discoveries implicate ghrelin as an

* Molecular Genetics, Institute of Life Science, Kurume University, Kurume, Fukuoka, Japan

essential hormone in the maintenance of GH release and energy homeostasis in vertebrates. © 2008 Elsevier Inc.

I. INTRODUCTION

Ghrelin is a peptide hormone originally isolated in 1999 from rat stomach (Kojima *et al.*, 1999). The name "ghrelin" derives from "ghre," a word root for "grow" in Proto-Indo-European languages, to indicate its ability to stimulate growth hormone (GH) release from the pituitary.

Ghrelin has two main physiological functions, GH-releasing activity and appetite-stimulating activity (Kojima and Kangawa, 2005). Ghrelin also exhibits cardiovascular effects (Nagaya *et al.*, 2001), mediates increases in gastric movement and gastric acid secretion (Masuda *et al.*, 2000), suppresses sympathetic nerve output (Nagaya *et al.*, 2004), and contributes to the regulation of glucose metabolism (Thorens and Larsen, 2004; van der Lely *et al.*, 2004).

Intravenous injection of ghrelin induces GH release in a dose-dependent manner (Arvat *et al.*, 2000; Kojima *et al.*, 1999; Peino *et al.*, 2000; Takaya *et al.*, 2000). Coadministration of ghrelin and growth hormone-releasing hormone (GHRH), another GH-releasing peptide, results in a synergistic effect on GH secretion; coadministration results in greater GH release than that following administration of either GHRH or ghrelin alone (Arvat *et al.*, 2001; Hataya *et al.*, 2001).

Ghrelin, produced primarily in gastrointestinal organs in response to hunger and starvation, circulates in the blood, serving as a peripheral signal telling the central nervous system to stimulate feeding (Ariyasu *et al.*, 2001; Cummings *et al.*, 2004; Date *et al.*, 2000; Tschop *et al.*, 2001). Intracerebroventricular (i.c.v.), intravenous, and subcutaneous injections of ghrelin have been demonstrated to increase food intake (Nakazato *et al.*, 2001; Tschop *et al.*, 2000; Wren *et al.*, 2000). Ghrelin secreted from the stomach stimulates the vagus nerve, transmitting a hunger signal to the brain that acts on the hypothalamic arcuate nucleus, an important region in the control of appetite (Date *et al.*, 2002a).

Ghrelin forms have been identified in almost all vertebrate species examined (Kojima and Kangawa, 2005). The characteristic structure of ghrelin is an acyl-modification at the third amino acid, Ser (Thr in the bullfrog and edible frog). This modification is essential for ghrelin's biological activity (Kojima *et al.*, 1999; Matsumoto *et al.*, 2001). Ser3 is modified with a medium-chain fatty acid (MCFA), typically *n*-octanoic acid.

This chapter surveys the known molecular forms of ghrelin, describes the structures of both mammalian and nonmammalian ghrelins, and discusses ghrelin's acyl-modification.

II. Mammalian Ghrelin

A. Molecular forms of ghrelin in rat stomach

1. n-Octanoyl ghrelin is the major form of gastric ghrelin

Ghrelin was initially purified from rat stomach by four sequential chromatography steps: gel filtration, two ion-exchange high performance liquid chromatography (HPLC) steps, and a final reverse-phase HPLC (RP-HPLC) procedure. The second ion-exchange HPLC step yielded two active peaks, later identified as ghrelin and des-Gln14-ghrelin (Hosoda et al., 2000b). The primary active peak, purified by RP-HPLC, was determined to be a 28-amino acid peptide, in which Ser3 was n-octanoylated, a modification essential for ghrelin activity (Fig. 1). Ghrelin was the first peptide hormone found to be modified by a fatty acid.

2. Des-Gln14-ghrelin

In rat stomach, a second type of ghrelin peptide was purified, identified as des-Gln14-ghrelin (Hosoda et al., 2000b). With the exception of the deletion of Gln14, des-Gln14-ghrelin is identical to ghrelin, retaining the n-octanoic acid modification. Des-Gln14-ghrelin has a similar potency of activities with that of as ghrelin. Thus, although two types of n-octanoyl-modified ghrelin peptides are produced in the rat stomach, ghrelin and des-Gln14-ghrelin, des-Gln14-ghrelin is only present in low amounts, indicating that ghrelin is the major active form.

The 28-amino acid functional ghrelin peptide is encoded by two exons of the ghrelin gene (Kanamoto et al., 2004; Tanaka et al., 2001). In the rat ghrelin gene, the codon for Gln14 (CAG) serves as an alternative splicing

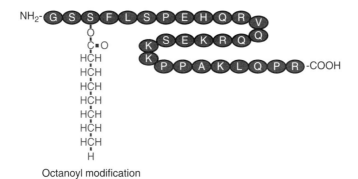

Figure 1 Structure of human ghrelin. Human ghrelin is 28-amino acid peptide, in which Ser3 is modified by a fatty acid, primarily n-octanoic acid. This modification is essential for ghrelin's activity.

site whose use leads to the generation of a second ghrelin messenger RNA (mRNA) that encodes a des-Gln14-ghrelin precursor.

Complementary DNA clone of human des-Gln14-ghrelin is also registered in GeneBank database (GenBank Accession number: AB035700). However, des-Gln14-ghrelin peptides have not yet been isolated from stomach tissue. Two cDNA clones from a fetal *Homo sapiens* library that encode human des-Gln14-ghrelin are deposited in the NCBI nucleotide database (AI338429 and BY149645). Two forms of porcine ghrelin cDNA, which encode ghrelin and des-Gln14-ghrelin, respectively, are present at an approximate ratio of 1:1 (GenBank Accession number: AB035703 and AB035704). In cows, only one ghrelin mRNA has been identified, encoding a 27-amino acid peptide that corresponds structurally to des-Gln14-ghrelin.

3. Desacyl ghrelin

A nonacylated form of ghrelin, desacyl ghrelin, is also found in significant quantities in both the stomach and the blood (Hosoda *et al.*, 2000a). In blood, desacyl ghrelin is detected in amounts far greater than those of acylated ghrelin.

Desacyl ghrelin does not replace radiolabeled ghrelin at binding sites for acylated ghrelin in the hypothalamus and pituitary. This peptide does not exhibit any GH-releasing and endocrine activities in rats (Baldanzi *et al.*, 2002). A specific receptor for desacyl ghrelin has not been identified; thus, it remains unclear if desacyl ghrelin has specific functions distinct from those of acylated ghrelin. Evidence suggests the existence of another ghrelin receptor in the cardiovascular system and fetal spinal cord (Cassoni *et al.*, 2004; Muccioli *et al.*, 2004; Sato *et al.*, 2006). Desacyl ghrelin shares with active acyl-modified ghrelin a number of nonendocrine actions, including stimulation of increased food intake (Toshinai *et al.*, 2006), modulation of cell proliferation (Baldanzi *et al.*, 2002; Filigheddu *et al.*, 2007; Sato *et al.*, 2006), and minor effects on adipogenesis (Muccioli *et al.*, 2004). Further study is required to determine if desacyl ghrelin is biologically active and binds to an as yet unidentified receptor.

B. Multiple forms of ghrelin in human stomach

Several minor forms of human ghrelin peptides have been isolated from human stomach (Hosoda *et al.*, 2003). These can be classified into four groups by the type of acylation observed at Ser3: nonacylated, *n*-octanoylated (C8:0), *n*-decanoylated (C10:0), and decenoylated (C10:1). All peptides identified thus far are either 27 or 28 amino acids in length. The 27-amino acid forms of ghrelin lack C-terminal Arg28, which may be removed by carboxypeptidase digestion. As in the rat, the major active form of human ghrelin is a 28-amino acid peptide with an *n*-octanoylated Ser3. Synthetic *n*-octanoylated and *n*-decanoylated ghrelins stimulate increases in intracellular Ca^{2+} in growth hormone secretagogue receptor (GHSR)-expressing cells and stimulate GH release to a similar degree in rats.

C. Molecular forms of ghrelin in several tissues

Gastric ghrelin was the first identified *n*-octanoyl-modified peptide. Previously, it was not known whether ghrelin is also modified by *n*-octanoic acid in tissues other than stomach.

The molecular forms of ghrelin in the jejunum (Date *et al.*, 2000), pancreas (Date *et al.*, 2002b), and hypothalamus (Sato *et al.*, 2005) have also been investigated. Peptide extracts from these tissues were analyzed and characterized by RP-HPLC and ghrelin-specific RIA. Two ghrelin forms were identified in these tissues, *n*-octanoyl ghrelin and desacyl ghrelin, as seen in gastric tissue. These results indicate that acyl-modification is not specific to stomach tissue, but occurs in tissues in which the ghrelin gene is expressed and ghrelin peptide produced.

D. Other mammalian ghrelin

In mammals, ghrelin has been identified in American bison (GenBank Accession number: AH013663), cats (GenBank Accession number: NM_001009853), cows (GenBank Accession number: AB035702), dogs (GenBank Accession number: AB060700), goats (GenBank Accession number: AH013721), humans, mongolian gerbils (GenBank Accession number: AF442491), mice (GenBank Accession number: AB035701), moose (GenBank Accession number: AH013724), pigs (GenBank Accession number: AB035703), pronghorns (GenBank Accession number: AY454075), pygmy sperm whales (GenBank Accession number: AH013719), rats, reindeer (GenBank Accession number: AH013722), rhesus monkeys (GenBank Accession number: AY371699), sheep (GenBank Accession number: AB060699), wapiti (GenBank Accession number: AH013723), water buffalo (GenBank Accession number: DQ118139), and white-tailed deer (GenBank Accession number: AY455987). Rat and human ghrelins have been discussed above. The amino acid sequences of mammalian ghrelins are well conserved; the seven N-terminal amino acids in particular are identical. The third amino acid, which is modified by a fatty acid and is essential for ghrelin activity, is Ser in all of these peptides (Fig. 2). This structural conservation and the universal requirement for acyl-modification of the third residue indicate that the N-terminal region is critical for peptide activity.

The ghrelin forms found in American bison, cows, goats, moose, pronghorns, pygmy sperm whales, reindeer, sheep, wapiti, water buffalo, and white-tailed deer are 27-amino acid peptides that, similar to rat des-Gln14 ghrelin, lack the corresponding Gln14 residue.

Figure 2 Sequence comparison of mammalian ghrelins. Identical amino acids are colored. The asterisk indicates acyl-modified third amino acid, Ser.

Mammalian ghrelin	
American bison	GSSFLSPEHQKLQ-RKEPKKPSGRLKPR
Cat	GSSFLSPEHQKVQ-RKESKKPPAKLQPR
Cow	GSSFLSPEHQKLQQRKEAKKPSGRLKPR
Dog	GSSFLSPEHQKLQQRKESKKPPAKLQPR
Goat	GSSFLSPEHQKLQ-RKEPKKPSGRLKPR
Human	GSSFLSPEHQRVQQRKESKKPPAKLQPR
Mongolian gerbil	GSSFLSPEHQKTQQRKESKKPPAKLQPR
Moose	GSSFLSPDHQKLQ-RKEPKKPSGRLKPR
Mouse	GSSFLSPEHQKAQQRKESKKPPAKLQPR
Pig	GSSFLSPEHQKVQQRKESKKPAAKLKPR
Pronghorn	GSSFLSPEHQKLQ-RKEPKKPSGM----
Pygmy sperm whale	GSSFLSPEHQKLQ-RKEAKKPSGRLKPR
Rat	GSSFLSPEHQKAQQRKESKKPPAKLQPR
Reindeer	GSSFLSPEHQKLQ-RKEPKKPSGRLKPR
Rhesus monkey	GSSFLSPEHQRAQQRKESKKPPAKLQPR
Sheep	GSSFLSPEHQKLQ-RKEPKKPSGRLKPR
Wapiti	GSSFLSPEHQKLQ-RKEPKKPSGRLKPR
Water buffalo	GSSFLSPEHQKLQ-RKEPKKPSGRLKPR
White-tailed deer	GSSFLSPEHQKLQ-RKEPKKPSGRLKPR

n-Octanoyl-modified ghrelin is the primary form of endogenous active ghrelin in all mammalian species. There exist, however, several minor ghrelin forms in the stomach. There are at least 14 different forms of ghrelin in cat stomach (Ida *et al.*, 2006). These are desacyl ghrelin, ghrelin-(1–27) with different acyl-modifications (C8:0, C10:0, C10:1, C10:2), n-octanoyl-modified des-Gln14-ghrelin-(1–27), ghrelin-(1–28) with different acyl-modifications (C8:0, C10:0, C10:1, C10:2, C13:0, C13:1), and n-octanoyl-modified des-Gln14-ghrelin-(1–28).

III. Nonmammalian Ghrelin

A. Bird ghrelin

1. Chicken ghrelin

Chicken (*Gallus gallus*) ghrelin is 26 amino acids in length and possesses 54% sequence identity with human ghrelin (Kaiya *et al.*, 2002). The Ser residue at position 3 (Ser3) is conserved between the chicken and mammalian species as is acylation by either n-octanoic or n-decanoic acid. Chicken ghrelin mRNA, predominantly expressed in the stomach, is present in the proventriculus, but absent from the gizzard. Reverse-transcriptase polymerase chain reaction (RT-PCR) analysis revealed low levels of expression in the brain, lung, and intestines.

Administration of chicken ghrelin to both rats and chicks increases plasma GH levels with a potency similar to that of rat or human ghrelin. Although ghrelin stimulates feeding in rats, both central and peripheral injection of ghrelin strongly suppressed feeding in neonatal chicks (Geelissen et al., 2006; Saito et al., 2002). This anorexic effect was also observed following injection of chicken or rat ghrelin. i.c.v. injection of GHRP-2 (KP-102), a synthetic GHS, also inhibited feeding, indicating that food intake is inhibited by GHSR agonists in neonatal chicks. The mechanism by which ghrelin suppresses rather than stimulates food intake in neonatal chicks remains to be elucidated.

2. Other bird ghrelins

Additional avian ghrelins have been identified in ducks (GenBank Accession number: AY338466), geese (GenBank Accession number: AY338465), emus (GenBank Accession number: AY338467), Japanese quail (GenBank Accession number: AB244056), and turkeys (GenBank Accession number: AY497549) (Fig. 3). The precursors of all avian ghrelins, except those found in turkeys, possess a pair of two basic amino acids, Arg-Arg, at the C-terminus. The Arg-Arg motif is removed by a carboxypeptide to produce the mature ghrelin peptide. Turkey ghrelin has a Pro-Arg-processing motif at this site, similar to mammalian homologues.

B. Fish ghrelins

Fish ghrelins have been identified both by purifying peptides from the stomach or by cDNA cloning analyses in black sea bream (Yeung et al., 2006), channel catfish (Kaiya et al., 2005), eel (Kaiya et al., 2003b), European sea bass (GenBank Accession number: DQ665912), goldfish (Unniappan et al., 2002), rainbow trout (Kaiya et al., 2003a), shark (*Carcharhinus melanopterus*; GenBank Accession number: AB254129, and *Sphyrna lewini*; GenBank Accession number: AB254128), tilapia (Kaiya et al., 2003c), and zebrafish (GenBank Accession number: AM055940) (Fig. 4). Multiple forms of fish ghrelin have been found that vary in amino acid length and specific acyl-modifications. The most characteristic feature, however, is a C-terminal

Bird ghrelin

Chicken	GSSFLSPTYKNIQQQKDTRKPTARLH
Duck	GSSFLSPEFKKIQQQNDPTKTTAKIH
Emu	GSSFLSPDYKKIQQRKDPRKPTTKLH
Goose	GSSFLSPEFKKIQQQNDPAKATAKIH
Japanese quail	GSSFLSPAYKNIQQQKNTRKPAARLH
Turkey	GSSFLSPAYKNIQQQKDTRKPTARLHPR

Figure 3 Sequence comparison of bird ghrelins. Identical amino acids are colored. The asterisk indicates acyl-modified third amino acid, Ser.

```
Fish ghrelin                                          *
Rainbow Trout 1                       GSSFLSPSQKPQVRQGKGK-PPRV
                                                    /
Rainbow Trout 2                       GSSFLSPSQKPQGKGK-PPRV
European sea bass                     GSSFLSPSQKPQSRGK-SSRV
Black sea bream                       GSSFLSPSQKPQNRGK-SSRV
Channel catfish                       GSSFLSPTQKPQNRGDRKPPRV
Goldfish                              GTSFLSPAQKPQ--GRRPPRM
Japanese eel                          GSSFLSPSQRPQGKDKKPPRV
Tilapia                               GSSFLSPSQKPQNKVK-SSRI
Zebrafish                             GTSFLSPTQKPQ--GRRPPRV
Shark (Carcharhinus melanopterus)     GVSFHPRLKEKDDNSSGNTRKFSP
Shark (Sphyrna lewini)                GVSFHPRLKEKDDNSSGNSRKSNP
```

Figure 4 Sequence comparison of fish ghrelins. Identical amino acids are colored. The asterisk indicates acyl-modified third amino acid, Ser.

amide structure observed in all fish ghrelins. All fish ghrelin precursors have a Gly residue just after the C-terminal amino acid residue of the mature peptide. This Gly residue serves as a donor for the C-terminal amide. Although this C-terminal amide structure is not necessary for ghrelin activity, it may have a role in modulation of the ghrelin turnover rate.

Injection of fish ghrelin also stimulates GH release and appetite, as seen for mammalian ghrelins.

1. Rainbow trout

Rainbow trout ghrelin was initially identified from the stomach (Kaiya et al., 2003a). Four isoforms of ghrelin peptide were isolated, including a 24-amino acid C-terminal-amidated form (rt ghrelin 1; GSSFLSPSQKPQ-VRQGKGKPPRV-amide), des-VRQ-rt ghrelin (rt ghrelin 2), from which three amino acids (V13R14Q15) have been deleted, and two additional forms that retain a glycine residue at their C-termini, rt ghrelin-Gly and des-VRQ-rt ghrelin-Gly. The third serine residue was modified by n-octanoic acid, n-decanoic acid, or the unsaturated forms of those fatty acids. High levels of ghrelin mRNA expression were detected in the stomach, while moderate levels were detected in the brain, hypothalamus, and intestines.

2. Eel ghrelin

Eel ghrelin was purified from stomach extracts of a teleost fish, the Japanese eel (*Anguilla japonica*). Purified peptide preparations also possessed an amide structure at the C-terminal end (Kaiya et al., 2003b). Two molecular forms of ghrelin, each composed of 21 amino acids, were identified by cDNA and mass spectrometric analyses, one modified by n-octanoic acid and the other by n-decanoic acid at the third Ser residue. Northern blot and RT-PCR analyses revealed high levels of gene expression in the stomach. RT-PCR analysis also identified low expression levels in the brain, intestines, kidney, and head kidney.

3. Tilapia ghrelin

Tilapia ghrelin was originally identified from the stomach of a euryhaline tilapia, *Oreochromis mossambicus* (Kaiya et al., 2003c). The third Ser residue was modified by *n*-decanoic acid or *n*-decenoic acid. The *n*-decanoyl modification generates the major ghrelin in tilapia; the content of *n*-octanoylated ghrelin is very low. The C-terminal end of the peptide also possesses an amide structure. RT-PCR analysis revealed high expression levels in the stomach and low levels in the brain, kidney, and gills.

4. Goldfish ghrelin

Goldfish ghrelin was originally identified by cDNA analyses (Unniappan et al., 2002). Mature goldfish ghrelin, a 19-amino acid peptide, contains a prohormone processing signal for cleavage and amidation (GRR). Use of this site produces a short ghrelin peptide of 12 amino acids with a C-terminal Gln-amide structure. Ghrelin mRNA expression could be detected in the brain, pituitary, intestine, liver, spleen, and gills by RT-PCR followed by Southern blot analysis. Northern blot could detect expression in the intestine.

5. Zebrafish ghrelin

A BLAST search of the zebrafish genomic database identified a zebrafish ghrelin gene encoding a peptide of the sequence GTSFLSPTQKPQ-GRRPPRV (GenBank Accession number: AL918922 and AM055940). The 19-amino acid zebrafish ghrelin is most similar to goldfish ghrelin, with which it shares a C-terminal Val-linked amide structure and a putative cleavage site for amidation (GRR) after amino acid 12. The second amino acid is a Thr, similar to goldfish ghrelin.

6. Channel catfish ghrelin

Channel catfish (*Ictalurus punctatus*) ghrelin is a 22-amino acid peptide bearing modification of the third Ser residue by *n*-decanoic acid (C10:0) or different unsaturated fatty acids such as C10:1 and C10:2 (Kaiya et al., 2005).

C. Frog ghrelin

Three molecular forms of ghrelin have been identified in bullfrog stomach (Kaiya et al., 2001), which contain either 27 or 28 amino acids with 29% sequence identity to human ghrelin. In frogs, the existence of two forms is not due to alternative splicing, but due to differential processing of the C-terminal Asn residue. The third residue (Thr3) of bullfrog ghrelin differs from the Ser3 seen in mammalian ghrelins. As Ser and Thr both possess

Amphibian ghrelin
Bullfrog   *
                 GLTFLSPADMQKIAERQSQNKLRHGNMN
Edible frog   GLTFLSPADMRKIAERQSQNKLRHGNMN

Reptile ghrelin
                          *
Red-eared slider turtle   GSSFLSPEYQNTQQRKDPKKHTKLN

Figure 5 Sequences of two amphibian ghrelins and reptile ghrelin. Identical amino acids in frog ghrelins are colored. The asterisks indicate acyl-modified third amino acids, Thr in amphibian ghrelins, and Ser in reptile ghrelin.

hydroxyl groups, both can be modified by fatty acids. Bullfrog Thr3 is modified with either n-octanoic or n-decanoic acid.

An additional form of frog ghrelin has been identified in the edible frog (GenBank Accession number: AM055941). Edible frog ghrelin is 28-amino acid peptide with only 1 amino acid change in comparison to bullfrog ghrelin (Fig. 5). The third amino acid, as in bullfrog ghrelin, is Thr, not Ser.

Northern blot analysis demonstrated that bullfrog ghrelin mRNA is predominantly expressed in the stomach. Low levels of gene expression were also observed in the heart, lung, small intestine, gall bladder, pancreas, and testis.

Bullfrog ghrelin stimulates the secretion of both GH and prolactin from dispersed bullfrog pituitary cells, indicating that the activity of ghrelin to induce GH secretion is conserved throughout evolution. Increases in both plasma and mRNA levels of ghrelin in the stomach after starvation suggest the critical involvement of ghrelin in energy homeostasis in the bullfrog.

D. Reptile ghrelin

Ghrelin has also been purified from the stomach of the red-eared slider turtle, *Trachemys scripta elegans* (Kaiya *et al.*, 2004). *Trachemys* ghrelin is a 25-amino acid peptide with acyl-modification of the third Ser by n-octanoic (C8:0), n-decanoic (C10:0), or unsaturated decanoic acid (C10:1) (Fig. 5).

IV. Ghrelin and Motilin Family

The amino acid sequence of ghrelin exhibits homology to motilin, another gastrically produced peptide with gastric contractile activity (Asakawa *et al.*, 2001; Smith *et al.*, 2001). Alignment of the 28-amino acid ghrelin peptide with 19-amino acid motilin reveals eight identical amino acids (Fig. 6). After discovery of ghrelin, Tomasetto *et al* (2000) reported the identification of a gastric peptide, which they named motilin-related peptide (MTLRP). That study used differential screening to isolate new

```
Ghrelin:  GSSFLSPEHQRVQQRKESKKPPAKLQPR
Motilin:  FVPIFTYGELQRMQE-KERNKGQ
```

Figure 6 Sequence comparison of human ghrelin and motilin. Identical amino acids are colored.

proteins whose expression was restricted to the gastric epithelium. The amino acid sequence of MTLRP later turned out to be identical to that of ghrelin [1–18]; the putative processing site of MTLRP, Lys-Lys, however, is not utilized in gastric cells. The sequence data alone did not reveal the acyl-modification, a critical alteration essential for ghrelin activity (Coulie and Miller, 2001; Del Rincon *et al.*, 2001; Folwaczny *et al.*, 2001).

The region of homology between ghrelin and motilin is not near the N-terminus, the site of ghrelin acyl-modification that also serves as the core region for ghrelin receptor activation, but in the central region.

Ghrelin and motilin play similar roles in the stomach; both induce gastric acid secretion and stimulate gastric movement.

Thus, ghrelin and motilin, which may have evolved from a common ancestral peptide, structurally and functionally define a new peptide superfamily.

V. Activity Change by Fatty Acid Chain Length of Ghrelin

Fatty acid chain length affects the activity of ghrelin (Matsumoto *et al.*, 2001). Beginning with desacyl rat ghrelin ($EC_{50} = 3500$ nM), we systematically elongated the adjacent fatty acid chain by two carbon units. Biological activity, which was substantially enhanced by the introduction of an acetyl (C2:0) group ($EC_{50} = 780$ nM), increased as a function of the length of the fatty acid chain (Table 1). A maximal response was observed in the presence of an *n*-octanoyl (C8:0) group, supporting the structural observations of natural ghrelin. Activity was maintained with the addition of as large as a *n*-palmitoyl (C16:0) group ($EC_{50} = 6.5$ nM) or longer fatty acid, although the activity gradually decreased with increasing side chain length. The observation of maximal activity for the *n*-octanoyl (C8:0) modification suggests that the C8:0 unit has been naturally selected during molecular evolution.

VI. Putative Ghrelin Acyl-Modifying Enzyme

While an enzyme that catalyzes the acyl-modification of ghrelin has not yet been identified, the universal incorporation of *n*-octanoic acid or *n*-decanoic acid in mammals, fish, birds, and amphibians suggests that this enzyme must be highly specific in its choice of MCFA substrates.

Table 1 Property of Ser3-side chain of ghrelin and activity

Carbon	Peptide	Structure	EC$_{50}$ (nM)
0	[desacyl]-rat ghrelin	GSSFLSPEHQKAQQR-KESKKPPAKLQPR	3500
2	[Ser3(acetyl)]-rat ghrelin	GSS(O-CO-CH$_3$)FLSPEHQKAQQRKESK-KPPAKLQPR	780
4	[Ser3(butyryl)]-rat ghrelin	GSS(O-CO-C$_3$H$_7$)FLSPEHQKAQQRK-ESKKPPAKLQPR	280
6	[Ser3(hexanoyl)]-rat ghrelin	GSS(O-CO-C$_5$H$_{11}$)FLSPEHQKAQQRKES-KKPPAKLQPR	16
8	Rat ghrelin	GSS(O-CO-C$_7$H$_{15}$)FLSPEHQKAQQRKESK-KPPAKLQPR	1.5
10	[Ser3(decanoyl)]-rat ghrelin	GSS(O-CO-C$_9$H$_{19}$)FLSPEHQKAQQRKESK-KPPAKLQPR	1.7
12	[Ser3(lauroyl)]-rat ghrelin	GSS(O-CO-C$_{11}$H$_{23}$)FLSPEHQKAQQRKESK-KPPAKLQPR	2.4
16	[Ser3(palmitoyl)]-rat ghrelin	GSS(O-CO-C$_{15}$H$_{31}$)FLSPEHQKAQRKESK-KPPAKLQPR	6.5

EC$_{50}$ is the concentration of peptides or peptide derivatives at [Ca^{2+}]$_i$ increase on GHSR-expressing cells.

The majority of the n-octanoic acid used for the modification of ghrelin is derived from *in vivo* synthesis. Ingestion of either MCFAs or medium-chain triacylglycerols (MCTs), however, specifically increases the production of acyl-modified ghrelin without changing the total (acyl- and desacyl) ghrelin levels (Nishi et al., 2005). When mice ingested either MCFAs or MCTs, the acyl group attached to nascent ghrelin molecules corresponded to that of the ingested MCFAs or MCTs (Nishi et al., 2005). These results indicate that ingested MCFAs are directly used for the acyl-modification of ghrelin.

Further investigations characterizing the putative ghrelin Ser O-acyltransferase will be required to elucidate the mechanism governing the unique acyl-modification of ghrelin.

VII. Conclusion

Ghrelin forms have been identified through peptide purification and cDNA cloning from multiple vertebrate species. Structural analyses of these endogenous ghrelins have demonstrated that acyl-modification by n-octanoic acid or n-decanoic acid is necessary for the biological activity of ghrelin. At present, ghrelin is the only known peptide hormone bearing such an acyl-modification. The study of ghrelin forces us to reevaluate the importance of identification and purification of natural substances. Although we can identify amino acid sequences by scanning genome databases, the natural substances may escape our notice. Ghrelin is such a peptide.

The diverse functions of ghrelin raise the possibility of clinical application to several diseases, such as GH deficiency, eating disorders, cardiovascular diseases, and gastrointestinal diseases. The structures of natural ghrelin in vertebrate species and synthetic unnatural ghrelin provide basic information necessary to design more potent and long-acting synthetic ghrelins and ghrelin derivatives.

REFERENCES

Ariyasu, H., Takaya, K., Tagami, T., Ogawa, Y., Hosoda, K., Akamizu, T., Suda, M., Koh, T., Natsui, K., Toyooka, S., Shirakami, G., Usui, T., *et al.* (2001). Stomach is a major source of circulating ghrelin, and feeding state determines plasma ghrelin-like immunoreactivity levels in humans. *J. Clin. Endocrinol. Metab.* **86**(10), 4753–4758.

Arvat, E., Di Vito, L., Broglio, F., Papotti, M., Muccioli, G., Dieguez, C., Casanueva, F. F., Deghenghi, R., Camanni, F., and Ghigo, E. (2000). Preliminary evidence that Ghrelin, the natural GH secretagogue (GHS)-receptor ligand, strongly stimulates GH secretion in humans. *J. Endocrinol. Invest.* **23**(8), 493–495.

Arvat, E., Maccario, M., Di Vito, L., Broglio, F., Benso, A., Gottero, C., Papotti, M., Muccioli, G., Dieguez, C., Casanueva, F. F., Deghenghi, R., Camanni, F., *et al.* (2001). Endocrine activities of ghrelin, a natural growth hormone secretagogue (GHS), in humans: Comparison and interactions with hexarelin, a nonnatural peptidyl GHS, and GH-releasing hormone. *J. Clin. Endocrinol. Metab.* **86**(3), 1169–1174.

Asakawa, A., Inui, A., Kaga, T., Yuzuriha, H., Nagata, T., Ueno, N., Makino, S., Fujimiya, M., Niijima, A., Fujino, M. A., and Kasuga, M. (2001). Ghrelin is an appetite-stimulatory signal from stomach with structural resemblance to motilin. *Gastroenterology* **120**(2), 337–345.

Baldanzi, G., Filigheddu, N., Cutrupi, S., Catapano, F., Bonissoni, S., Fubini, A., Malan, D., Baj, G., Granata, R., Broglio, F., Papotti, M., Surico, N., *et al.* (2002). Ghrelin and desacyl ghrelin inhibit cell death in cardiomyocytes and endothelial cells through ERK1/2 and PI 3-kinase/Akt. *J. Cell Biol.* **159**(6), 1029–1037.

Cassoni, P., Ghe, C., Marrocco, T., Tarabra, E., Allia, E., Catapano, F., Deghenghi, R., Ghigo, E., Papotti, M., and Muccioli, G. (2004). Expression of ghrelin and biological activity of specific receptors for ghrelin and desacyl ghrelin in human prostate neoplasms and related cell lines. *Eur. J. Endocrinol.* **150**(2), 173–184.

Coulie, B. J., and Miller, L. J. (2001). Identification of motilin-related peptide. *Gastroenterology* **120**(2), 588–589.

Cummings, D. E., Frayo, R. S., Marmonier, C., Aubert, R., and Chapelot, D. (2004). Plasma ghrelin levels and hunger scores in humans initiating meals voluntarily without time- and food-related cues. *Am. J. Physiol. Endocrinol. Metab.* **287**(2), E297–E304.

Date, Y., Kojima, M., Hosoda, H., Sawaguchi, A., Mondal, M. S., Suganuma, T., Matsukura, S., Kangawa, K., and Nakazato, M. (2000). Ghrelin, a novel growth hormone-releasing acylated peptide, is synthesized in a distinct endocrine cell type in the gastrointestinal tracts of rats and humans. *Endocrinology* **141**(11), 4255–4261.

Date, Y., Murakami, N., Toshinai, K., Matsukura, S., Niijima, A., Matsuo, H., Kangawa, K., and Nakazato, M. (2002a). The role of the gastric afferent vagal nerve in ghrelin-induced feeding and growth hormone secretion in rats. *Gastroenterology* **123**(4), 1120–1128.

Date, Y., Nakazato, M., Hashiguchi, S., Dezaki, K., Mondal, M. S., Hosoda, H., Kojima, M., Kangawa, K., Arima, T., Matsuo, H., Yada, T., and Matsukura, S. (2002b). Ghrelin is present in pancreatic alpha-cells of humans and rats and stimulates insulin secretion. *Diabetes* **51**(1), 124–129.

Del Rincon, J. P., Thorner, M. O., and Gaylinn, B. G. (2001). Motilin-related peptide and ghrelin: Lessons from molecular techniques, peptide chemistry, and receptor biology. *Gastroenterology* **120**(2), 587–588; author reply 589.

Filigheddu, N., Gnocchi, V. F., Coscia, M., Cappelli, M., Porporato, P. E., Taulli, R., Traini, S., Baldanzi, G., Chianale, F., Cutrupi, S., Arnoletti, E., Ghe, C., et al. (2007). Ghrelin and desacyl ghrelin promote differentiation and fusion of C2c12 skeletal muscle cells. *Mol. Biol. Cell* **18**(3), 986–994.

Folwaczny, C., Chang, J. K., and Tschop, M. (2001). Ghrelin and motilin: Two sides of one coin? *Eur. J. Endocrinol.* **144**(4), R1–R3.

Geelissen, S. M., Swennen, Q., Geyten, S. V., Kuhn, E. R., Kaiya, H., Kangawa, K., Decuypere, E., Buyse, J., and Darras, V. M. (2006). Peripheral ghrelin reduces food intake and respiratory quotient in chicken. *Domest. Anim. Endocrinol.* **30**(2), 108–116.

Hataya, Y., Akamizu, T., Takaya, K., Kanamoto, N., Ariyasu, H., Saijo, M., Moriyama, K., Shimatsu, A., Kojima, M., Kangawa, K., and Nakao, K. (2001). A low dose of ghrelin stimulates growth hormone (GH) release synergistically with GH-releasing hormone in humans. *J. Clin. Endocrinol. Metab.* **86**(9), 4552.

Hosoda, H., Kojima, M., Matsuo, H., and Kangawa, K. (2000a). Ghrelin and desacyl ghrelin: Two major forms of rat ghrelin peptide in gastrointestinal tissue. *Biochem. Biophys. Res. Commun.* **279**(3), 909–913.

Hosoda, H., Kojima, M., Matsuo, H., and Kangawa, K. (2000b). Purification and characterization of rat des-Gln14-Ghrelin, a second endogenous ligand for the growth hormone secretagogue receptor. *J. Biol. Chem.* **275**(29), 21995–22000.

Hosoda, H., Kojima, M., Mizushima, T., Shimizu, S., and Kangawa, K. (2003). Structural divergence of human ghrelin. Identification of multiple ghrelin-derived molecules produced by post-translational processing. *J. Biol. Chem.* **278**(1), 64–70.

Ida, T., Miyazato, M., Naganobu, K., Nakahara, K., Sato, M., Lin, X. Z., Kaiya, H., Doi, K., Noda, S., Kubo, A., Murakami, N., and Kangawa, K. (2006). Purification and characterization of feline ghrelin and its possible role. *Domest. Anim. Endocrinol.* **32**(2), 93–105.

Kaiya, H., Kojima, M., Hosoda, H., Koda, A., Yamamoto, K., Kitajima, Y., Matsumoto, M., Minamitake, Y., Kikuyama, S., and Kangawa, K. (2001). Bullfrog ghrelin is modified by n-octanoic acid at its third threonine residue. *J. Biol. Chem.* **276**(44), 40441–40448.

Kaiya, H., Van Der Geyten, S., Kojima, M., Hosoda, H., Kitajima, Y., Matsumoto, M., Geelissen, S., Darras, V. M., and Kangawa, K. (2002). Chicken ghrelin: Purification, cDNA cloning, and biological activity. *Endocrinology* **143**(9), 3454–3463.

Kaiya, H., Kojima, M., Hosoda, H., Moriyama, S., Takahashi, A., Kawauchi, H., and Kangawa, K. (2003a). Peptide purification, complementary deoxyribonucleic acid (DNA) and genomic DNA cloning, and functional characterization of ghrelin in rainbow trout. *Endocrinology* **144**(12), 5215–5226.

Kaiya, H., Kojima, M., Hosoda, H., Riley, L. G., Hirano, T., Grau, E. G., and Kangawa, K. (2003b). Amidated fish ghrelin: Purification, cDNA cloning in the Japanese eel and its biological activity. *J. Endocrinol.* **176**(3), 415–423.

Kaiya, H., Kojima, M., Hosoda, H., Riley, L. G., Hirano, T., Grau, E. G., and Kangawa, K. (2003c). Identification of tilapia ghrelin and its effects on growth hormone and prolactin release in the tilapia, Oreochromis mossambicus. *Comp. Biochem. Physiol. B. Biochem. Mol. Biol.* **135**(3), 421–429.

Kaiya, H., Sakata, I., Kojima, M., Hosoda, H., Sakai, T., and Kangawa, K. (2004). Structural determination and histochemical localization of ghrelin in the red-eared slider turtle, Trachemys scripta elegans. *Gen. Comp. Endocrinol.* **138**(1), 50–57.

Kaiya, H., Small, B. C., Bilodeau, A. L., Shepherd, B. S., Kojima, M., Hosoda, H., and Kangawa, K. (2005). Purification, cDNA cloning, and characterization of ghrelin in channel catfish, Ictalurus punctatus. *Gen. Comp. Endocrinol.* **143**(3), 201–210.

Kanamoto, N., Akamizu, T., Tagami, T., Hataya, Y., Moriyama, K., Takaya, K., Hosoda, H., Kojima, M., Kangawa, K., and Nakao, K. (2004). Genomic structure and characterization of the 5′-flanking region of the human ghrelin gene. *Endocrinology* **145**(9), 4144–4153.

Kojima, M., and Kangawa, K. (2005). Ghrelin: Structure and function. *Physiol. Rev.* **85**(2), 495–522.

Kojima, M., Hosoda, H., Date, Y., Nakazato, M., Matsuo, H., and Kangawa, K. (1999). Ghrelin is a growth-hormone-releasing acylated peptide from stomach. *Nature* **402**(6762), 656–660.

Masuda, Y., Tanaka, T., Inomata, N., Ohnuma, N., Tanaka, S., Itoh, Z., Hosoda, H., Kojima, M., and Kangawa, K. (2000). Ghrelin stimulates gastric acid secretion and motility in rats. *Biochem. Biophys. Res. Commun.* **276**(3), 905–908.

Matsumoto, M., Hosoda, H., Kitajima, Y., Morozumi, N., Minamitake, Y., Tanaka, S., Matsuo, H., Kojima, M., Hayashi, Y., and Kangawa, K. (2001). Structure-activity relationship of ghrelin: Pharmacological study of ghrelin peptides. *Biochem. Biophys. Res. Commun.* **287**(1), 142–146.

Muccioli, G., Pons, N., Ghe, C., Catapano, F., Granata, R., and Ghigo, E. (2004). Ghrelin and desacyl ghrelin both inhibit isoproterenol-induced lipolysis in rat adipocytes via a non-type 1a growth hormone secretagogue receptor. *Eur. J. Pharmacol.* **498**(1–3), 27–35.

Nagaya, N., Uematsu, M., Kojima, M., Ikeda, Y., Yoshihara, F., Shimizu, W., Hosoda, H., Hirota, Y., Ishida, H., Mori, H., and Kangawa, K. (2001). Chronic administration of ghrelin improves left ventricular dysfunction and attenuates development of cardiac cachexia in rats with heart failure. *Circulation* **104**(12), 1430–1435.

Nagaya, N., Moriya, J., Yasumura, Y., Uematsu, M., Ono, F., Shimizu, W., Ueno, K., Kitakaze, M., Miyatake, K., and Kangawa, K. (2004). Effects of ghrelin administration on left ventricular function, exercise capacity, and muscle wasting in patients with chronic heart failure. *Circulation* **110**(24), 3674–3679.

Nakazato, M., Murakami, N., Date, Y., Kojima, M., Matsuo, H., Kangawa, K., and Matsukura, S. (2001). A role for ghrelin in the central regulation of feeding. *Nature* **409**(6817), 194–198.

Nishi, Y., Hiejima, H., Hosoda, H., Kaiya, H., Mori, K., Fukue, Y., Yanase, T., Nawata, H., Kangawa, K., and Kojima, M. (2005). Ingested medium-chain fatty acids are directly utilized for the acyl modification of ghrelin. *Endocrinology* **146**(5), 2255–2264.

Peino, R., Baldelli, R., Rodriguez-Garcia, J., Rodriguez-Segade, S., Kojima, M., Kangawa, K., Arvat, E., Ghigo, E., Dieguez, C., and Casanueva, F. F. (2000).

Ghrelin-induced growth hormone secretion in humans. *Eur. J. Endocrinol.* **143**(6), R11–R14.
Saito, E. S., Kaiya, H., Tachibana, T., Tomonaga, S., Denbow, D. M., Kangawa, K., and Furuse, M. (2002). Chicken ghrelin and growth hormone-releasing peptide-2 inhibit food intake of neonatal chicks. *Eur. J. Pharmacol.* **453**(1), 75–79.
Sato, M., Nakahara, K., Goto, S., Kaiya, H., Miyazato, M., Date, Y., Nakazato, M., Kangawa, K., and Murakami, N. (2006). Effects of ghrelin and desacyl ghrelin on neurogenesis of the rat fetal spinal cord. *Biochem. Biophys. Res. Commun.* **350**(3), 598–603.
Sato, T., Fukue, Y., Teranishi, H., Yoshida, Y., and Kojima, M. (2005). Molecular forms of hypothalamic ghrelin and its regulation by fasting and 2-deoxy-d-glucose administration. *Endocrinology* **146**(6), 2510–2516.
Smith, R. G., Leonard, R., Bailey, A. R., Palyha, O., Feighner, S., Tan, C., McKee, K. K., Pong, S. S., Griffin, P., and Howard, A. (2001). Growth hormone secretagogue receptor family members and ligands. *Endocrine* **14**(1), 9–14.
Takaya, K., Ariyasu, H., Kanamoto, N., Iwakura, H., Yoshimoto, A., Harada, M., Mori, K., Komatsu, Y., Usui, T., Shimatsu, A., Ogawa, Y., Hosoda, K., *et al.* (2000). Ghrelin strongly stimulates growth hormone release in humans. *J. Clin. Endocrinol. Metab.* **85**(12), 4908–4911.
Tanaka, M., Hayashida, Y., Iguchi, T., Nakao, N., Nakai, N., and Nakashima, K. (2001). Organization of the mouse ghrelin gene and promoter: Occurrence of a short noncoding first exon. *Endocrinology* **142**(8), 3697–3700.
Thorens, B., and Larsen, P. J. (2004). Gut-derived signaling molecules and vagal afferents in the control of glucose and energy homeostasis. *Curr. Opin. Clin. Nutr. Metab. Care* **7**(4), 471–478.
Tomasetto, C., Karam, S. M., Ribieras, S., Masson, R., Lefebvre, O., Staub, A., Alexander, G., Chenard, M. P., and Rio, M. C. (2000). Identification and characterization of a novel gastric peptide hormone: The motilin-related peptide. *Gastroenterology* **119**(2), 395–405.
Toshinai, K., Yamaguchi, H., Sun, Y., Smith, R. G., Yamanaka, A., Sakurai, T., Date, Y., Mondal, M. S., Shimbara, T., Kawagoe, T., Murakami, N., Miyazato, M., *et al.* (2006). Des-acyl ghrelin induces food intake by a mechanism independent of the growth hormone secretagogue receptor. *Endocrinology* **147**(5), 2306–2314.
Tschop, M., Smiley, D. L., and Heiman, M. L. (2000). Ghrelin induces adiposity in rodents. *Nature* **407**(6806), 908–913.
Tschop, M., Wawarta, R., Riepl, R. L., Friedrich, S., Bidlingmaier, M., Landgraf, R., and Folwaczny, C. (2001). Post-prandial decrease of circulating human ghrelin levels. *J. Endocrinol. Invest.* **24**(6), RC19–RC21.
Unniappan, S., Lin, X., Cervini, L., Rivier, J., Kaiya, H., Kangawa, K., and Peter, R. E. (2002). Goldfish ghrelin: Molecular characterization of the complementary deoxyribonucleic acid, partial gene structure and evidence for its stimulatory role in food intake. *Endocrinology* **143**(10), 4143–4146.
van der Lely, A. J., Tschop, M., Heiman, M. L., and Ghigo, E. (2004). Biological, physiological, pathophysiological, and pharmacological aspects of ghrelin. *Endocr. Rev.* **25**(3), 426–457.
Wren, A. M., Small, C. J., Ward, H. L., Murphy, K. G., Dakin, C. L., Taheri, S., Kennedy, A. R., Roberts, G. H., Morgan, D. G., Ghatei, M. A., and Bloom, S. R. (2000). The novel hypothalamic peptide ghrelin stimulates food intake and growth hormone secretion. *Endocrinology* **141**(11), 4325–4328.
Yeung, C. M., Chan, C. B., Woo, N. Y., and Cheng, C. H. (2006). Seabream ghrelin: cDNA cloning, genomic organization and promoter studies. *J. Endocrinol.* **189**(2), 365–379.

CHAPTER FOUR

The Growth Hormone Secretagogue Receptor

Conrad Russell Young Cruz* *and* Roy G. Smith*,[†,‡,§]

Contents

I. Introduction: A Case of Reverse Pharmacology	48
II. Genetics and Molecular Biology	49
A. The GHSR gene	49
B. Structure of GHSR1a	52
III. Regulation of Gene Expression	55
A. Promoter sequences	55
B. Hormonal regulation	56
C. Ligand-independent activity	57
IV. Signal Transduction	58
A. Pathways leading to GH secretion	58
B. Other signaling systems	61
V. GHSR1a: Tissue Distribution and Functions in Different Organ Systems	62
A. The hypothalamic-pituitary axis and energy balance	62
B. Energy balance	64
C. Immune system modulation and anti-inflammatory properties of ghrelin	67
D. Ghrelin, GHSR, and aging	71
E. Expression and effects in other tissues	74
VI. Other Ligands	74
A. Alternative forms of ghrelin	74
B. Adenosine	75
C. Cortistatin	77
VII. Evidence for Other Receptors	77
VIII. Future Directions	78
References	79

* Translational Biology and Molecular Medicine Program, Baylor College of Medicine, One Baylor Plaza, Houston, Texas 77030
[†] Huffington Center on Aging, Baylor College of Medicine, One Baylor Plaza, Houston, Texas 77030
[‡] Department of Molecular and Cellular Biology, Baylor College of Medicine, One Baylor Plaza, Houston, Texas 77030
[§] Department of Medicine, Baylor College of Medicine, One Baylor Plaza, Houston, Texas 77030

Abstract

The neuroendocrine hormone ghrelin, a recently discovered acylated peptide with numerous activities in various organ systems, exerts most of its known effects on the body through a highly conserved G-protein–coupled receptor, the growth hormone secretagogue receptor (GHSR) type 1a. The GHSR's wide expression in different tissues reflects activity of its ligands in the hypothalamic-pituitary, cardiovascular, immune, gastrointestinal, and reproductive systems. Its extensive cellular distribution along with its important actions on the growth hormone (GH)/insulin-like growth factor-1 (IGF-1) axis and other neuroendocrine and metabolic systems suggest a pivotal role in governing the mechanisms of aging. A more comprehensive characterization of the receptor, and a more thorough identification of its various agonists and antagonists, will undoubtedly introduce important clinical applications in age-related states like anorexia, cardiovascular pathology, cancer, impaired energy balance, and immune dysfunction.

Although present knowledge points to a single functional receptor and a single endogenous ligand, recent investigations suggest the existence of additional GHSR subtypes, as well as other endogenous agonists.

It has been more than a decade since the landmark cloning of this ubiquitous, highly conserved receptor, and the considerable extent of its effects on normal physiology and disease states have filled the literature with incredible insights on how organisms regulate various functions through subtle signaling processes. But science has barely scratched the surface, and we can be assured that the mysteries surrounding the precise nature of ghrelin and its receptor(s) are only beginning to unravel. © 2008 Elsevier Inc.

I. Introduction: A Case of Reverse Pharmacology

The discovery of the hormone ghrelin by Kojima *et al.* (1999) was preceded by the cloning and identification of its receptor, the growth hormone secretagogue receptor (GHSR)—the result of expression cloning studies undertaken by investigators from Merck Research Laboratories (Howard *et al.*, 1996). Both ligand and receptor characterization were consequences of a series of studies aimed primarily at identifying substances that could amplify the amplitude of episodic growth hormone (GH) release from the anterior pituitary gland.

The search for secretagogues able to rejuvenate pulsatile GH release for potential use in deficiency states was modeled on the work of Bowers *et al.* (1981), who identified enkephalin-derived peptides able to stimulate GH secretion *in vitro*, prior to the identification of the GH releasing hormone (GHRH). Subsequent studies (Bowers *et al.*, 1991; Cheng *et al.*, 1989; Malozowski *et al.*, 1991) indicated that the mechanism involved is distinct from what is now recognized as the two primary regulators of GH release: GHRH activation and somatostatin antagonism.

Growth hormone secretagogue (GHS) activity during GH secretion was associated with increases in intracellular calcium in somatotrophs, increases in the activity of protein kinase C (PKC), magnesium-dependent binding to a high-affinity, low-abundance receptor, and inhibition by guanosine triphosphate analogues, all suggesting that the then uncharacterized GHSR was G-protein coupled (Howard et al., 1996). An expression cloning strategy, used to overcome the extraordinarily low abundance of a GHS (in this experiment, the MK0677 non-peptidyl mimetic)-binding site (Smith, 2005), was utilized to identify the G-protein–coupled receptor (GPCR). A cRNA pool prepared from a cDNA expression library of porcine pituitary gland messenger RNAs, a cRNA encoding the G-protein subunit $G_{\alpha 11}$ (which, in previous experiments, showed an ability to improve the sensitivity of the cloning assay), and a cRNA-encoding aequorin (a protein that gives off a bioluminescent signal in response to calcium, whose intracellular concentration is expected to increase during binding of GHS to the putative GHSR) were injected in *Xenopus* oocytes, and their light responses to MK0677 were measured. A clone was subsequently isolated after stepwise fractionations, and nucleotide sequencing revealed it to be a novel GPCR (Howard et al., 1996; Fig. 1).

The identification of an endogenous ligand came a few years later when a group led by Kojima et al. (1999) expressed the rat GHSR in a Chinese hamster ovary cell line and screened various tissue preparations for the characteristic increase in intracellular calcium concentrations induced by the GHSs, eventually finding a peptide from stomach extracts. The isolated hormone, christened "ghrelin" (after the Proto-Indo-European root of "grow"), featured a hitherto undiscovered modification: an octanoyl moiety in its third amino acid Ser that was vital for its activity (Fig. 2).

To date, ghrelin's documented activities include: the release of various hormones like GH, ACTH, cortisol, and prolactin (Hosoda et al., 2006); modulation of cell proliferation and survival; influences on food intake, sleep, behavior, glucose metabolism, pancreatic function, and gastric acid secretion (van der Lely et al., 2004); and inhibition of proinflammatory cascades and modulation of immune function that play important roles in (among other systems) aging and the cardiovascular system (Smith et al., 2005b).

II. Genetics and Molecular Biology

A. The GHSR gene

The gene encoding GHSR resides in chromosome 3q26.2 (McKee et al., 1997a) and presents sequence homology with only two other identified proteins receptors: the motilin receptor, whose ligand is an important signal

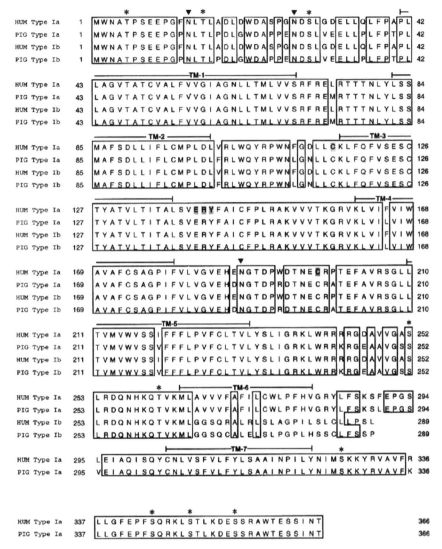

Figure 1 Amino acid sequence of the GHSR (aka ghrelin receptor) highlighting residues involved in forming various TM domains. Also shown is the homology between human and porcine receptors, for both GHSR transcripts 1a and 1b. Reprinted from *Science* (Howard et al., 1996).

involved in interdigestive migrating contractions (Itoh, 1997), with 52% sequence identity (Feighner et al., 1999), and the neurotensin receptor, a putative mediator of antipsychotic drug action and the neuronal networks serving the reward/sensitization pathways involved in drug abuse (Caceda et al., 2006), with 35% sequence identity (McKee et al., 1997b).

O=C(CH2)$_6$CH$_3$
|
GSSFLSPEHQKAQQRKESKKPPAKLQPR

Figure 2 Structure of ghrelin, showing the 28-amino acid peptide octanoylated on Ser3.

The gene is highly conserved in humans, chimpanzees, pigs, cows, rats, and mice (Howard *et al.*, 1996). It contains a single, ~2-kb intron separating two exons and encodes a GPCR with seven transmembrane (TM) domains (McKee *et al.*, 1997a; Fig. 3). Its ligand activation domain—key amino acid residues in the second TM domain (TM2) and third TM domain (TM3) essential for binding and activation by different ligands—has been evolutionarily conserved for 400 million years, highlighting its importance in fundamental physiological processes (Palyha *et al.*, 2000).

Several studies have sought to link mutations and polymorphisms in the ghrelin receptor gene to specific physiologic mechanisms involved in metabolism and in the development of obesity, but no consensus has ever been reached. Baessler *et al.* (2005) identified single nucleotide polymorphisms and haplotypes within the GHSR gene associate with obesity, while Wang *et al.* (2004) obtained no conclusive evidence for such involvement of GHSR gene variation. Vartiainen *et al.* (2004) found that genetic variations do not seem to associate with insulin-like growth factor-1 (IGF-1) levels, though they did see hints that IGFBP-1 concentrations may provide a better link.

Some genotypes seem to associate with specific disease processes. Baessler *et al.* (2006) showed a significantly higher percentage of subjects with left ventricular hypertrophy presented with one of the two most common single nucleotide polymorphism haplotypes in a linkage disequilibrium analysis. Another group led by Miyasaka *et al.* (2006) found higher frequencies of a certain GHSR genotype (the CC type) in patients suffering from bulimia. The actions of ghrelin on the cardiovascular system and appetite tend to support such associations with pathologies in these systems, though again further studies are needed to confirm whether such associations are clinically relevant.

Interestingly, the genetic locus of the GHSR gene also maps near the locus for Brachmann-de Lange syndrome (McKee *et al.*, 1997a), a disease characterized by growth delays during pre- and postnatal development, along with a typical facies and a constellation of other visceral and morphological defects (Boog *et al.*, 1999). Currently, however, no studies have explored the relationship between the two.

The functional ghrelin receptor—GHSR type 1a (GHSR1a)—is one of the transcripts produced by the GHSR gene. GHSR1b appears to be derived from an unspliced mRNA that terminates at a stop codon in the intron. Type 1a encodes a 366-amino acid polypeptide containing all seven

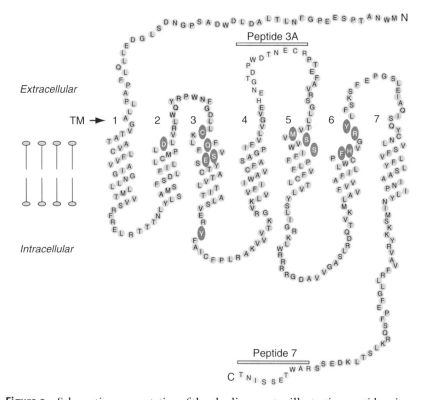

Figure 3 Schematic representation of the ghrelin receptor illustrating peptide epitopes (green) used to generate antibodies to extracellular loop 2 (peptide 3A) and the C-terminus (peptide 7) for determining orientation of the GHSR in the plasma membrane. Also highlighted are residues mutated to investigate the binding pocket occupied by MK0677. Reprinted from *Molecular Endocrinology*, Copyright 1998, *The Endocrine Society* (Feighner et al., 1998). (See Color Insert.)

TM domains, while type 1b encodes a truncated 289-amino acid polypeptide with only five TM domains. Both sequences are identical from the Met translation site to Leu at the 265th amino acid position (Howard et al., 1996). Expression studies in HEK293 cells, tagging each transcript to the fluorescent protein GFP, showed that GHSR1a localized to the plasma membrane. Curiously, however, GHSR1b was shown to localize in the nucleus (Smith et al., 2005b; Fig. 4). The significance of such a finding is still the subject of further inquiry.

B. Structure of GHSR1a

Studies of rat GHSR1a showed that it contains features consistent with most GPCRs: (1) the characteristic sequence Glu-Arg-Tyr adjacent to TM3, Cys115 and Cys197 in the first two extracellular loops that confer

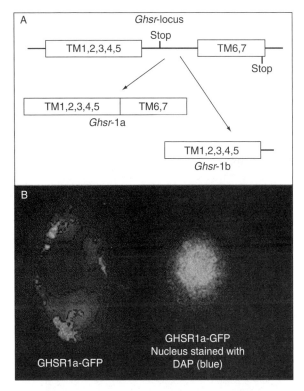

Figure 4 (A) Processing of GHSR primary transcript illustrating formation of GHSR subtypes. (B) Cellular localization of GHSR1a and GHSR1b following expression of GHSR1a and GHSR1b cDNAs tagged with green fluorescent protein (GFP) in HEK293 cells. GHSR1a is shown localized on the plasma membrane and GHSR1b is localized to the nucleus. Reprinted from *Trends in Endocrinology and Metabolism*, Copyright 2005, with permission from *Elsevier* (Smith et al., 2005). (See Color Insert.)

possible disulfide bond formation; (2) putative N-terminal N-glycosylation sites; (3) C-terminus and third cytoplasmic loop phosphorylation sites; and (4) conserved Pro residues in TM4–7 (McKee et al., 1997a). Mutagenesis of TM sites contributed much to analysis of functional domains within the GHSR. Amino acids like E124, D99, and R102 of TM2 and TM3 seem to comprise part of the ligand activation site for the receptor, while E124, F119, and Q120 of TM3 appear to mediate binding activity (Palyha et al., 2000).

Using antibody-binding studies directed at different regions of the ghrelin receptor, Feighner et al. (1998) confirmed that its membrane orientation resembled typical GPCR topology. Like the previously mentioned study by Palyha et al. (2000), the group identified the amino acid residue E124 as an important binding site for GHSs, mediating a salt bridge interaction with basic moieties that appear to act independent of receptor activation (Feighner *et al.*, 1998). Meanwhile, Bednarek *et al.* (2000) demonstrated the importance of the hydrophobic octanoyl moiety attached to ghrelin's Ser3 residue to GHSR1a binding, and identified an "active core" sequence of Gly-Ser-Ser (octanoyl)-Phe necessary for receptor activation.

Using the fragmental prediction strategy, a group led by Pedretti *et al.* (2006) proposed a model for GHSR1a (Fig. 5). The approach entailed dividing the GHSR1a primary structure into several fragments and predicting (using various algorithms) the folding for each. Combinations of the fragments were obtained and scored—the model subsequently chosen represented the best combination with the highest score. The N-terminal domain assumes a β-hairpin, and the TM domains form a round calyx-like structure which is attributed to Pro residues in the center of the TM helices. TM3 occupies the

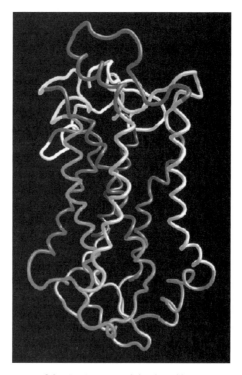

Figure 5 Tube structure of the GHSR1a model colored by segment. Color legend (NT, N-terminal domain; TM1–7, transmembrane domains 1–7; CL1–3, cytoplasmic loops 1–3; EL1–3, extracellular loops 1–3; CT, C-terminal domain): NT =white, TM1 = red, CL1 = green, TM2 = azure, EL1 =yellow, TM3 = dark red, CL2 = violet, TM4 = pink, EL2 = indigo, TM5 = gray, CL3 = orange, TM6 = dark green, EL3 = dark yellow, TM7 = brown, CT = blue. Reprinted with permission, Copyright 2006, *American Chemical Society* (Pedretti *et al.*, 2006). (See Color Insert.)

central position in the TM segment and TM5 is the most peripheral. Interestingly, based on interhelix distances, Pedretti et al. (2006) concluded that the rhodopsin structure cannot be used to model GHSR1a in its open space. Docking analysis with the FlexX program, an automated docking algorithm that "considers ligand conformational flexibility by an incremental fragment placing technique," was consistent with previous mutagenesis studies, lending credence to the model (Pedretti *et al.*, 2006).

While the International Union of Pharmacology recommends naming GHSR after its endogenous agonist and abbreviating it as GRLN receptor (Davenport *et al.*, 2005), this chapter still abbreviates the receptor as GHSR (which will represent the functional receptor unless otherwise written), or the more specific GHSR1a when necessary, and uses the terms "ghrelin receptor," and "growth hormone secretagogue receptor" interchangeably to maintain consistency with previously published papers.

III. REGULATION OF GENE EXPRESSION

Flanking the GHSR gene is an important upstream sequence that presumably confers tissue specificity, responsiveness to hormonal regulation, and constitutive activation.

A. Promoter sequences

In a series of experiments that illustrate the functions of the GHSR's regulatory elements, the 5′-untranslated region was inserted in a vector containing bacterial luciferase. Significant expression was observed in rat pituitary cells, but not in COS7 monkey kidney cells, human endometrium Skut-1B cells, mouse hypothalamic LHRH neuronal GT1–7 cells, or mouse corticotroph pituitary AtT20 cells. The monkey kidney COS7 cells, however, increased luciferase expression with coadministration of Pit-1. Pit-1 is a pituitary-specific transcription factor, suggesting that its binding elements control GHSR gene expression (Petersenn *et al.*, 2001). Sequence analysis of this region confirms a putative consensus binding site for Pit-1, along with the other POU domain transcription factors Oct-1 and Ptx1 (Kaji *et al.*, 1998).

Reported stimulatory effects of GHRH on GHSR mRNA expression levels (Kineman *et al.*, 1999; Yan *et al.*, 2004) may be mediated by an increase in Pit-1 expression. Soto *et al.* (1995) demonstrated an increase in transcription of the transcription factor Pit-1 in cultured rat pituitary cells treated with GHRH in an earlier study. Luque *et al.* (2004), however, reported quite the opposite: cells treated with a single dose of GHRH (equivalent to the dose that induces maximal release in porcine somatotrophs) diminished GHSR transcript levels by half.

B. Hormonal regulation

Hormonal regulation of the gene was suggested by increased GHSR gene expression in rat pituitary cells when treated with β-estradiol and the thyroid hormone T3 and decreased expression with hydrocortisone (Petersenn et al., 2001). These findings were supported by additional evidence from other experiments. Rat pituitary cells in culture treated with triiodothyronine (T3) exhibited elevated GHSR mRNA levels, even in the presence of an RNA synthesis inhibitor: the mechanism employed by the thyroid hormone involves increasing the stability of the mRNA transcript, and this is confirmed by an examination of its decay rates (Kamegai et al., 2001). GH3 cells transfected with the 5'-untranslated region of the GHSR gene ligated to luciferase showed a weak but significant inhibition of activity with the addition of hydrocortisone (Kaji et al., 2001).

In addition, GH, the secretion of which is increased by GHSR activity in somatotrophs, appears to modulate the expression of GHSR-expressing neurons. Increased levels of the receptor are seen in the arcuate and ventromedial nuclei, and hippocampal CA1 and CA2 neurons of GH-deficient, dwarf rats. These levels were reduced (significantly below normal in ARC and VMN but merely decreased in hippocampal neurons) by the administration of bovine GH, suggesting a possible mechanism for feedback regulation involving the ghrelin receptor (Bennett et al., 1997). Katayama et al. (2000) confirmed a similar increase in GHSR mRNA levels in the pituitary gland of GH-deficient rats. Ghrelin itself has been suggested to regulate the expression of its own receptor in somatotrophs—a single dose of the hormone that usually elicits maximal GH secretion was found to decrease GHSR mRNA levels in porcine pituitary cells by as much as 62% compared to controls (Luque et al., 2004). Ghrelin-mediated homologous desensitization has been examined more closely by Camiña et al. (2004). In studies using radioligand binding, intracellular calcium measurements, and confocal microscopy, the investigators show that prolonged stimulation with ghrelin, while initially showing the expected increase in intracellular calcium levels, subsequently decreased over time. Cells transfected with GHSR1a receptors conjugated to the fluorescent protein GFP showed internalization of membrane receptors after induction by ghrelin, through clathrin-mediated endocytosis. Fluorescence was noted to move from the plasma membrane to cytoplasmic vesicles, though the absence of lysosomal fusion (indicated by failure of the GHSR1a proteins to colocalize with cathepsin D, a marker of lysosome compartments) suggested that the receptors are not degraded, and may eventually get recycled back to the plasma membrane. Indeed, within 360 min of ghrelin treatment, radioligand binding and intracellular calcium steadily increased, while fluorescent GHSR1a began appearing under the plasma membrane (Camiña et al., 2004). However, we should be cautious in generalizing these results because the studies were conducted in cells that do not normally express the GHSR; therefore, the kinetics might be quite different in cells that express the GHSR endogenously.

For example, the biological responsiveness of pituitary cells is recovered within 60 min following desensitization with a saturating dose of the GHSR agonist GHRP-6 (Blake and Smith, 1991).

In other GPCRs, homologous desensitization is thought to occur through the actions of GPCR kinases, which specifically phosphorylate agonist-occupied receptors, minimally desensitizing it and marking it for arrestin binding. This proceeds to a subsequent sequestration of the receptor in endosomal vesicles, where it is targeted for destruction or mere inactivation (for later resensitization) (Iacovelli *et al.*, 1999). Recent bioluminescence resonance energy transfer studies between stably expressed luciferase-labeled arrestin stably transfected and transiently expressed GFP-labeled GHSR suggest such that ghrelin receptor interactions with arrestin do occur (Holst *et al.*, 2005). The precise mechanisms involved in GHSR-arrestin-mediated desensitization need to be further elucidated.

C. Ligand-independent activity

The promoter region of the GHSR lacks typical initiation sites like the TATA box, the CAAT box, or the GC-rich region. Instead, it contains TATA-less promoter elements that are distinctive features of constitutively expressed housekeeping genes, as well as other GPCR genes. Possible binding sites for bHLH and AP2 transcription factors may be involved in such purported GHSR basal activity (Kaji *et al.*, 1998).

Transient transfections of COS7 cells with the ghrelin receptor showed gene dose-dependent, ligand-independent increases in inositol phosphate accumulation, a measure of G-protein signaling through the phospholipase C (PLC) pathway, when compared with cultured cells injected with empty target vectors. These cells were pretreated with adenosine deaminase, mitigating concerns that endogenous production of this supposed GHSR agonist might interfere with the interpretation of results. Moreover, comparisons with transfection assays using the closely related motilin receptor showed no such ligand-independent activity: the measured inositol phosphate was not significantly different from cells treated with empty vectors (Holst *et al.*, 2003). The structural basis for such constitutive signaling may lie within the sixth and seventh TM domains: a systematic mutational approach showed that three significant aromatic residues within these domains appear to stabilize the ghrelin receptor in an active state (Holst *et al.*, 2004). However, the dose-dependent constitutive activity of the GHSR observed in these studies may not translate to physiology because the GHSR1a is expressed endogenously at extraordinarily low concentrations. For example, the most abundant endogenous source of GHSR1a is on membranes of the anterior pituitary gland, and Scatchard analyses on rat pituitary membranes indicate a concentration of 2.3 fmol receptor/mg protein (Pong *et al.*, 1996). All ligand binding is eliminated from pituitary membranes by 1-μM GTPγS; but when the GHSR is stably expressed in HEK293 cells, higher expression levels are observed and coupling is apparently different because agonist binding is far more resistant to GTPγS.

The importance of this basal activity is made evident in a rare mutation of the GHSR found in two Moroccan patients with familial short stature. (One patient, diagnosed with isolated GH deficiency, had a heterozygous mutation, while the other, with idiopathic short stature, had a homozygous one.) This C → A nucleotide substitution in a conserved residue of the first exon resulted in decreased cell surface expression of the receptor and decreased ligand-independent activity without a decrease in affinity to ghrelin (Pantel *et al.*, 2006), and may account for the abnormalities in the height of the patients. This result is in general agreement with the phenotype of $Ghsr^{-/-}$ mice, which exhibit modest but significant reduction in serum IGF-1 (Sun *et al.*, 2004).

IV. Signal Transduction

The most characterized signal transduction mechanism employed by the GHSR is involved in its stimulation of GH release. Indeed, the increase in intracellular calcium that is the hallmark of this pathway allowed the expression cloning of the receptor (Howard *et al.*, 1996).

A. Pathways leading to GH secretion

The earliest studies that examined the intracellular signaling events catalyzed by GHSR used the synthetic GHSs; at the time, the endogenous ligand ghrelin was not yet identified. Numerous studies indicated that the PLC-signaling pathway was the primary signaling mechanism involved. The PLC pathway (Fig. 6), stimulated by GPCRs, involves the substitution of GTP for GDP on receptor-associated G-proteins on ligand binding. This induces the dissociation of the $G_{q/\alpha 11}$-subunit and its subsequent stimulation of PLC. PLC cleaves the membrane lipid phosphoinositol 4,5 diphosphate (PtdIns (4,5) P_2) into diacylglycerol (DAG) and inositol (1,4,5) triphosphate (IP$_3$). IP$_3$, representing one arm of the cascade, mediates the release of calcium from stores inside the endoplasmic reticulum. DAG, representing the other arm, activates PKC, which inhibits potassium channels. The resulting membrane depolarization triggers voltage-gated calcium channels to open, and extracellular calcium rushes inside (Balla, 2006). The high intracellular calcium levels trigger endocytic release of GH into the extracellular milieu (Anderson *et al.*, 2004).

The GHSR stimulated by GHSs like GHRP-1 mediated elevations in intracellular calcium (Akman *et al.*, 1993). That this receptor induced two mechanisms of calcium release was shown by Herrington and Hille (1994)

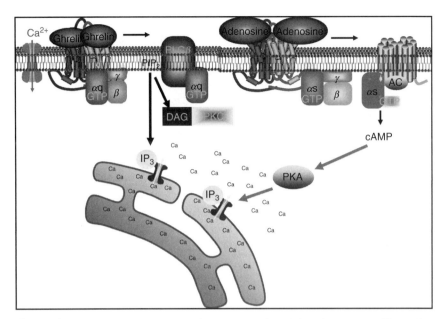

Figure 6 Signal transduction mechanisms employed by the GHSR illustrating the dominant PLC-PKC-inositol triphosphate (PLC/PKC/IP$_3$) pathway activated by ghrelin and ghrelin mimetics such as MK-0677 and GHRP-6, and the adenylate cyclase/PKA/cAMP (AC/PKA/cAMP) pathway activated by the partial agonist adenosine. The GHSR (ghrelin receptor) is represented as homodimers. (See Color Insert.)

in patch clamp and calcium photometry experiments done in rat somatotropes stimulated by another GHS, GHRP-6. An initial, rapid, but transient increase in intracellular calcium appears to originate within cells (the removal of extracellular calcium, the addition of the calcium channel blocker nitrendipine, and the induction of hyperpolarization through removal of extracellular sodium all failed to block this observed calcium spike), and a persistent elevation (this spike, in contrast, IS blocked by the removal of external calcium or sodium, or the addition of nitrendipine). PKC, whose effects include voltage-gated entry of extracellular calcium, was implicated in studies made using the same GHS on similar cells (Cheng *et al.*, 1991). Other studies on rat pituitary cells treated with GHRP-6 showed increased intracellular concentrations of sodium, suggesting that an increase in membrane permeability to the cation is an important factor contributing to cell depolarization (Kato and Sakuma, 1999). This extracellular calcium entry that preceded GH secretion in somatotrophs appeared to be mediated by a depolarization event involving potassium channels as well (Smith *et al.*, 1997). More studies on GHRP-6 further support the PLC hypothesis: human pituitary somatotrophs treated with GHRP-6 exhibited dose-dependent increases in the rate of phosphatidylinositol turnover (Lei *et al.*, 1995). A later study's observation that the $G_{\alpha 11}$-subunit augments intracellular calcium increase in pituitary cells stimulated with another GHS, MK0677, provided important confirmatory evidence (Howard *et al.*, 1996).

It is important to note that these actions were studied using non-endogenous agonists of the receptor, and a better appreciation of normal physiology necessarily involves characterization of ghrelin-stimulated GHSRs. The observations made on the ghrelin receptor itself actually lead to a more complex picture of GHSR signaling (Fig. 6).

GHSR dependence on both intracellular and extracellular calcium stores apparently holds true for ghrelin-induced activation as well (Yamazaki *et al.*, 2004). Experiments by Glavaski-Joksimovic *et al.* (2003) on porcine somatotrophs treated with ghrelin also confirmed the involvement of the PLC pathway in GHSR signaling; much like the GHSs, ghrelin-associated increase in intracellular calcium concentrations was attenuated in low extracellular calcium, blocked by nifedipine (a calcium channel blocker), reduced by U73122 (a PLC inhibitor), and decreased in the absence of extracellular sodium. The group, however, also documented a decrease in the actions of ghrelin on its receptor when the adenylate cyclase inhibitor SQ-22536 was added (Glavaski-Joksimovic *et al.*, 2003), implicating contributions of another pathway to its signaling transduction system. Similar findings were obtained by Malagon *et al.* (2003), who also used pig pituitary cells.

The proposed involvement of PKC in the inhibition of potassium currents was questioned by results obtained by Han *et al.* (2005), who suggested that the cGMP pathway may be involved after observing that reductions in the potassium current in GH3 cells treated with ghrelin were unaffected by both PKA and PKC inhibitors.

Carreira *et al.* (2004) speculated that the GHSR responds through different signaling mechanisms depending on the endogenous ligand, and undertook studies that aimed to characterize the intracellular cascade resulting from adenosine binding (Fig. 6). Adenosine has been previously reported as a possible endogenous ligand for the GHSR (Smith *et al.*, 2000). Adenosine-activated GHSR increased the levels of calcium as expected, but, in contrast to ghrelin, failed to increase the concentration of inositol phosphates. Moreover, the inhibitory effects of preincubation with the G_s-subunit activator cholera toxin (CTX), the adenylate cyclase inhibitor MDL-12,330 A, and the PKA inhibitor H-89 on these processes strongly suggest the utilization of the PKA/cAMP pathway via G_s-proteins (Carreira *et al.*, 2004).

The PKA/cAMP pathway is the proposed mechanism by which GHRH stimulates GH secretion (Root and Root, 2002). Although the GHSR has been known to show no such increase in intracellular cAMP levels when activated by ghrelin alone, coadministration with GHRH potentiated GHRH-induced increase in cAMP. In HeLa T4 cells cotransfected with both receptors, inhibition of PLC and PKC had no effect on GHSR potentiation of the GHRH cAMP increase (Cunha and Mayo, 2002). However, this result is in direct contrast to that observed in pituitary cells, which coexpress GHSR and GHRH receptors endogenously. In primary cultures of rat pituitary cells, potentiation of GHRH-induced cAMP accumulation and GHRH-induced GH release by GHSR agonists are mimicked by phorbol ester activation of PKC, inhibited by a PKC inhibitor, and inhibited by downregulation of PKC (Cheng et al., 1991; Smith et al., 1993). These results illustrate the need for caution regarding the physiological significance of extrapolating results derived from cell lines that express receptors exogenously.

B. Other signaling systems

Ghrelin stimulates other signaling systems as well, depending on the cell. In hepatoma cells expressing GHSR, ghrelin stimulated the intracellular signaling mechanisms characterized by Tyr phosphorylation of insulin receptor substrate-1 (IRS-1), binding of growth factor receptor-bound protein 2 (GRB2) to IRS-1, and the rise of mitogen-activated protein kinase (MAPK)—a downstream substrate of GRB2—activity (Murata et al., 2002). In neuropeptide Y (NPY)-containing neurons of the hypothalamus, ghrelin induced calcium influx via N-type calcium channels (as opposed to the L-type calcium channels shown to be stimulated in pituitary cells) using the PKA/cAMP pathway (Kohno et al., 2003). Whether GHSR mediates this ghrelin-activated system remains to be fully elucidated. In Chinese hamster ovary cells engineered to overexpress GHSR, ghrelin administration resulted in activation of the ERK1/2 pathway via PLC signaling (Mousseaux et al., 2006). A retroviral expression vector was used to express GHSR in nontransformed colonic epithelial cells (NCM460); ghrelin treatment caused phosphorylation and subsequent degradation of IκBα, allowing uninhibited activity of nuclear factor kappa B (NF-κB) to bind to interleukin-8 (IL-8) promoters and catalyze the latter's expression (Zhao et al., 2006).

Multiple-signaling properties is not unprecedented in GPCR biology, as GPCRs are proposed to adopt different active states that may mediate different G-protein-signaling cascades, depending on its interactions with the agonist (Perez and Karnik, 2005). Many other pathways are currently being identified in the literature. A review of the different possible signal transduction mechanisms employed by the GHSR is given by Camiña (2006). However, in reviewing this literature and for consideration of future reports, it is most important to question the physiological relevance of signaling pathways characterized in cell types that do not normally express the GHSR endogenously.

V. GHSR1a: Tissue Distribution and Functions in Different Organ Systems

A. The hypothalamic-pituitary axis and energy balance

Most of the GHSR1a receptors are concentrated in pituitary cells (Gnanapavan et al., 2002; Guan et al., 1997; Howard et al., 1996), and GHSR mRNA levels were greatly expressed in various nuclei of the hypothalamus demonstrated by *in situ* hybridization studies of the rat brain (Guan et al., 1997), consistent with its neuroendocrine roles in GH secretion and energy balance.

Ghrelin is known to induce a specific, dose-dependent release of GH when added to primary cultured pituitary cells, and when injected intravenously into anesthetized rats (Kojima et al., 1999). That such effects of ghrelin on GH release are mediated by GHSR was conclusively demonstrated in a study by Sun et al. (2004). Here, GHSR-null mice failed to show the ghrelin-induced GH secretion evident in wild-type mice.

Investigations of the endocrine functions of ghrelin on human GHSR revealed this ligand's tremendous potency: GH levels were significantly higher after ghrelin administration compared to levels obtained after treatment with either GHRH or the GHSR agonist hexarelin. Moreover, ghrelin was found to induce prolactin, adrenocorticotropin hormone (ACTH), cortisol, and aldosterone release (the last two probably result from ACTH stimulation) in humans (Arvat et al., 2001).

The pituitary-specific transcription factor Pit-1 was previously shown to promote the transcription of the gene for GH; this gene fails to express its product in the absence of Pit-1. This factor is also implicated in the embryological development of pituitary cells; Snell dwarf mice, who do not express Pit-1, fail to produce somatotrophs (Harvey et al., 2000). In an interesting report, ghrelin and the GHS GHRP-6 were shown to increase the expression of Pit-1 in an HEK293 cell line overexpressing the GHSR (Garcia et al., 2001). It is tempting to speculate putative roles for ghrelin in not just the secretion but the production of GH as well, but a closer examination of the aforementioned study ultimately showed failure of GHRP-6 to stimulate Pit-1 expression in pituitary cells derived from adult rats; only neonatal rats, with lower concentrations of GHSR exhibited Pit-1 transcription in response to GHRP-6 challenge (Garcia et al., 2001). Perhaps a better hypothesis lies in a role for ghrelin in the development of the pituitary, although clearly much more evidence is needed to pursue this conjecture.

GHSR in hypothalamic neurons exert important roles in GH endocrinology. Arcuate nuclei cells, when genetically altered to express antisense mRNA for GHSR (under tyrosine hydroxylase promoter control) so as to specifically inhibit receptor expression in just this group of cells, have been studied in transgenic rats. Here, the amplitude of endogenous GH pulsatility was markedly suppressed in female, but not male rats, and was associated with the anticipated reduction in serum IGF-1 (Shuto et al., 2002). Food intake, body weight, and adipose tissue weight were reduced in both male and female rats. These results support speculation that ghrelin regulates food intake and fat deposition; however, although $Ghsr^{-/-}$ mice exhibit reduced IGF-1, the ablation of the ghrelin receptor does not reduce food intake or fat mass, indicative of species differences (Sun et al., 2004).

The synergistic actions of ghrelin on GHRH-mediated GH release in humans (Hataya et al., 2001) suggest that the hormone may be acting on GHRH-expressing neurons. A ghrelin-induced GH response observed in rats was essentially eliminated by immunoneutralization of endogenous GHRH, encouraging speculation that ghrelin serves as a signal to link energy metabolism and GH secretion (Tannenbaum et al., 2003). Moreover, previously discussed stimulation of cAMP levels induced by ghrelin helps augment similarly induced increases by GHRH (Cunha and Mayo, 2002).

Experiments in transgenic mice modified to overexpress GHSR1a in their GHRH neurons resulted in several interesting observations when these animals were compared to their wild-type littermates. While transgenic mice exhibited slightly higher growth rates after weaning (which became significant after 6 weeks), the growth rate difference waned when the two groups of mice reached adulthood. Additionally, whereas increased production of GH contents was measured in the transgenic mice, no differences in basal secretion or ligand-induced GH release were noted (Lall et al., 2004).

Even though the stimulatory effect of ghrelin receptor activation on GH secretion is well established, there is still considerable debate on what the precise physiological role GHSR plays in the GH axis. Despite its potent induction of GH release, GHSR-null mice showed no phenotypic differences with wild-type mice, suggesting that ghrelin and its receptor do not exert dominant effects in this neuroendocrine process (Sun et al., 2004). Mice where the ghrelin gene was knocked out also do not exhibit any noticeable phenotypic abnormality—no changes in their size, growth rate, food intake, body composition, reproduction, gross behavior, and tissue pathology were detected (Sun et al., 2003). GHSR null mice do exhibit modest reductions in body weight and exhibit biochemical alterations in IGF-1 levels (Sun et al., 2004). IGF-1 generates most of the perceptible, anabolic effects of GH (like linear growth and increase in skeletal muscle mass) (Root and Root, 2002); unlike GH, which is secreted in a pulsatile manner, changes in IGF-1 levels are easier to measure as they do not fluctuate markedly (Sun et al., 2004). There exists, of course, the possibility that compensatory mechanisms circumvent the GH's requirement for

GHSR stimulation, but at the very least, these results indicate that ghrelin may play a more modulatory role in GH release (Sun et al., 2004).

Besides GH secretion, ghrelin and the GHSR mediate appetite, feeding behavior, and energy balance through their actions on various neurons in the hypothalamus.

B. Energy balance

Soon after its discovery, intracerebroventricular administration of ghrelin was observed to induce feeding and weight gain in rats. That these effects occurred through stimulation of the GHSR was demonstrated by suppression of the response by the GHSR antagonist [D-Lys3] GHRP-6. Measurements of c-fos expression (a marker of neuronal activity) showed immunoreactive neuronal populations in areas associated with the regulation of feeding behavior (Nakazato et al., 2001). The pivotal role of the arcuate nucleus (ARC) in ghrelin/GHSR-stimulated feeding behavior can be inferred from reports that show abolition of this response in rats neonatally treated with monosodium glutamate (a substance that destroys hypothalamic arcuate nuclei) (Tamura et al., 2002). NPY- and AgRP-containing cells in the ARC were further revealed to serve important roles in this GHSR-mediated mechanism: pretreatment with IgG antibodies to NPY, the NPY-associated receptors Y1 and Y5, or AgRP all abolished ghrelin-induced feeding (Nakazato et al., 2001). A series of experiments where wild-type $Ghsr^{-/-}$, $npy^{-/-}$, $agrp^{-/-}$, and $agrp^{-/-}.npy^{-/-}$ were treated with ghrelin to determine which downstream mediators of ghrelin regulated appetite showed that both AgRP and NPY had to be ablated to recapitulate the phenotype of $Ghsr^{-/-}$ mice (Chen et al., 2004).

Evidence for the role of GHSR in the regulation of feeding is seen in transgenic animals modified to selectively suppress GHSR expression in the ARC. Genetically modified male rats were shown to eat less than age-matched normal rats from weaning to 8 weeks of age, while transgenic females ate less than the control rats at 3, 4, 5, 6, 8, and 9 weeks of age. In addition, the administration of the GHS GHRP-2 to these rats failed to stimulate feeding (Shuto et al., 2002). GHSR-null mice also failed to increase food intake after intraperitoneal ghrelin injection. Wild-type counterparts showed increased feeding after intraperitoneal ghrelin during the first half-hour immediately following treatment, while no such activity was observed in saline-treated, wild-type mice (Sun et al., 2004).

Regulation of energy metabolism through appetite stimulation is an important process whose pathophysiological consequences include the development of obesity. Earlier findings of the appetite of mice whose gene for GHSR was knocked out showed no significant difference from wild-type animals: cumulative food intake and biweekly food intake were similar in both groups (Sun et al., 2004). Although ghrelin knockout mice failed to show decreases in food intake (Sun et al., 2004; Wortley et al., 2004), the observed preferences for metabolic substrates of these mice (the ghrelin null animals, like their GHSR knockout counterparts) showed decreases in their respiratory quotients (RQ; that is, they overused fat as a fuel source when administered a high-fat diet), indicating that a physiological role for ghrelin in energy balance involves the modulation of fuel utilization (Wortley et al., 2004; Zigman et al., 2005). More recently, Wortley et al. (2005) showed increased energy expenditure in young N3F2 $ghrelin^{-/-}$ mice when high-fat (HF) feeding was initiated immediately after weaning, but failed to detect a change in RQ. This observation contradicted their previous report in 14- to 16-week-old N2F2 mice, which indicated that HF feeding reduces RQ, but has no effect on energy expenditure (Wortley et al., 2004).

Recent reports suggest that ghrelin- and $Ghsr^{-/-}$ mice are resistant to DIO when fed HF diets immediately after weaning (Wortley et al., 2005; Zigman et al., 2005), which contrasts with results from adult congenic $ghrelin^{-/-}$ and $Ghsr^{-/-}$ mice, where DIO was not attenuated. One explanation for the apparent discrepancy is that as mice reach adulthood, they develop compensatory pathways to adjust for the loss of a ghrelin/GHSR signal (Sun et al., 2007). A more likely explanation emerges from consideration of the genetic backgrounds of the mice used. Attenuation of DIO was described in $Ghsr^{-/-}$ mice from a mixed (N2F1) genetic background (Zigman et al., 2005), whereas Sun and colleagues used C57BL/6J congenic $Ghsr^{-/-}$ mice. These differences in background could undoubtedly influence outcome of metabolic experiments. The N2 generation mice, generated by backcrossing F1 animals (genetic background 50% identical to C57BL/6 and 50% identical to 129Sv) to C57BL/6 for one generation, are 75% identical to C57BL/6 and 25% identical to 129Sv. Since the 129Sv mouse strain has a lean phenotype, whereas C57BL/6 is highly susceptible to DIO, the observed differences in the two studies might be explained by traits carried with the 129Sv genome.

The anatomical circuit for the energy regulatory roles of ghrelin and its receptor was deduced from electrophysiological studies of hypothalamic neurons. Ghrelin appears to be robustly expressed in axons which innervated several hypothalamic nuclei, including the ARC, dorsomedial hypothalamus (DMH), lateral hypothalamus (LH), and paraventricular hypothalamic nucleus (PVH). GHSR, identified through binding studies with biotinylated ghrelin, primarily localizes in the presynaptic boutons of the ARC, LH, and PVH. Ghrelin mediates depolarization/activation of NPY and AgRP neurons that stimulate appetite and hyperpolarization/inhibition of (POMC) neurons that are implicated in anorexigenic signaling (Cowley et al., 2003; Fig. 7).

Leptin, an anorexigenic hormone, acts on this anatomic circuit as well. Leptin depolarizes the POMC neurons while hyperpolarizing NPY cells (Cowley et al., 2001), causing satiety and decreased hyperphagia. In the

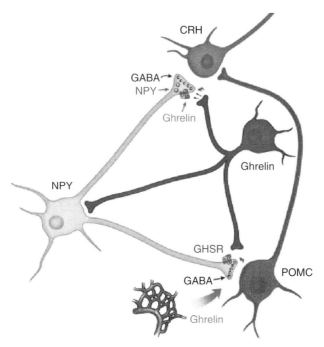

Figure 7 Hypothalamic circuits employed by the ghrelin/GHSR system during energy homeostasis illustrating ghrelin activation of NPY-containing neurons and inhibition of pro-opiomelanocortin (POMC) neurons. Corticotrophin-releasing hormone, CRH; γ-aminobutyric acid, GABA. Reprinted from Neuron, Copyright 2003, with permission from *Elsevier* (Cowley *et al.*, 2003). (See Color Insert.)

homeostatic mechanisms governing appetite and food intake, leptin serves as ghrelin's functional antagonist.

How ghrelin comes to act on the central nervous system in living organisms, to act on the GH-secretory and energy homeostatic pathways, is still a mystery. Circulating plasma ghrelin comes chiefly from the endocrine cells of the stomach (Ariyasu *et al.*, 2001), and there is still considerable debate on what mechanisms comprise the activation of hypothalamic and pituitary GHSRs by this hormone. This is because the blood–brain barrier presents a significant obstacle to the transport of ghrelin from the blood to brain tissues.

Studies made by Banks *et al.* (2002) using radiolabeled substrates suggested the existence of certain saturable transport systems that bind ghrelin in the blood and facilitate its transfer into the neuronal environment; data from these experiments recognized the importance of the unique octanoylation of ghrelin in this transport process. The presence of ghrelin mRNA in pituitary cells (Korbonits *et al.*, 2001), meanwhile, suggests the presence of paracrine/autocrine ghrelin pathways in the central nervous system, which may actually be the primary signals involved. Staining of ghrelin immunoreactive axons in various hypothalamic networks (caudal part of the supraoptic nucleus, the periventricular and suprachiasmatic nuclei, the periventricular part of the paraventricular nucleus, the ventral perifornical region, the rostral part of the supraoptic nucleus) in postmortem human brain samples (Menyhert *et al.*, 2006) indicates that the control of energy homeostasis is, at least in part, mediated by ghrelin/neuronal signaling from within the hypothalamus. On the other hand, the involvement of the afferent vagal nerve in such a process is advocated by Date *et al.* (2002) following a series of investigations that revealed (1) the presence of GHSRs in vagal afferents and (2) the abolition of ghrelin-induced GH secretion, increase in feeding behavior, and NPY cell activity on vagotomy or capsaicin (an afferent neuron inhibitor) blockade.

C. Immune system modulation and anti-inflammatory properties of ghrelin

Growth and metabolic control represent but a facet of the neuroendocrine system's role in maintaining the organism; the powerful, diverse, and highly regulated neuroendocrine hormones are often utilized by the body in numerous other processes. The GH/IGF-1 axis, for example, has been documented to possess stimulatory functions on immune cells—ranging from enhancement of T cell proliferation to induction of immunoglobulin production (Murphy *et al.*, 1995).

In one of the many investigations seeking to extend the physiological effects of ghrelin into other organ systems, a search for other sites of GHSR cellular localization revealed possible involvement with the body's mechanism for self-defense. The expression of GHSR in cells of the immune system was detected by reverse-transcriptase polymerase chain reaction (RT-PCR), and subsequently confirmed by DNA sequencing. The human leukemic cell lines Raji, Daudi, Jurkat, K-562, HL-60, and Hut-78, as well as T cells, B cells, and neutrophils from normal individuals, all demonstrated GHSR mRNA expression. Transcript levels, however, varied widely among individuals (Hattori *et al.*, 2001). Such widespread expression of GHSR supports a role for ghrelin in the regulation of immune-related functions. That the GHSR is actually expressed on the surface of T cells and monocytes was demonstrated by Dixit *et al.* (2004), who also observed (1) increases in intracellular calcium (an important characteristic response of GHSR on activation) in human T cells treated with ghrelin and ghrelin fragments, and (2) a subcellular reorganization event involving the receptors in activated T cells: the GHSRs aggregated and colocalized in GM1 + lipid rafts during T cell receptor ligation (and consequent cell activation). Activated T cells also exhibit the formation of lamellipodia and actin cytoskeleton remodeling—events that lead to polarization and directional migration, and, as the researchers subsequently found, events that were induced in cultured lymphocytes by the ligation of GHSR (Dixit et al., 2004). These findings validate inferences of a functional GHSR in T cells.

Influences of the GHSR on immune cells were suggested by the actions of a GHS, an MK0677 analogue. The GHS consistently increased the numbers of peripheral white blood cells in young mice; the secretagogue also maintained lymphopoiesis at times that mice normally have decreased white blood, and lymphoid cells. Its effects are particularly more dramatic in older mice: a statistically significant increase in thymic cellularity was observed in GHS-treated animals, as well as a remarkable resistance against tumor development. The transplantable tumor EL4 (which did not exhibit any effects when treated with GHS *in vitro*) was injected subcutaneously into elderly mice, a group of which received GHS, while another received empty vehicle. All eight vehicle-treated, 22-month-old mice developed tumors, two of which developed metastases in the pleural and peritoneal cavities; on the other hand, only five (out of seven) of their ghrelin-treated counterparts developed tumors, and none developed metastases. Enhanced production of cytotoxic T lymphocytes against EL4 cells in treated mice suggests that its receptor mediates important immune-enhancing effects (Koo et al., 2001).

The same group also observed that the MK0677 analogue enhanced antibody production, promoted the engraftment of bone marrow cells in severe, combined immunodeficiency (SCID) mice, and increased cycling cells in the spleens of 9- and 10-month-old mice (Koo et al., 2001).

Studies on human umbilical vein endothelial cells (HUVECs) show a significant effect of ghrelin against inflammatory mechanisms induced by tumor necrosis factor-α (TNF-α); HUVECs pretreated with ghrelin had lower levels of IL-8 and MCP1 compared to untreated cells. The transcription of NF-κB, a key regulator involved in the production of chemotactic cytokines and the expression of adhesion molecules, was also decreased by ghrelin in HUVECs transduced to express NF-κB-dependent luciferase— both in the presence and absence of TNF-α (Li et al., 2004). Li et al. (2004) also found other anti-inflammatory properties brought about by ghrelin: the reduction of TNF-α-mediated mononuclear cell adhesion (a crucial factor in vascular inflammation and atherosclerosis) and the inhibition of hydrogen peroxide promoted release of inflammatory mediators IL-8 (through mechanisms independent of reactive oxygen species scavenging).

GHSR expression on monocytes may explain some of the anti-inflammatory effects of ghrelin. Monocytes are important sources of proinflammatory cytokines, and initial studies show that ghrelin acts to inhibit the production of IL-1β and IL-6 via the GHSR (evidence for the involvement of GHSR came from observations that this cytokine inhibition is blunted in the presence of GHSR antagonists) (Dixit et al., 2004).

Zhao et al. (2006) reported proinflammatory effects of ghrelin in mice with experimentally induced colitis based on their observation that levels of ghrelin and GHSR mRNA were elevated during the acute phase of colitis. In direct contrast, using the same model of colitis, Gonzalez-Rey et al. concluded that ghrelin showed novel anti-inflammatory actions in the gastrointestinal tract. The discrepancy is probably explained by the fact that Zhao et al. concluded that ghrelin plays a proinflammatory role in the gastrointestinal tract by stimulating the NF-κB pathway and increasing IL-8. However, the results in support of this conclusion were derived from studies in a nonphysiological *in vitro* system, where a colonic epithelial cell line was transfected with GHSR rather than from the *in vivo* model (Zhao et al., 2006). Confirming that the results from the *in vitro* model were misleading, in the *in vivo* model Gonzalez-Rey et al. demonstrated that exogenously administered ghrelin deactivated the intestinal inflammatory response and restored mucosal immune tolerance at multiple levels. Ghrelin reduced serum levels of the proinflammatory cytokines IL-6, TNF-α, IFN-γ and increased the anti-inflammatory cytokine, IL-10. Indeed, these authors concluded that ghrelin administration provided a novel therapeutic approach for treating Crohn's disease and other Th1-mediated inflammatory diseases (Gonzalez-Rey et al., 2006).

Sepsis is a deleterious consequence of excessive immune system activation after systemic infection and remains a major cause of death (Bach et al., 2006). Positive functions mediated by the GHSR in this pathological state have been shown in several studies. Rats intravenously injected with bacterial lipopolysaccharide (LPS), a potent inducer of sepsis, developed severe hypotension, heart failure, and critical metabolic disturbances that all led to a significant mortality rate. The administration of ghrelin either during or 12 h after administration of LPS decreased mortality rates (25% in LPS-treated mice given ghrelin at the same time, 47% in LPS-treated rats given ghrelin at a later time, and 55% in LPS-treated rats without ghrelin). Improved hemodynamic parameters (mean arterial blood pressure, heart rate, cardiac systolic, and diastolic function) were also seen in rats who received ghrelin early (Chang et al., 2003).

Ghrelin may be involved in the inhibition of TNF-α-induced activation of NF-κB and may also be involved in the downregulation of endothelin-1, the most potent vasoconstrictor produced by endothelial cells (Wu et al., 2005). Ghrelin was also shown to dramatically reduce the production of the inflammatory cytokines IL-8, TNF-α, and MCP-1 in LPS-treated rats (Li et al., 2004). Dixit et al. (2004) reported reductions in the expression of TNF-α, IL-6, and IL-1β in LPS-induced endotoxemia *in vivo*. There are findings that GHSRs in the cardiovascular system are upregulated during the early phase of sepsis. Protein levels (measured by Western blots) of GHSR1a were increased in the aorta, heart, and the small intestine during the hyperdynamic phase of a cecal ligation and puncture-induced polymicrobial sepsis in rats; whether this represents effects caused by the cytokines involved in the systemic inflammatory response is yet to be determined (Wu et al., 2004).

The actions of ghrelin and its receptor on energy balance and immunity present an opportunity for the body to coordinate the immune and metabolic systems, allowing appropriate responses to various physiological and pathological conditions: such an integrative role may provide the rationale for GHSR expression in immune cells (Dixit and Taub, 2005).

Cachexia is an excellent example of a disease linked with both metabolism and immunity. It is characterized by unintentional weight loss, often with greater wasting from muscle tissue. There is evidence of cytokine excess, and proinflammatory cytokines are currently recognized as the primary cause of cachexia (Morley *et al.*, 2006). Leptin, previously mentioned in discussions of the hypothalamic circuits governing appetite and food intake, appears to play an important role in metabolism and immunity as well. Besides acting as a functional antagonist for ghrelin in energy balance, it appears that leptin acts as an opposing hormone in inflammation (Fig. 8). Leptin induces the production of IL-6 and TNF-α in PBMCs and monocytes (Zarkesh-Esfahani *et al.*, 2001) and—if it may be conjectured—probably contributes to the pathogenesis of cachexia.

Again acting through the GHSR, ghrelin significantly inhibits the production of IL-1-β, IL-6, and TNF-α by peripheral blood mononuclear cells. But more importantly, ghrelin was able to inhibit leptin-induced

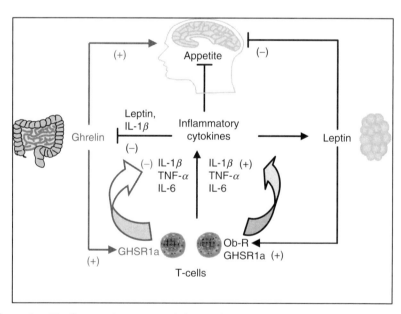

Figure 8 Ghrelin as an immunomodulator. Ghrelin inhibits the production of proinflammatory cytokines such as IL-1β, TNF-α, IL-6, and leptin; ghrelin stimulates appetite, whereas leptin and the proinflammatory cytokines inhibit food intake. (See Color Insert.)

cytokine expression (Dixit *et al.*, 2004). Its unique place at the interface of metabolism and immunity give it importance as a therapeutic option for the debilitating disease.

D. Ghrelin, GHSR, and aging

Pituitary/endocrine hormone secretion, energy metabolism, and immunity represent homeostatic mechanisms in the body that possess complex regulatory networks. In a sense, such nonlinear pathways allow a robustness that facilitates adaptability to different environments: biological systems existing "at the edge of chaos" are essentially more resistant to disruption than systems that are more repetitive, linear, or random. Alterations in the components of such a system that lead to reduced complexity result in a state of increased susceptibility and decreased adaptability to environmental stress, trauma, and insult: the phenotype we associate with aging (Smith *et al.*, 2005a).

Levels of GH, and its liver-derived primary effector IGF-1, decline with age, both in its basal state and in response to stimulation, causing decreases in protein synthesis, lean body mass, bone mass, and immune function (Chahal and Drake, 2007). The pituitary reserve of GH, however, appears to be intact (Arlt and Hewison, 2004).

The anorexia of aging is an important cause of morbidity in the elderly, and it is believed to result from a host of different factors. A decline in taste sensitivity, decreased hunger and increased satiety, delayed gastric emptying, and the altered responsiveness of gut hormones (including leptin and ghrelin) all play a role in the elderly phenotype manifested as poor appetite and poor food intake (Hays and Roberts, 2006).

Immune dysfunction is another unfortunate hallmark of aging. There is an increase of cytokines, primarily IL-6 and TNF-α, and inflammatory mediators that—while not achieving levels that are typically observed in disease states—confer a chronic low-grade inflammation. This state appears to predispose the elderly to a number of diseases, including atherosclerosis, frailty, and cognitive decline (Krabbe *et al.*, 2004).

The GHSR's important effects in the endocrine, metabolic, and immune processes—the networks that comprise the complex systems affected in aging—suggest that it plays a pivotal role in the maintenance of a youthful phenotype and aging itself, a pathophysiological consequence of decreased GHSR sensitivity to its endogenous ligands (Smith *et al.*, 2005b). While a decrease in the expression of GHSR and a decrease in the production of ghrelin are equally acceptable reasons for ghrelin pathology in the elderly, recent findings tend to refute these theories and tend to support the GHSR resistance hypothesis.

In mice, both the expression of GHSR mRNA and the circulating levels of ghrelin remain stable during aging; ghrelin levels even increased slightly (a finding that also supports increasing ghrelin resistance) (Sun et al., 2007). Other studies appear to validate these observations. One group found that fasting and postprandial levels of ghrelin do not significantly differ between the young and the old (Di Francesco et al., 2006). Another, while also demonstrating no significant differences in the levels of acylated and unacylated ghrelin with age, noted a decrease in orexigenic effect (Akimoto-Takano et al., 2006).

There are reports of a decrease in the acylated fraction of total ghrelin in the elderly when compared to their younger counterparts (Baranowska et al., 2006).

It is no coincidence, then, that the history of ghrelin is interwoven with the search for the mythical fountain of youth. The ghrelin receptor's identification came in the wake of research aimed at restoring the youthful phenotype through increasing GH pulsatile amplitude in the elderly. And at the moment, accumulating data point to more and more physiological systems associated with, and influenced by, ghrelin and its receptor in the reversal of age-related decline.

Dopamine provides a classic example of age-related change resulting in a system with reduced complexity. During aging, there is a decline in the number of dopaminergic neurons, as well as a decrease in the production of the neurotransmitter. The consequent reduction in complexity and signaling effectiveness of the neurons could then lead to failures in adaptive mechanisms, causing, for example, vulnerability to ischemic injury (Smith et al., 2005a).

The possible role of the GHSR in ghrelin's modulation of dopamine signaling was illustrated in an elegant set of experiments by Jiang et al. (2006). Briefly, the researchers described coexpression of the receptors for ghrelin and dopamine D1R in neurons of the cortex, hippocampal structures, substantia nigra, midbrain, and ventral tegmental area; they also observed synergistic increases in intracellular cAMP levels (the main signaling molecule involved in dopamine activation) with coadministration of ghrelin and dopamine—a response not seen with the administration of ghrelin alone or the absence of the GHSR. An *in vitro* HEK293 cell model expressing both receptors was then used to explore the mechanisms whereby ghrelin amplifies dopamine signaling. Co-immunoprecipitation and BRET experiments suggest a formation of heterotrimeric complexes; experiments using pertussis toxin (an inhibitor of $G_{\alpha i/o}$), as well as phorbol 12-myristate 13-acetate and bisindoylmaleimide I (activator and inhibitor of PKC, respectively), meanwhile, provide evidence that cross talk activation involves a different G_α-subunit—$G_{\alpha i/o}$ in place of the usual GHSR-associated $G_{q/\alpha 11}$—and consequently, a different signal transduction pathway (Jiang et al., 2006; Fig. 9).

Such an interaction allows one to envisage an important effect ghrelin may bring about in elderly subjects: the improvement of cognitive function

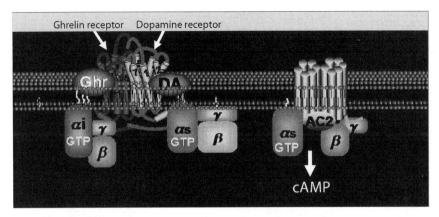

Figure 9 Model of ghrelin amplification of dopamine signaling involving GHSR/D1R heterodimerization causing a switch in GHSR G-protein coupling from $G_{\alpha q11}$ to $G_{\alpha i}$, resulting in synergistic activation of adenylyl cyclase-2 via $G_{\alpha s}$ (D1R coupling) and $\beta\gamma$-subunits derived from $G_{\alpha i}$ (GHSR coupling). Modified from *Molecular Endocrinology*, Copyright 2006, *The Endocrine Society* (Jiang et al., 2006). (See Color Insert.)

(Jiang et al., 2006). There is increasing evidence that dopamine plays a crucial role in higher cognitive processes (Backman et al., 2006). Dopamine increases intracellular cAMP levels, an event which was shown to induce changes in ionic conductance and synthesis of new proteins—efforts linked to the molecular mechanisms involved in learning and memory (Kandel et al, 2001).

Like the augmentation of dopaminergic signaling, the ideal, comprehensive solution to age-related decline should involve restoring the nonlinear, deterministic chaos that governs the neuronal and hormonal networks disrupted by this "natural" process. Agonists to the GHSR, an apparently central signal that is involved in most systems affected by aging, may represent such an intervention. There is, in fact, ample evidence that activating the ghrelin receptor can provide substantial improvements in these age-related processes. And even if no single system or function is dramatically improved, subtle enhancement of a multitude of physiological processes is likely to have a significant impact on the quality of life (Smith et al., 2005a).

MK0677, a GHSR agonist, has shown promise in different investigations (Smith, 2005). It has been observed to (1) increase peak amplitude and the 24-h area under the curve (AUC) for GH in older men and women (Chapman et al., 1996); (2) increase IGF-1 levels by up to 60% in the elderly (Chapman et al., 1996); (3) effect a reversal of diet-induced nitrogen wasting in humans (Murphy et al., 1998); (4) improved functional performance of elderly hip fracture patients in a placebo-controlled, randomized, double-blind trial (Bach et al., 2004); and (5) partially reverse thymic shrinkage and age-related reduction in the manufacture of T cells in mice, as well as reduce tumor growth rate and metastasis, in mice (Koo et al., 2001). Ultimately, better agonists with more desirable pharmacological properties will be discovered, paving the way for a revolution in the way we care for the elderly.

The growing evidence that functional plasticity remains in older individuals (Smith et al., 2005a) gives cause for optimism; and providing the adequate signal to restart still-operable functions that have merely lost their stimuli will be an exciting field of future research.

E. Expression and effects in other tissues

Besides the hypothalamic nuclei, pituitary somatotrophs, and cells of the immune system, the GHSR has been found elsewhere. Detectable levels of GHSR transcripts have been documented in other areas of the central nervous system, including the CA2 and CA3 regions of the hippocampus, the substantia nigra, the ventral tegmental area, the dentate gyrus of the hippocampal formation, and the dorsal and median raphe nuclei (Guan et al., 1997). In peripheral tissue, receptor-binding assays with the radiolabeled [^{125}I] Tyr-Ala-hexarelin detected putative ghrelin receptors in the myocardium, adrenal glands, gonads, arteries and veins, lungs, liver, skeletal muscle, kidney, thyroid, adipose tissue, uterus, skin, and lymph nodes (Papotti et al., 2000).

There is great amount of literature supporting the role of GHSR in various systems, and a number of these explore its responsibilities in the gastrointestinal system (Dass et al., 2003; Sibilia et al., 2006), the cardiovascular system (Kleinz et al., 2006), the gonadotropin hormones (Fernandez-Fernandez et al., 2004) and other organs of the reproductive system (Sirotkin et al., 2006; Tena-Sempere et al., 2002), the adrenal axis (Stevanovic et al., 2007), and the sleep–wake cycle (Tolle et al., 2002). Pituitary adenomas, not surprisingly, express GHSR (Kim et al., 2001), but, remarkably, various other tumors have also been documented to express transcripts of the ghrelin receptor and/or ghrelin, including astrocytoma (Dixit et al., 2006), highly differentiated Leydig cell tumors (Gaytan et al., 2004), and benign serous tumors of the ovary (Gaytan et al., 2005).

The physiological processes influenced by the receptor in these systems are the subjects of reviews found elsewhere in this volume.

VI. OTHER LIGANDS

A. Alternative forms of ghrelin

A few months after the identification of ghrelin as the GHSR's endogenous ligand, an alternatively spliced modification of this protein was shown to act as a second natural ligand for the receptor. Des-Gln14 ghrelin, a 27-amino acid peptide similar to ghrelin except for one Gln, shows similar abilities to increase intracellular calcium and stimulate GH release (Hosoda et al., 2000).

A particularly common peptide found in significant levels in the blood is desacyl ghrelin, the unacylated form of the hormone. Desacyl ghrelin, initially thought of as an inactive form of ghrelin, is now recognized to exhibit biological activity. Studies by Toshinai et al. (2006) showed that desacyl ghrelin increased food intake through a receptor and hypothalamic circuit distinct from the GHSR. To establish the mechanism involved, the effects of ghrelin and desacyl ghrelin were compared in $Ghsr^{-/-}$ and $orexin^{-/-}$ mice. Ghrelin increased food intake when administered to $orexin^{-/-}$, but not $Ghsr^{-/-}$ mice, whereas desacyl ghrelin stimulated feeding in $Ghsr^{-/-}$, but not $orexin^{-/-}$ mice. Hence, it was concluded that the orexigenic properties of desacyl ghrelin were mediated by activation of orexin neurons. This was confirmed by studies in isolated orexin neurons. Chen et al. (2005), meanwhile, showed that desacyl ghrelin decreases 1- and 2-h cumulative food intake in food-deprived rats by acting on the corticotrophin-releasing hormone (CRH) receptor (Chen et al., 2005).

The unacylated form of ghrelin produced varying effects in different prostatic cell lines: at high concentrations, it was able to inhibit cell proliferation in PC3 and DU-145 cells, while at low concentrations, it significantly increased the numbers of PC3 cells (and had no effect on DU 145 cells). A similar profile was seen with ghrelin (Cassoni et al., 2004). Desacyl ghrelin was also found to stimulate fetal spinal cord proliferation in rat cells *in vitro* (Sato et al., 2006) and to prevent isoproterenol-induced myocardial injury (Li et al., 2006) in rat models *in vivo*. However, the receptor involved has not been identified.

Over the years, a number of other modifications of ghrelin have been found, including octanoylated, decanoylated, and decenoylated ghrelin. These peptides were identified after ESI-mass spectrometry analysis of stomach tissue extracts; their ability to stimulate GH secretion through the GHSR was verified using synthetic versions of the ghrelins injected into anesthetized rats (Hosoda et al., 2003).

B. Adenosine

Smith et al. (2000) first identified adenosine as a putative ligand for the GHSR after testing fractionated hypothalamic fractions for their capacity to activate Ca^{2+}-mediated aequorin bioluminescence in HEK293 cells stably expressing the GHSR and aqueorin. Microspray tandem mass spectrometry identified the agonist as adenosine, and it was shown to act specifically at the GHSR as a partial agonist (Smith et al., 2000). Tullin et al. independently concluded that adenosine was a GHSR agonist by showing that adenosine induced increases in intracellular calcium in BHK and HEK293 cells transiently transfected with GHSR; the nontransfected cells failed to exhibit any calcium response (Tullin *et al.*, 2000). Despite this ability to induce intracellular calcium increase, adenosine is unable to stimulate the release of GH (Smith *et al.*, 2000; Tullin *et al.*, 2000), and this might relate to the partial agonist properties of adenosine (Smith *et al.*, 2000). Scatchard plot analysis of binding to membranes isolated from GHSR expressing cells using ^{35}S-MK0677 and ^{3}H-adenosine HH showed that the concentration of binding sites was similar irrespective of which radiolabeled ligand was used (Tullin *et al.*, 2000). However, based on competitive binding assays using ^{35}S-MK0677, it was clear that the adenosine-binding pocket was distinct from that of the GHSs (Smith *et al.*, 2000). Adenosine apparently acts via a different signal transduction pathway to increase calcium levels, which is not surprising since the adenosine-binding site does not overlap with that of the GHSs; in contrast to the PLC-PKC actions of ghrelin on GHSR1a, adenosine was found to stimulate cAMP levels, most likely leading to increased PKA and calcium (Carreira *et al.*, 2004).

In spite of these independent reports identifying adenosine as a GHSR agonist, there are claims in the most recent literature arguing that adenosine is not a GHSR agonist. Reexamining the previously observed properties by adenosine, Johansson *et al.* (2005) hypothesized that adenosine's effects on intracellular calcium levels may be mediated by other receptors. In experiments using adenosine and adenosine analogues, HEK293-EBNA cells transfected to transiently express GHSR and nontransfected HEK293 cells exhibited similar increases in intracellular calcium. However, this obvious control was not overlooked by the Smith and Tullin groups, who failed to see such effects in non-GHSR expressing control cells (Smith *et al.*, 2000; Tullin *et al.*, 2000). Johansson *et al.* (2005) also reported that in GHSR-transfected CHO cells known to lack adenosine receptors, no such intracellular calcium effects were noted, suggesting that the agonist properties of adenosine on the GHSR was a result of artifactual activation via the adenosine receptor. The lack of adenosine-induced GHSR signaling in CHO cells could be explained by the lack of appropriate G-protein coupling. Finally, Johansson *et al.* (2005) cited the fact that binding by adenosine and its analogue 2-chloroadenosine cannot be displaced by different GHSR1a agonists. This was not unanticipated because Smith *et al.* (2000) reported that adenosine activated the GHSR by binding to a different binding pocket than that for ghrelin and the GHS. Carreira *et al.* (2006) reported that adenosine binding to nontransfected HEK293 cells corresponded to endogenous adenosine receptors type 2b and 3. The researchers detected an increase in adenosine-binding affinity to these endogenous receptors, which was associated with overexpression of GHSR1a. They suggest that adenosine receptors may partially use the signaling transduction mechanisms of the ghrelin receptor (Carreira *et al.*, 2006). However, the evidence from these recent publications is not sufficiently compelling to refute the claim that adenosine is a partial agonist of the GHSR.

C. Cortistatin

The somatostatin analogue cortistatin, while able to bind the different somatostatin subtypes (Broglio et al., 2002a) and exhibit endocrine actions similar to somatostatin (mainly the inhibition of GH secretion) (Gottero et al., 2004), has recently been discovered as another possible endogenous ligand of GHSR1a (Deghenghi et al., 2001). Cortistatin, like somatostatin, is able to inhibit both GHRH- and ghrelin-stimulated release of GH (Broglio et al., 2002a), and one report found that it may inhibit the release of ghrelin itself (Broglio et al., 2002b). The hormone does not antagonize all activities of ghrelin; however, studies show that it failed to influence the ghrelin-activated release of prolactin, ACTH, and cortisol (Broglio et al., 2002a).

To what extent the natural physiological actions of cortistatin are mediated by GHSR1a remains uncertain and is currently an area of considerable research.

VII. Evidence for Other Receptors

Most of the functions identified with the ghrelin receptor have so far been attributed to GHSR1a, but there have been several studies attempting to characterize the precise role of its truncated counterpart, GHSR1b.

GHSR1b is more widely expressed than GHSR1a, being found in tissues where no detectable levels of GHSR1a receptor transcripts were seen (Gnanapavan et al., 2002).

Overexpression inhibits constitutive activity of GHSR1a. HEK293 cells cotransfected with GHSR1a and GHSR1b-containing vectors in a 1:5 ratio, respectively, significantly increased the functional receptor's K_d without affecting expression of GHSR1a. Observed decreases in the log concentration–response curves seen in GHSR1b-transfected cells without changes in EC_{50} and increase in labeled inositol phosphate accumulation suggest that the truncated receptor alters the basal activity of GHSR1a (Chu et al., 2007). Such decreases in constitutive activation were also observed by Leung et al. (2007). They also found that the expression of GHSR1a is decreased and appears to translocate into the nucleus; this phenomenon is thought to result from GHSR1a–GHSR1b heterodimer interactions and to cause the observed decrease in basal activity (Leung et al., 2007).

Several studies increasingly recognize that other GHSRs exist in order to account for the different actions of ghrelin and other endogenous GHSR ligands that do not seem to involve the functional ghrelin receptor.

In the cardiovascular system, for instance, the GHS hexarelin's actions on the coronary vasculature are apparently mediated by CD36. Activation by hexarelin causes an increase in coronary perfusion pressure, an effect absent in CD36-deficient mice (Bodart et al., 2002). In adipocytes, desacyl ghrelin and ghrelin caused antagonistic effects on lipolysis induced by the administration of isoproterenol. But RT-PCR of adipose tissue and receptor assays of adipocytes did not reveal the presence of GHSR (Muccioli et al., 2004). Additionally, ghrelin and desacyl ghrelin exhibited stimulatory effects on bone marrow adipogenesis, while the GHSR1a agonist L163,255 did not (Thompson et al., 2004). In osteoblasts, ghrelin stimulated increases in cellular DNA content and thymidine uptake—suggesting the stimulation of proliferation—again not through the GHSR: real-time PCR of osteoblasts revealed no GHSR1a mRNA expression (Delhanty et al., 2006). In hepatocytes, different effects on glucose output are catalyzed by ghrelin and desacyl ghrelin; ghrelin increased glucose output, while unacylated ghrelin inhibited glucose release—both in a dose-dependent fashion. However, it is generally accepted that GHSR1a mRNA is undetectable in hepatocytes.

These studies provide exciting evidence that the explication of the ghrelin-GHSR axis is still in its infancy and that important contributions—in the form of new receptor components and new endogenous agonists/antagonists—to present knowledge of what appears to be a distinct endocrine system will undoubtedly fill the pages of scientific literature in the years to come.

VIII. Future Directions

This versatile hormone receptor system has already opened many avenues in various fields of basic science and clinical research. Still, the future remains full of exciting opportunities for ghrelin and GHSR research.

We may yet see a more intricate regulatory network governing body physiology and pathophysiology. The emerging picture is that of a neuroendocrine system with fundamental roles in metabolic homeostasis, aging, immune modulation, and integration of complex physiological systems. Ultimately, the understanding of these networks at the molecular and whole organism levels will provide us a better blueprint of how fundamental processes are disrupted and how we can intervene. The multiple alterations involved in the aging process, in the metabolic syndrome and obesity, and in cardiovascular disease are examples of pathologies in these networks, pathologies that we may soon have a better understanding of.

We may yet possess powerful new drugs that reverse the aging process, treat the pathology and complications of cancer, provide better treatments for cardiovascular disease, replace deficient endocrine hormones, and counter the obesity epidemic: improving the quality of life of an increasingly older population. Studies are currently underway, examining the potential applications of ghrelin agonists or antagonists in both diagnosis and treatment of complicated disorders.

We are certain to discover more functions for GHSR or other as yet unidentified members of its family. A hormone that was first identified in the stomach through its actions on the pituitary is now known to exert multiple influences over multiple organ systems.

We may yet move into an era featuring a different kind of research: the complex activity of ghrelin and its receptor(s) will force us to investigate biological systems using a more holistic perspective—away from the reductionist approach that has served the exponential research of the past century well, yet finds itself inadequate for the deluge of information in, and interdisciplinary nature of, twenty-first century scientific inquiry.

REFERENCES

Akimoto-Takano, S., Sakurai, C., Kanai, S., Hosoya, H., Ohta, M., and Miyasaka, K. (2006). Differences in the appetite-stimulating effect of orexin, neuropeptide Y and ghrelin among young, adult and old rats. *Neuroendocrinology* **82**(5–6), 256–263 (Abstract).

Akman, M. S., Girard, M., O'Brien, L. F., Ho, A. K., and Chik, C. L. (1993). Mechanisms of action of a second generation growth hormone-releasing peptide (Ala-His-D-beta Nal-Ala-Trp-D-Phe-Lys-NH2) in rat anterior pituitary cells. *Endocrinology* **132**(3), 1286–1291.

Anderson, L. L., Jeftinija, S., and Scanes, C. G. (2004). Growth hormone secretion: Molecular and cellular mechanisms and *in vivo* approaches. *Exp. Biol. Med. (Maywood)* **299**(4), 291–302.

Ariyasu, H., Takaya, K., Tagami, T., Ogawa, Y., Hosoda, K., Akamizu, T., Suda, M., Koh, T., Natsui, K., Toyooka, S., Shirakami, G., Usui, T., *et al.* (2001). Stomach is a major source of circulating ghrelin, and feeding state determines plasma ghrelin-like immunoreactivity levels in humans. *J. Clin. Endocrinol. Metab.* **86**(10), 4753–4758.

Arlt, W., and Hewison, M. (2004). Hormones and immune function: Implications of aging. *Aging Cell* **3**(4), 209–216.

Arvat, E., Maccario, M., Di Vito, L., Broglio, F., Benso, A., Gottero, C., Papotti, M., Muccioli, G., Dieguez, C., Casanueva, F. F., Deghenghi, R., Camanni, F., *et al.* (2001). Endocrine activities of ghrelin, a natural growth hormone secretagogue (GHS), in humans: Comparison and interactions with hexarelin, a nonnatural peptidyl GHS, and GH-releasing hormone. *J. Clin. Endocrinol. Metab.* **86**(3), 1169–1174.

Bach, F., Bause, H., Kaisers, U., and Mertzlufft, F. (2006). Sepsis—an ever new challenge. *Anasthesiol. Intensivmed. Notfallmed. Schmerzther.* **41**(1), 27–31.

Bach, M. A., Rockwood, K., Zetterberg, C., Thamsborg, G., Hebert, R., Devogelaer, J. P., Christiansen, J. S., Rizzoli, R., Ochsner, J. L., Beisaw, N., Gluck, O., Yu, L., *et al.* (2004). The effects of MK-0677, an oral growth hormone secretagogue, in patients with hip fracture. *J. Am. Geriatr. Soc.* **52**(4), 516–523.

Backman, L., Nyberg, L., Lindenberger, U., Li, S. C., and Farde, L. (2006). The correlative triad among aging, dopamine, and cognition: Current status and future prospects. *Neurosci. Biobehav. Rev.* **30**(6), 791–807.

Baessler, A., Hasinoff, M. J., Fischer, M., Reinhard, W., Sonnenberg, G. E., Olivier, M., Erdmann, J., Schunkert, H., Doering, A., Jacob, H. J., Comuzzie, A. G., Kissebah, A. H., *et al.* (2005). Genetic linkage and association of the growth hormone secretagogue receptor (ghrelin receptor) gene in human obesity. *Diabetes* **54**, 259–267.

Baessler, A., Kwitek, A. E., Fischer, M., Koehler, M., Reinhard, W., Erdmann, J., Riegger, G., Doering, A., Schunkert, H., and Hengstenberg, C. (2006). Association of the ghrelin receptor gene region with left ventricular hypertrophy in the general population: Results

of the MONICA/KORA Augsburg Echocardiographic Substudy. *Hypertension* **47**(5), 920–927.
Balla, T. (2006). Phosphoinositide-derived messengers in endocrine signaling. *J. Endocrinol.* **188**(2), 135–153.
Banks, W. A., Tschöp, M., Robinson, S. M., and Heiman, M. L. (2002). Extent and direction of ghrelin transport across the blood-brain barrier is determined by its unique primary structure. *J. Pharmacol. Exp. Ther.* **302**, 822–827.
Baranowska, B., Bik, W., Baranowska-Bik, A., Wolinska-Witort, E., Szybinska, A., Martynska, L., and Chmielowska, M. (2006). Neuroendocrine control of metabolic homeostasis in Polish centenarians. *J. Physiol. Pharmacol.* **57**(Suppl. 6), 55–61.
Bednarek, M. A., Feighner, S. D., Pong, S. S., McKee, K. K., Hreniuk, D. L., Silva, M. V., Warren, V. A., Howard, A. D., van der Ploeg, L. H. Y., and Heck, J. V. (2000). Structure-function studies on the new growth hormone-releasing peptide ghrelin: Minimal sequence of ghrelin necessary for activation of growth hormone secretagogue receptor 1a. *J. Med. Chem.* **43**, 4370–4376.
Bennett, P. A., Thomas, G. B., Howard, A. D., Feighner, S. D., Van der Ploeg, L. H. T., Smith, R. G., and Robinson, I. C. A. F. (1997). Hypothalamic growth hormone secretagogue receptor (GHS-R) expression is regulated by growth hormone in the rat. *Endocrinology* **138**, 4552–4557.
Blake, A. D., and Smith, R. G. (1991). Desensitization studies using perifused rat pituitary cells show that growth hormone releasing hormone and His-D-Trp-Ala-Trp-D-Phe-Lys-NH2 stimulates growth hormone release through distinct receptor sites. *J. Endocrinol.* **129**, 11–19.
Bodart, V., Febbraio, M., Demers, A., McNicoll, N., Pohankova, P., Perreault, A., Sejlitz, T., Escher, E., Silverstein, R. L., Lamontagne, D., and Ong, H. (2002). CD36 mediates the cardiovascular action of growth hormone-releasing peptides in the heart. *Circ. Res.* **90**(8), 844–849.
Boog, G., Sagot, F., Winer, N., David, A., and Nomballais, M. F. (1999). Brachmann-de Lange syndrome: A cause of early symmetric fetal growth delay. *Eur. J. Obstet. Gynecol. Reprod. Biol.* **85**(2), 173–177.
Bowers, C. Y., Momany, F., Reynolds, G. A., Chang, D., Hong, A., and Chang, K. (1981). Structure-activity relationships of a synthetic pentapeptide that specifically releases growth hormone *in vitro*. *Endocrinology* **106**(3), 663–667.
Bowers, C. Y., Sartor, A. O., Reynolds, G. A., and Badger, T. M. (1991). On the actions of the growth hormone-releasing hexapeptide, GHRP. *Endocrinology* **128**(4), 2027–2035.
Broglio, F., Arvat, E., Benso, A., Gottero, C., Prodam, F., Grottoli, S., Papotti, M., Muccioli, G., Van der Lely, A. J., Deghenghi, R., and Ghigo, E. (2002a). Endocrine activities of cortistatin-14 and its interaction with GHRH and ghrelin in humans. *J. Clin. Endocrinol. Metab.* **87**(8), 3783–3790.
Broglio, F., Van Koetsveld, P., Benso, A., Gottero, C., Prodam, F., Papotti, M., Muccioli, G., Gauna, C., Hofland, L., Deghenghi, R., Arvat, E., Van der Lely, E., *et al.* (2002b). Ghrelin secretion is inhibited by either somatostatin or cortistatin in humans. *J. Clin. Endocrinol. Metab.* **87**(10), 4823–4832.
Caceda, R., Kinkead, B., and Nemeroff, C. B. (2006). Neurotensin: Role in psychiatric and neurological diseases. *Peptides* **27**(10), 2385–2404.
Camiña, J. P. (2006). Cell biology of the ghrelin receptor. *J. Neuroendocrinol.* **18**(1), 65–76.
Camiña, J. P., Carreira, M. C., Messari, S. E., Llorens-Cortes, C., Smith, R. G., and Casanueva, F. F. (2004). Desensitization and endocytosis mechanisms of ghrelin-activated growth hormone secretagogue receptor 1a. *Endocrinology* **145**(2), 930–940.
Carreira, M. C., Camiña, J. P., Smith, R. G., and Casanueva, F. F. (2004). Agonist-specific coupling of growth hormone secretagogue receptor type 1a to different intracellular signaling systems. *Neuroendocrinology* **79**, 13–25.

Carreira, M. C., Camiña, J. P., Diaz-Rodriguez, E., Alvear-Perez, R., Llorens-Cortes, C., and Casanueva, F. F. (2006). Adenosine does not bind to the growth hormone secretagogue receptor type 1a (GHS-R1a). *J. Endocrinol.* **191,** 147–157.

Cassoni, P., Ghe, C., Marrocco, T., Tarabra, E., Allia, E., Catapano, F., Deghenghi, R., Ghigo, E., Papotti, M., and Muccioli, G. (2004). Expression of ghrelin and biological activity of specific receptors for ghrelin and des-acyl ghrelin in human prostate neoplasms and related cell lines. *Eur. J. Endocrinol.* **150,** 173–184.

Chahal, H., and Drake, W. (2007). The endocrine system and ageing. *J. Pathol.* **211**(2), 173–180.

Chang, L., Zhao, J., Yang, J., Zhang, Z., Du, J., and Tang, C. (2003). Therapeutic effects of ghrelin on endotoxic shock in rats. *Eur. J. Pharmacol.* **473**(2–3), 171–176.

Chapman, I. M., Bach, M. A., Van Cauter, E., Farmer, M., Krupa, D., Taylor, A. M., Schilling, L. M., Cole, K. Y., Skiles, E. H., Pezzoli, S. S., Hartman, M. L., Veldhuis, J. D., *et al.* (1996). Stimulation of the growth hormone (GH)-insulin-like growth factor I axis by daily oral administration of a GH secretagogue (MK-677) in healthy elderly subjects. *J. Clin. Endocrinol. Metab.* **81**(12), 4249–4257.

Chen, C. Y., Inui, A., Asakawa, A., Fujino, K., Kato, I., Chen, C. C., Ueno, N., and Fujimiya, M. (2005). Des-acyl ghrelin acts by CRF type 2 receptors to disrupt fasted stomach motility in conscious rats. *Gastroenterology* **129**(1), 8–25.

Chen, H. Y., Trumbauer, M. E., Chen, A. S., Weingarth, D. T., Adams, J. R., Frazier, E. G., Shen, Z., Marsh, D. J., Feighner, S. D., Guan, X. M., Ye, Z., Nargund, R. P., *et al.* (2004). Orexigenic action of peripheral ghrelin is mediated by Neuropeptide Y (NPY) and Agouti-Related Protein (AgRP). *Endocrinology* **145,** 2607–2612.

Cheng, K., Chan, W. W., Barreto, A., Jr., Convey, E. M., and Smith, R. G. (1989). The synergistic effects of His-D-Trp-Ala-Trp-D-Phe-Lys-NH2 on growth hormone (GH)-releasing factor-stimulated GH release and intracellular adenosine $3'$, $5'$-monophosphate accumulation in rat primary pituitary cell culture. *Endocrinology* **124**(6), 2791–2798.

Cheng, K., Chan, W. W., Butler, B., Barreto, A., Jr., and Smith, R. G. (1991). Evidence for a role of protein kinase-C in His-D-Trp-Ala-Trp-D-Phe-Lys-NH2-induced growth hormone release from rat primary pituitary cells. *Endocrinology* **129**(6), 3337–3342.

Chu, K. M., Chow, K. B., Leung, P. K., Lau, P. N., Chan, C. B., Cheng, C. H., and Wise, H. (2007). Over-expression of the truncated ghrelin receptor polypeptide attenuates the constitutive activation of phosphatidylinositol-specific phospholipase C by ghrelin receptors but has no effect on ghrelin-stimulated extracellular signal-regulated kinase 1/2 activity. *Int. J. Biochem. Cell. Biol.* **39,** 752–764.

Cowley, M. A., Smart, J. L., Rubinstein, M., Cerdan, M. G., Diano, S., Horvath, T. L., Cone, R. D., and Low, M. J. (2001). Leptin activates anorexigenic POMC neurons through a neural network in the arcuate nucleus. *Nature* **411**(6836), 480–484.

Cowley, M. A., Smith, R. G., Diano, S., Tschop, M., Pronchuk, N., Grove, K. L., Strasburger, C. J., Bidlingmaier, M., Esterman, M., Heiman, M. L., Garcia-Segura, L. M., Nillni, E. A., *et al.* (2003). The distribution and mechanism of action of ghrelin in the CNS demonstrates a novel hypothalamic circuit regulating energy homeostasis. *Neuron* **37**(4), 649–661.

Cunha, S. R., and Mayo, K. E. (2002). Ghrelin and growth hormone (GH) secretagogues potentiate GH-releasing hormone (GHRH)-induced cyclic adenosine $3',5'$-monophosphate production in cells expressing transfected GHRH and GH secretagogue receptors. *Endocrinology* **143**(12), 4570–4582.

Dass, N. B., Munonyara, M., Bassil, A. K., Hervieu, G. J., Osbourne, S., Corcoran, S., Morgan, M., and Sanger, G. J. (2003). Growth hormone secretagogue receptors in rat and human gastrointestinal tract and the effects of ghrelin. *Neuroscience* **120,** 443–453.

Date, Y., Murakami, N., Toshinai, K., Matsukura, S., Niijima, A., Matsuo, H., Kangawa, K., and Nakazato, M. (2002). The role of the gastric afferent vagal nerve in ghrelin-induced feeding and growth hormone secretion in rats. *Gastroenterology* **123**(4), 1120–1128.

Davenport, A. P., Bonner, T. I., Foord, S. M., Harmar, A. J., Neubig, R. R., Pin, J. P., Spedding, M., Kojima, M., and Kangawa, K. (2005). International union of pharmacology. LVI. Ghrelin receptor nomenclature, distribution, and function. *Pharmacol. Rev.* **51**(4), 541–546.

Deghenghi, R., Papotti, M., Ghigo, E., and Muccioli, G. (2001). Cortistatin, but not somatostatin, binds to growth hormone secretagogue (GHS) receptors of human pituitary gland. *J. Endocrinol. Invest.* **24**(1), RC1–RC3.

Delhanty, P. J., van der Eerden, B. C., van der Velde, M., Gauna, C., Pols, H. A., Jahr, H., Chiba, H., van der Lely, A. J., and van Leeuwen, J. P. (2006). Ghrelin and unacylated ghrelin stimulate human osteoblast growth via mitogen-activated protein kinase (MAPK)/phosphoinositide 3-kinase (PI3K) pathways in the absence of GHSR1a. *J. Endocrinol.* **188**(1), 37–47.

Di Francesco, V., Zamboni, M., Zoico, E., Mazzali, G., Dioli, A., Omizzolo, F., Bissoli, L., Fantin, F., Rizzotti, P., Solerte, S. B., Micciolo, R., and Bosello, O. (2006). Unbalanced serum leptin and ghrelin dynamics prolong postprandial satiety and inhibit hunger in healthy elderly: Another reason for the "anorexia of aging." *Am. J. Clin. Nutr.* **83**(5), 1149–1152.

Dixit, V. D., and Taub, D. D. (2005). Ghrelin and immunity: A young player in an old field. *Exp. Gerontol.* **40**(11), 900–910.

Dixit, V. D., Schaffer, E. M., Pyle, R. S., Collins, G. D., Sakthivel, S. K., Palaniappan, R., Lillard, J. W., and Taub, D. D. (2004). Ghrelin inhibits leptin- and activation-induced proinflammatory cytokine expression by human monocytes and T cells. *J. Clin. Invest.* **114**(1), 57–66.

Dixit, V. D., Weeraratna, A. T., Yang, H., Bertak, D., Cooper-Jenkins, A., Riggins, G. J., Eberhart, C. G., and Taub, D. D. (2006). Ghrelin and the growth hormone secretagogue receptor constitute a novel autocrine pathway in astrocytoma motility. *J. Biol. Chem.* **281**(24), 16681–16690.

Feighner, S. D., Howard, A. D., Prendergast, K., Palyha, O. C., Hreniuk, D. L., Nargund, R., Underwood, D., Tata, J. R., Dean, D. C., Tan, C. P., McKee, K. K., Woods, J. W., *et al.* (1998). Structural requirements for the activation of the human growth hormone secretagogue receptor by peptide and nonpeptide secretagogues. *Mol. Endocrinol.* **12,** 137–145.

Feighner, S. D., Tan, C. P., McKee, K. K., Palyha, O. C., Hreniuk, D. L., Pong, S. S., Austin, C. P., Figueroa, D., MacNeil, D., Cascieri, M. A., Nargund, R., Bakshi, R., *et al.* (1999). Receptor for motilin identified in the human gastrointestinal system. *Science* **284**(5423), 2184–2188.

Fernandez-Fernandez, R., Tena-Sempere, M., Aguilar, E., and Pinilla, L. (2004). Ghrelin effects on gonadotropin secretion in male and female rats. *Neurosci. Lett.* **362**(2), 103–107.

Garcia, A., Alvarez, C. V., Smith, R. G., and Dieguez, C. (2001). Regulation of PIT-1 expression by ghrelin and GHRP-6 through the GH secretagogue receptor. *Mol. Endocrinol.* **15**(9), 1484–1495.

Gaytan, F., Barreiro, M. L., Caminos, J. E., Chopin, L. K., Herington, A. C., Morales, C., Pinilla, L., Paniagua, R., Nistal, M., Casanueva, F. F., Aguilar, E., Dieguez, C., *et al.* (2004). Expression of ghrelin and its functional receptor, the type 1a growth hormone secretagogue receptor, in normal human testis and testicular tumors. *J. Clin. Endocrinol. Metab.* **89,** 400–409.

Gaytan, F., Morales, C., Barreiro, M. L., Jeffery, P., Chopin, L. K., Herington, A. C., Casanueva, F. F., Aguilar, E., Diegez, C., and Tena-Sempere, M. (2005). Expression of growth hormone secretagogue receptor type 1a, the functional ghrelin receptor, in human ovarian surface epithelium, mullerian duct derivatives, and ovarian tumors. *J. Clin. Endocrinol. Metab.* **90**(3), 1798–1804.

Glavaski-Joksimovic, A., Jeftinija, K., Scanes, C. G., Anderson, L. L., and Jeftinija, S. (2003). Stimulatory effect of ghrelin on isolated porcine somatotropes. *Neuroendocrinology* **77**(6), 367–379.

Gnanapavan, S., Kola, B., Bustin, S. A., Morris, D. G., McGee, P., Fairclough, P., Bhattacharya, S., Carpenter, R., Grossman, A. B., and Korbonits, M. (2002). The tissue distribution of the mRNA of ghrelin and subtypes of its receptor, GHS-R, in humans. *J. Clin. Endocrinol. Metab.* **87**(6), 2988–2991.

Gonzalez-Rey, E., Chorny, A., and Delgado, M. (2006). Therapeutic action of ghrelin in a mouse model of colitis. *Gastroenterology* **130,** 1707–1720.

Gottero, C., Prodam, F., Destefanis, S., Benso, A., Gauna, C., Me, E., Filtri, L., Riganti, F., Van Der Lely, A. J., Ghigo, E., and Broglio, F. (2004). Cortistatin-17 and -14 exert the same endocrine activities as somatostatin in humans. *Growth Horm. IGF Res.* **14**(5), 382–387 (Abstract).

Guan, X. M., Yu, H., Palyha, O. C., McKee, K. K., Feighner, S. D., Sirinathsinghji, D. J. S., Smith, R. G., Van der Ploeg, L. H. T., and Howard, A. D. (1997). Distribution of mRNA encoding the growth hormone secretagogue receptor in brain and peripheral tissues. *Brain Res. Mol. Brain Res.* **48,** 23–29.

Han, X. F., Zhu, Y. L., Hernandez, M., Keating, D. J., and Chen, C. (2005). Ghrelin reduces voltage-gated potassium currents in GH3 cells via cyclic GMP pathways. *Endocrine* **28**(2), 217–224. (Abstract).

Harvey, S., Azumaya, Y., and Hull, K. L. (2000). Pituitary and extrapituitary growth hormone: Pit-1 dependence? *Can. J. Physiol. Pharmacol.* **78**(12), 1013–1028.

Hataya, Y., Akamizu, T., Takaya, K., Kanamoto, N., Ariyasu, H., Saijo, M., Moriyama, K., Shimatsu, A., Kojima, M., Kangawa, K., and Nakao, K. (2001). A low dose of ghrelin stimulates growth hormone (GH) release synergistically with GH-releasing hormone in humans. *J. Clin. Endocrinol. Metab.* **86**(9), 4552–4555.

Hattori, N., Saito, T., Yagyu, T., Jiang, B. H., Kitagawa, K., and Inagaki, C. (2001). GH, GH receptor, GH secretagogue receptor, and ghrelin expression in human T cells, B cells, and neutrophils. *J. Clin. Endocrinol. Metab.* **86**(9), 4284–4291.

Hays, N. P., and Roberts, S. B. (2006). The anorexia of aging in humans. *Physiol. Behav.* **88**(3), 257–266.

Herrington, J., and Hille, B. (1994). Growth hormone-releasing hexapeptide elevates intracellular calcium in rat somatotropes by two mechanisms. *Endocrinology* **135**(3), 1100–1108.

Holst, B., Cygankiewicz, A., Jensen, T. H., Ankersen, M., and Schwartz, T. W. (2003). High constitutive signaling of the ghrelin receptor—identification of a potent inverse agonist. *Mol. Endocrinol.* **17**(11), 2201–2210.

Holst, B., Holliday, N. D., Bach, A., Elling, C. E., Cox, H. M., and Schwartz, T. W. (2004). Common structural basis for constitutive activity of the ghrelin receptor family. *J. Biol. Chem.* **279**(51), 53806–53817.

Holst, B., Brandt, E., Bach, A., Heding, A., and Schwartz, T. W. (2005). Nonpeptide and peptide growth hormone secretagogues act both as ghrelin receptor agonist and as positive or negative allosteric modulators of ghrelin signaling. *Mol. Endocrinol.* **19**(9), 2400–2411.

Hosoda, H., Kojima, M., Matsuo, H., and Kangawa, K. (2000). Purification and characterization of rat des-Gln14-Ghrelin, a second endogenous ligand for the growth hormone secretagogue receptor. *J. Biol. Chem.* **275**(29), 21995–22000.

Hosoda, H., Kojima, M., Mizushima, T., Shimizu, S., and Kangawa, K. (2003). Structural divergence of human ghrelin. Identification of multiple ghrelin-derived molecules produced by post-translational processing. *J. Biol. Chem.* **278**(1), 64–70.

Hosoda, H., Kojima, M., and Kangawa, K. (2006). Biological, physiological, and pharmacological aspects of ghrelin. *J. Pharmacol. Sci.* **100,** 398–410.

Howard, A. D., Feighner, S. D., Cully, D. F., Arena, J. P., Liberator, P. A., Rosenblum, C. I., Hamelin, M., Hreniuk, D. L., Palyha, O. C., Anderson, J., Paress, P. S., Diaz, C., et al. (1996). A receptor in pituitary and hypothalamus that functions in growth hormone release. *Science* **273**, 974–977.

Iacovelli, L., Sallese, M., Mariggiò, S., and De Blasi, A. (1999). Regulation of G-protein-coupled receptor kinase subtypes by calcium sensor proteins. *FASEB J.* **13**, 1–8.

Itoh, Z. (1997). Motilin and clinical application. *Peptides* **18**(4), 593–608.

Jiang, H., Betancourt, L., and Smith, R. G. (2006). Ghrelin amplifies dopamine signaling by cross talk involving formation of growth hormone secretagogue receptor/dopamine receptor subtype 1 heterodimers. *Mol. Endocrinol.* **20**(8), 1772–1785.

Johansson, S., Fredholm, B. B., Hjort, C., Morein, T., Kull, B., and Hu, P. S. (2005). Evidence against adenosine analogues being agonists at the growth hormone secretagogue receptor. *Biochem. Pharmacol.* **70**(4), 598–605.

Kaji, H., Tai, S., Okimura, Y., Iguchi, G., Takahashi, Y., Abe, H., and Chihara, K. (1998). Cloning and characterization of the 5′-flanking region of the human growth hormone secretagogue receptor gene. *J. Biol. Chem.* **273**(51), 33885–33888.

Kaji, H., Kishimmoto, M., Kirimura, T., Iguchi, G., Murata, M., Yoshioka, S., Iida, K., Okimura, Y., Yoshimoto, Y., and Chihara, K. (2001). Hormonal regulation of the human ghrelin receptor gene transcription. *Biochem. Biophys. Res. Commun.* **284**, 660–666.

Kamegai, J., Ishii, H., Ishii, S., Sugihara, H., and Wakabayashi, I. (2001). Thyroid hormones regulate pituitary growth hormone secretagogue receptor gene expression. *J. Neuroendocrinol.* **13**, 275–278.

Kandel, E. R. (2001). The molecular biology of memory storage: A dialog between genes and synapses. *Science* **294**(5544), 1030–1038.

Katayama, M., Nogami, H., Nishiyama, J., Kawase, T., and Kawamura, K. (2000). Developmentally and regionally regulated expression of growth hormone secretagogue receptor mRNA in rat brain and pituitary gland. *Neuroendocrinology* **72**, 333–340.

Kato, M., and Sakuma, Y. (1999). The effect of GHRP-6 on the intracellular Na^+ concentration of rat pituitary cells in primary culture. *J. Neuroendocrinol.* **11**(10), 795–800.

Kim, K., Arai, K., Sanno, N., Osamura, R. Y., Teramoto, A., and Shibasaki, T. (2001). Ghrelin and growth hormone (GH) secretagogue receptor (GHSR) mRNA expression in human pituitary adenomas. *Clin. Endocrinol. (Oxf.).* **54**(6), 759–768.

Kineman, R. D., Kamegai, J., and Frohman, L. A. (1999). Growth hormone (GH)-releasing hormone (GHRH) and the GH secretagogue (GHS), L692,585, differentially modulate rat pituitary GHS receptor and GHRH receptor messenger ribonucleic acid levels. *Endocrinology* **140**(8), 3581–3586.

Kleinz, M. J., Maguire, J. J., Skepper, J. N., and Davenport, A. P. (2006). Functional and immunocytochemical evidence for a role of ghrelin and des-octanoyl ghrelin in the regulation of vascular tone in man. *Cardiovasc. Res.* **69**(1), 227–235.

Kohno, D., Gao, H. Z., Muroya, S., Kikuyama, S., and Yada, T. (2003). Ghrelin directly interacts with neuropeptide Y-containing neurons in the rat arcuate nucleus: Calcium signaling via protein kinase A and N-type channel-dependent mechanisms and cross-talk with leptin and orexin. *Diabetes* **52**, 948–956.

Kojima, M., Hosoda, H., Date, Y., Nakazato, M., Matsuo, H., and Kangawa, K. (1999). Ghrelin is a growth-hormone releasing acylated peptide from stomach. *Nature* **402**, 656–660.

Koo, G. C., Huang, C., Camacho, R., Trainor, C., Blake, J. T., Sirotina-Meisher, A., Schleim, K. D., Wu, T. J., Cheng, K., Nargund, R., and McKissick, G. (2001). Immune enhancing effect of a growth hormone secretagogue. *J. Immunol.* **166**, 4195–4201.

Korbonits, M., Bustin, S. A., Kojima, M., Jordan, S., Adams, E. F., Lowe, D. G., Kangawa, K., and Grossman, A. B. (2001). The expression of the growth hormone secretagogue receptor ligand ghrelin in normal and abnormal human pituitary and other neuroendocrine tumors. *J. Clin. Endocrinol. Metab.* **86**(2), 881–887.

Krabbe, K. S., Pedersen, M., and Bruunsgaard, H. (2004). Inflammatory mediators in the elderly. *Exp. Gerontol.* **39**(5), 687–699.

Lall, S., Balthasar, N., Carmignac, D., Magoulas, C., Sesay, A., Houston, P., Mathers, K., and Robinson, I. (2004). Physiological studies of transgenic mice overexpressing growth hormone (GH) secretagogue receptor 1a in GH-releasing hormone neurons. *Endocrinology* **145**(4), 1602–1611.

Lei, T., Buchfelder, M., Fahlbusch, R., and Adams, E. F. (1995). Growth hormone releasing peptide (GHRP-6) stimulates phosphatidylinositol (PI) turnover in human pituitary somatotroph cells. *J. Mol. Endocrinol.* **14**(1), 135–138.

Leung, P. K., Chow, K. B., Lau, P. N., Chu, K. M., Chan, C. B., Cheng, C. H., and Wise, H. (2007). The truncated ghrelin receptor polypeptide (GHS-R1b) acts as a dominant-negative mutant of the ghrelin receptor. *Cell Signal.* **19**, 1011–1022.

Li, L., Zhang, L. K., Pang, Y. Z., Pan, C. S., Qi, Y. F., Chen, L., Wang, X., Tang, C. S., and Zhang, J. (2006). Cardioprotective effects of ghrelin and des-octanoyl ghrelin on myocardial injury induced by isoproterenol in rats. *Acta Pharmacol. Sin.* **19**(5), 527–535.

Li, W. G., Gavrila, D., Liu, X., Wang, L., Gunnlaugsson, S., Stoll, L. L., McCormick, M. L., Sigmund, C. D., Tang, C., and Weintraub, N. L. (2004). Ghrelin inhibits proinflammatory responses and nuclear factor-kappa B activation in human endothelial cells. *Circulation* **109**(18), 2221–2226.

Luque, R. M., Kineman, R. D., Park, S., Peng, X. D., Gracia-Navarro, F., Castano, J. P., and Malagon, M. M. (2004). Homologous and heterologous regulation of pituitary receptors for ghrelin and growth hormone-releasing hormone. *Endocrinology* **145**(7), 3182–3189.

Malagon, M. M., Luque, R. M., Ruiz-Guerrero, E., Rodriguez-Pacheco, F. R., Garcia-Navarro, S., Casanueva, F. F., Gracia-Navarro, F., and Castaño, J. P. (2003). Intracellular signaling mechanisms mediating ghrelin-stimulated growth hormone release in somatotropes. *Endocrinology* **144**(12), 5372–5380.

Malozowski, S., Hao, E. H., Ren, S. G., Marin, G., Liu, L., Southers, J. L., and Merriam, G. R. (1991). Growth hormone (GH) responses to the hexapeptide GH-releasing peptide and GH-releasing hormone (GHRH) in the cynomolgus macaque: Evidence for non-GHRH-mediated responses. *J. Clin. Endocrinol. Metab.* **73**, 314–317.

McKee, K. K., Palyha, O. C., Feighner, S. D., Hreniuk, D. L., Tan, C. P., Phillips, M. S., Smith, R. G., Van der Ploeg, L. H. T., and Howard, A. D. (1997a). Molecular analysis of rat pituitary and hypothalamic growth hormone secretagogue receptors. *Mol. Endocrinol.* **11**, 415–423.

McKee, K. K., Tan, C. P., Palyha, O. C., Liu, J., Feighner, S. D., Hreniuk, D. L., Smith, R. G., Howard, A. D., and Van der Ploeg, L. H. T. (1997b). Cloning and characterization of two human G protein-coupled receptor genes (GPR38 and GPR39) related to the growth hormone secretagogue and neurotensin receptors. *Genomics* **46**, 426–434.

Menyhert, J., Wittmann, G., Hrabovsky, E., Szlavik, N., Keller, E., Tschop, M., Liposits, Z., and Fekete, C. (2006). Distribution of ghrelin-immunoreactive neuronal networks in the human hypothalamus. *Brain Res.* **1125**(1), 31–36.

Miyasaka, K., Hosoya, H., Sekime, A., Ohta, M., AMono, H., Matsushita, S., Suzuki, K., Higuchi, S., and Funakoshi, A. (2006). Association of ghrelin receptor gene polymorphism with bulimia nervosa in a Japanese population. *J. Neural. Transm.* **113**(9), 1279–1285.

Morley, J. E., Thomas, D. R., and Wilson, M. M. (2006). Cachexia: Pathophysiology and clinical relevance. *Am. J. Clin. Nutr.* **83**(4), 735–743.

Mousseaux, D., Gallic, L. L., Ryan, J., Oiry, C., Gagne, D., Fehrentz, J. A., Galleyrand, J. C., and Martinez, J. (2006). Regulation of ERK1/2 activity by ghrelin-activated growth hormone secretagogue receptor 1A involves a PLC/PKCε pathway. *Br. J. Pharmacol.* **148,** 350–365.

Muccioli, G., Pons, N., Ghe, C., Catapano, F., Granata, R., and Ghigo, E. (2004). Ghrelin and des-acyl ghrelin both inhibit isoproterenol-induced lipolysis in rat adipocytes via a non-type 1a growth hormone secretagogue receptor. *Eur. J. Pharmacol.* **498**(1–3), 27–35.

Murata, M., Okimura, Y., Iida, K., Matsumoto, M., Sowa, H., Kaji, H., Kojima, M., Kangawa, K., and Chihara, K. (2002). Ghrelin modulates the downstream molecules of insulin signaling in hepatoma cells. *J. Biol. Chem.* **277**(7), 5667–5674.

Murphy, M. G., Plunkett, L. M., Gertz, B. J., He, W., Wittreich, J., Polvino, W. M., and Clemmons, D. R. (1998). MK-677, an orally active growth hormone secretagogue, reverses diet-induced catabolism. *J. Clin. Endocrinol. Metab.* **83**(2), 320–325.

Murphy, W. J., Rui, H., and Longo, D. L. (1995). Effects of growth hormone and prolactin immune development and function. *Life Sci.* **57**(1), 1–14.

Nakazato, M., Murakamim, N., Date, Y., Kojima, M., Matsuo, H., Kangawa, K., and Matsukura, S. (2001). A role for ghrelin in the central regulation of feeding. *Nature* **409**(6817), 194–198.

Palyha, O. C., Feighner, S. D., Tan, C. P., McKee, K. K., Hreniuk, D. L., Gao, Y. D., Schleim, K. D., Yang, L., Morriello, G. J., Nargund, R., Patchett, A. A., Howard, A. D., *et al.* (2000). Ligand activation domain of human orphan growth hormone (GH) secretagogue receptor (GHS-R) conserved from pufferfish to humans. *Mol. Endocrinol.* **14,** 160–169.

Pantel, J., Legendre, M., Cabrol, S., Hilal, L., Hajaji, Y., Morisset, S., Nivot, S., Vie-Luton, M. P., Grouselle, D., de Kerdanet, M., Kadiri, A., Epelbaum, J., *et al.* (2006). Loss of constitutive activity of the growth hormone secretagogue receptor in familial short stature. *J. Clin. Invest.* **116**(3), 760–768.

Papotti, M., Ghè, C., Cassoni, P., Catapano, F., Deghenghi, R., Ghigo, E., and Muccioli, G. (2000). Growth hormone secretagogue binding sites in peripheral human tissues. *J. Clin. Endocrinol. Metab.* **85,** 3803–3807.

Pedretti, A., Villa, M., Pallavicini, M., Valoti, E., and Vistoli, G. (2006). Construction of human ghrelin receptor (hGHS-R1a) model using a fragmental prediction approach and validation through docking analysis. *J. Med. Chem.* **49,** 3077–3085.

Perez, D. M., and Karnik, S. S. (2005). Multiple signaling states of G-protein-coupled receptors. *Pharmacol. Rev.* **57,** 147–161.

Petersenn, S., Rasch, A. C., Penshorn, M., Beil, F. U., and Schulte, H. M. (2001). Genomic structure and transcriptional regulation of the human growth hormone secretagogue receptor. *Endocrinology* **142**(6), 2649–2659.

Pong, S.-S., Channg, L.-Y., Dean, D. C., Nargund, R. P., Patchett, A. A., and Smith, R. G. (1996). Identification of a new G-protein-linked receptor for growth Hormone secretagogues. *Mol. Endocrinol.* **10,** 57–61.

Root, A. W., and Root, M. J. (2002). Clinical pharmacology of human growth hormone and its secretagogues. *Curr. Drug Targets Immune Endocr. Metabol. Disord.* **2**(1), 27–52.

Sato, M., Nakahara, K., Goto, S., Kaiya, H., Miyazato, M., Date, Y., Nakazato, M., Kangawa, K., and Murakami, M. (2006). Effects of ghrelin and des-acyl ghrelin on neurogenesis of the rat fetal spinal cord. *Biochem. Biophys. Res. Commun.* **350**(3), 598–603.

Shuto, Y., Shibasaki, T., Otagiri, A., Kuriyama, H., Ohata, H., Tamura, H., Kamegai, J., Sugihara, H., Oikawa, S., and Wakabayashi, I. (2002). Hypothalamic growth hormone secretagogue receptor regulates growth hormone secretion, feeding, and adiposity. *J. Clin. Invest.* **109,** 1429–1436.

Sibilia, V., Muccioli, G., Deghenghi, R., Pagani, F., De Luca, V., Rapetti, D., Locatelli, V., and Netti, C. (2006). Evidence for a role of the GHS-R1a receptors in ghrelin inhibition of gastric acid secretion in the rat. *J. Neuroendocrinol.* **18**(2), 122–128.

Sirotkin, A. V., Grossmann, R., Maria-Peon, M. T., Roa, J., Tena-Sempere, M., and Klein, S. (2006). Novel expression and functional role of ghrelin in chicken ovary. *Mol. Cell. Endocrinol.* **257–258,** 15–25.
Smith, R. G. (2005). Development of growth hormone secretagogues. *Endocr. Rev.* **26**(3), 346–360.
Smith, R. G., Cheng, K., Schoen, W. R., Pong, S.-S., Hickey, G. J., Jacks, T. M., Butler, B. S., Chan, W. W.-S., Chaung, L.-Y. P., Judith, F., Taylor, A. M., Wyvratt, M., Jr., et al. (1993). A nonpeptidyl growth hormone secretagogue. *Science* **260,** 1640–1643.
Smith, R. G., Van der Ploeg, L. H., Howard, A. D., Feighner, S. D., Cheng, K., Hickey, G. J., Wyvratt, M. J., Jr., Fisher, M. H., Nargund, R. P., and Patchett, A. A. (1997). Peptidomimetic regulation of growth hormone secretion. *Endocr. Rev.* **18**(5), 621–645.
Smith, R. G., Griffin, P. R., Xu, Y., Smith, A. G. A., Liu, K., Calacay, J., Feighner, S. D., Pong, C. S., Leong, D., Pomés, A., Cheng, K., Van der Ploeg, L. H. T., et al. (2000). Adenosine: A partial agonist of the growth hormone secretagogue receptor. *Biochem. Biophys. Res. Commun.* **276,** 1306–1313.
Smith, R. G., Betancourt, L., and Sun, Y. (2005a). Molecular endocrinology and physiology of the aging central nervous system. *Endocr. Rev.* **26**(2), 203–250.
Smith, R. G., Jiang, H., and Sun, Y. (2005b). Developments in ghrelin biology and potential clinical relevance. *Trends. Endocrinol. Metab.* **16**(9), 436–442.
Soto, J. L., Castrillo, J. L., Dominguez, F., and Dieguez, C. (1995). Regulation of the pituitary-specific transcription factor GHF-1/Pit-1 messenger ribonucleic acid levels by growth hormone-secretagogues in rat anterior pituitary cells in monolayer culture. *Endocrinology* **136**(9), 3863–3870.
Stevanovic, D., Milosevic, V., Starcevic, V. P., and Severs, W. B. (2007). The effect of centrally administered ghrelin on pituitary ACTH cells and circulating ACTH and corticosterone in rats. *Life Sci.* **80,** 867–872.
Sun, Y., Ahmed, S., and Smith, R. G. (2003). Deletion of ghrelin impairs neither growth nor appetite. *Mol. Cell. Biol.* **23**(22), 7973–7981.
Sun, Y., Wang, P., Zheng, H., and Smith, R. G. (2004). Ghrelin stimulation of growth hormone release and appetite is mediated through the growth hormone secretagogue receptor. *Proc. Natl. Acad. Sci. USA* **101**(13), 4679–4684.
Sun, Y., Garcia, J. M., and Smith, R. G. (2007). Ghrelin and growth hormone secretagogue receptor (GHS-R) expression in mice during aging. *Endocrinology* **148,** 1323–1329.
Tamura, H., Kamegai, J., Shimizu, T., Ishii, S., Sugihara, H., and Oikawa, S. (2002). Ghrelin stimulates GH but not food intake in arcuate nucleus ablated rats. *Endocrinology* **143**(9), 3268–3275.
Tannenbaum, G. S., Epelbaum, J., and Bowers, C. Y. (2003). Interrelationship between the novel peptide ghrelin and somatostatin/growth hormone-releasing hormone in regulation of pulsatile growth hormone secretion. *Endocrinology* **144**(3), 967–974.
Tena-Sempere, M., Barreiro, M. L., Gonzales, L. C., Gaytan, F., Zhang, F. P., Caminos, J. E., Pinilla, L., Casanueva, F. F., Dieguez, C., and Aguilar, E. (2002). Novel expression and functional role of ghrelin in rat testis. *Endocrinology* **143**(2), 717–725.
Thompson, N. M., Gill, D. A., Davies, R., Loveridge, N., Houston, P. A., Robinson, I. C., and Wells, T. (2004). Ghrelin and des-octanoyl ghrelin promote adipogenesis directly *in vivo* by a mechanism independent of the type 1a growth hormone secretagogue receptor. *Endocrinology* **145**(1), 234–242.
Tolle, V., Bassant, M. H., Zizzari, P., Poindessous-Jazat, F., Tomasetto, C., Epelbaum, J., and Bluet-Pajot, M. T. (2002). Ultradian rhythmicity of ghrelin secretion in relation with GH, feeding behavior, and sleep-wake patterns in rats. *Endocrinology* **143**(4), 1353–1361.

Toshinai, K., Yamaguchi, H., Sun, Y., Smith, R. G., Yamanaka, A., Sakurai, T., Date, Y., Mondal, M. S., Shimbara, T., Kawagoe, T., Murakami, N., Miyazato, M., et al. (2006). Des-acyl ghrelin induces food intake by a mechanism independent of the growth hormone secretagogue receptor. *Endocrinology* **147**(5), 2306–2314.

Tullin, S., Hansen, B. S., Ankersen, M., Møller, J., von Cappelen, K. A., and Thim, L. (2000). Adenosine is an agonist of the growth hormone secretagogue receptor. *Endocrinology* **141**(9), 3397–3402.

van der Lely, A. J., Tschöp, M., Heiman, M. L., and Ghigo, E. (2004). Biological, physiological, pathophysiological, and pharmacological aspects of ghrelin. *Endocr. Rev.* **25**(3), 426–457.

Vartiainen, J., Pöykkö, S. M., Räisänen, T., Kesäniemi, Y. A., and Ukkola, O. (2004). Sequencing analysis of the ghrelin receptor (growth hormone secretagogue receptor type 1a) gene. *Eur. J. Endocrinol.* **150**, 457–463.

Wang, H. J., Geller, F., Dempfle, A., Schäuble, N., Friedel, S., Lichtner, P., Fontenla-Horro, F., Wudy, S., Hagemann, S., Gortner, L., Huse, K., Remschmidt, H., et al. (2004). Ghrelin receptor gene: Identification of several sequence variants in extremely obese children and adolescents, healthy normal-weight and underweight students, and children with short normal stature. *J. Clin. Endocrinol. Metab.* **89**(1), 157–162.

Wortley, K. E., Anderson, K. D., Garcia, K., Murray, J. D., Malinova, L., Liu, R., Moncrieffe, M., Thabet, K., Cox, H. J., Yancopoulos, G. D., Wiegand, S. J., and Sleeman, M. W. (2004). Genetic deletion of ghrelin does not decrease food intake but influences metabolic fuel preference. *Proc. Natl. Acad. Sci. USA* **101**(21), 8227–8232.

Wortley, K. E., Del Rincon, J. P., Murray, J. D., Garcia, K., Ilda, K., Thorner, M. O., and Sleeman, M. W. (2005). Absence of ghrelin protects against early-onset obesity. *J. Clin. Invest.* **115**, 3573–3578.

Wu, R., Zhou, M., Cui, X., Simms, H. H., and Wang, P. (2004). Upregulation of cardiovascular ghrelin receptor occurs in the hyperdynamic phase of sepsis. *Am. J. Physiol. Heart Circ. Physiol.* **287**, 1296–1302.

Wu, R., Dong, W., Zhou, M., Cui, X., Hank Simms, H., and Wang, P. (2005). Ghrelin improves tissue perfusion in severe sepsis via downregulation of endothelin-1. *Cardiovasc. Res.* **68**(2), 318–326.

Yamazaki, M., Kobayashi, H., Tanaka, T., Kangawa, K., Inoue, K., and Sakai, T. (2004). Ghrelin-induced growth hormone release from isolated rat anterior pituitary cells depends on intracellular and extracellular Ca^{2+} stores. *J. Neuroendocrinol.* **16**(10), 825–831.

Yan, M., Hernandez, M., Xu, R., and Chen, C. (2004). Effect of GHRH and GHRP-2 treatment *in vitro* on GH secretion and levels of GH, pituitary transcription factor-1, GHRH-receptor, GH-secretagogue-receptor and somatostatin receptor mRNAs in ovine pituitary cells. *Eur. J. Endocrinol.* **150**, 235–242.

Zhao, D., Zhan, Y., Zeng, H., Moyer, M. P., Mantzoros, C. S., and Pothoulakis, C. (2006). Ghrelin stimulates interleukin-8 gene expression through protein kinase C-mediated NF-kappa B pathway in human colonic epithelial cells. *J. Cell. Biochem.* **97**(6), 1317–1327.

Zigman, J. M., Nakano, Y., Coppari, R., Balthasar, N., Marcus, J. N., Lee, C. E., Jones, J. E., Deysher, A. E., Waxman, A. R., White, R. D., Williams, T. D., Lachey, J. L., et al. (2005). Mice lacking ghrelin receptors resist the development of diet-induced obesity. *J. Clin. Invest.* **115**(12), 3564–3572.

Zarkesh-Esfahani, H., Pockley, G., Metcalfe, R. A., Bidlingmaier, M., Wu, Z., Ajami, A., Weetman, A. P., Strasburger, C. J., and Ross, R. J. (2001). High-dose leptin activates human leukocytes via receptor expression on monocytes. *J. Immunol.* **167**(8), 4593–4599.

CHAPTER FIVE

Basic Aspects of Ghrelin Action

Yolanda Pazos,[*,‡] Felipe F. Casanueva,[*,†,‡] and Jesus P. Camiña[*,‡]

Contents

I. Introduction	90
II. Structure of GHSR1a: A G-Protein–Coupled Receptor	91
III. How to Define the Role of the System Ghrelin/GHSR1a?	96
IV. Are There Alternative Ligands for the GHSR1a?	97
V. Endocytosis of GHSR1a	98
VI. Homo- or Heteromeric Complexes for GHSR1a	99
VII. GHSR1a: G-Protein-Signaling Pathways	101
VIII. A Brief Commentary: "New" Receptors for Ghrelin and Desacyl Ghrelin	107
IX. Concluding Remarks	108
Acknowledgments	109
References	109

Abstract

The identification of the natural ligand for growth hormone secretagogue receptor 1a (GHSR1a) added a new element to the complex machinery of the physiological regulators for both growth hormone (GH) secretion and food intake. Initially, the incorporation of this "novel system" contributes to clarify some aspects of the regulation of GH that previously were not fully understood. However, this system is not as simple as it was thought at first. Ghrelin and its receptor became recognized not only for stimulating GH release but also by the discovery that this system appeared to exert an important role on several aspects of energy homeostasis. In this way, GHSR1a becomes a potential therapeutic target for the treatment of wasting syndromes. One of the important features of GHSR1a is the basal activity in the absence of an agonist,

---

[*] Laboratory of Molecular Endocrinology, Research Area, Complejo Hospitalario Universitario de Santiago (CHUS), Santiago de Compostela, Spain
[†] Department of Medicine, School of Medicine, University of Santiago de Compostela, Santiago de Compostela, Spain
[‡] CIBER de Fisiopatología, Obesidad y Nutrición (CB06/03) Instituto de Salud Carlos III, Spain

resulting in a high degree of receptor internalization as well as of signaling activity. This constitutive activity seems to provide a tonic signal required for the development of normal height, probably through an effect on the GH axis. Additionally, GHSR1a might function as homo- or heteromeric complexes in living cells which introduce a key concept that could have significant implications in different aspects of receptor biogenesis and function. At molecular level, GHSR1a regulates the activation of the downstream mitogen-activated protein kinase, Akt, nitric oxide synthase, and AMPK cascades in different cellular systems. Added to this complexity, the idea that GHSR1a is not the single receptor for ghrelin has been progressively more recognized. In this sense, the available data are quite ambiguous and many fundamental questions need to be clarified. The purpose of this chapter is to summarize the most recent characteristics of GHSR1a as the features to define the action of ghrelin and its physiological implication. © 2008 Elsevier Inc.

I. INTRODUCTION

To talk about ghrelin, it is necessary to look back over till early 1980 when Bowers discovered that certain molecules from a series of synthetic opioid-like peptides showed the capacity to release growth hormone (GH) from isolated pituitary cells (Bowers *et al.*, 1980). GHRP-6 was one of the first peptides that showed to induce GH secretion independently from the GH-releasing hormone receptor (GHRH-R). Thus, several pharmaceutical companies initiated several projects to develop new drugs based on this compound. Thus, a series of peptidic and nonpeptidic compounds, termed growth hormone secretagogues (GHSs), were described up through the 1990s (Bowers *et al.*, 1984; Patchett *et al.*, 1995; Smith *et al.*, 1993). These GHSs act at pituitary and hypothalamus level through an orphan receptor, which was cloned in the late 1990s and so-called the growth hormone secretagogue receptor 1a (GHSR1a) (Howard *et al.*, 1996; McKee *et al.*, 1997). This fact indicated the existence of an endogenous ligand that had not been identified yet. It was not till 1999, when using cells expressing the mentioned receptor to monitor intracellular changes of second messengers, the endogenous ligand was identified and named ghrelin (Kojima *et al.*, 1999). Ghrelin is a 28-amino acid peptide with the peculiarity of an *n*-octanoic acid covalently linked to the hydroxy group of the serine 3 (Ser3) residue. Like many other peptidic hormones, ghrelin is processed from a 94-amino acid precursor through limited proteolytic cleavage at a single arginine. *In vivo*, the enzyme responsible for this event is the prohormone convertase 1/3 (Zhu *et al.*, 2006). Surprisingly, ghrelin is expressed in large amounts in endocrine cells in the stomach, and it was soon recognized that besides its capacity to release GH, ghrelin was much more than the endogenous GHS since other central and peripheral actions, such as energy homeostasis, reproduction, sleep regulation, cardiovascular actions, corticotrope secretion, stimulation of lactotrope, and influence on gastroenteropancreatic functions, were continuously emerging (Korbonits *et al.*, 2004; van der Lely *et al.*, 2004). The fact that ghrelin powerfully increases food intake and body weight (Nakazato *et al.*, 2001; Tschop *et al.*, 2000; Wren *et al.*, 2000) shifted the focus of the research on ghrelin/ GHSR1a system to its role in energy homeostasis. Among the attributes of this hormone, its capacity to stimulate short-term food intake is remarkable, which is more effective than any known molecule with the exception of neuropeptide Y (NPY) (Asakawa *et al.*, 2001; Wren *et al.*, 2001a), an effect that is equally exerted when ghrelin is central or peripherally administrated. Ghrelin showed to increase food intake when is delivered in humans (Druce *et al.*, 2005; Laferrere *et al.*, 2005; Neary *et al.*, 2004; Schmid *et al.*, 2005; Wren *et al.*, 2001a; Wynne *et al.*, 2005). Furthermore, chronic administration of ghrelin increased body weight as a result of anabolic effects on food intake, energy expenditure, and fuel utilization (Asakawa *et al.*, 2001; Nagaya *et al.*, 2004; Nakazato *et al.*, 2001; Tschop *et al.*, 2000; Wren *et al.*, 2001b; Xie *et al.*, 2004). Additionally, the ghrelin/GHSR1a system exerts other central and peripheral actions, including effects on hormone secretion, glucose homeostasis, pancreatic function, reproduction, gastrointestinal motility, cardiovascular function, cell proliferation and cell survival, immunity, inflammation, bone metabolism, memory, sleep, and much more (van der Lely *et al.*, 2004). Therefore, the regulation of food intake as well as the diverse array of processes involved direct or indirectly in energy expenditure and fuel utilization have revitalized the interest of the ghrelin/GHSR1a system as a potential target for the treatment of obesity and wasting disorders (Druce *et al.*, 2005). These processes are regulated by complex mechanisms that involve multiple intracellular signaling pathways. As consequence, efficient organized mechanisms are required for the appropriate biological and physiological situations. The efficacy of this organization is orchestrated, in a first approach, by the ghrelin receptor GHSR1a. By this reason, the purpose of this chapter is to review the classical and the most recent aspects of the basic features of the ghrelin receptor, as well as its inferences for receptor biology and endocrinology.

II. STRUCTURE OF GHSR1A: A G-PROTEIN–COUPLED RECEPTOR

The ghrelin receptor (GHSR) is expressed by a single conserved gene located at chromosomal location 3q26.2 (Mckee *et al.*, 1997). This gene is composed by two exons separated by one intron. Exon 1 is composed of the 5′-untranslated region and encodes for the first 265 amino acids from the transmembrane (TM) regions I to V. Exon II encodes for 101 amino acids from the TM regions VI and VII, and includes the 3′-untranslated region.

The GHSR gene encodes two types of GHSR mRNA, noted as 1a and 1b, by alternative mRNA processing (Mckee et al., 1997). The GHSR1a mRNA is encoded by the exons 1 and 2 where about 2152 nucleotides of the intronic sequence are removed from the pre-RNA by splicing (Petersenn et al., 2001). As a result, the GHSR1a is a protein of 366 amino acids with seven TM domains. By contrast, the GHSR1b mRNA is encoded by the exon 1 and the intron, probably due to the use of an alternative splice site and an alternative polyadenylation site (Howard et al., 1996; Mckee et al., 1997), which determines a polypeptide of 289 amino acids with only 5 TM domains plus 24 differing amino acids at the C-terminal region. Only the GHSR1a subtype is activated by ghrelin and the GHSs. It is important to note that the expression of each subtype, tagged with the green fluorescent protein (GFP) in HEK293 cells, confirmed that GHSR1a–GFP was localized in the plasma membrane, but curiously, GHSR1b–GFP was confined to the nucleus (Smith et al., 2005). To date, the functional activity of GHSR1b remains unclear.

Regarding the expression of GHSR1a, two cellular areas in the brain were found to be relevant. One is the pituitary gland where GHSR1a is mainly expressed in somatotrope pituitary cells of the anterior lobe, consistent with the role of ghrelin in regulating GH release. The second area was found in the hypothalamus, where the GHSR1a is mainly expressed in the arcuate nucleus, an area that is crucial for the neuroendocrine and appetite-stimulating activity of ghrelin (Korbonits et al., 2004; Shuto et al., 2002; Tannenbaum et al., 1998; van der Lely et al., 2004; Willesen et al., 1999), with other hypothalamic nuclei as the suprachiasmatic, the anteroventral preoptic, the anterior hypothalamic, the paraventricular, and the tubero-mammillary (Guan et al., 1997; Howard et al., 1996). Moreover, a discrete expression was displayed in other brain areas such as the CA2 and CA3 regions of the hippocampus, the substantia nigra, the ventral tegmental area (VTA), and the dorsal and median raphe nuclei. This wide distribution suggests a role for ghrelin in extrahypothalamic actions (Guan et al., 1997; Katayama et al., 2000; Muccioli et al., 1998); however, its physiological relevance remains to be determined. Additionally, a widespread expression was demonstrated in a great variety of peripheral tissues: thyroid gland, pancreas, spleen, myocardium, adrenal gland, testis, ovary, stomach, and in the neuronal cells of the gut (Dass et al., 2003; Gaytan et al., 2003, 2004, 2005; Gnanapavan et al., 2002; Shuto et al., 2001). This peripheral distribution strengthens the growing functions of ghrelin, other than the regulation of GH secretion and energy homeostasis.

A heterologous group of endocrine signals regulate pituitary and hypothalamic expression of the GHSR gene. In consequence, GHSR mRNA expression is inhibited by GH (Bennett et al., 1997; Horikawa et al., 2000; Kamegai et al., 1998; Nass et al., 2000), while it is upregulated by GHRH (Kineman et al., 1999). Additional evidences suggest that the anorexigenic hormone, leptin, reduces the level of GHSR expression in the arcuate nuclei (Nogueiras *et al.*, 2004). Therefore, ghrelin and leptin act as complementary players of one regulatory system which inform the central nervous system (CNS) about the current status of acute and chronic energy balance. Similarly, pituitary GHSR mRNA levels are inhibited by the insulin-like growth factor type-1 (IGF-1) (Kamegai *et al.*, 2005; Peng *et al.*, 2001; Wallenius *et al.*, 2001; Zhou *et al.*, 1997). Other additional regulatory factors are glucocorticoids (Kaji *et al.*, 2001; Petersenn *et al.*, 2001; Tamura *et al.*, 2000; Thomas *et al.*, 2000), thyroid hormones (Kaji *et al.*, 2001; Kamegai *et al.*, 2001; Petersenn *et al.*, 2001), and sex steroids (Kamegai *et al.*, 1999). Regarding homologous regulation, GHSR1a synthesis is susceptible to be downregulated by GHSs in the pituitary (Kineman *et al.*, 1999). In contrast, the level of GHSR expression in the arcuate nucleus is increased by ghrelin, an effect that appears to be GH dependent (Bresciani *et al.*, 2004; Nogueiras *et al.*, 2004). The homologous regulation appears to be tissue-specific and might reflect different physiological roles of GHSR in the target tissues.

GHSR1a is a member of the small family of receptors for peptidic hormones and neuropeptides, which includes the motilin receptor, neurotensin receptors 1 and 2, the GPR39, the neuromedin receptors 1 and 2. These receptors share a common molecular architecture consisting of seven TM spanning α-helices connected by alternating intra- and extracellular loops with the N-terminus located on the extracellular side and the C-terminus on the intracellular side. GHSR1a transduces information provided by ghrelin and by the GHSs not related structurally to ghrelin. This property was explained by the existence of a common binding domain imposed by the orientation of the TM segments that determine the stereo- and geometric specificity on the ligand's entry and binding to the TM core (Feighner *et al.*, 1998). This binding site allows that a conserved structure of the agonists recognizes a complementary conserved binding pocket accommodating the variable part of the ligand which interacts with specific agonist-associated regions determining an overlapping in the agonist-binding site (Bondensgaard *et al.*, 2004). A recent mutational analysis in the main ligand-binding pocket suggested that ghrelin interacts only with residues in the middle part of the pocket (TM III, VI, and VII) (Holst *et al.*, 2006). In addition to these proposed contact residues, the peptide will have additional contact points with the receptor, probably in the loop regions. Based on this model, it was suggested that the receptor activation was associated with a vertical seesaw movement of mainly TM VI and VII around a pivotal conserved Pro residue in the middle of each helix. TM VI and VII are thought to move inward toward TM III in the main binding pocket, whereas the intracellular loops of these helices are shifting the opposite way facilitating its coupling to G-proteins (Holst *et al.*, 2006). One of the most important features of GHSR1a is the basal activity in the absence of an agonist, resulting in a high degree of receptor internalization as well as of signaling activity. Based on gene-dosing experiments, the constitutive activity was confirmed by the inositol phosphate production, cAMP-responsive element (CRE), and serum-responsive element (SRE) transcriptional activity (Holst *et al.*, 2003, 2004), showing different degree of signaling activity. Regarding the inositol phosphate production, the G_q/phosphoinositide-specific phospholipase C (PI-PLC) pathway was 50% of its maximal activity. With respect to CRE-mediated transcription, the GHSR1a showed a solid ligand-independent activity which might be mediated through different downstream kinase mediated through of a G_q-protein-dependent pathway. A lower constitutive activity was observed for the SRE pathway, which suggests that a G-protein other than $G_{\alpha q}$ might be responsible for this constitutive activation. However, this constitutive activity is not extended to the entire GHSR1a-associated signaling pathways since this receptor did not show any constitutive activity in mitogen-activated protein kinase (MAPK) activation, at least through the extracellular signal-regulated kinase-1/2 (ERK1/2) signaling pathway (Holst *et al.*, 2004). Regarding the structural basis, the aromatic cluster on the inner face of the GHSR1a seems to determine this signaling activity (Holst *et al.*, 2004). This cluster is established by the positions of three Phe on the inner face of the extracellular ends of TMS VI and VII (PheVI:16, PheVII:06, and PheVII:09). In the model described, these hydrophobic groups ensure a favorable docking of the extracellular end of TM VI and VII through the formation of a hydrophobic core between these helices. This structural characteristic confers stabilization to the active conformation of this receptor providing a signaling activity independent of the ligand. In addition, inverse agonists seem to prevent the inward movement of TM VI and VII in particular, and thereby blocking receptor activation (Holst *et al.*, 2006). In line with these findings, the binding cavity of GHSR1a is considered to be formed by two subpockets: a polar cavity bearing the key residues involved in receptor activation and an aromatic/apolar subpocket, which plays a crucial role in determining the high constitutive activity. From these studies, a model of the two possible situations for GHSR1a was proposed: an open state, involved in agonist recognition, in which the two subpockets form two different binding sites; and a close one, involved in the constitutive activity, in which the aromatic cluster approaches to the polar subpocket. Recent docking studies confirmed the relevance of these subpockets (Pedretti *et al.*, 2006). Regarding the physiological importance of such constitutive signaling, it is difficult to differentiate between an effect of constitutive receptor signaling and an effect of the endogenous ligand *in vivo*. About this point, it has been suggested that the GHSR1a might play an important role in modulating the orexigenic signals in the regulatory pathways that are integrating anorexigenic signals such as leptin and insulin. The upregulation observed for GHSR1a in the hypothalamus during fasting might provoke an increase in the agonist-independent stimulation of GHSR1a, and so in the receptor-associated signaling tone (Holst et al., 2004). A recent work provides a very strong evidence for a physiological role of the constitutive activity of the GHSR1a in the development of normal height in humans (Pantel et al., 2006). A missense mutation in the ghrelin receptor (Ala204Glu) was found in two individuals that segregate with short stature. Importantly, this mutation resulted in a decrease in the cell-surface expression of the GHSR1a and selectively impaired the characteristic of constitutive signaling activity, but the receptor ability to respond to ghrelin was preserved. Other missense variant (Phe279Leu) was detected in child with short stature, but without any functional characterization (Wang et al., 2004). However, an analysis of the residues located in the main ligand-binding pocket of the GHSR1a demonstrated that Phe279 exerts an important role for the constitutive activity (Holst et al., 2004). Thus, this mutation probably shows a similar molecular pharmacology profile to that of Ala204Glu. These findings suggest that the ghrelin receptor constitutive activity provides a tonic signal required for the development of normal height, possibly through an effect on the GH axis (Holst et al., 2006; Pantel et al., 2006). Although there is considerable evidence that the GHSR1a-associated constitutive activity is related to a pathogenic mechanism of growth failure, additional questions raise about the physiological significance of the decreased cell-surface expression of the mutated receptors.

Regarding the bioactive conformation of ghrelin and mode of interaction with the binding pocket of GHSR1a, there is no data at present. This is fundamentally determined by the high conformational flexibility introduced by a 28-amino acid peptide in which Ser3 is further modified by *n*-octanoylation. There is some information about the structural features of ghrelin necessary for its interaction with the receptor, based on binding assays and activation of GHSR1a by measuring intracellular calcium mobilization using HEK293 cells transfected with the GHSR1a (Bednarek et al., 2000). The role of the *n*-octanoyl group in binding and activation of GHSR1a was carried out by means of ghrelin analogues in which the hydroxyl group of Ser3 was acylated by diverse aliphatic and aromatic acids. Modified peptides with longer aliphatic chains showed to activate the receptor as efficiently as the parent compound. By contrast, the analogue with the small acetyl group replacing the *n*-octanoyl group was a poor agonist for receptor activation. This fact reveals a significant role of the hydrophobic interaction in the recognition and activation of this receptor. An analogue in which the *n*-octanoyl group was replaced by the 1-adamantane acetyl group allowed to confirm this point as this bulky and rigid hydrophobic substituent induced a similar response to that of analogues with flexible extended hydrocarbon groups. This result might give an idea of the size of the GHSR1a-binding site. The substitution of the ester bond in the side chain of residue 3 by an amide bond did not show important modifications in the activity regarding the native ghrelin. The octanoylation at the other three Ser residues (Ser2, 6, and 18) showed a decrease in the activity for these analogues, only the Ser2 analogue was as potent as the parent peptide. Furthermore, the first five residues of the ghrelin sequence [Gly-Ser-Ser(n-octanoyl)-Phe-Leu] are required to activate the GHSR1a, which determines the active core for ghrelin. However, these results did not fit well with the results obtained with the short truncated analogues (encompassing the first 4-, 8-, or 14-amino acid residues) which were ineffective in stimulating GH secretion in neonatal rats and did not displace radiolabeled ghrelin from binding sites in membranes of human hypothalamus or pituitary (Torsello et al., 2002). Although it is difficult to explain these discrepancies, it would be possible to postulate that the affinity of the truncated analogues for GHSR1a binding might determine the grade of activation of the GHSR1a-associated intracellular machinery. In this sense, the intracellular calcium mobilization might not be a reflex of a complete activation of the signal transduction systems required, for example, to GH release. It is important to note that calcium mobilization is only a component of the complex signal transduction system associated to GHSR1a.

III. How to Define the Role of the System Ghrelin/GHSR1a?

Ghrelin and its receptor became recognized not only for stimulating GH release but also by the discovery that this system appeared to exert an important role on several aspects of energy homeostasis. In this way, the ghrelin/GHSR1a system became a potential therapeutic target for the treatment of wasting syndromes. However, in an attempt to demonstrate the relevance of this system, several studies developed with transgenic animals, in which the genes encoding ghrelin and GHSR1a were knocked out, did not display the expected results specially related to the feeding behavior and body fat deposition. Mainly, these animals showed a normal growth, energy expenditure, and food intake (Sun et al., 2003, 2004; Wortley et al., 2004). Moreover, mice lacking ghrelin fed with a high-fat diet showed no increase in body weight or adiposity. Thus, these loss-of-function studies raised doubts regarding the physiological significance of the ghrelin/GHSR1a system in regulating these processes. However, two recent works report that deleting ghrelin and GHSR1a protects against diet-induced obesity in mice started on a high-fat diet in the early post-weaning period (Wortley et al., 2005; Zigman et al., 2005). The diminution in weight gain was associated with a decrease in adiposity and an increase in energy expenditure and locomotor activity as the animals aged. Two main questions rose from these studies: why adult ghrelin-deficient mice become obese on a high-fat diet, whereas young, postweaning animals are resistant to the effects of this diet; and why these ghrelin-deficient mice preserve the GH/IGF-1 axis. As previously indicated (Grove and Cowley, 2005), the deletion of a single orexigenic system does not always lead to a leaner phenotype. An example of this affirmation is the most well-known melanin-concentrating hormone-null mice (Shimada et al., 1998). On the contrary, the deletion of a single anorexigenic signal results in obesity (leptin and melanocortin). Consequently, it is obvious that these animals are capable of developing compensatory mechanisms that regulate the body weight in the absence of orexigenic systems and, in this way, to distort the functional significance of the ghrelin/GHSR1a system. This fact might give an idea about the complexity of the system that regulates the body weight. Therefore, more investigations are needed to understand not only the role of the ghrelin/GHSR1a system in the body weight homeostasis but also the compensatory mechanisms that are able to balance its action.

IV. Are There Alternative Ligands for the GHSR1a?

GHSR1a shows the characteristic to transduce the information provided by ghrelin and the group of GHSs not related structurally to ghrelin. This property is explained by the existence of a common binding domain demonstrated by molecular modeling and site-directed mutagenesis studies developed with peptidic and nonpeptidic GHSs (Feighner et al., 1998). This binding site might determine that a conserved structure of the agonists recognizes a complementary conserved binding pocket, which directs the variable part of the ligand and interacts with specific agonist-associated regions establishing an overlapping in the agonist-binding site (Bondensgaard et al., 2004). Additionally, GHSR1a seems to have other alternative ligands. This piece of information was revealed by studies carried out with adenosine (Smith et al., 2000; Tullin et al., 2000) and also with cortistatin (Deghenghi et al., 2001). Regarding adenosine, administration of this nucleoside to cells not expressing the GHSR1a is apparently devoid of action, but when administered to cells expressing this receptor, adenosine triggers intracellular calcium rise, although it fails to stimulate GH secretion (Tullin et al., 2000). Concerning intracellular calcium mobilization, a signaling pathway was proposed involving adenylate cyclase (AC)/protein kinase A (PKA)/inositol 1,4,5-triphosphate (IP_3)-receptors, different from ones activated by ghrelin (Carreira et al., 2004). Based on such results, it appeared that the GHSR1a was able to activate different intracellular second messenger systems that regulate disconnected agonist-dependent mechanisms. However, the fact that adenosine did not show cross competition with ghrelin in binding to GHSR1a (Camina et al., 2004; Carreira et al., 2004) remained as a discordant data, considering that this fact is a requisite to define alternative ligands for a specific receptor. However, two recent works report that adenosine is not a ligand of the GHSR1a, and its action is mediated by specific adenosine receptors (Carreira et al., 2006; Johansson et al., 2005). Actually, this conclusion is clearly endorsed by the reduction in the functionality of adenosine after depleting cellular levels of adenosine receptors, A2b adenosine receptor (A2b-R) and A3 adenosine receptor (A3-R), by transfecting siRNA specially directed against each isoform (Carreira et al., 2006). It thus seems that the overexpression of GHSR1a modifies the efficacy of G-protein-coupling activities for A2b-R/A3-R with the consequent modulation of the GHSR1a response. Indeed, the expression of the GHSR1a in HEK293 cells modifies adenosine receptor properties changing a nonoperative receptor in a receptor capable "to use" or "to modulate" the GHSR1a activity. In consequence, the interaction between both receptors, probably by heterodimerization, is a mechanism that helps to aggregate the signal transduction machinery facilitating the modulation of signaling. This is a clear example of how the elevated receptor expression levels are able to lead to unexpected interactions with other receptors, change the receptor properties and, inevitably, their functions. Furthermore, this represents a lesson of how a classical and simple biochemist assay (binding assay) can resolve "artifacts" introduced by the use of molecular biology techniques.

Unlike adenosine, the idea that cortistatin might be a ligand of GHSR1a emerges from binding studies that demonstrate a cross competition with ghrelin binding in membranes from human hypothalamus and pituitary gland (Deghenghi et al., 2001; Muccioli et al., 2001). Despite that this is requisite to define alternative ligands for a specific receptor, the functional ability of cortistatin to bind GHSR1a is not evident yet and the endocrine interaction of cortistatin with this receptor requires further studies.

V. Endocytosis of GHSR1a

The binding of ghrelin to GHSR1a promotes a series of molecular interactions that determine the receptor endocytosis, signaling through G-protein-independent signal transduction pathways and feedback regulation of G-protein coupling. In this sense, the GHSR1a activity represents a synchronized balance among molecular mechanisms regulating receptor signaling, desensitization, and resensitization. An important aspect of ghrelin activity and regulation is the internalization of ghrelin-activated GHSR1a as part of the mechanism that controls the responsiveness in the target cells. This cellular mechanism was evaluated in HEK293 and CHO cells stably expressing GHSR1a by means of radioligand binding, confocal microscopy, and intracellular calcium rise after receptor activation (Camina *et al.*, 2004). The exposure of GHSR1a to ghrelin results in a rapid attenuation of the receptor response due to a combination of the uncoupling from heterotrimeric G-proteins and the internalization to the intracellular compartments. The homologous desensitization is not mediated by the second messenger-stimulated protein kinase C (PKC), at least in these cellular models. However, ghrelin-activated PKC mediates the agonist-independent desensitization of surrounding receptors susceptible to be regulated by PKC and, in this way, it represents a heterologous desensitization. The kinetic studies of GHSR1a is internalization demonstrate that GHSR1a was internalized by a clathrin-mediated mechanism through an endosomal trafficking pathway (Camina *et al.*, 2004). Once the ligand-receptor complex is internalized into vesicles, GHSR1a is sorted into endosomes to be recycled back to the plasma membrane. The level of receptors on the cell surface rises once again close to 100% after 6 h of agonist removal, where GHSR1a came from endosomes rather than *de novo* receptor synthesis. This slow recycling might be determined by the stability of the complex GHSR1a/β-arrestin during clathrin-mediated endocytosis since this complex appears to dictate the profile of the receptor resensitization (Luttrell and Lefkowitz, 2002). Indeed, receptor binding of β-arrestin 2 to GHSR1a in response to ghrelin has been described (Holst *et al.*, 2005). Interestingly, physiological experiments show that the GH response after two consecutive pulses of ghrelin is blunted when both pulses are separated by a short interval (60 min), but are restored to its initial amplitude when the second pulse is administered after 180, 240, or 360 min, which fits in well with the kinetics of receptor recycling. Therefore, GHSR1a mediates ghrelin uptake and degradation through an internalization pathway essentially involving clathrin-coated pits, dissociation of ligand-receptor complex in early endosomes, and reappearance of the receptor by a recycling pathway. In this way, receptor desensitization and endocytosis prevent an over stimulation and determine the frequency of ghrelin response. Thus, alterations of this mechanism lead to an uncontrolled or defective stimulation of target cells with the consequent physiological alterations.

VI. Homo- or Heteromeric Complexes for GHSR1a

Whereas most of the membrane receptors, including tyrosine kinase receptor, cytokine receptors, guanylate cyclase receptors, and ligand-gated channels, form oligomeric entities to transduce the signal from extracellular to intracellular side, the G-protein–coupled receptors (GPCR) have long been thought to function as monomers. This concept is basically built on the idea that most of GPCRs have a membrane core domain composed of seven TM-spanning helices, responsible for both, the ligand recognition and the intracellular effectors activation. In this way, a GPCR shows diverse conformations that oscillate from active to inactive states determined by an agonist or an inverse agonist. However, this "simple" model is unable to explain certain cooperative phenomena observed in ligand binding. This led to the demonstration that GPCRs can form dimers or even higher ordered oligomers, although the functional significance of this event is not yet clear. Interestingly, a work suggested the existence of GHSR1a homodimers based on titration curves obtained from a bioluminescence resonance energy transfer (BRET) assay (Jiang et al., 2006). This finding might endorse a dimeric receptor model proposed to explain the pharmacological properties of GHS compounds on their own as agonists and their influence on ghrelin signaling (Holst et al., 2005), that is, the action of some GHSs as positive and negative allosteric modulators of ghrelin action. The model for GHSR1a involves a homodimeric unit, where ghrelin is bound only to one subunit avoiding almost certainly, by negative cooperativity, the binding of a second ghrelin molecule in the other subunit. Thus, this receptor model prevents the binding of two ghrelin molecules at the same time; however, the structure requirements for the negative cooperativity might not prevent that a small nonpeptidic agonist was bound to the other subunit. In this way, the occupancy of both receptor subunits should result in a variation of signaling efficacy. The small nonpeptidic compound that binds relatively deep in the main ligand-binding pocket may exert a positive cooperative effect (i.e., L-692,429) or a neutral cooperative effect, although with an increase in the maximal signaling capacity (i.e., MK0677). For other agonists, as in the case of GHRP-6, the conformational changes of the binding pocket from the second subunit lead to a decrease of binding effectiveness which might exert a negative cooperative effect with a subsequent diminution in the ghrelin potency. Such complex mechanism of this homodimeric unit offers extraordinary possibilities for allosterically regulating ghrelin activity using compounds that operate at different binding sites of the homodimeric unit. This state could be more complex taking into account that GHRP-6 shows the capacity to bind and activate the TM protein CD36 introducing the possibility of heterodimerization between CD36 and GHSR1a (Bodart et al., 2002). Concerning the possibility of heterodimerization, it was reported that ghrelin can amplify dopamine signaling through a mechanism that involves the GHSR1a/dopamine D1 receptor (D1R) heterodimer (Jiang et al., 2006). As a result, activation of GHSR1a by ghrelin amplifies dopamine/D1R-induced cAMP accumulation through a switch in G-protein coupling of GHSR1a from $G_{\alpha 11/q}$ to $G_{\alpha i/0}$. It was therefore proposed that ghrelin provides temporal control over the magnitude of dopamine signaling in neurons where both receptors are expressed. Furthermore, a work showed that GHSR1a is present in VTA, where it participates in the dopamine neuronal activity, synapse formation, and dopamine turnover in the nucleus accumbens (Abizaid et al., 2006). In this area, GHSR1a seems to be playing as a physiological mediator of appetite.

The fact that GHSR1a might function as homo- or heteromeric complexes in living cells introduces a new important concept that could have significant implications in different aspects of receptor biogenesis and function. The existence of homodimers raises fundamental questions about the molecular mechanisms involved in signal transduction. Besides, the formation of heterodimers enlarges fascinating combinatorial possibilities that could underlie an unexpected level of pharmacological diversity, and contribute to a cross-talk regulation between transmission systems. The existence of dimers introduces a new point of view for the development and screening of new drugs that not only enhance the efficacy and potency of the endogenous ligand but also modulate its effectiveness.

VII. GHSR1A: G-PROTEIN-SIGNALING PATHWAYS

Despite the long time since the discovery of ghrelin, the GHSR1a-activated intracellular calcium mobilization turns to be the best characterized intracellular function associated to this hormone. GHSs activate an intracellular calcium mobilization mediated by a PI-PLC through a $G_{\alpha q/11}$-protein (Howard et al., 1996; McKee et al., 1997). In somatotrope cells, this enzyme acts on phosphatidylinositol 4,5-biphosphate (PIP_2) at the internal leaflet of the plasma membrane generating IP_3 and diacylglycerol (DAG) (Adams et al., 1995). IP_3 triggers the release of calcium from IP_3 intracellular sensitive stores, while DAG is responsible for the activation of PKC. The IP_3-dependent calcium rise triggers voltage-independent K^+ channels of the small (SK) conductance type, transiently hyperpolarizing the cells (Chen et al., 1990). The first calcium current is followed by a sustained calcium entry through L- and T-type calcium channels via a membrane depolarization (Chen and Clarke, 1995; Chen et al., 1996, 1990; Herrington and Hille, 1994; Smith et al., 1993). Although the mechanisms involved in the depolarization have not been fully defined, an inhibition of K^+ channels, probably mediated through the activation of PKC together with an increase in the membrane Na^+ conductance, may play a key role in this effect (Kato and Sakuma, 1999). However, an alternative transduction system appears to be implicated in the ghrelin-induced calcium mobilization. In the NPY-containing neurons, the ghrelin-induced calcium rise depends on the calcium influx through N-type calcium channels via the cAMP/PKA-signaling pathway coupled to GHSR1a

(Kohno et al., 2003). Furthermore, three distinct signaling systems, including the AC/PKA, PLC/PKC, and the extracellular calcium system, were involved in the response of porcine somatotropes to ghrelin, operating in a sequential manner (Kohno et al., 2003). The implication of AC/PKA is also reported in ovine but not rat pituitary somatotropes (Wu et al., 1996). Thus, the existence of different signaling pathways to trigger intracellular calcium mobilization raises fundamental questions about the molecular mechanisms involved in these signal transductions. Regarding this point, it is possible to postulate about the existence of different GHSR1a subtypes that exhibit essential differences in their coupling to G_s and their dependence on the AC/PKA system. This diversity is established by sequence variations within the intracellular loops of GHSR1a that act as a key in facilitating its coupling to G_q and/or G_s. Nevertheless, it is evident that GHSR1a is capable of coupling to multiple G-protein/effectors in a species- and/or tissue-specific manner.

In addition to the ghrelin-activated ionic currents, the proliferative activity of ghrelin involves the activation of the downstream MAPK in different cellular systems. An attempt to delimitate the signaling machinery involved in the ghrelin-mediated ERK1/2 activation has been reported (Mousseaux et al., 2006). To this end, CHO cells that stably express GHSR1a were selected as the cellular model in which GHSR1a controls ERK1/2 activity through a PLC- and PKCepsilon (PKCε)-dependent pathway. No evidence for β-arrestins, G_i, phosphatidylinositol-3 kinase (PI-3K), or tyrosine kinases is found in ERK1/2 activation. However, these data are not consistent with previous findings. In 3T3-L1 preadipocytes, the mitogenic effect of ghrelin is mediated via activation of the PI-3K/Akt and MAPK pathways through a pertussis toxin (PTX)-sensitive G-protein ($G_{i/o}$) (Kim et al., 2004). In addition, the IRS-1-associated PI-3K activity and Akt phosphorylation are also stimulated. The PI-3K/Akt pathway has also been implicated in the mitogenic action on pancreatic adenocarcinoma cells (Duxbury et al., 2003). On the other hand, in human and rat adrenal zona glomerulosa cells, the proliferative action of ghrelin involves the activation of ERK1/2 and seems to be independent of PKA and PKC (Andreis et al., 2003; Mazzocchi et al., 2004). Additionally, in hepatoma cells, ghrelin stimulates cell proliferation through the tyrosine phosphorylation of IRS-1 and association with the adapter molecule growth factor receptor-bound protein 2 (GRB2) with the consequent activation of MAPK (Murata et al., 2002). Therefore, the downstream molecules of GHSR1a can cross talk with the insulin-signaling pathway in these cells. Indeed, combination of ghrelin and insulin results in an additive increase in the downstream signaling of IRS-1. However, ghrelin suppresses Akt kinase activity despite the presence of insulin, as well as upregulates the amount of phosphoenolpyruvate carboxykinase (PEPCK) mRNA expression, a gluconeogenic enzyme that catalyzes the conversion of oxaloacetate to phosphoenolpyruvate. This anti-insulin action might be affecting the glucogenesis by attenuating the effect of insulin on the expression of PEPCK. The mitogenic activity of ghrelin through the GHSR1a has also been described in osteoblastic cells (Maccarinelli et al., 2005) and the prostate cancer cell line (PC3) (Jeffery et al., 2002), although no data are available about the signaling pathways implicated. In general, these results allow us to evaluate the real complexity of the cell regulatory network in which the existence of different cascades to give a common response varies as a function of the cell type. In this way, the response depends on the signal transduction proteins, their isoforms, on the positive and negative feedbacks and feedforward controls, and so on (Dumont et al., 2001). For this reason, the use of cell models with the implicit assumption that networks of ERK1/2 regulation are "universal" is not adequate as they are, in fact, cell specific.

Activation of PI-3K/Akt pathway has been implicated in the neuroprotection exerted by the GHSs GHRP-6, and hexarelin. GHRP-6 stimulates phosphorylation of Akt and proapoptotic bcl-2 family member BAD in specific areas of the CNS of adult rats, such as hypothalamus, hippocampus, and cerebellum, with no change in MAPK or glycogen synthase kinase-3β (Gsk-3β) (Frago et al., 2002). Furthermore, this activity is coincident with IGF-1 mRNA expression, although the IGF receptor and IGF-binding protein 2 (IGFBP-2) were not affected. Therefore, activation of intracellular signaling mechanisms involving Akt might be directly activated by GHRP-6, or via increased IGF-1. In several in vitro experiments developed in an embryonic rat hypothalamic neuronal cell line, RCA-6, GHRP-6 shows to stimulate IGF-1 and NPY mRNA synthesis and activates PI-3K/Akt-signaling pathway, although in an independent way (Frago et al., 2005). In this cellular model, no significant increase to IGF-1 is detected until 2 h posttreatment, a delayed response that excludes the implication of IFG-I in the rapid response to GHRP-6 in NPY synthesis. Additionally, GHRP-6 shows to reverse glutamate-induced cell death by decreasing activation of caspases 9 and 7 and poly(ADP-ribose) polymerase fragmentation (Delgado-Rubin de Celix et al., 2006). On the other hand, hexarelin reduces neonatal brain injury via phosphorylation of Akt, indicating involvement of PI-3K, reduction of caspase-3 activity, and attenuation of Gsk-3β, with no effect in ERK activation (Brywe et al., 2005). However, this study shows no significant increase in IGF-1, probably due to the experimental conditions. Curiously, an increase in the IGF-1R was detected which opens the possibility of a transactivation of the IGF-1R, although it is not possible to distinguish between an endogenous factor and a cross signaling. Unfortunately, the implication of GHSR1a has not been determined yet, although this receptor might mediate this action as it is expressed in all areas where a response is observed (Kojima et al., 2001).

$5'$-AMP-activated protein kinase (AMPK) is the downstream component of a protein kinase cascade that plays a pivotal role in the regulation of energy metabolism acting as an energy sensor. When activated, AMPK stimulates glucose uptake and lipid oxidation to produce energy, while turning off energy-consuming processes to restore energy balance. In this sense, AMPK works as a signal integrator among peripheral tissues and the hypothalamus in the control of body energy balance in response to hormones and nutrients (Long and Zierath, 2006). Recently, it was demonstrated that *in vivo* administration of ghrelin, which leads to an increase in food intake, stimulates AMPK activity (Andersson *et al.*, 2004); conversely, administration of anorexigenic peptides including leptin, which decreases food intake, decreases its activity in the hypothalamus (Andersson *et al.*, 2004). The complete mechanisms by which ghrelin regulates AMPK are not delineated yet, although the tumor suppressor LKB1 serine/threonine kinase may be an upstream mediator that activates AMPK kinase (Hardie, 2004). Furthermore, the regulatory action of ghrelin on AMPK activity is related to fat distribution and metabolism in a tissue-specific appearance. Ghrelin stimulates AMPK activity in the heart; but, this activity is inhibited in the liver and the adipose tissue (Kola *et al.*, 2005). In rat liver, ghrelin decreases AMPK activity, showing a lipogenic and glucogenic pattern of gene expression, and an increase in triglyceride content with unchanged mitochondrial oxidative enzyme activities (Barazzoni *et al.*, 2005). The tissue-specific AMPK activity mediated by ghrelin is determined by the upstream regulator and downstream effector signals which need to be defined.

In relation with the control of feeding behavior, ghrelin activates the nitric oxide synthase (NOS) in the hypothalamus (Gaskin *et al.*, 2003) as other orexigenic peptides, such as NPY (Morley *et al.*, 1999); on the contrary, this activity is decreased by several anorexigenic peptides including leptin (Calapai *et al.*, 1998, 1999). Although the GHSR1a-associated signaling pathway is not delineated, it is thought that this action is mediated through NPY to modify the nitric oxide (NO) levels. Consequently, ghrelin may stimulate the NPYergic neurons to produce NPY which in turn activates NOS converting L-arginine into NO affecting the food intake. Diverse effects of ghrelin have been related with the ghrelin-activated NO pathway, including the control of insulin and glucagon secretion (Qader *et al.*, 2005). Studies in rat pancreatic islets revealed that the effects of ghrelin on glucagon and insulin secretion are in relation with the rise in NO derived from calcium/calmodulin-dependent NOS (neuronal NO). A recent work demonstrates that ghrelin stimulates NO production in endothelium using a signaling pathway that involves PI-3K, Akt, and endothelial NOS (eNOS) (Iantorno *et al.*, 2006). In these cells, ghrelin also stimulates phosphorylation of MAPK cascade, although, unlike insulin, does not stimulate MAPK-dependent secretion of the vasoconstrictor endothelin-1 (ET-1). This novel vascular action of ghrelin allows the development of therapeutic strategies to treat diabetes and related diseases characterized by reciprocal relationships between endothelial dysfunction and insulin resistance.

The ghrelin/GHSR1a system shows the property to amplify the cAMP levels stimulated by GHRH through GHRH-R, an effect evaluated in a homogenous population of cells expressing cloned GHRH and GHS receptors (Cunha and Mayo, 2002). Although the activation of the GHSR1a alone has no effect on cAMP production, coactivation of the GHSR1a and GHRH receptors produces a cAMP response approximately twofold enhanced of that observed after GHRH receptor activation alone. The inhibition of signaling molecules associated to GHSR1a activation, including PKC, PI-PLC, G-protein subunits, does not modify the potentiation of GHRH-activated cAMP. Furthermore, GHSR1a activation does not potentiate the forskoline-induced cAMP production. These results suggest the heterodimerization of GHSR1a/GHRH-R as a possible explanation for the potentiation of the second messenger pathways. This finding endorses the idea that ghrelin/GHS actions on the pituitary to release GH are dependent on a functional GHRH receptor. This is not an isolated fact as GHSR1a showed to amplify dopamine/D1R-induced cAMP accumulation through a switch in G-protein coupling of GHSR1a from $G_{\alpha 11/q}$ to $G_{\alpha i/0}$ (Jiang et al., 2006). It was therefore proposed that GHSR1a regulates the dopamine-signaling pathway in neurons expressing both receptors involving formation of heterodimers. Interestingly, different signal transductions are involved in the control of GHRH- and dopamine-stimulated cAMP biosynthesis. While the amplification of GHRH-R signaling is through a $G_{\alpha 11/q}$/PKC-dependent pathway, independent of $G_{i/0}$-protein, the regulation of D1R signaling is dependent on $G_{i/0}$-protein and independent of PKC (Jiang et al., 2006). In light to these effects, it was proposed that the amplification of $G_{\alpha s}$-D1R-mediated cAMP accumulation is through the $\beta\gamma$-subunit dissociated from G_i-protein that amplifies the accumulation of cAMP via activation of AC2. Amplification of GHRH-induced cAMP accumulation might be also explained via activation of AC2, but in this case via PKC-mediated phosphorylation of AC2. The main question that rose from these studies is why the target protein, AC2, is activated by the use of different signaling pathways through the same receptor. In the case that the synergic effect of ghrelin on GHRH signaling was due to a heterodimeric unit, this type of association might cause specific changes in the TMα-helices of GHSR1a, which would modify the conformation of intracellular loops and uncover different masked G-protein-binding sites determining the interaction with these proteins. As a result, the heterodimer unit determines the G-protein subunits and the intracellular signaling responses.

Additionally, ghrelin seems to exert an inhibitory effect on angiogenic factors. Assays developed with endothelial HUVEC cells showed that the fibroblast growth factor-2 (FGF-2)-induced cell proliferation and the Matrigel tube formation were inhibited by ghrelin (Conconi et al., 2004). This effect seems to be restricted to proliferating endothelial cells, indicative of a high selectivity as an antivascular agent, and is modulated by an autocrine–paracrine mechanism since ghrelin/GHSR1a are expressed in this cell line. This angiostatic property was also observed in neuromicrovascular endothelial cells, in which ghrelin inhibited the tyrosine kinase and ERK1/2 activities associated to FGF-2 (Baiguera et al., 2004). In addition, ghrelin showed to inhibit the angiotensin II-induced migration of human aorta endothelial cells (Rossi et al., 2006). Curiously, this inhibitory action involves dependence on the AC/PKA system and probably of the G_s-protein, although the implication of GHSR1a as ghrelin receptor was not determined.

The expression of ghrelin and its receptor is significantly increased during acute experimental colitis in mice degradation (Zhao et al., 2006). In human colon epithelial NCM460 cells stably transfected with GHSR1a, ghrelin stimulated IκBα phosphorylation and its subsequent degradation. Furthermore, ghrelin showed to stimulate nuclear factor-kappa B (NF-κB)-binding activity and NF-κB p65 subunit phosphorylation in a calcium/PKC-dependent pathway. This process was followed by an increase in interleukin (IL)-8 promoter activity and IL-8 protein secretion (Zhao et al., 2006). This finding opens up the exciting possibility that ghrelin may participate in the pathophysiology of colonic inflammation through the PKC-dependent NF-κB activation and the IL-8 production. These results stand in contrast to the potential protective function suggested in some types of inflammatory processes. In this regard, ghrelin was shown to inhibit leptin and endotoxin LPS-induced expression of proinflammatory cytokines such as IL-1β, IL-6, and tumor necrosis factor-α (TNF-α) in isolated monocytes and in T cells, as well as LPS-cytokine expression *in vivo* (Dixit et al., 2004; Li et al., 2004). Additionally, ghrelin prevented the endotoxin-induced release of IL-6 from peritoneal macrophages isolated from adjuvant-induced arthritic rats (Granado et al., 2005). In endothelial cells, ghrelin inhibited IL-8 and monocyte chemoattractant protein-1 (MCP-1) production in response to TNF-α, which showed to be mediated by attenuation of NF-κB nuclear translocation and transcriptional activity (Li et al., 2004). More recently, GHSR1a has been involved in the regulation of the peroxisome proliferator-activated receptor γ (PPARγ)-liver X receptor α (LXRα)-ATP-binding cassette transporters A1/G1 (ABCA1/ABCG1) cascade in macrophages, and such modulation is associated with enhanced cholesterol efflux by macrophages (Avallone et al., 2006). On view of these results and considering the implication of the role of PPARs in mediating anti-inflammatory processes (Castrillo and Tontonoz, 2004; Lee et al., 2003; Welch et al., 2003), it was proposed that the activation of GHSR1a might regulate the inflammatory response in macrophages through PPAR isoforms (Avallone et al., 2006). Even though the exact intracellular mechanism has not been delineated in most of the cases, the differences between proinflammatory and anti-inflammatory effects associated to ghrelin might provide further evidence to validate the concept of transducing cascade networks, cell type specific and physiologically dependent. In this sense, it is important to mention that the activation of PI-3K, MAPK, and PKC have been shown to promote NF-κB pathways in different systems, signaling pathways associated to ghrelin action (Lilienbaum and Israel, 2003; Zhao et al., 2001, 2005). Probably, the implication of PI-3K, MAPK, and/or PKC in the ghrelin signaling network should be evaluated in the different systems in concert to their implications in the ghrelin action.

VIII. A Brief Commentary: "New" Receptors for Ghrelin and Desacyl Ghrelin

The impression that GHSR1a is not the single receptor for ghrelin and/or GHSs has been progressively more accepted. Since the finding that ghrelin and peptidic/nonpeptidic GHSs showed different binding and biological activities in the same systems, the puzzle has grown to include binding sites for desacyl ghrelin, different receptor subtypes for ghrelin, and even common binding sites for ghrelin and desacyl ghrelin. The presence of a specific uncharacterized ghrelin receptor, different from the GHSR1a, has been described in chondrocytes (Caminos et al., 2005), human erythroleukemic HEL cells (De Vriese et al., 2005), and cardiomyocytes (Iglesias et al., 2004). In addition, the existence of a common uncharacterized receptor that mediates shared biological effects for ghrelin and desacyl ghrelin has been described in breast cancer cell lines (Cassoni et al., 2001), cardiomyocytes (Baldanzi et al., 2002), human prostatic neoplasms, and related cell lines (Cassoni et al., 2004). The involvement of a common receptor was described in the promotion of rat bone marrow adipogenesis *in vivo* (Thompson et al., 2004) as well as in the antilipolytic action in rat adipose tissue (Muccioli et al., 2004). Further evidence for a shared binding site was described in HT-T15 β-cells (Granata et al., 2006). Both ghrelin and desacyl ghrelin activate similar signaling pathways (cAMP/PKA, ERK1/2, and PI-3K/Akt) and promote proliferation and inhibition of apoptosis in this β-cell model. The same effect on cell proliferation for ghrelin and desacyl ghrelin was also described in human osteoblast cells via PI-3K/MAPK pathway in absence of GHSR1a (Delhanty et al., 2006). Despite the shared binding site for ghrelin and desacyl ghrelin, the latter seems to act as an independent hormone through other receptor(s) (Ariyasu et al., 2005; Asakawa et al., 2005; Baldanzi et al., 2002; Bedendi et al., 2003; Broglio et al., 2001; Cassoni et al., 2001, 2004; Chen et al., 2005; Gauna et al., 2004, 2005; Muccioli et al., 2004). Recently, it was suggested that desacyl ghrelin may directly enter the brain through the blood–brain barrier and act through corticotropin-releasing factor receptor type 2 (CRF-R2) to induce the anorectic effects and the disruptive change in fasted motor activity in the stomach (Chen *et al.*, 2005). Most recently, central administration of desacyl ghrelin has shown to increase feeding through activation of the orexin pathway. Furthermore, central administration of desacyl ghrelin to GHSR1a-deficient mice stimulated feeding, suggesting that this peptide acts on a target protein that is specific for it (Toshinai *et al.*, 2006). Added to the complexity of diverse ghrelin receptor subtypes, the presence of receptors that show a specific binding for synthetic peptidic GHSs has been described in anterior pituitary (Ong *et al.*, 1998), lung tissue and CALU-1 cells (Ghe *et al.*, 2002), and cardiomyocytes and microvascular endothelial cells (Bodart *et al.*, 2002). Interestingly, a unique peptidic GHS-binding site has been identified in the heart as CD36, a multifunctional B-type scavenger receptor. The data strongly suggest that CD36 mediates the increase of the coronary perfusion pressure induced by hexarelin, a hexapeptide of the GHS family (Bodart *et al.*, 2002). Although CD36 might be a candidate to mediate the effects described for peptidic GHSs, this receptor appears to be distinct from those identified in the pituitary (Ong *et al.*, 1998). CD36 has been implicated in the regulation of PPARγ-LXRα-ABC cascade in macrophages that provide a novel mechanism by which this receptor might regulate cholesterol metabolism in this system (Avallone *et al.*, 2006). In light of these results, there is adequate amount of data sustaining the presence of "alternative" receptors, different from GHSR1a, which seem to be involved in mediating the major nonendocrine actions of ghrelin, GHSs, and probably of desacyl ghrelin. However, these data are quite ambiguous and many fundamental questions need to be clarified, for instance, the common and antagonized actions described for ghrelin and desacyl ghrelin in the peripheral systems, the existence of a common receptor, or, by contrast, the existence of specific receptors for both factors, and, in this particular occurrence, the cross-affinity grade for each factor. The characterization of these emerging receptors as well as their functional significance is required as the starting point.

IX. Concluding Remarks

Although the regulation of food intake and energy homeostasis is far from being understood, it is evident that our insight on this problem advances unrelently. In addition to leptin and other orexigenic and anorexigenic peptides, ghrelin appears as a key regulator. At this moment, several pieces of a complex puzzle are available, though it needs to be completed by means of more complete progressive knowledge of the structure and signaling of the GHSR1a as well as its supposed subtypes. This will make possible to get a better understanding of *in vivo* actions of ghrelin with the consequent development of new treatment strategies for either wasting diseases or the obesity pandemia.

ACKNOWLEDGMENTS

We would like to express our gratitude to Mary Lage (Department of Medicine, University of Santiago de Compostela and Complejo Hospitalario Universitario de Santiago, Spain) for expert technical assistance. Y.P. and J.P.C. are recipients of a Research Contract from the Spanish Ministry of Health, Fondo de Investigacion Sanitaria (FIS), Instituto de Salud Carlos III in the Research Area of the C.H.U.S. This work was performed with financial support provided by Instituto de Salud Carlos III, FIS, Ministerio Español de Sanidad y Consumo [PI042251, PI050382, PI050319, PI050798, PI060705, PI060239, CIBER 03: Physiopathology of obesity and nutrition] and Xunta de Galicia [PGIDIT05BTF20802PR, PGIDIT06PXIB918360PR, PGIDIT06PXIB918322PR]. We would like to note that not all published data were discussed in this chapter. Therefore, we really apologize to those whose works were not mentioned.

REFERENCES

Abizaid, A., Liu, Z. W., Andrews, Z. B., Shanab rough, M., Borok, E., Elsworth, J. D., Roth, R. H., Sleeman, M. W., Picciotto, M. R., Tschop, M. H., Gao, X. B., and Horvath, T. L. (2006). Ghrelin modulates the activity and synaptic input organization of midbrain dopamine neurons while promoting appetite. *J. Clin. Invest.* **116**(12), 3229–3239.

Adams, E. F., Petersen, B., Lei, T., Buchfelder, M., and Fahlbusch, R. (1995). The growth hormone secretagogue, L-692,429, induces phosphatidylinositol hydrolysis and hormone secretion by human pituitary tumors. *Biochem. Biophys. Res. Commun.* **208**, 555–561.

Andersson, U., Filipsson, K., Abbott, C. R., Woods, A., Smith, K., Bloom, S. R., Carling, D., and Small, C. J. (2004). AMP-activated protein kinase plays a role in the control of food intake. *J. Biol. Chem.* **279**, 12005–12008.

Andreis, P. G., Malendowicz, L. K., Trejter, M., Neri, G., Spinazzi, R., Rossi, G. P., and Nussdorfer, G. G. (2003). Ghrelin and growth hormone secretagogue receptor are expressed in the rat adrenal cortex: Evidence that ghrelin stimulates the growth, but not the secretory activity of adrenal cells. *FEBS Lett.* **536**, 17317–17319.

Ariyasu, H., Takaya, K., Iwakura, H., Hosoda, H., Akamizu, T., Arai, Y., Kangawa, K., and Nakao, K. (2005). Transgenic mice overexpressing des-acyl ghrelin show small phenotype. *Endocrinology* **146**, 355–364.

Asakawa, A., Inui, A., Kaga, T., Yuzuriha, H., Nagata, T., Ueno, N., Makino, S., Fujimiya, M., Niijima, A., Fujino, M. A., and Kasuga, M. (2001). Ghrelin is an appetite-stimulatory signal from stomach with structural resemblance to motilin. *Gastroenterology* **120**, 337–345.

Asakawa, A., Inui, A., Fujimiya, M., Sakamaki, R., Shinfuku, N., Ueta, Y., Meguid, M. M., and Kasuga, M. (2005). Stomach regulates energy balance *via* acylated ghrelin and desacyl ghrelin. *Gut* **54**, 18–24.

Avallone, R., Demers, A., Rodrigue-Way, A., Bujold, K., Harb, D., Anghel, S., Wahli, W., Marleau, S., Ong, H., and Tremblay, A. (2006). A Growth hormone-releasing peptide that binds scavenger receptor CD36 and ghrelin receptor upregulates ABC sterol transporters and cholesterol efflux in macrophages through a PPAR{gamma}-dependent pathway. *Mol. Endocrinol.* **20**, 3165–3178.

Baiguera, S., Conconi, M. T., Guidolin, D., Mazzocchi, G., Malendowicz, L. K., Parnigotto, P. P., Spinazzi, R., and Nussdorfer, G. G. (2004). Ghrelin inhibits *in vitro* angiogenic activity of rat brain microvascular endothelial cells. *Int. J. Mol. Med.* **14,** 849–854.

Baldanzi, G., Filigheddu, N., Cutrupi, S., Catapano, F., Bonissoni, S., Fubini, A., Malan, D., Baj, G., Granata, R., Broglio, F., Papotti, M., Surico, N., *et al.* (2002).Ghrelin and desacyl ghrelin inhibit cell death in cardiomyocytes and endothelial cells through ERK1/2 and PI 3-kinase/AKT. *J. Cell Biol.* **159,** 1029–1037.

Barazzoni, R., Bosutti, A., Stebel, M., Cattin, M. R., Roder, E., Visintin, L., Cattin, L., Biolo, G., Zanetti, M., and Guarnieri, G. (2005). Ghrelin regulates mitochondrial-lipid metabolism gene expression and tissue fat distribution in liver and skeletal muscle. *Am. J. Physiol. Endocrinol. Metab.* **288,** E228–E235.

Bedendi, I., Alloatti, G., Marcantoni, A., Malan, D., Catapano, F., Ghe, C., Deghenghi, R., Ghigo, E., and Muccioli, G. (2003). Cardiac effects of ghrelin and its endogenous derivatives des-octanoyl ghrelin and des-Gln14-ghrelin. *Eur. J. Pharmacol.* **476,** 87–95.

Bednarek, M. A., Feighner, S. D., Pong, S. S., McKee, K. K., Hreniuk, D. L., Silva, M. V., Warren, V. A., Howard, A. D., Van Der Ploeg, L. H., and Heck, J. V. (2000). Structure-function studies on the new growth hormone-releasing peptide, ghrelin: Minimal sequence of ghrelin necessary for activation of growth hormone secretagogue receptor 1a. *J. Med. Chem.* **43,** 4370–4376.

Bennett, P. A., Thomas, G. B., Howard, A. D., Feighner, S. D., van der Ploeg, L. H., Smith, R. G., and Robinson, I. C. (1997). Hypothalamic growth hormone secretagogue-receptor (GHS-R) expression is regulated by growth hormone in the rat. *Endocrinology* **138,** 4552–4557.

Bodart, V., Febbraio, M., Demers, A., McNicoll, N., Pohankova, P., Perreault, A., Sejlitz, T., Escher, E., Silverstein, R. L., Lamontagne, D., and Ong, H. (2002). CD36 mediates the cardiovascular action of growth hormone-releasing peptides in the heart. *Circ. Res.* **90,** 844–849.

Bondensgaard, K., Ankersen, M., Thogersen, H., Hansen, B. S., Wulff, B. S., and Bywater, R. P. (2004). Recognition of privileged structures by G-protein coupled receptors. *J. Med. Chem.* **47,** 888–899.

Bowers, C. Y., Momany, F., Reynolds, G. A., Chang, D., Hong, A., and Chang, K. (1980). Structure-activity relationships of a synthetic pentapeptide that specifically releases growth hormone *in vitro*. *Endocrinology* **10,** 663–667.

Bowers, C. Y., Momany, F. A., Reynolds, G. A., and Hong, A. (1984). On the *in vitro* and *in vivo* activity of a new synthetic hexapeptide that acts on the pituitary to specifically release growth hormone. *Endocrinology* **114,** 1537–1545.

Bresciani, E., Nass, R., Torsello, A., Gaylinn, B., Avallone, R., Locatelli, V., Thorner, M. O., and Muller, E. E. (2004). Hexarelin modulates the expression of growth hormone secretagogue receptor type 1a mRNA at hypothalamic and pituitary sites. *Neuroendocrinology* **80,** 52–59.

Broglio, F., Arvat, E., Benso, A., Gotero, C., Muccioli, G., Papotti, M., van der Lely, A. J., Deghenghi, R., and Ghigo, E. (2001). Ghrelin, a natural GH secretagogue produced by the stomach, induces hyperglycemia and reduces insulin secretion in humans. *J. Clin. Endocrinol. Metab.* **86,** 5083–5086.

Brywe, K. G., Leverin, A. L., Gustavsson, M., Mallard, C., Granata, R., Destefanis, S., Volante, M., Hagberg, H., Ghigo, E., and Isgaard, J. (2005). Growth hormone-releasing peptide hexarelin reduces neonatal brain injury and alters Akt/glycogen synthase kinase-3β phosphorylation. *Endocrinology* **146,** 4665–4672.

Calapai, G., Corica, F., Allegra, A., Corsonello, A., Sautebin, L., De Gregorio, T., Di Rosa, M., Costantino, G., Buemi, M., and Caputi, A. P. (1998). Effects of intracerebroventricular leptin administration on food intake, body weight gain and diencephalic nitric oxide synthase activity in the mouse. *Br. J. Pharmacol.* **125,** 798–802.

Calapai, G., Corica, F., Corsonello, A., Sautebin, L., Di Rosa, M., Campo, G. M., Buemi, M., Mauro, V. N., and Caputi, A. P. (1999). Leptin increases serotonin turnover by inhibition of brain nitric oxide synthesis. *J. Clin. Invest.* **104,** 975–982.

Camina, J. P., Carreira, M. C., El Messari, S., Llorens-Cortes, C., Smith, R. G., and Casanueva, F. F. (2004). Desensitization and endocytosis mechanisms of ghrelin-activated growth hormone secretagogue receptor 1a. *Endocrinology* **145,** 930–940.

Caminos, J. E., Gualillo, O., Lago, F., Otero, M., Blanco, M., Gallego, R., Garcia-Caballero, T., Goldring, M. B., Casanueva, F. F., Gomez-Reino, J. J., and Dieguez, C. (2005). The endogenous growth hormone secretagogue (ghrelin) is synthesized and secreted by chondrocytes. *Endocrinology* **146,** 1285–1292.

Carreira, M. C., Camina, J. P., Smith, R. G., and Casanueva, F. F. (2004). Agonist-specific coupling of growth hormone secretagogue receptor type 1a to different intracellular signaling systems. Role of adenosine. *Neuroendocrinology* **79,** 13–25.

Carreira, M. C., Camina, J. P., Diaz-Rodriguez, E., Alvear-Perez, R., Llorens-Cortes, C., and Casanueva, F. F. (2006). Adenosine does not bind to the growth hormone secretagogue receptor type-1a (GHS-R1a). *J. Endocrinol.* **191,** 147–157.

Cassoni, P., Papotti, M., Ghe, C., Catapano, F., Sapino, A., Graziani, A., Deghenghi, R., Reissmann, T., Ghigo, E., and Muccioli, G. (2001). Identification, characterization, and biological activity of specific receptors for natural (ghrelin) and synthetic growth hormone secretagogues and analogs in human breast carcinomas and cell lines. *J. Clin. Endocrinol. Metab.* **84,** 1738–1745.

Cassoni, P., Ghe, C., Marrocco, T., Tarabra, E., Allia, E., Catapano, F., Deghenghi, R., Ghigo, E., Papotti, M., and Muccioli, G. (2004). Expression of ghrelin and biological activity of specific receptors for ghrelin and des-acyl ghrelin in human prostate neoplasms and related cell lines. *Eur. J. Endocrinol.* **150,** 173–184.

Castrillo, A., and Tontonoz, P. (2004). Nuclear receptors in macrophage biology: At the crossroads of lipid metabolism and inflammation. *Annu. Rev. Cell Dev. Biol.* **20,** 455–480.

Chen, C., and Clarke, I. J. (1995). Effects of growth hormone-releasing peptide-2 (GHRP-2) on membrane Ca^{2+} permeability in cultured ovine somatotrophs. *J. Neuroendocrinol.* **7,** 179–186.

Chen, C., Zhang, J., Vincent, J. D., and Israel, J. M. (1990). Two types of voltage-dependent calcium current in rat somatotrophs are reduced by somatostatin. *J. Physiol.* **425,** 29–42.

Chen, C., Wu, D., and Clarke, I. J. (1996). Signal transduction systems employed by synthetic GH-releasing peptides in somatotrophs. *J. Endocrinol.* **148,** 381–386.

Chen, C. Y., Inui, A., Asakawa, A., Fujino, K., Kato, I., Chen, C. C., Ueno, N., and Fujimiya, M. (2005). Des-acyl ghrelin acts by CRF type 2 receptors to disrupt fasted stomach motility in conscious rats. *Gastroenterology* **129,** 8–25.

Conconi, M. T., Nico, B., Guidolin, D., Baiguera, S., Spinazzi, R., Rebuffat, P., Malendowicz, L. K., Vacca, A., Carraro, G., Parnigotto, P. P., Nussdorfer, G. G., and Ribatti, D. (2004). Ghrelin inhibits FGF-2-mediated angiogenesis *in vitro* and *in vivo*. *Peptides* **25,** 2179–2185.

Cunha, S. R., and Mayo, K. E. (2002). Ghrelin and growth hormone (GH) secretagogues potentiate GH-releasing hormone (GHRH)-induced cyclic adenosine $3',5'$-monophosphate production in cells expressing transfected GHRH and GH secretagogue receptors. *Endocrinology* **143,** 4570–4582.

Dass, N. B., Munonyara, M., Bassil, A. K., Hervieu, G. J., Osbourne, S., Corcoran, S., Morgan, M., and Sanger, G. J. (2003). Growth hormone secretagogue receptors in rat and human gastrointestinal tract and the effects of ghrelin. *Neuroscience* **120,** 443–453.

Deghenghi, R., Papotti, M., Ghigo, E., and Muccioli, G. (2001). Cortistatin, but not somatostatin, binds to growth hormone secretagogue (GHS) receptors of human pituitary gland. *J. Endocrinol. Invest.* **24,** RC1–RC3.

Delgado-Rubin de Celix, A., Chowen, J. A., Argente, J., and Frago, L. M. (2006). Growth hormone releasing peptide-6 acts as a survival factor in glutamate-induced excitotoxicity. *J. Neurochem.* **99,** 839–849.

Delhanty, P. J., van der Eerden, B. C., van der Velde, M., Gauna, C., Pols, H. A., Jahr, H., Chiba, H., van der Lely, A. J., and van Leeuwen, J. P. (2006). Ghrelin and unacylated ghrelin stimulate human osteoblast growth *via* mitogen-activated protein kinase (MAPK)/phosphoinositide 3-kinase (PI3K) pathways in the absence of GHS-R1a. *J. Endocrinol.* **188,** 37–47.

De Vriese, C., Gregoire, F., De Neef, P., Robberecht, P., and Delporte, C. (2005). Ghrelin is produced by the human erythroleukemic HEL cell line and involved in an autocrine pathway leading to cell proliferation. *Endocrinology* **146,** 1514–1522.

Dixit, V. D., Schaffer, E. M., Pyle, R. S., Collins, G. D., Sakthivel, S. K., Palaniappan, R., Lillard, J. W., Jr., and Tabú, D. D. (2004). Ghrelin inhibits leptin- and activation-induced proinflammatory cytokine expression by human monocytes and T cells. *J. Clin. Invest.* **114,** 57–66.

Dumont, J. E., Pecasse, F., and Maenhaut, C. (2001). Crosstalk and specificity in signalling. Are we crosstalking ourselves into general confusion? *Cell. Signal.* **13,** 457–463.

Duxbury, M. S., Waseem, T., Ito, H., Robinson, M. K., Zinner, M. J., Ashley, S. W., and Whang, E. E. (2003). Ghrelin promotes pancreatic adenocarcinoma cellular proliferation and invasiveness. *Biochem. Biophys. Res. Commun.* **309,** 464–468.

Druce, M. R., Wren, A. M., Park, A. J., Milton, J. E., Patterson, M., Frost, G., Ghatei, M. A., Small, C., and Bloom, S. R. (2005). Ghrelin increases food intake in obese as well as lean subjects. *Int. J. Obes. (Lond.)* **29,** 1130–1136.

Feighner, S. D., Howard, A. D., Prendergast, K., Palyha, O. C., Hreniuk, D. L., Nargund, R., Underwood, D., Tata, J. R., Dean, D. C., Tan, C. P., McKee, K. K., Woods, J. W., *et al.* (1998). Structural requirements for the activation of the human growth hormone secretagogue receptor by peptide and nonpeptide secretagogues. *Mol. Endocrinol.* **12,** 137–145.

Frago, L. M., Paneda, C., Dickson, S. L., Hewson, A. K., Argente, J., and Chowen, J. A. (2002). Growth hormone (GH) and GH-releasing peptide-6 increase brain insulin-like growth factor-I expression and activate intracellular signaling pathways involved in neuroprotection. *Endocrinology* **143,** 4113–4122.

Frago, L. M., Paneda, C., Argente, J., and Chowen, J. A. (2005). Growth hormone-releasing peptide-6 increases insulin-like growth factor-I mRNA levels and activates Akt in RCA-6 cells as a model of neuropeptide Y neurones. *J. Neuroendocrinol.* **17,** 701–710.

Gauna, C., Meyler, F. M., Janssen, J. A., Delhanty, P. J., Abribat, T., van Koetsveld, P., Hofland, L. J., Broglio, F., Ghigo, E., and van der Lely, A. J. (2004). Administration of acylated ghrelin reduces insulin sensitivity, whereas the combination of acylated plus unacylated ghrelin strongly improves insulin sensitivity. *J. Clin. Endocrinol. Metab.* **89,** 5035–5042.

Gauna, C., Delhanty, P. J., Hofland, L. J., Janssen, J. A., Broglio, F., Ross, R. J., Ghigo, E., and van der Lely, A. J. (2005). Ghrelin stimulates, whereas des-octanoyl ghrelin inhibits, glucose output by primary hepatocytes. *J. Clin. Endocrinol. Metab.* **90,** 1055–1060.

Gaskin, F. S., Farr, S. A., Banks, W. A., Kumar, V. B., and Morley, J. E. (2003). Ghrelin-induced feeding is dependent on nitric oxide. *Peptides* **24,** 913–918.

Gaytan, F., Barreiro, M. L., Chopin, L. K., Herington, A. C., Morales, C., Pinilla, L., Casanueva, F. F., Aguilar, E., Dieguez, C., and Tena-Sempere, M. (2003). Immunolocalization of ghrelin and its functional receptor, the type 1a growth hormone secretagogue receptor, in the cyclic human ovary. *J. Clin. Endocrinol. Metab.* **88,** 879–887.

Gaytan, F., Barreiro, M. L., Caminos, J. E., Chopin, L. K., Herington, A. C., Morales, C., Pinilla, L., Paniagua, R., Nistal, M., Casanueva, F. F., Aguilar, E., Dieguez, C., et al. (2004). Expression of ghrelin and its functional receptor, the type 1a growth hormone secretagogue receptor, in normal human testis and testicular tumors. *J. Clin. Endocrinol. Metab.* **89,** 400–409.

Gaytan, F., Morales, C., Barreiro, M. L., Jeffery, P., Chopin, L. K., Herington, A. C., Casanueva, F. F., Aguilar, E., Dieguez, C., and Tena-Sempere, M. (2005). Expression of growth hormone secretagogue receptor type 1a, the functional ghrelin receptor, in human ovarian surface epithelium, mullerian duct derivatives and ovarian tumors. *J. Clin. Endocrinol. Metab.* **90,** 1798–17804.

Ghe, C., Cassoni, P., Catapano, F., Marrocco, T., Deghenghi, R., Ghigo, E., Muccioli, G., and Papotti, M. (2002). The antiproliferative effect of synthetic peptidyl GH secretagogues in human CALU-1 lung carcinoma cells. *Endocrinology* **143,** 484–491.

Gnanapavan, S., Kola, B., Bustin, S. A., Morris, D. G., McGee, P., Fairclough, P., Bhattacharya, S., Carpenter, R., Grossman, A. B., and Korbonits, M. (2002). The tissue distribution of the mRNA of ghrelin and subtypes of its receptor, GHS-R, in humans. *J. Clin. Endocrinol. Metab.* **87,** 2988–2991.

Granado, M., Priego, T., Martín, A. I., Villanua, M. A., and Lopez-Calderon, A. (2005). Ghrelin receptor agonist GHRP-2 prevents arthritis-induced increase in E3 ubiquitin-ligating enzymes MuRF1 and MAFbx gene expression in skeletal muscle. *Am. J. Physiol. Endocrinol. Metab.* **289,** E1007–E10014.

Granata, R., Settanni, F., Biancone, L., Trovato, L., Nano, R., Bertuzzi, F., Destefanis, S., Annunziata, M., Martinetti, M., Catapano, F., Ghe, C., Isgaard, J., et al. (2007). Acetylated and unacylated ghrelin promote proliferation and inhibit apoptosis of pancreatic beta cells and human islets. Involvement of cAMP/PKA, ERK1/2 and PI3K/AKT signaling. *Endocrinology* **148,** 512–529.

Grove, K. L., and Cowley, M. A. (2005). Is ghrelin a signal for the development of metabolic systems? *J. Clin. Invest.* **115,** 3393–3397.

Guan, X. M., Yu, H., Palyha, O. C., McKee, K. K., Feighner, S. D., Sirinathsinghji, D. J., Smith, R. G., Van der Ploeg, L. H., and Howard, A. D. (1997). Distribution of mRNA encoding the growth hormone secretagogue receptor in brain and peripheral tissues. *Brain Res. Mol. Brain Res.* **48,** 23–29.

Hardie, D. G. (2004). The AMP-activated protein kinase pathway-new players upstream and downstream. *J. Cell. Sci.* **117,** 5479–5487.

Herrington, J., and Hille, B. (1994). Growth hormone-releasing hexapeptide elevates intracellular calcium in rat somatotropes by two mechanisms. *Endocrinology* **135,** 1100–1108.

Holst, B., and Schwartz, T. W. (2004). Constitutive ghrelin receptor activity as a signaling set-point in appetite regulation. *Trends Pharmacol. Sci.* **25,** 113–117.

Holst, B., and Schwartz, T. W. (2006). Ghrelin receptor mutations–too little height and too much hunger. *J. Clin. Invest.* **116,** 637–641.

Holst, B., Cygankiewicz, A., Jensen, T. H., Ankersen, M., and Schwartz, T. W. (2003). High constitutive signaling of the ghrelin receptor-identification of a potent inverse agonist. *Mol. Endocrinol.* **17,** 2201–2210.

Holst, B., Holliday, N. D., Bach, A., Elling, C. E., Cox, H. M., and Schwartz, T. W. (2004). Common structural basis for constitutive activity of the ghrelin receptor family. *J. Biol. Chem.* **279,** 53806–53817.

Holst, B., Brandt, E., Bach, A., Heding, A., and Schwartz, T. W. (2005). Nonpeptide and peptide growth hormone secretagogues act both as ghrelin receptor agonist and as positive or negative allosteric modulators of ghrelin signaling. *Mol. Endocrinol.* **19,** 2400–2411.

Holst, B., Lang, M., Brandt, E., Bach, A., Howard, A., Frimurer, T. M., Beck-Sickinger, A., and Schwartz, T. W. (2006). Ghrelin receptor inverse agonists: Identification of an active peptide core and its interaction epitopes on the receptor. *Mol. Pharmacol.* **70**, 936–946.

Howard, A. D., Feighner, S. D., Cully, D. F., Arena, J. P., Liberator, P. A., Rosenblum, C. I., Hamelin, M., Hreniuk, D. L., Palyha, O. C., Anderson, J., Paress, P. S., Diaz, C., *et al.* (1996). A receptor in pituitary and hypothalamus that functions in growth hormone release. *Science* **273**, 974–977.

Horikawa, R., Tachibana, T., Katsumata, N., Ishikawa, H., and Tanaka, T. (2000). Regulation of pituitary growth hormone-secretagogue and growth hormone-releasing hormone receptor RNA expression in young Dwarf rats. *Endocr. J.* **47**, S53–S56.

Iantorno, M., Chen, H., Kim, J. A., Tesauro, M., Lauro, D., Cardillo, C., and Quon, M. J. (2007). Ghrelin has novel vascular actions that mimic only PI 3-kinase-dependent insulin-stimulated production of nitric oxide (NO) but not MAP-kinase-dependent insulin-stimulated secretion of ET-1 from endothelial cells. *Am. J. Physiol. Endocrinol. Metab.* **292**, E756–E764.

Iglesias, M. J., Pineiro, R., Blanco, M., Gallego, R., Dieguez, C., Gualillo, O., Gonzalez-Juanatey, J. R., and Lago, F. (2004). Growth hormone releasing peptide (ghrelin) is synthesized and secreted by cardiomyocytes. *Cardiovasc. Res.* **62**, 481–488.

Jeffery, P. L., Herington, A. C., and Chopin, L. K. (2002). Expression and action of the growth hormone releasing peptide ghrelin and its receptor in prostate cancer cell lines. *J. Endocrinol.* **172**, R7–R11.

Jiang, H., Betancourt, L., and Smith, R. G. (2006). Ghrelin amplifies dopamine signaling by cross talk involving formation of growth hormone secretagogue receptor/dopamine receptor subtype 1 heterodimers. *Mol. Endocrinol.* **20**, 1772–1785.

Johansson, S., Fredholm, B. B., Hjort, C., Morein, T., Kull, B., and Hu, P. S. (2005). Evidence against adenosine analogues being agonists at the growth hormone secretagogue receptor. *Biochem. Pharmacol.* **70**, 598–605.

Kaji, H., Kishimoto, M., Kirimura, T., Iguchi, G., Murata, M., Yoshioka, S., Iida, K., Okimura, Y., Yoshimoto, Y., and Chihara, K. (2001). Hormonal regulation of the human ghrelin receptor gene transcription. *Biochem. Biophys. Res. Commun.* **284**, 660–666.

Kamegai, J., Wakabayashi, I., Miyamoto, K., Unterman, T. G., Kineman, R. D., and Frohman, L. A. (1998). Growth hormone-dependent regulation of pituitary GH secretagogue receptor (GHS-R) mRNA levels in the spontaneous dwarf rat. *Neuroendocrinology* **68**, 312–318.

Kamegai, J., Wakabayashi, I., Kineman, R. D., and Frohman, L. A. (1999). Growth hormone-releasing hormone receptor (GHRH-R) and growth hormone secretagogue receptor (GHS-R) mRNA levels during postnatal development in male and female rats. *J. Neuroendocrinol.* **11**, 299–306.

Kamegai, J., Tamura, H., Ishii, S., Sugihara, H., and Wakabayashi, I. (2001). Thyroid hormones regulate pituitary growth hormone secretagogue receptor gene expression. *J. Neuroendocrinol.* **13**, 275–278.

Kamegai, J., Tamura, H., Shimizu, T., Ishii, S., Sugihara, H., and Oikawa, S. (2005). Insulin-like growth factor-I down-regulates ghrelin receptor (growth hormone secretagogue receptor) expression in the rat pituitary. *Regul. Pept.* **127**, 203–206.

Katayama, M., Nogami, H., Nishiyama, J., Kawase, T., and Kawamura, K. (2000). Developmentally and regionally regulated expression of growth hormone secretagogue receptor mRNA in rat brain and pituitary gland. *Neuroendocrinology* **72**, 333–340.

Kato, M., and Sakuma, Y. (1999). The effect of GHRP-6 on the intracellular Na^+ concentration of rat pituitary cells in primary culture. *J. Neuroendocrinol.* **11**, 795–800.

Kim, M. S., Yoon, C. Y., Jang, P. G., Park, Y. J., Shin, C. S., Park, H. S., Ryu, J. W., Pak, Y. K., Park, J. Y., Lee, K. U., Kim, S. Y., Lee, H. K., *et al.* (2004). The mitogenic

and antiapoptotic actions of ghrelin in 3T3-L1 adipocytes. *Mol. Endocrinol.* **18**, 2291–2301.
Kineman, R. D., Kamegai, J., and Frohman, L. A. (1999). Growth hormone (GH)-releasing hormone (GHRH) and the GH secretagogue (GHS), L692,585, differentially modulate rat pituitary GHS receptor and GHRH receptor messenger ribonucleic acid levels. *Endocrinology* **140**, 3581–3586.
Kojima, M., Hosoda, H., Date, Y., Nakazato, M., Matsuo, H., and Kangawa, K. (1999). Ghrelin is a growth-hormone-releasing acylated peptide from stomach. *Nature* **402**, 656–660.
Kojima, M., Hosoda, H., Matsuo, H., and Kangawa, K. (2001). Ghrelin: Discovery of the natural endogenous ligand for the growth hormone secretagogue receptor. *Trends Endocrinol. Metab.* **12**, 118–122.
Kola, B., Hubina, E., Tucci, S. A., Kirkham, T. C., Garcia, E. A., Mitchell, S. E., Williams, L. M., Hawley, S. A., Hardie, D. G., Grossman, A. B., and Korbonits, M. (2005). Cannabinoids and ghrelin have both central and peripheral metabolic and cardiac effects via AMP-activated protein kinase. *J. Biol. Chem.* **280**, 25196–25201.
Kohno, D., Gao, H. Z., Muroya, S., Kikuyama, S., and Yada, T. (2003). Ghrelin directly interacts with neuropeptide-Y-containing neurons in the rat arcuate nucleus: Ca^{2+} signaling via protein kinase A and N-type channel-dependent mechanisms and crosstalk with leptin and orexin. *Diabetes* **52**, 948–956.
Korbonits, M., Goldstone, A. P., Gueorguiev, M., and Grossman, A. B. (2004). Ghrelin: A hormone with multiple functions. *Front. Neuroendocrinol.* **25**, 27–68.
Laferrere, B., Abraham, C., Russell, C. D., and Bowers, C. Y. (2005). Growth hormone releasing peptide-2 (GHRP-2), like ghrelin, increases food intake in healthy men. *J. Clin. Endocrinol. Metab.* **90**, 611–614.
Lee, C. H., Chawla, A., Urbiztondo, N., Liao, D., Boisvert, W. A., Evans, R. M., and Curtiss, L. K. (2003). Transcriptional repression of atherogenic inflammation: Modulation by PPARdelta. *Science* **302**, 453–457.
Li, W. G., Gavrila, D., Liu, X., Wang, L., Gunnlaugsson, S., Stoll, L. L., McCormick, M. L., Sigmund, C. D., Tang, C., and Weintraub, N. L. (2004). Ghrelin inhibits proinflammatory responses and nuclear factor-kappaB activation in human endothelial cells. *Circulation* **109**, 2221–2226.
Long, Y. C., and Zierath, J. R. (2006). AMP-activated protein kinase signaling in metabolic regulation. *J. Clin. Invest.* **116**, 1776–1783.
Luttrell, L. M., and Lefkowitz, R. J. (2002). The role of beta-arrestins in the termination and transduction of G-protein-coupled receptor signals. *J. Cell. Sci.* **115**, 455–465.
Maccarinelli, G., Sibilia, V., Torsello, A., Raimondo, F., Pitto, M., Giustina, A., Netti, C., and Cocchi, D. (2005). Ghrelin regulates proliferation and differentiation of osteoblastic cells. *J. Endocrinol.* **184**, 249–256.
Mazzocchi, G., Neri, G., Rucinski, M., Rebuffat, P., Spinazzi, R., Malendowicz, L. K., and Nussdorfer, G. G. (2004). Ghrelin enhances the growth of cultured human adrenal zona glomerulosa cells by exerting MAPK-mediated proliferogenic and antiapoptotic effects. *Peptides* **25**, 1269–1277.
McKee, K. K., Palyha, O. C., Feighner, S. D., Hreniuk, D. L., Tan, C. P., Phillips, M. S., Smith, R. G., Van der Ploeg, L. H., and Howard, A. D. (1997). Molecular analysis of rat pituitary and hypothalamic growth hormone secretagogue receptors. *Mol. Endocrinol.* **11**, 415–423.
Morley, J. E., Alshaher, M. M., Farr, S. A., Flood, J. F., and Kumar, V. B. (1999). Leptin and neuropeptide Y (NPY) modulate nitric oxide synthase: Further evidence for a role of nitric oxide in feeding. *Peptides* **20**, 595–600.
Mousseaux, D., Le Gallic, L., Ryan, J., Oiry, C., Gagne, D., Fehrentz, J. A., Galleyrand, J. C., and Martinez, J. (2006). Regulation of ERK1/2 activity by ghrelin-

activated growth hormone secretagogue receptor 1A involves a PLC/PKCvarepsilon pathway. *Br. J. Pharmacol.* **148,** 350–365.

Muccioli, G., Ghe, C., Ghigo, M. C., Papotti, M., Arvat, E., Boghen, M. F., Nilsson, M. H., Deghenghi, R., Ong, H., and Ghigo, E. (1998). Specific receptors for synthetic GH secretagogues in the human brain and pituitary gland. *J. Endocrinol.* **157,** 99–106.

Muccioli, G., Papotti, M., Locatelli, V., Ghigo, E., and Deghenghi, R. (2001). Binding of 125I-labeled ghrelin to membranes from human hypothalamus and pituitary gland. *J. Endocrinol. Invest.* **24,** RC7–RC9.

Muccioli, G., Pons, N., Ghe, C., Catapano, F., Granata, R., and Ghigo, E. (2004). Ghrelin and des-acyl ghrelin both inhibit isoproterenol-induced lipolysis in rat adipocytes *via* a non-type 1a growth hormone secretagogue receptor. *Eur. J. Pharmacol.* **498,** 27–35.

Murata, M., Okimura, Y., Iida, K., Matsumoto, M., Sowa, H., Kaji, H., Kojima, M., Kangawa, K., and Chihara, K. (2002). Ghrelin modulates the downstream molecules of insulin signaling in hepatoma cells. *J. Biol. Chem.* **277,** 5667–5674.

Nakazato, M., Murakami, N., Date, Y., Kojima, M., Matsuo, H., Kangawa, K., and Matsukura, S. (2001). A role for ghrelin in the central regulation of feeding. *Nature* **409,** 194–198.

Nagaya, N., Moriya, J., Yasumura, Y., Uematsu, M., Ono, F., Shimizu, W., Ueno, K., Kitakaze, M., Miyatake, K., and Kangawa, K. (2004). Effects of ghrelin administration on left ventricular function, exercise capacity, and muscle wasting in patients with chronic heart failure. *Circulation* **110,** 3674–3679.

Nass, R., Gilrain, J., Anderson, S., Gaylinn, B., Dalkin, A., Day, R., Peruggia, M., and Thorner, M. O. (2000). High plasma growth hormone (GH) levels inhibit expression of GH secretagogue receptor messenger ribonucleic acid levels in the rat pituitary. *Endocrinology* **141,** 2084–2089.

Neary, N. M., Small, C. J., Wren, A. M., Lee, J. L., Druce, M. R., Palmieri, C., Frost, G. S., Ghatei, M. A., Combes, R. C., and Bloom, S. R. (2004). Ghrelin increases energy intake in cancer patients with impaired appetite: Acute, randomized, placebo-controlled trial. *J. Clin. Endocrinol. Metab.* **89,** 2832–2836.

Nogueiras, R., Tovar, S., Mitchell, S. E., Rayner, D. V., Archer, Z. A., Dieguez, C., and Williams, L. M. (2004). Regulation of growth hormone secretagogue receptor gene expression in the arcuate nuclei of the rat by leptin and ghrelin. *Diabetes* **53,** 2552–2558.

Ong, H., McNicoll, N., Escher, E., Collu, R., Deghenghi, R., Locatelli, V., Ghigo, E., Muccioli, G., Boghen, M., and Nilsson, M. (1998). Identification of a pituitary growth hormone-releasing peptide (GHRP) receptor subtype by photoaffinity labeling. *Endocrinology* **139,** 432–435.

Pantel, J., Legendre, M., Carbol, S., Hilal, L., Hajaji, Y., Moriste, S., Nivot, S., Vie-Luton, M. P., Grouselle, D., de Kerdanet, M., Kadiri, A., Epelbaum, J., *et al.* (2006). Loss of constitutive activity of the growth hormone secretagogue receptor in familial short stature. *J. Clin. Invest.* **116,** 760–768.

Patchett, A. A., Nargund, R. P., Tata, J. R., Chen, M. H., Barakat, K. J., Johnston, D. B., Cheng, K., Chan, W. W., Butler, B., Hickey, G., Jacks, T., Schleim, K., *et al.* (1995). Design and biological activities of L-163,191 (MK-0677): A potent, orally active growth hormone secretagogue. *Proc. Natl. Acad. Sci. USA* **92,** 7001–7005.

Pedretti, A., Villa, M., Pallavicini, M., Valoti, E., and Vistoli, G. (2006). Construction of human ghrelin receptor (hGHS-R1a) model using a fragmental prediction approach and validation through docking analysis. *J. Med. Chem.* **49,** 3077–3085.

Peng, X.-D., Park, S., Gadelha, M. R., Coschigano, K. C., Kopchick, K. C., and Frohman, K. C. (2001). The growth hormone (GH)-axis of GH receptor/binding protein gene-disrupted and metallothionein-human GH-releasing hormone transgenic mice: Hypothalamic neuropeptide and pituitary receptor expression in the absence and presence of GH feedback. *Endocrinology* **142,** 1117–1123.

Petersenn, S., Rasch, A. C., Penshorn, M., Beil, F. U., and Schulte, H. M. (2001). Genomic structure and transcriptional regulation of the human growth hormone secretagogue receptor. *Endocrinology* **142,** 2649–2659.

Qader, S. S., Lundquist, I., Ekelund, M., Hakanson, R., and Salehi, A. (2005). Ghrelin activates neuronal constitutive nitric oxide synthase in pancreatic islet cells while inhibiting insulin release and stimulating glucagon release. *Regul. Pept.* **128,** 51–56.

Rossi, F., Bertone, C., Petricca, S., and Santiemma, V. (2006). Ghrelin inhibits angiotensin II-induced migration of human aortic endothelial cells. *Atherosclerosis* **192**(2), 291–297.

Schmid, D. A., Held, K., Ising, M., Uhr, M., Weikel, J. C., and Steiger, A. (2005). Ghrelin stimulates appetite, imagination of food, GH, ACTH, and cortisol, but does not affect leptin in normal controls. *Neuropsychopharmacology* **30,** 1187–1192.

Shimada, M., Tritos, N. A., Lowell, B. B., Flier, J. S., and Maratos-Flier, E. (1998). Mice lacking melanin-concentrating hormone are hypophagic and lean. *Nature* **396,** 670–674.

Shuto, Y., Shibasaki, T., Otagiri, A., Kuriyama, H., Ohata, H., Tamura, H., Kamegai, J., Sugihara, H., Oikawa, S., and Wakabayashi, I. (2002). Hypothalamic growth hormone secretagogue receptor regulates growth hormone secretion, feeding, and adiposity. *J. Clin. Invest.* **109,** 1429–1436.

Smith, R. G., Cheng, K., Schoen, W. R., Pong, S. S., Hickey, G., Jacks, T., Butler, B., Chan, W. W., Chaung, L. Y., Judith, F., Taylor, J., Wyvratt, J., et al. (1993). A nonpeptidyl growth hormone secretagogue. *Science* **260,** 1640–1643.

Smith, R. G., Griffin, P. R., Xu, Y., Smith, A. G., Liu, K., Calacay, J., Feighner, S. D., Pong, C., Leong, D., Pomes, A., Cheng, K., Van der Ploeg, L. H., et al. (2000). Adenosine: A partial agonist of the growth hormone secretagogue receptor. *Biochem. Biophys. Res. Commun.* **276,** 1306–1313.

Smith, R. G., Jiang, H., and Sun, Y. (2005). Developments in ghrelin biology and potential clinical relevance. *Trends Endocrinol. Metab.* **16,** 436–442.

Sun, Y., Ahmed, S., and Smith, R. G. (2003). Deletion of ghrelin impairs neither growth nor appetite. *Mol. Cell Biol.* **23,** 7973–7981.

Sun, Y., Wang, P., Zheng, H., and Smith, R. G. (2004). Ghrelin stimulation of growth hormone release and appetite is mediated through the growth hormone secretagogue receptor. *Proc. Natl. Acad. Sci. USA* **101,** 4679–4684.

Tannenbaum, G. S., Lapointe, M., Beaudet, A., and Howard, A. D. (1998). Expression of growth hormone secretagogue-receptors by growth hormone-releasing hormone neurons in the mediobasal hypothalamus. *Endocrinology* **139,** 4420–4423.

Tamura, H., Kamegai, J., Sugihara, H., Kineman, R. D., Frohman, L. A., and Wakabayashi, I. (2000). Glucocorticoids regulate pituitary growth hormone secretagogue receptor gene expression. *J. Neuroendocrinol.* **12,** 481–485.

Thomas, G. B., Bennett, P. A., Carmignac, D. F., and Robinson, I. C. (2000). Glucocorticoid regulation of growth hormone (GH) secretagogue-induced growth responses and GH secretagogue receptor expression in the rat. *Growth Horm. IGF Res.* **10,** 45–52.

Thompson, N. M., Gill, D. A., Davies, R., Loveridge, N., Houston, P. A., Robinson, I. C., and Wells, T. (2004). Ghrelin and des-octanoyl ghrelin promote adipogenesis directly *in vivo* by a mechanism independent of the type 1a growth hormone secretagogue receptor. *Endocrinology* **145,** 234–242.

Torsello, A., Ghe, C., Bresciani, E., Catapano, F., Ghigo, E., Deghenghi, R., Locatelli, V., and Muccioli, G. (2002). Short ghrelin peptides neither displace ghrelin binding *in vitro* nor stimulate GH release *in vivo*. *Endocrinology* **143,** 1968–1971.

Toshinai, K., Yamaguchi, H., Sun, Y., Smith, R. G., Yamanaka, A., Sakurai, T., Date, Y., Mondal, M. S., Shimbara, T., Kawagoe, T., Murakami, N., Miyazato, M., et al. (2006).

Des-acyl ghrelin induces food intake by a mechanism independent of the growth hormone secretagogue receptor. *Endocrinology* **147,** 2306–2314.
Tullin, S., Hansen, B. S., Ankersen, M., Moller, J., Von Cappelen, K. A., and Thim, L. (2000). Adenosine is an agonist of the growth hormone secretagogue receptor. *Endocrinology* **141,** 3397–3402.
Tschop, M., Smiley, D. L., and Heiman, M. L. (2000). Ghrelin induces adiposity in rodents. *Nature* **407,** 908–913.
van der Lely, A. J., Tschop, M., Heiman, M. L., and Ghigo, E. (2004). Biological, physiological, pathophysiological, and pharmacological aspects of ghrelin. *Endocr. Rev.* **25,** 426–457.
Wallenius, K., Sjögren, K., Peng, X. D., Park, S., Wallenius, V., and Liu, V. (2001). Liver-derived IGF-I regulates GH secretion at the pituitary level in mice. *Endocrinology* **142,** 114762–114770.
Wang, H. J., Geller, F., Dempfle, A., Schauble, N., Friedel, S., Lichtner, P., Fontenla-Horro, F., Wudy, S., Hagemann, S., Gortner, L., Huse, K., Remschmidt, H., et al. (2004). Ghrelin receptor gene: Identification of several sequence variants in extremely obese children and adolescents, healthy normal-weight and underweight students, and children with short normal stature. *J. Clin. Endocrinol. Metab.* **89,** 157–162.
Welch, J. S., Ricote, M., Akiyama, T. E., Gonzalez, F. J., and Glass, C. K. (2003). PPARgamma and PPARdelta negatively regulate specific subsets of lipopolysaccharide and IFN-gamma target genes in macrophages. *Proc. Natl. Acad. Sci. USA* **100,** 6712–6717.
Willesen, M. G., Kristensen, P., and Romer, J. (1999). Co-localization of growth hormone secretagogue receptor and NPY mRNA in the arcuate nucleus of the rat. *Neuroendocrinology* **70,** 306–316.
Wortley, K. E., Anderson, K. D., Garcia, K., Murria, J. D., Malinova, L., Liu, R., Moncrieffe, M., Thabet, K., Cox, H. J., Yancopoulos, G. D., Wiegand, S. J., and Sleeman, M. W. (2004). Genetic deletion of ghrelin does not decrease food intake but influences metabolic fuel preference. *Proc. Natl. Acad. Sci. USA* **101,** 8227–8232.
Wortley, K. E., del Rincón, J. P., Murria, J. D., Garcia, K., Iida, K., Thorner, M. O., and Sleeman, M. W. (2005). Absence of ghrelin protects against early-onset obesity. *J. Clin. Invest.* **115,** 3573–3578.
Wren, A. M., Small, C. J., Ward, H. L., Murphy, K. G., Dakin, C. L., Taheri, S., Kennedy, A. R., Roberts, G. H., Morgan, D. G., Ghatei, M. A., and Bloom, S. R. (2000). The novel hypothalamic peptide ghrelin stimulates food intake and growth hormone secretion. *Endocrinology* **141,** 4325–4328.
Wren, A. M., Seal, L. J., Cohen, M. A., Brynes, A. E., Frost, G. S., Murphy, K. G., Dhillo, W. S., Ghatei, M. A., and Bloom, S. R. (2001a). Ghrelin enhances appetite and increases food intake in humans. *J. Clin. Endocrinol. Metab.* **86,** 5992–5995.
Wren, A. M., Small, C. J., Abbott, C. R., Dhillo, W. S., Seal, L. J., Cohen, M. A., Batterham, R. L., Taheri, S., Stanley, S. A., Ghatei, M. A., and Bloom, S. R. (2001b). Ghrelin causes hyperphagia and obesity in rats. *Diabetes* **50,** 2540–2547.
Wu, D., Chen, C., Zhang, J., Bowers, C. Y., and Clarke, I. J. (1996). The effects of GH-releasing peptide-6 (GHRP-6) and GHRP-2 on intracellular adenosine $3',5'$-monophosphate (cAMP) levels and GH secretion in ovine and rat somatotrophs. *J. Endocrinol.* **148,** 197–205.
Wynne, K., Giannitsopoulou, K., Small, C. J., Patterson, M., Frost, G., Ghatei, M. A., Brown, E. A., Bloom, S. R., and Choi, P. (2005). Subcutaneous ghrelin enhances acute food intake in malnourished patients who receive maintenance peritoneal dialysis: A randomized, placebo-controlled trial. *J. Am. Soc. Nephrol.* **16,** 2111–2118.

Xie, Q. F., Wu, C. X., Meng, Q. Y., and Li, N. (2004). Ghrelin and truncated ghrelin variant plasmid vectors administration into skeletal muscle augments long-term growth in rats. *Domest. Anim. Endocrinol.* **27,** 155–164.

Zhao, D., Keates, A. C., Kuhnt-Moore, S., Moyer, M. P., Kelly, C. P., and Pothoulakis, C. (2001). Signal transduction pathways mediating neurotensin-stimulated interleukin-8 expression in human colonocytes. *J. Biol. Chem.* **276,** 44464–44471.

Zhao, D., Zhan, Y., Zeng, H., Koon, H. W., Moyer, M. P., and Pothoulakis, C. (2005). Neurotensin stimulates interleukin-8 expression through modulation of I kappa. B alpha phosphorylation and p65 transcriptional activity: Involvement of protein kinase C alpha. *Mol. Pharmacol.* **67,** 2025–2031.

Zhao, D., Zhan, Y., Zeng, H., Moyer, M. P., Mantzoros, C. S., and Pothoulakis, C. (2006). Ghrelin stimulates interleukin-8 gene expression through protein kinase C-mediated NF-kappaB pathway in human colonic epithelial cells. *J. Cell Biochem.* **97,** 1317–1327.

Zhou, Y., Xu, B. C., Maheshwari, H. G., He, L., Reed, M., and Lozykowski, M. (1997). A mammalian model for Laron syndrome produced by targeted disruption of the mouse growth hormone receptor/binding protein gene (the Laron mouse). *Proc. Natl. Acad. Sci. USA* **94,** 13215–13220.

Zigman, J. M., Nakano, Y., Coppari, R., Baltasar, N., Marcus, J. N., Lee, C. E., Jones, J. E., Deysher, A. E., Waxman, A. R., White, R. D., Williams, T. D., Lachey, J. L., *et al.* (2005). Mice lacking ghrelin receptors resist the development of diet-induced obesity. *J. Clin. Invest.* **115,** 3564–3572.

Zhu, X., Cao, Y., Voodg, K., and Steiner, D. F. (2006). On the processing of proghrelin to ghrelin. *J. Biol. Chem.* **281,** 38867–38870.

CHAPTER SIX

Appetite and Metabolic Effects of Ghrelin and Cannabinoids: Involvement of AMP-Activated Protein Kinase

Hinke van Thuijl,* Blerina Kola,* *and* Márta Korbonits*

Contents

I. Introduction	122
II. Ghrelin	122
A. Function	122
B. Appetite effects	123
C. Metabolic effects	125
D. Effects on cardiovascular system	126
III. Cannabinoids	127
A. Function	127
B. Appetite effects	128
C. Metabolic effects	129
D. Effects on cardiovascular system	130
IV. Adenosine Monophosphate-Activated Protein Kinase	131
A. Structure	131
B. Role on appetite	133
C. Role on peripheral metabolism: Glucose and lipids	133
D. Role on cardiovascular function	134
V. The Effects of Ghrelin and Cannabinoids on AMPK	135
A. Human studies with rimonabant	137
References	138

Abstract

Obesity is one of the most important health threats to the Western world, and the physiology of appetite-regulating hormones has become a major interest in the last decades. One of the orexigenic hormones, ghrelin is the stomach-derived "brain-gut" peptide, which stimulates energy consumption and

* Department of Endocrinology, Barts and The London School of Medicine, Queen Mary University of London, London EC1M 6BQ, United Kingdom

storage. Ghrelin promotes gluconeogenesis and adipose tissue deposition. Endocannabinoids, such as anandamide and 2-arachydoglycerol, are lipid-like neurotransmitter molecules activating the cannabinoid receptors. Endocannabinoids, apart from the well-known psychological effects, cause an increase in appetite, and they peripherally promote *de novo* fatty acid synthesis and gluconeogenesis. Adenosine monophosphate-activated protein kinase (AMPK) is an energy-sensing kinase, which responds to changes in the energy levels of the cell and the whole body in order to maintain adequate ATP levels in the cell. Recently, several hormones have been shown to regulate AMPK activity, and interestingly in a strictly tissue-specific manner. Orexigenic agents such as ghrelin and cannabinoids stimulate hypothalamic AMPK leading to increase in appetite while inhibiting AMPK activity in the liver and adipose tissue, therefore leading to lipogenic and diabetogenic effects. Here we summarize the recent data on hormonal AMPK regulation. © 2008 Elsevier Inc.

I. INTRODUCTION

The focus on appetite regulation and the physiology of orexigenic and anorectic compounds has increased considerably in the last few years. The reason for this is the fact that the average BMI of the people in the Western world has reached an alarmingly high level, and data from the United States predicts that this is unlikely to improve in the near future. Here we will summarize the relationship between appetite-inducing endogenous compounds ghrelin and cannabinoids and adenosine monophosphate-activated protein kinase (AMPK), a cellular "fuel gauge".

II. GHRELIN

A. Function

Ghrelin is a member of the brain–gut peptide family and is an endogenous ligand for the growth hormone secretagogue receptor (GHSR), a G-protein–coupled receptor (Kojima *et al.*, 1999). The ghrelin gene is located on chromosome 3p26. The prepropeptide codes for the 28-amino acid ghrelin and for a separate protein, the 23-amino acid obestatin. Interestingly, the former has orexigenic properties while the latter has anorexigenic properties (Gil-Campos *et al.*, 2006). Ghrelin is expressed in largest amount in the stomach, but mRNA and protein expression has been detected in a very wide range of tissues (Ariyasu *et al.*, 2001; Krsek *et al.*, 2002). While most orexigenic neuropeptides [neuropeptide Y (NPY), agouti-related protein (AgRP), orexin, and melanocortin-concentrating hormone (MCH)] originate from the central nervous system (CNS), ghrelin is mainly derived from the periphery and has effects both on the hypothalamus as well as direct effects on several peripheral organs (Tschop et al., 2000). Ghrelin has been named the "ultimate anabolic hormone" (Cummings, 2006) because ghrelin stimulation causes the body to consume and store energy.

B. Appetite effects

The stimulating effects of synthetic GHSs on food intake were first discovered in 1995 (Locke et al., 1995; Okada et al., 1996; Shibasaki et al., 1998; Torsello et al., 1998) and a similar effect was reported for ghrelin in 2000 (Tschop et al., 2000). Short-term fasting results in preprandial rise in ghrelin levels and hunger scores corresponded to ghrelin levels in a study involving subjects initiating meals voluntarily without time- and food-related clues (Cummings et al., 2004). This implicates that ghrelin might be an important factor for meal initiation (Ariyasu et al., 2001; Cummings et al., 2001; Faulconbridge et al., 2003; Shiiya et al., 2002; Tschop et al., 2001a). A study following ghrelin levels after different calorie intake found that the level of postprandial ghrelin suppression is proportional to the ingested caloric load but that recovery of plasma ghrelin is not a critical determinant of the intermeal interval (Callahan et al., 2004). Longer fasting after 3 days or even 44 days results in low ghrelin levels (Chan et al., 2004; Korbonits et al., 2005; Fig. 1).

The effects of ghrelin on appetite, and consequently body weight, can be divided into short- and long-term effects. The rapid and short-lived effects

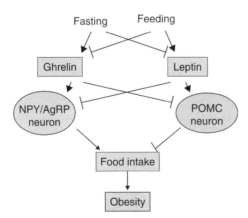

Figure 1 Simplified mechanism of appetite regulation. Leptin, an adipocyte hormone, downregulates appetite via an inhibitory effect on the NPY/AgRP neurons, and a stimulatory effect on proopiomelanocortin (POMC) neurons. Ghrelin causes an activation of NPY/AgRP cells in the ARC and an inhibition of POMC neurons resulting in an increase in food intake.

of ghrelin fit its function to signal and influence meal-related behavior, for example, administration of exogenous ghrelin triggers eating even at times of minimal spontaneous food intake (Asakawa et al., 2001; Tschop et al., 2000; Wren et al., 2000). In addition, the search for food is stimulated as well, as has been shown in ghrelin-stimulated animals. The number of meals seems to be more important than meal size as ghrelin reduces the interval between the meals (Cummings, 2006). Both suggest that ghrelin increases appetite rather than consummatory behavior. Food intake inhibits ghrelin levels. The extent of downregulation is partly dependent on the type of food and drinks consumed. Lipids suppress ghrelin less than carbohydrates or proteins do (Foster-Schubert et al., 2005; Overduin et al., 2005) and fructose-sweetened drinks also have less effect compared to glucose-sweetened drinks probably because they are less capable of increasing insulin levels (Teff et al., 2004). Ghrelin receptor (GHSR) gene expression in the arcuate nucleus (ARC) and ventromedial nucleus of the seasonal Siberian hamster increases after food restriction. The Siberian hamster is known for its seasonal changes in body weight and fat mass, and the ghrelin levels do not follow the same pattern, implying that it acts as a short-term regulator (Tups et al., 2004). Ghrelin responds to changes in body weight. Its levels increase after weight loss and go down after weight gain (Cummings and Shannon, 2003). This decrease cannot be attributed to additional food in the gastrointestinal tract (Williams et al., 2006). The sensitivity to ghrelin is reduced in obese mice, which improves after weight loss (Perreault et al., 2004). Obese people have low ghrelin levels; but when they are fasting, their ghrelin levels increase more than in lean people. Also when they lose weight, ghrelin goes up. After each meal, ghrelin is less suppressed and as a result their weight gain is supported (Cummings, 2006; English et al., 2002; Weigle et al., 2003). Ghrelin decreases during weight gain, not only in overfed people but also in anorexia nervosa patients who tend to have high ghrelin levels (Ariyasu et al., 2001; Hansen et al., 2002; Otto et al., 2001; Ravussin et al., 2001; Shiiya et al., 2002; Tschop et al., 2001b). Patients suffering from anorexia nervosa of the restricted type have lower basal ghrelin levels after treatment compared to a control group, despite still having a lower BMI, suggesting that there is a connection between ghrelin levels and pathological feeding behavior (Janas-Kozik et al., 2006).

Ghrelin can reach the hypothalamus via three different ways: via the blood stream into the ARC, via peripheral ghrelin receptors on the vagus and the dorsal vagal complex, and via local hypothalamic ghrelin-secreting cells in the hypothalamus. Recent data show that central ghrelin could be of importance in the pathogenesis of obesity in humans. Anterior hypothalamic material at time of autopsy showed elevated ghrelin and NPY expression (protein and mRNA) in obese people compared to lean ones (Couce et al., 2006). Earlier data on the role of the vagus in peripheral ghrelin effects suggested its crucial importance (Date et al., 2002); however, a carefully conducted study suggests that the acute eating-stimulatory effect of intraperitoneal ghrelin does not require vagal afferent signaling (Arnold et al., 2006). The hyperphagic fatty Zucker rats have an increased sensitivity to ghrelin due to a higher expression of ghrelin receptor mRNA in the hypothalamus (Brown et al., 2007).

Mice lacking the gene for either ghrelin or its receptor do not show a striking phenotype in basal circumstances probably due to compensatory changes during embryonic development, but they are more resistant to high-fat diet-induced obesity (Sun et al., 2003, 2004; Wortley et al., 2004, 2005). The blockade of ghrelin using anti-ghrelin antibodies, ghrelin-receptor antagonists, or antisense oligonucleotides results in a decrease in spontaneous food intake, leading to weight loss in longer studies (Asakawa et al., 2003; Bagnasco et al., 2003; Murakami et al., 2002; Nakazato et al., 2001; Shuto et al., 2002).

Ghrelin could also be involved in the mechanisms for alcohol craving as high circulating ghrelin levels are seen in subjects with an increased craving (Addolorato et al., 2006). This psychostimulating effect has also been shown in rats as ghrelin augments the rewarding effects of cocaine (Davis et al., 2007).

To conclude, ghrelin is of great importance in the regulation of appetite. Hypothalamic ghrelin is increased in obese patients, while their circulating ghrelin levels are lower. During fasting and weight loss, these levels increase more compared to "normal" people leading to an increase in appetite and making it very difficult for obese patients to lose weight. Still, it is promising that ghrelin levels, which are increased during weight loss, return to baseline with sustained weight maintenance (Garcia et al., 2006).

C. Metabolic effects

Ghrelin affects glucose metabolism in several ways. In the brain, there are two types of glucose-sensing neurons: those which are excited and those which are inhibited as glucose levels rise. In response to these changes, the brain exerts effects, which modulate food intake and peripheral organ activity via autonomic output. Ghrelin has a modulating effect on the sensitivity of these glucose sensors (Penicaud et al., 2006).

Glucose production in hepatocytes is directly stimulated by ghrelin, and this is inhibited by its unacylated form (Gauna et al., 2005). Ghrelin also blocks the inhibitory effect of insulin on the gluconeogenic enzymes in a hepatoma cell line (Murata et al., 2002b). The effect of ghrelin on insulin production is unclear as an inhibiting effect has been found in mice and the opposite in rats (Heijboer et al., 2006).

An increase in ghrelin has a positive effect on adipose tissue deposition. In rats, ghrelin increases bone marrow adipogenesis via a direct peripheral action in which the GHSR is not involved (Thompson et al., 2004).

In vitro studies show that ghrelin promotes the differentiation of preadipocytes and antagonizes lipolysis (Choi *et al.*, 2003). This mechanism acts independently of food intake and weight gain because ghrelin-treated animals pair-fed to saline-treated ones showed an increase in fat tissue content and no change in body weight (Wren *et al.*, 2004). Ghrelin reduces isoprotenerol-induced lipolysis (Muccioli *et al.*, 2004). Ageing is associated with a decline in appetite and weight loss. Elderly people have an increased fat mass and a reduced fat-free mass (FFM). Data show that increased ghrelin levels signal FFM depletion in elderly subjects and therefore promote lipogenesis (Bertoli *et al.*, 2006). Undernourished older women have higher ghrelin levels compared to well-nourished women (Sturm *et al.*, 2003). There are controversial data whether ghrelin levels change with age. Some, but not all, studies suggest a negative correlation between ghrelin and age in mouse and humans (reviewed in Korbonits *et al.*, 2004). According to studies where multiple regression analysis was performed taking a number of parameters into account [BMI, insulin sensitivity, body composition, high-density lipoproteins (HDLs), sex (Purnell *et al.*, 2003), or insulin and fat distribution (Goldstone *et al.*, 2004)], age was not an independent factor in ghrelin levels, but was in another (Chan *et al.*, 2004). Ghrelin's function seems to become impaired with age (Broglio *et al.*, 2003; Chan *et al.*, 2004; Schutte *et al.*, 2007).

Experimental data suggest that ghrelin preferentially promotes lipid deposition in visceral fat tissue and not in subcutaneous fat tissue (Kola *et al.*, 2005). It has been shown that ghrelin interacts with plasma lipid particles: desoctanoylation of acylated ghrelin occurs when it is in contact with low-density lipoproteins (LDLs). Recent data found that triglyceride-rich lipids mostly transport AG, while HDLs and very high-density lipoproteins (VHDLs) transport both acylated and unacylated ghrelin (De Vriese *et al.*, 2007). This suggests that ghrelin desoctanoylation may occur in the LDLs and lipoprotein-poor plasma fractions. The enzymes butyrylcholinesterase and platelet-activating factor acetylhydrolase have been implicated in the process (De Vriese *et al.*, 2007). The positive effect ghrelin exerts on adipose tissue deposition fits its orexigenic properties.

D. Effects on cardiovascular system

Ghrelin has an inhibitory effect on the cardiovascular sympathetic activity (Matsumura *et al.*, 2002). This causes vasodilatation (Okumura *et al.*, 2002) and a decrease in blood pressure (Nagaya *et al.*, 2001) in humans. Ghrelin has been recognized to have a long-term vasodilatative effect (Wiley and Davenport, 2002). Some studies, however, show a vasoconstrictive effect on rat coronary vasculature without altering cardiac peptide secretion (Pemberton *et al.*, 2004). Other studies using pharmacological instead of physiological doses of ghrelin have shown beneficial cardiovascular effects

(Chang *et al.*, 2004; Frascarelli *et al.*, 2003; Korbonits *et al.*, 2004). These effects are independent of growth hormone (Broglio *et al.*, 2001), and *in vitro* antiapoptotic effects on embryonic (H9c2) and adult (HL-1) heart muscle cell lines were observed (Baldanzi *et al.*, 2002; Pettersson *et al.*, 2002). Diastolic dysfunction associated with myocardial stunning improves with ghrelin analogue treatment (Weekers *et al.*, 2000). Furthermore, it attenuates the development of cardiac cachexia; humans with cachexia have elevated ghrelin levels. This could be a compensatory mechanism in response to the catabolic/anabolic imbalance (Iglesias *et al.*, 2004).

Some GHSs have been shown to have an inhibitory effect on coronary disease acting via the CD36 scavenger receptor, but ghrelin does not share this activity (Bodart *et al.*, 2002; Demers *et al.*, 2004; Marleau *et al.*, 2005). GHSR density shows upregulation in atherosclerotic plaques (Katugampola *et al.*, 2001). Presumably as a consequence, ghrelin levels are positively associated with carotid artery atherosclerosis in males. This is controversial because obese people are known to have low ghrelin levels, but these findings remained reliable after adjusting for the common risk factors for atherosclerosis, making it an interesting future subject (Poykko *et al.*, 2006). Ghrelin does not have an effect on the vascular endothelial cell barrier function (Elbatarny *et al.*, 2006).

III. Cannabinoids

The natural compound Δ^9-tetrahydrocannabinol (Δ^9-THC) is derived from *Cannabis sativa*, while there are several endogenous cannabinoids including arachidonylethanolamide (anandamide) (Devane *et al.*, 1992) and 2-arachydol-glycerol (2-AG) (Mechoulam *et al.*, 1995; Sugiura *et al.*, 1995). These are lipid-like molecules derived from arachidonic acid. They are synthesized on demand and are degraded rapidly so they are not "classical" hormones, rather neurotransmitters or cytokines.

A. Function

Cannabis (marijuana) has been used for centuries for various medical and pleasure purposes; and from the 1980s, it has been used for cancer patients to prevent nausea (Walsh *et al.*, 2003) and more recently in AIDS patients to stimulate their appetite. However, it was only in 1990 that the first cannabinoid receptor, CB1, was discovered (Matsuda *et al.*, 1990). The CB1 receptor is a presynaptic G-protein–coupled membrane receptor (Fig. 2). It is located in the CNS, adipose tissue, muscle, heart, gastrointestinal tract, and the liver (Pacher *et al.*, 2006). The CB2 receptor is present in immune

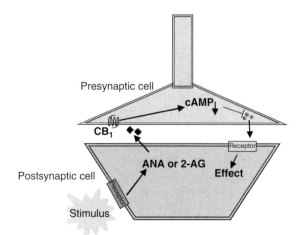

Figure 2 The CB1 receptor is located on the presynaptic cell. Stimuli cause endocannabinoids anandamide and/or 2-AG to be released from the postsynaptic cell and activate the CB1 receptor located on the presynaptic cell. This usually results in downregulation of cAMP leading to inhibition of neurotransmitter release from the presynaptic cell. As a result, the postsynaptic cell receives less neurotransmitter input. CB1 receptor inhibits also Ca^{2+} channels and activates K^+ channels.

cells where it is possibly involved in cytokine release (Howlett, 2002), but also in the CNS (Van Sickle et al., 2005). Endocannabinoids bind to both types of receptors with slightly different characteristics. A third important endocannabinoid called noladin ether has been identified. It promotes food intake after systemic treatment with very low doses (1 μg/kg i.p.) in food-restricted mice. Its effect is mediated via the CB1 receptor as it is reversed by rimonabant (Avraham et al., 2004). Although it has been questioned whether this endocannabinoid is naturally present in the brain (Oka et al., 2003), it can be a useful tool because of its relative stability compared to anandamide and 2-AG which degrade more quickly.

B. Appetite effects

Cannabinoids exert their orexigenic effects at two levels in the brain. First, via the mesolimbic pathway, which is involved in reward mechanisms, to stimulate the finding and consuption of food. Second, in the hypothalamus, which is activated after a short period of food deprivation and then transiently regulates the level and action of other orexigenic or anorectic mediators to induce appetite. This assumption has been confirmed in experiments in which endocannabinoids were injected in the limbic forebrain resulting in an increase in food intake (Jamshidi and Taylor, 2001; Kirkham et al., 2002).

CB1 receptors in the brain have been identified in the hypothalamus, the nucleus accumbens, the vagus nerve, and the nodos ganglion (Matias and Di Marzo, 2006). In the hypothalamus, CB1 is present in cocaine- and amphetamine-regulated transcript (CART) neurons in the ARC, in neurons containing orexin and MCH in the lateral hypothalamus, and those expressing corticotrophin-releasing hormone (CRH) in the paraventricular nucleus (PVN). The mechanisms via which cannabinoids exert their orexigenic effect are still a matter of debate and investigation. NPY, CRH, CART, orexin have been proposed as downstream mediators of the cannabinoid action in the hypothalamus. Hypothalamic endocannabinoid content seems also to be important in food regulation mechanisms and varies with the nutritional state. Fasting increases cannabinoid content, in particular 2-AG, which then declines as the animals are refed and return to normal levels in satiated animals (Kirkham et al., 2002).

Leptin has an inhibitory effect on appetite via downregulation of hypothalamic endocannabinoid levels (Di Marzo et al., 2001).

Cannabinoids stimulate the sensitivity to palatable foods. Rats treated with the cannabinoid antagonist rimonabant showed a decrease in intake of palatable food, while the intake of chow was not affected, except in the first few days when food intake in general was suppressed (Arnone et al., 1997). In marmosets, a similar effect has been found: rimonabant reduced the intake of cane sugar without suppressing the intake of standard pellets (Simiand et al., 1998). The effect is confirmed with the use of CB1 agonists, after administration, rats work harder to obtain palatable ingesta (beer compared to sucrose solutions), while antagonists attenuate this response (Gallate and McGregor, 1999; Gallate et al., 1999). $CB1^{-/-}$ mice show lower levels of responding to sucrose (Sanchis-Segura et al., 2004). The stimulation by CB1 agonists to chow is not related to satiation or energy status, this implicates that they directly activate eating motivation (Kirkham et al., 2002; Williams and Kirkham, 2002). Intestinal anandamide has been pointed out to give the "hunger signal" because stimulation or blocking of peripheral CB1 receptors influences central motivation (Jamshidi and Taylor, 2001; Kirkham et al., 2002; Koch, 2001). The satiety hormone cholecystokinin (CCK) naturally inhibits anandamide, which could activate the vagus nerve.

C. Metabolic effects

Cota et al. (2003) showed that cannabinoids and their antagonists affect body weight not only via an acute effect on food intake but also via a direct peripheral effect on adipose tissue. A High-fat diet activates the endocannabinoid system in the white adipose tissue causing a downregulation of adiponectin levels and increased lipogenesis (Di Marzo and Matias, 2005). The role of the liver in *de novo* lipogenesis is major compared to adipose tissue

(Diraison et al., 2003). Endocannabinoids directly target the liver and the CB1 receptor is present in hepatocytes (Osei-Hyiaman et al., 2005). Anandamide activates CB1 in hepatocytes, which results in an increase in *de novo* fatty acid synthesis through an induction in the hepatic gene expression of the lipogenic transcription factor sterol regulatory element-binding protein-1 (SREBP1) (Brown and Goldstein, 1998) and its target enzymes acetyl CoA carboxylase (ACC) and FAS. There is an increase of CB1 in adipose tissue in obese versus lean rats (Bensaid et al., 2003). There are data confirming the upregulation of the endocannabinoid system in obesity in humans too. 2-AG levels in adipose tissue are significantly higher in overweight and obese than in normoweight patients (Matias et al., 2006). Blood levels of endocannabinoids in obese women with binge-eating disorders are also significantly elevated (Monteleone et al., 2005), despite the high levels of leptin that are normally present in obese patients. Similarly, circulating endocannabinoid levels were found to be increased in obese postmenopausal women and to be inversely correlated with decreased fatty acid amide hydrolase (FAAH) expression in adipose tissue (Engeli et al., 2005). This would suggest that the high plasma anandamide and 2-AG levels are due to increased cannabinoid levels in this tissue as a consequence of decreased endocannabinoid degradation. Nevertheless, this would be in contrast with the general accepted view that endocannabinoids are not normally released from tissues into the bloodstream to act as hormones or circulating mediators. Matias and Di Marzo (2006) suggested that this is either due to a widespread endocannabinoid "hyperactivity" in obesity, which is reflected in circulating cannabinoid levels, or due to overproduction and/or reduced degradation of endocannabinoids in blood cells as well. The finding of a strong correlation between overweight/obesity and a phenotypic missense mutation of FAAH (Sipe et al., 2005) might confirm a dysregulation of the endocannabinoid system in obesity. In obese men, plasma 2-AG, but not anandamide, levels correlated positively with BMI, waist girth, intraabdominal obesity, fasting plasma triglyceride, and insulin levels, and negatively with HDL cholesterol and adiponectin levels (Cote et al., 2007).

D. Effects on cardiovascular system

Cannabinoids and their endogenous and synthetic analogues exert important cardiovascular effects. The underlying mechanisms are complex, involving direct effects on the vasculature and myocardium, and modulating the autonomic outflow through action on the central and the peripheral system (Pacher et al., 2006). Endocannabinoids play a role in several cardiovascular disorders such as hypertension, circulatory shock, and myocardial ischemia reperfusion (I/R) injury. The vasodilatating effect of histamine is induced by endocannabinoids and cannabinoid agonists (Cenni et al., 2006). The vascular effects of cannabinoids have largely been studied, while less is known about their direct cardiac effects. The endocannabinoid system has been implicated in endotoxin-induced preconditioning against myocardial I/R injury (Lagneux and Lamontagne, 2001). Endotoxin pretreatment enhances functional recovery on reperfusion and reduces infarct size. Perfusion of isolated rat hearts with 2-AG but not with anandamide provided protection against ischemia by improving myocardial recovery and decreasing myocardial damage and infarct size (Underdown et al., 2005). Experiments in whole animal models of I/R injury have provided further evidence for a protective role of cannabinoids in myocardial ischemic damage and that this effect is most probably mediated by CB2 receptors (Pacher et al., 2006). This receptor is also involved in the inhibiting effect cannabinoids have on disease progression in atherosclerosis (Steffens and Mach, 2006).

IV. Adenosine Monophosphate-Activated Protein Kinase

A. Structure

AMPK is an energy-sensing kinase, which responds to changes in the energy levels of the cell and the whole body in order to maintain adequate ATP levels in the cell (Fig. 3). AMPK is a heterotrimeric protein with a catalytic α-subunit and two regulatory subunits (β and γ) (Hardie, 2004). Multiple forms of each subunit exist; these are the products of different genes ($\alpha1$, $\alpha2$, $\beta1$, $\beta2$, $\gamma1$, $\gamma2$, $\gamma3$). The α-subunits have a protein kinase catalytic domain at the N-terminus and a regulatory domain at the C-terminus binding β and γ together. It also contains an auto-inhibitory region. In this way, it controls enzyme activity and stability (Crute et al., 1998). The β-subunit contains a glycogen-binding domain (GBD/KIS), which is closely related to enzymes that metabolize glucans such as starch and glycogen. Another domain on the β-subunits is on the C-terminus which functions to form complexes with the α- and γ-subunits. The γ-subunit has a variable N-terminus followed by two pairs of CBS (cystathionine-β-synthase) tandem repeats, which bind ATP. ATP is synthesized from two ADP molecules which form ATP and AMP (Hardie and Hawley, 2001).

AMP, which activates AMPK, should not be confused with cyclic AMP (cAMP), a second messenger commonly involved in hormone effects and which can activate different enzymes including protein kinase A (PKA). Fig. 4 shows the difference in structure of AMP and cAMP.

When AMPK is activated, it is phosphorylated on Thr172 by LKB1, a serine/threonine kinase multitasking enzyme, which is also involved in cell polarity, tumor growth, and energy metabolism (Hawley et al., 1996),

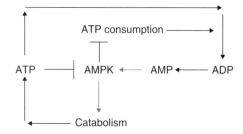

Figure 3 AMPK and its effects on the energy metabolism.

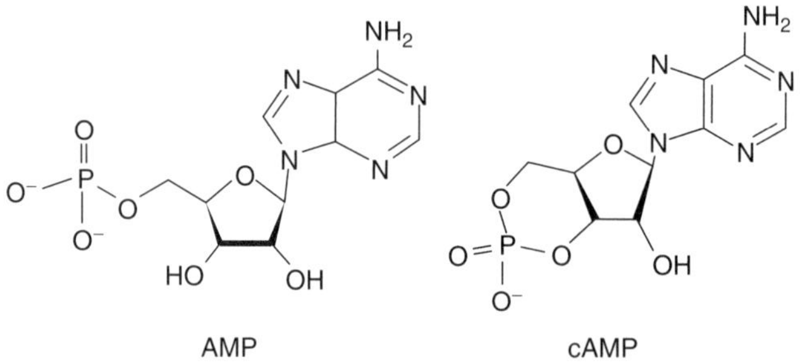

Figure 4 The molecular structure of AMP and cAMP.

and by CaMKK, a calmodulin kinase kinase (Hawley et al., 2005; Hurley et al., 2005; Woods et al., 2005). AMP promotes AMPK activation in three ways. First is the allosteric activation by binding of AMP. Second, the binding makes AMPK more accessible for upstream kinase LKB1. The phosphorylation on Thr172 starts the activation loop, making the reaction up to a 100-fold more active. Third, the binding of AMP prevents phosphatases to dephosphorylate Thr172 (Davies et al., 1995). These three actions are inhibited by high levels of ATP. The advantage of having three different mechanisms is that it makes the AMPK activation ultrasensitive; only a small change in AMP can have a large effect on AMPK (Hardie et al., 1999).

AMPK has a regulating function in the energy metabolism in both peripheral tissues and in the CNS (Minokoshi et al., 2002, 2004). The activation of AMPK is initiated when pathological or physiological stresses cause a fall in the ATP levels. Examples of pathological stresses are hypoxia, metabolic poisoning, oxidative stress, and nutrient deprivation. A physiological stress can be the ATP depletion after muscle contraction during exercise (Winder and Hardie, 1996). The beneficial effect of exercise in relation to AMPK activation has given new perspectives concerning the treatment of obesity and type 2 diabetes. This suggestion has been reinforced by the fact that AMPK is also stimulated by biguanides (e.g., metformin) and thiazolidinediones (e.g., rosiglitazone, pioglitazone), which are drugs used for the treatment of type 2 diabetes (Hardie, 2004). One of the main enzyme targets of AMPK is ACC, which is phosphorylated and therefore inactivated by AMPK, and the two isoforms of this enzyme respectively stimulate lipogenesis and inhibit lipid oxidation. In brown and white adipose tissue, AMPK activity is upregulated under conditions of chronic cold exposure, showing that AMPK has a thermogenic potential (Mulligan et al., 2007). In response to chronic energy deprivation, AMPK activation provides energy in skeletal muscle via mitochondrial biogenesis (Zong et al., 2002).

B. Role on appetite

It has emerged that AMPK, apart from its role at cellular level, is also involved in the energy regulation of the whole body by influencing food (i.e., energy) intake (Minokoshi et al., 2002, 2004). The regions in the hypothalamus involved in satiety and food intake, the ventromedial (VMH) and arcuate nuclei, are negatively regulated in conditions of high leptin (Minokoshi et al., 2004), α-melanocyte stimulating hormone, insulin, FAS inhibitor C75 (Kim et al., 2004), and glucose (Lee et al., 2005).

On the other hand, during energy depletion or when orexigenic compounds are administered (AgRP or the antimetabolite 2-deoxyglucose) to the brain, AMPK activity is stimulated and the food intake increases. This effect is mediated through regulation of NPY and AgRP expression (Fryer and Carling, 2005).

C. Role on peripheral metabolism: Glucose and lipids

In general, the effects of AMPK are focused on maintaining the appropriate energy level in the cell. In times of energy deprivation, it switches off anabolic, ATP-consuming pathways, for example, gluconeogenesis and fatty acid and cholesterol biosynthesis. At the same time, it activates the catabolic pathways that generate ATP such as glycolysis, glucose uptake, and fatty acid oxidation. AMPK inhibits glucose output in the liver resulting in a decrease in hyperglycemia (Zhou et al., 2001). AMPK's effect on lipid metabolism involves a downregulation of SREBP1 and its target enzymes ACC and FAS, resulting in a reduction of fatty acid synthesis and an increase in fatty acid oxidation (Zhou et al., 2001). Interesting data from Foretz et al. showed that short-term overexpression of constitutively active AMPKα2 in liver decreased blood glucose levels and hepatic gluconeogenesis gene expression. This results in low availability of glucose and causes a switch from glucose utilization to fatty acid utilization and the generation of ketone bodies. This is associated with a decrease in white adipose tissue mass, lipid uptake from peripheral tissue, and lipid accumulation in liver (Foretz et al., 2005). Data from longer term AMPKα2 activation is now awaited. Further complexity to the liver AMPK regulation is that hypothalamic AMPK activation stimulates liver gluconeogenesis (McCrimmon et al., 2004).

In skeletal muscle, AMPK is activated during exercise, promoting glucose uptake, by stimulating GLUT4 (a glucose transporter) via an insulin-independent pathway (Vavvas et al., 1997). In insulin-resistant type 2 diabetic patients, basal glucose uptake is impaired while exercise-induced glucose uptake is intact. The beneficial effects of metformin and glitazones on AMPK activity are prominent in skeletal muscle. CaMKK, an upstream regulator of AMPK, exerts its effect on glucose uptake in skeletal muscle independent of AMPK (Witczak et al., 2007).

Chronic and acute exercise stimulates AMPK activity in rat adipocytes and this effect is mediated by noradrenalin (Koh et al., 2007). In order to supply energy, lipogenesis is inhibited by AMPK and lipid oxidation is stimulated. During fasting, energy is derived from the breakdown of triglycerides to provide fuel for the peripheral tissues. This process is called lipolysis. The effects of AMPK on lipolysis remain unclear because different research groups have published contradictory results. Stimulation would be expected and has been described by Yin et al. (2003), but it has been suggested that lipolysis is an energy-consuming process for adipocytes and therefore is inhibited in times of energy depletion (Daval et al., 2006). The glucose-dependent insulinotropic polypeptide (GIP) has an inhibiting effect on AMPK, which results in activation of lipoprotein lipase and triglyceride accumulation, increasing fat storage in adipocytes (Kim et al., 2007). To conclude, AMPK activation increases fatty acid oxidation and inhibits lipogenesis and triacylglycerol synthesis, contributing therefore to improved insulin sensitivity.

D. Role on cardiovascular function

AMPK is a key regulator of energy metabolism also in the heart. The heart, which is an organ with high energy demand, utilizes both fatty acids and glucose as energy resources. AMPK influences heart glucose metabolism targeting glucose transporters with an overall effect of stimulating glucose uptake, glycolysis, glycogenolysis, GLUT4 transport, and decreasing glycogen synthesis. Pharmacological activation of AMPK intervenes with cardiac growth and thereby inhibits cardiac hypertrophy (Li et al., 2007). In cardiac muscle, AMPK activates heart-specific phosphofructokinase-2 and therefore stimulates glycolytic flux during ischemia, suggesting a protective, beneficial role of AMPK in the heart (Marsin et al., 2002). However, overactivation of AMPK can be detrimental as increased fatty acid oxidation could predominate and provide most of the acetyl CoA to the Krebs cycle leading to accumulation of glycolytic products protons and lactate (Dyck and Lopaschuk, 2002).

V. THE EFFECTS OF GHRELIN AND CANNABINOIDS ON AMPK

We have hypothesized that orexigenic compounds could change the AMPK activity of the hypothalamus. We have established that ghrelin and cannabinoids stimulate AMPK activity in rat hypothalamus after intracerebroventricular (i.c.v.) and intraperitoneal (i.p.) administration (Kola et al., 2005; Fig. 5). Ghrelin and cannabinoids are synthesized peripherally as well as in the hypothalamus so both approaches seem to have physiological relevance. An interaction was shown between ghrelin and the cannabinoid system when subanorectic doses of rimonabant, a CB1 antagonist, inhibited the orexigenic effect of ghrelin (Tucci et al., 2004; Fig. 6).

We then studied CB1 receptor knockout animals and found that endocannabinoids lose their effect in the hypothalamus in CB1$^{-/-}$ mice, suggesting that the AMPK-stimulating effect is via the CB1 receptor (our unpublished data). We have established that ghrelin and cannabinoids affect hypothalamic AMPK activity and probably interact with each other.

Both cannabinoids and ghrelin have direct peripheral effects and the lean CB1$^{-/-}$ animals have a lower body weight than their pair-fed wild-type littermates, suggesting that purely food intake is not the only cause of the weight difference (Cota et al., 2003). We have suggested that the direct peripheral effect involves the inhibitory effect of cannabinoids and ghrelin on AMPK activity in the liver and in the adipose tissue. Active AMPK is known to increase fatty acid oxidation, while inhibition leads to an increase in adipose tissue mass and glucose and fatty acid synthesis. ACC, a target enzyme of SREBP1, which in turn is regulated by AMPK, has been linked to be a target of CB1 activation (Osei-Hyiaman et al., 2005). Ghrelin is known to have a positive effect on gluconeogenesis; this has been shown with the use of human hepatoma cell lines (Murata et al., 2002a), possibly this occurs via the inhibition of AMPK. Our data is supported by a study of 4-day ghrelin treatment showing reduced pAMPK content in rat liver, which usually correlates with AMPK activity, although kinase activity was not directly assessed (Barazzoni et al., 2005).

Ghrelin and cannabinoids are known to increase lipid stores, while AMPK promotes lipid oxidation. We have shown that ghrelin and cannabinoids inhibit adipose tissue AMPK activity, especially in mesenteric fat. As rimonabant promotes adiponectin synthesis (Bensaid et al., 2003), which in turn stimulates AMPK, this could be an indirect effect. AMPK activity is

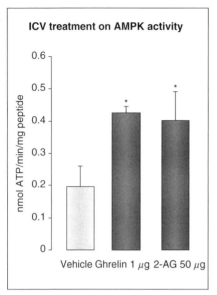
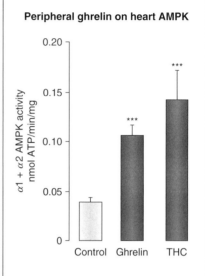
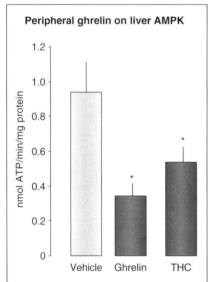
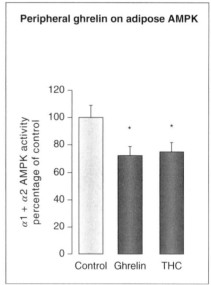

Figure 5 Rats were treated with intracerebroventricular ghrelin or 2-AG (left upper panel) or intraperitoneal ghrelin and THC (other panels), and AMPK activity was measured in the tissues 1 h later (Kola *et al.*, 2005).

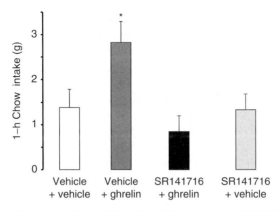

Figure 6 The effect of ghrelin is inhibited by CB1 antagonist SR141716 (rimonabant) (Tucci et al., 2004).

decreased after treatment with a CB1 agonist; this supports the suggestion of a mechanism acting via the CB1 receptor (Bensaid et al., 2003).

Cannabinoids play an important role in limiting damage to the heart in shock and in myocardial infarction. They also improve atherosclerosis via a CB2-mediated effect on immune cells (Steffens et al., 2005). Similarly, AMPK protects ATP levels in the heart, thereby reducing infarct size and damage to myocytes during ischemia. Adiponectin knockout animals show pressure overload and cardiac hypertrophy, and adiponectin is known to stimulate AMPK activity (Russell et al., 2004). The beneficial effects of cannabinoids could be mediated via activation of AMPK, although excessive AMPK activation can be harmful (Altarejos et al., 2005; Arad et al., 2002). The positive effects of ghrelin on cardiovascular function including improved heart stroke volume and attenuation of cardiac cachexia could also possibly act via AMPK as AMPK phosphorylation and activation is stimulated by ghrelin.

To conclude, the stimulating effects of ghrelin and cannabinoids on AMPK activity centrally cause an increase in appetite and facilitate restoring fuel levels after deprivation of energy. On the other hand, AMPK is inhibited peripherally to promote energy storage, especially in fat tissues. These two different effects of ghrelin and cannabinoids have the combined effect to stimulate lipid deposition.

A. Human studies with rimonabant

Rimonabant (SR141716A) is a cannabinoid receptor antagonist recently approved for the treatment of obesity and associated metabolic disorders in Europe. Phase III clinical trials of rimonabant have shown sustained effects on weight loss in humans through effects on glucose and fat metabolism

(Despres et al., 2005; Pi-Sunyer et al., 2006; Scheen et al., 2006; Van Gaal et al., 2005). The Rimonabant in Obesity (RIO) program of Sanofi-Aventis consisted of four 1- to 2-year Phase III trials designed to assess the efficacy and safety of rimonabant in over 6600 patients in the treatment of overweight/obesity, hyperlipidemia, and type 2 diabetes. The four published studies have a similar setup of 5- and 20-mg rimonabant and placebo. The weight-reducing effect of 20-mg rimonabant reached 6–8 kg by 1 year and was not different from other available therapies (orlistat, sibutramin, and well-supported diet and exercise programs), even though there are no head-to-head comparisons available. In addition to weight loss, the results of these studies showed statistically significant and sustained mean reductions in waist circumference and weight, LDL particle size, TG, HOMA-IR, leptin, and CRP, and significant increase for HDL-C and adiponectin for rimonabant 20 mg/day compared with placebo. To note, the improvements in lipid profile (HDL-C and TG), in insulin resistance (fasting insulin and HOMA-IR), and the increase in adiponectin secretion were partly independent of weight loss (Van Gaal et al., 2005), suggesting a direct effect of rimonabant. The data obtained from these four trials suggest that therapeutic targeting of the CB1 receptor is a promising approach for the comprehensive treatment of multiple cardiometabolic abnormalities. We believe that some of these effects could have resulted from the inhibition of rimonabant on hypothalamic AMPK and stimulation on liver and adipose tissue AMPK activity.

REFERENCES

Addolorato, G., Capristo, E., Leggio, L., Ferrulli, A., Abenavoli, L., Malandrino, N., Farnetti, S., Domenicali, M., D'Angelo, C., Vonghia, L., Mirijello, A., Cardone, S., et al. (2006). Relationship between ghrelin levels, alcohol craving, and nutritional status in current alcoholic patients. *Alcohol Clin. Exp. Res.* **30**, 1933–1937.

Altarejos, J. Y., Taniguchi, M., Clanachan, A. S., and Lopaschuk, G. D. (2005). Myocardial ischemia differentially regulates LKB1 and an alternate 5′-AMP-activated protein kinase kinase. *J. Biol. Chem.* **280**, 183–190.

Arad, M., Benson, D. W., Perez-Atayde, A. R., McKenna, W. J., Sparks, E. A., Kanter, R. J., McGarry, K., Seidman, J. G., and Seidman, C. E. (2002). Constitutively active AMP kinase mutations cause glycogen storage disease mimicking hypertrophic cardiomyopathy. *J. Clin. Invest.* **109**, 357–362.

Ariyasu, H., Takaya, K., Tagami, T., Ogawa, Y., Hosoda, K., Akamizu, T., Suda, M., Koh, T., Natsui, K., Toyooka, S., Shirakami, G., Usui, T., et al. (2001). Stomach is a major source of circulating ghrelin, and feeding state determines plasma ghrelin-like immunoreactivity levels in humans. *J. Clin. Endocrinol. Metab.* **86**, 4753–4758.

Arnold, M., Mura, A., Langhans, W., and Geary, N. (2006). Gut vagal afferents are not necessary for the eating-stimulatory effect of intraperitoneally injected ghrelin in the rat. *J. Neurosci.* **26**, 11052–11060.

Arnone, M., Maruani, J., Chaperon, F., Thiebot, M. H., Poncelet, M., Soubrie, P., and Le Fur, G. (1997). Selective inhibition of sucrose and ethanol intake by SR 141716, an antagonist of central cannabinoid (CB1) receptors. *Psychopharmacology (Berl.)* **132,** 104–106.

Asakawa, A., Inui, A., Kaga, T., Yuzuriha, H., Nagata, T., Ueno, N., Makino, S., Fujimiya, M., Niijima, A., Fujino, M. A., and Kasuga, M. (2001). Ghrelin is an appetite-stimulatory signal from stomach with structural resemblance to motilin. *Gastroenterology* **120,** 337–345.

Asakawa, A., Inui, A., Kaga, T., Katsuura, G., Fujimiya, M., Fujino, M. A., and Kasuga, M. (2003). Antagonism of ghrelin receptor reduces food intake and body weight gain in mice. *Gut* **52,** 947–952.

Avraham, Y., Ben Shushan, D., Breuer, A., Zolotarev, O., Okon, A., Fink, N., Katz, V., and Berry, E. M. (2004). Very low doses of delta 8-THC increase food consumption and alter neurotransmitter levels following weight loss. *Pharmacol. Biochem. Behav.* **77,** 675–684.

Bagnasco, M., Tulipano, G., Melis, M. R., Argiolas, A., Cocchi, D., and Muller, E. E. (2003). Endogenous ghrelin is an orexigenic peptide acting in the arcuate nucleus in response to fasting. *Regul. Pept.* **111,** 161–167.

Baldanzi, G., Filigheddu, N., Cutrupi, S., Catapano, F., Bonissoni, S., Fubini, A., Malan, D., Baj, G., Granata, R., Broglio, F., Papotti, M., Surico, N., *et al.* (2002). Ghrelin and desacyl ghrelin inhibit cell death in cardiomyocytes and endothelial cells through ERK1/2 and PI 3-kinase/AKT. *J. Cell Biol.* **159,** 1029–1037.

Barazzoni, R., Bosutti, A., Stebel, M., Cattin, M. R., Roder, E., Visintin, L., Cattin, L., Biolo, G., Zanetti, M., and Guarnieri, G. (2005). Ghrelin regulates mitochondrial-lipid metabolism gene expression and tissue fat distribution in liver and skeletal muscle. *Am. J. Physiol. Endocrinol. Metab.* **288,** E228–E235.

Bensaid, M., Gary-Bobo, M., Esclangon, A., Maffrand, J. P., Le Fur, G., Oury-Donat, F., and Soubrie, P. (2003). The cannabinoid CB1 receptor antagonist SR141716 increases Acrp30 mRNA expression in adipose tissue of obese fa/fa rats and in cultured adipocyte cells. *Mol. Pharmacol.* **63,** 908–914.

Bertoli, S., Magni, P., Krogh, V., Ruscica, M., Dozio, E., Testolin, G., and Battezzati, A. (2006). Is ghrelin a signal of decreased fat-free mass in elderly subjects? *Eur. J. Endocrinol.* **155,** 321–330.

Bodart, V., Febbraio, M., Demers, A., McNicoll, N., Pohankova, P., Perreault, A., Sejlitz, T., Escher, E., Silverstein, R. L., Lamontagne, D., and Ong, H. (2002). CD36 mediates the cardiovascular action of growth hormone-releasing peptides in the heart. *Circ. Res.* **90,** 844–849.

Broglio, F., Benso, A., Valetto, M. R., Gottero, C., Quaranta, L., Podio, V., Arvat, E., Bobbio, M., Bisi, G., and Ghigo, E. (2001). Growth hormone-independent cardiotropic activities of growth hormone- releasing peptides in normal subjects, in patients with growth hormone deficiency, and in patients with idiopathic or ischemic dilated cardiomyopathy. *Endocrine* **14,** 105–108.

Broglio, F., Benso, A., Castiglioni, C., Gottero, C., Prodam, F., Destefanis, S., Gauna, C., van der Lely, A. J., Deghenghi, R., Bo, M., Arvat, E., and Ghigo, E. (2003). The endocrine response to ghrelin as a function of gender in humans in young and elderly subjects. *J. Clin. Endocrinol. Metab.* **88,** 1537–1542.

Brown, M. S., and Goldstein, J. L. (1998). Sterol regulatory element binding proteins (SREBPs): Controllers of lipid synthesis and cellular uptake. *Nutr. Rev.* **56,** S1–S3.

Brown, L. M., Benoit, S. C., Woods, S. C., and Clegg, D. J. (2007). Intraventricular (i3vt) ghrelin increases food intake in fatty Zucker rats. *Peptides* **28,** 612–616.

Callahan, H. S., Cummings, D. E., Pepe, M. S., Breen, P. A., Matthys, C. C., and Weigle, D. S. (2004). Postprandial suppression of plasma ghrelin level is proportional to ingested caloric load but does not predict intermeal interval in humans. *J. Clin. Endocrinol. Metab.* **89,** 1319–1324.

Cenni, G., Blandina, P., Mackie, K., Nosi, D., Formigli, L., Giannoni, P., Ballini, C., Della, C. L., Francesco, M. P., and Beatrice, P. M. (2006). Differential effect of cannabinoid agonists and endocannabinoids on histamine release from distinct regions of the rat brain. *Eur. J. Neurosci.* **24,** 1633–1644.

Chan, J. L., Bullen, J., Lee, J. H., Yiannakouris, N., and Mantzoros, C. S. (2004). Ghrelin levels are not regulated by recombinant leptin administration and/or three days of fasting in healthy subjects. *J. Clin. Endocrinol. Metab.* **89,** 335–343.

Chang, L., Ren, Y., Liu, X., Li, W. G., Yang, J., Geng, B., Weintraub, N. L., and Tang, C. (2004). Protective effects of ghrelin on ischemia/reperfusion injury in the isolated rat heart. *J. Cardiovasc. Pharmacol.* **43,** 165–170.

Choi, K., Roh, S. G., Hong, Y. H., Shrestha, Y. B., Hishikawa, D., Chen, C., Kojima, M., Kangawa, K., and Sasaki, S. (2003). The role of ghrelin and growth hormone secretagogues receptor on rat adipogenesis. *Endocrinology* **144,** 754–759.

Cota, D., Marsicano, G., Tschop, M., Grubler, Y., Flachskamm, C., Schubert, M., Auer, D., Yassouridis, A., Thone-Reineke, C., Ortmann, S., Tomassoni, F., Cervino, C., et al. (2003). The endogenous cannabinoid system affects energy balance via central orexigenic drive and peripheral lipogenesis. *J. Clin. Invest.* **112,** 423–431.

Cote, M., Matias, I., Lemieux, I., Petrosino, S., Almeras, N., Despres, J. P., and Di Marzo, V. (2007). Circulating endocannabinoid levels, abdominal adiposity and related cardiometabolic risk factors in obese men. *Int. J. Obes. (Lond.)* **31**(4), 692–699.

Couce, M. E., Cottam, D., Esplen, J., Teijeiro, R., Schauer, P., and Burguera, B. (2006). Potential role of hypothalamic ghrelin in the pathogenesis of human obesity. *J. Endocrinol. Invest.* **29,** 599–605.

Crute, B. E., Seefeld, K., Gamble, J., Kemp, B. E., and Witters, L. A. (1998). Functional domains of the alpha1 catalytic subunit of the AMP-activated protein kinase. *J. Biol. Chem.* **273,** 35347–35354.

Cummings, D. E. (2006). Ghrelin and the short- and long-term regulation of appetite and body weight. *Physiol. Behav.* **89,** 71–84.

Cummings, D. E., and Shannon, M. H. (2003). Roles for ghrelin in the regulation of appetite and body weight. *Arch. Surg.* **138,** 389–396.

Cummings, D. E., Purnell, J. Q., Frayo, R. S., Schmidova, K., Wisse, B. E., and Weigle, D. S. (2001). A preprandial rise in plasma ghrelin levels suggests a role in meal initiation in humans. *Diabetes* **50,** 1714–1719.

Cummings, D. E., Frayo, R. S., Marmonier, C., Aubert, R., and Chapelot, D. (2004). Plasma ghrelin levels and hunger scores in humans initiating meals voluntarily without time- and food-related cues. *Am. J. Physiol. Endocrinol. Metab.* **287,** E297–E304.

Date, Y., Murakami, N., Toshinai, K., Matsukura, S., Niijima, A., Matsuo, H., Kangawa, K., and Nakazato, M. (2002). The role of the gastric afferent vagal nerve in ghrelin-induced feeding and growth hormone secretion in rats. *Gastroenterology* **123,** 1120–1128.

Daval, M., Foufelle, F., and Ferre, P. (2006). Functions of AMP-activated protein kinase in adipose tissue. *J. Physiol.* **574,** 55–62.

Davies, S. P., Helps, N. R., Cohen, P. T., and Hardie, D. G. (1995). 5′-AMP inhibits dephosphorylation, as well as promoting phosphorylation, of the AMP-activated protein kinase. Studies using bacterially expressed human protein phosphatase-2C alpha and native bovine protein phosphatase-2AC. *FEBS Lett.* **377,** 421–425.

Davis, K. W., Wellman, P. J., and Clifford, P. S. (2007). Augmented cocaine conditioned place preference in rats pretreated with systemic ghrelin. *Regul. Pept.* **140**(3), 148–152.

De Vriese, C., Hacquebard, M., Gregoire, F., Carpentier, Y., and Delporte, C. (2007). Ghrelin interacts with human plasma lipoproteins. *Endocrinology* **148**(5), 2355–2362.

Demers, A., McNicoll, N., Febbraio, M., Servant, M., Marleau, S., Silverstein, R., and Ong, H. (2004). Identification of the growth hormone-releasing peptide binding site in CD36: A photoaffinity cross-linking study. *Biochem. J.* **382**, 417–424.

Despres, J. P., Golay, A., and Sjostrom, L. (2005). Effects of rimonabant on metabolic risk factors in overweight patients with dyslipidemia. *N. Eng. J. Med.* **353**, 2121–2134.

Devane, W. A., Hanus, L., Breuer, A., Pertwee, R. G., Stevenson, L. A., Griffin, G., Gibson, D., Mandelbaum, A., Etinger, A., and Mechoulam, R. (1992). Isolation and structure of a brain constituent that binds to the cannabinoid receptor. *Science* **258**, 1946–1949.

Di Marzo, V., Goparaju, S. K., Wang, L., Liu, J., Batkai, S., Jarai, Z., Fezza, F., Miura, G. I., Palmiter, R. D., Sugiura, T., and Kunos, G. (2001). Leptin-regulated endocannabinoids are involved in maintaining food intake. *Nature* **410**, 822–825.

Di Marzo, V., and Matias, L. (2005). Endocannabinoid control of food intake and energy balance. *Nat. Neurosci.* **8**, 585–589.

Diraison, F., Yankah, V., Letexier, D., Dusserre, E., Jones, P., and Beylot, M. (2003). Differences in the regulation of adipose tissue and liver lipogenesis by carbohydrates in humans. *J. Lipid Res.* **44**, 846–853.

Dyck, J. R., and Lopaschuk, G. D. (2002). Malonyl CoA control of fatty acid oxidation in the ischemic heart. *J. Mol. Cell. Cardiol.* **34**, 1099–1109.

Elbatarny, H. S., Netherton, S. J., Ovens, J. D., Ferguson, A. V., and Maurice, D. H. (2007). Adiponectin, ghrelin, and leptin differentially influence human platelet and human vascular endothelial cell functions: Implication in obesity-associated cardiovascular diseases. *Eur. J. Pharmacol.* **558**(1–3), 7–13.

Engeli, S., Bohnke, J., Feldpausch, M., Gorzelniak, K., Janke, J., Batkai, S., Pacher, P., Harvey-White, J., Luft, F. C., Sharma, A. M., and Jordan, J. (2005). Activation of the peripheral endocannabinoid system in human obesity. *Diabetes* **54**, 2838–2843.

English, P. J., Ghatei, M. A., Malik, I. A., Bloom, S. R., and Wilding, J. P. (2002). Food fails to suppress ghrelin levels in obese humans. *J. Clin. Endocrinol. Metab.* **87**, 2984–2987.

Faulconbridge, L. F., Cummings, D. E., Kaplan, J. M., and Grill, H. J. (2003). Hyperphagic effects of brainstem ghrelin administration. *Diabetes* **52**, 2260–2265.

Foretz, M., Ancellin, N., Andreelli, F., Saintillan, Y., Grondin, P., Kahn, A., Thorens, B., Vaulont, S., and Viollet, B. (2005). Short-term overexpression of a constitutively active form of AMP-activated protein kinase in the liver leads to mild hypoglycemia and fatty liver. *Diabetes* **54**, 1331–1339.

Foster-Schubert, K. E., McTiernan, A., Frayo, R. S., Schwartz, R. S., Rajan, K. B., Yasui, Y., Tworoger, S. S., and Cummings, D. E. (2005). Human plasma ghrelin levels increase during a one-year exercise program. *J. Clin. Endocrinol. Metab.* **90**, 820–825.

Frascarelli, S., Ghelardoni, S., Ronca-Testoni, S., and Zucchi, R. (2003). Effect of ghrelin and synthetic growth hormone secretagogues in normal and ischemic rat heart. *Basic Res. Cardiol.* **98**, 401–405.

Fryer, L. G., and Carling, D. (2005). AMP-activated protein kinase and the metabolic syndrome. *Biochem. Soc. Trans.* **33**, 362–366.

Gallate, J. E., and McGregor, I. S. (1999). The motivation for beer in rats: Effects of ritanserin, naloxone and SR 141716. *Psychopharmacology (Ber.)* **142**, 302–308.

Gallate, J. E., Saharov, T., Mallet, P. E., and McGregor, I. S. (1999). Increased motivation for beer in rats following administration of a cannabinoid CB1 receptor agonist. *Eur. J. Pharmacol.* **370**, 233–240.

Garcia, J. M., Iyer, D., Poston, W. S., Marcelli, M., Reeves, R., Foreyt, J., and Balasubramanyam, A. (2006). Rise of plasma ghrelin with weight loss is not sustained during weight maintenance. *Obesity (Silver Spring)* **14**, 1716–1723.

Gauna, C., Delhantyz, P. J., Hofland, L. J., Janssen, J. A., Broglio, F., Ross, R. J., Ghigo, E., and van der Lely, A. J. (2005). Ghrelin stimulates, whereas des-octanoyl ghrelin inhibits, glucose output by primary hepatocytes. *J. Clin. Endocrinol. Metab.* **90,** 1055–1060.

Gil-Campos, M., Aguilera, C. M., Canete, R., and Gil, A. (2006). Ghrelin: A hormone regulating food intake and energy homeostasis. *Br. J. Nutr.* **96,** 201–226.

Goldstone, A. P., Thomas, E. L., Brynes, A. E., Castroman, G., Edwards, R., Ghatei, M. A., Frost, G., Holland, A. J., Grossman, A. B., Korbonits, M., Bloom, S. R., and Bell, J. D. (2004). Elevated fasting plasma ghrelin in Prader-Willi syndrome adults is not solely explained by their reduced visceral adiposity and insulin resistance. *J. Clin. Endocrinol. Metab.* **89,** 1718–1726.

Hansen, T. K., Dall, R., Hosoda, H., Kojima, M., Kangawa, K., Christiansen, J. S., and Jorgensen, J. O. (2002). Weight loss increases circulating levels of ghrelin in human obesity. *Clin. Endocrinol. (Oxf.)* **56,** 203–206.

Hardie, D. G. (2004). The AMP-activated protein kinase pathway–new players upstream and downstream. *J. Cell Sci.* **117,** 5479–5487.

Hardie, D. G., and Hawley, S. A. (2001). AMP-activated protein kinase: The energy charge hypothesis revisited. *Bioessays* **23,** 1112–1119.

Hardie, D. G., Salt, I. P., Hawley, S. A., and Davies, S. P. (1999). AMP-activated protein kinase: An ultrasensitive system for monitoring cellular energy charge. *Biochem. J.* **338** (Pt. 3), 717–722.

Hawley, S. A., Davison, M., Woods, A., Davies, S. P., Beri, R. K., Carling, D., and Hardie, D. G. (1996). Characterization of the AMP-activated protein kinase kinase from rat liver and identification of threonine 172 as the major site at which it phosphorylates AMP-activated protein kinase. *J. Biol. Chem.* **271,** 27879–27887.

Hawley, S. A., Pan, D. A., Mustard, K. J., Ross, L., Bain, J., Edelman, A. M., Frenguelli, B. G., and Hardie, D. G. (2005). Calmodulin-dependent protein kinase kinase-beta is an alternative upstream kinase for AMP-activated protein kinase. *Cell Metab.* **2,** 9–19.

Heijboer, A. C., Pijl, H., Van den Hoek, A. M., Havekes, L. M., Romijn, J. A., and Corssmit, E. P. (2006). Gut-brain axis: Regulation of glucose metabolism. *J. Neuroendocrinol.* **18,** 883–894.

Howlett, A. C. (2002). The cannabinoid receptors. *Prostaglandins Other Lipid Mediat.* **68-69,** 619-631.

Hurley, R. L., Anderson, K. A., Franzone, J. M., Kemp, B. E., Means, A. R., and Witters, L. A. (2005). The Ca^{2+}/calmodulin-dependent protein kinase kinases are AMP-activated protein kinase kinases. *J. Biol. Chem.* **280,** 29060–29066.

Iglesias, M. J., Pineiro, R., Blanco, M., Gallego, M., Dieguez, C., Gualillo, O., Gonzalez-Juanatey, J. R., and Lago, F. (2004). Growth hormone releasing peptide (ghrelin) is synthesized and secreted by cardiomyocytes. *Cardiovasc. Res.* **62,** 481–488.

Jamshidi, N., and Taylor, D. A. (2001). Anandamide administration into the ventromedial hypothalamus stimulates appetite in rats. *Br. J. Pharmacol.* **134,** 1151–1154.

Janas-Kozik, M., Krupka-Matuszczyk, I., Malinowska-Kolodziej, I., and Lewin-Kowalik, J. (2007). Total ghrelin plasma level in patients with the restrictive type of anorexia nervosa. *Regul. Pept.* **140,** 43–46.

Katugampola, S. D., Pallikaros, Z., and Davenport, A. P. (2001). [125I-His(9)]-ghrelin, a novel radioligand for localizing GHS orphan receptors in human and rat tissue: Up-regulation of receptors with atherosclerosis. *Br. J. Pharmacol.* **134,** 143–149.

Kim, E. K., Miller, I., Aja, S., Landree, L. E., Pinn, M., McFadden, J., Kuhajda, F. P., Moran, T. H., and Ronnett, G. V. (2004). C75, a fatty acid synthase inhibitor, reduces food intake via hypothalamic AMP-activated protein kinase. *J. Biol. Chem.* **279,** 19970–19976.

Kim, S. J., Nian, C., and McIntosh, C. H. (2007). Activation of lipoprotein lipase (LPL) by glucose-dependent insulinotropic polypeptide (GIP) in adipocytes: A role for a protein kinase B (PKB), LKB1 and AMP-activated protein kinase (AMPK) cascade. *J. Biol. Chem.* **282,** 8557–8567.

Kirkham, T. C., Williams, C. M., Fezza, F., and Di Marzo, V. (2002). Endocannabinoid levels in rat limbic forebrain and hypothalamus in relation to fasting, feeding and satiation: Stimulation of eating by 2-arachidonoyl glycerol. *Br. J. Pharmacol.* **136,** 550–557.

Koch, J. E. (2001). Delta(9)-THC stimulates food intake in Lewis rats: Effects on chow, high-fat and sweet high-fat diets. *Pharmacol. Biochem. Behav.* **68,** 539–543.

Koh, H. J., Hirshman, M. F., He, H., Li, Y., Manabe, Y., Balschi, J. A., and Goodyear, L. J. (2007). Epinephrine is a critical mediator of acute exercise-induced AMP-activated protein kinase activation in adipocytes. *Biochem. J.* **403(3),** 473–481.

Kojima, M., Hosoda, H., Date, Y., Nakazato, M., Matsuo, H., and Kangawa, K. (1999). Ghrelin is a growth-hormone-releasing acylated peptide from stomach. *Nature* **402,** 656–660.

Kola, B., Hubina, E., Tucci, S. A., Kirkham, T. C., Garcia, E. A., Mitchell, S. E., Williams, L. M., Hawley, S. A., Hardie, D. G., Grossman, A. B., and Korbonits, M. (2005). Cannabinoids and ghrelin have both central and peripheral metabolic and cardiac effects via AMP-activated protein kinase. *J. Biol. Chem.* **280,** 25196–25201.

Korbonits, M., Goldstone, A. P., Gueorguiev, M., and Grossman, A. B. (2004). Ghrelin—a hormone with multiple functions. *Front. Neuroendocrinol.* **25,** 27–68.

Korbonits, M., Blaine, D., Elia, M., and Powell-Tuck, J. (2005). Refeeding David Blaine—studies after a 44-day fast. *N. Eng. J. Med.* **353,** 2306–2307.

Krsek, M., Rosicka, M., Haluzik, M., Svobodova, J., Kotrlikova, E., Justova, V., Lacinova, Z., and Jarkovska, Z. (2002). Plasma ghrelin levels in patients with short bowel syndrome. *Endocr. Res.* **28,** 27–33.

Lagneux, C., and Lamontagne, D. (2001). Involvement of cannabinoids in the cardioprotection induced by lipopolysaccharide. *Br. J. Pharmacol.* **132,** 793–796.

Lee, W. J., Koh, E. H., Won, J. C., Kim, M. S., Park, J. Y., and Lee, K. U. (2005). Obesity: The role of hypothalamic AMP-activated protein kinase in body weight regulation. *Int. J. Biochem. Cell Biol.* **37,** 2254–2259.

Li, H. L., Yin, R., Chen, D., Liu, D., Wang, D., Yang, Q., and Dong, Y. G. (2007). Long-term activation of adenosine monophosphate-activated protein kinase attenuates pressure-overload-induced cardiac hypertrophy. *J. Cell Biochem.* **100,** 1086–1099.

Locke, W., Kirgis, H. D., Bowers, C. Y., and Abdoh, A. A. (1995). Intracerebroventricular growth-hormone-releasing peptide-6 stimulates eating without affecting plasma growth hormone responses in rats. *Life Sci.* **56,** 1347–1352.

Marleau, S., Harb, D., Bujold, K., Avallone, R., Iken, K., Wang, Y., Demers, A., Sirois, M. G., Febbraio, M., Silverstein, R. L., Tremblay, A., and Ong, H. (2005). EP 80317, a ligand of the CD36 scavenger receptor, protects apolipoprotein E-deficient mice from developing atherosclerotic lesions. *FASEB J.* **19,** 1869–1871.

Marsin, A. S., Bouzin, C., Bertrand, L., and Hue, L. (2002). The stimulation of glycolysis by hypoxia in activated monocytes is mediated by AMP-activated protein kinase and inducible 6-phosphofructo-2-kinase. *J. Biol. Chem.* **277,** 30778–30783.

Matias, I., and Di Marzo, V. (2006). Endocannabinoid synthesis and degradation, and their regulation in the framework of energy balance. *J. Endocrinol. Invest.* **29,** 15–26.

Matias, I., Gonthier, M. P., Orlando, P., Martiadis, V., De Petrocellis, L., Cervino, C., Petrosino, S., Hoareau, L., Festy, F., Pasquali, R., Roche, R., Maj, M., *et al.* (2006). Regulation, function, and dysregulation of endocannabinoids in models of adipose and beta-pancreatic cells and in obesity and hyperglycemia. *J. Clin. Endocrinol. Metab.* **91,** 3171–3180.

Matsuda, L. A., Lolait, S. J., Brownstein, M. J., Young, A. C., and Bonner, T. I. (1990). Structure of a cannabinoid receptor and functional expression of the cloned cDNA. *Nature* **346**, 561–564.

Matsumura, K., Tsuchihashi, T., Fujii, K., Abe, I., and Iida, M. (2002). Central ghrelin modulates sympathetic activity in conscious rabbits. *Hypertension* **40**, 694–699.

McCrimmon, R. J., Fan, X., Ding, Y., Zhu, W., Jacob, R. J., and Sherwin, R. S. (2004). Potential role for AMP-activated protein kinase in hypoglycemia sensing in the ventromedial hypothalamus. *Diabetes* **53**, 1953–1958.

Mechoulam, R., Ben Shabat, S., Hanus, L., Ligumsky, M., Kaminski, N. E., Schatz, A. R., Gopher, A., Almog, S., Martin, B. R., Compton, D. R., Pertwee, R. G., Griffin, G., et al. (1995). Identification of an endogenous 2-monoglyceride, present in canine gut, that binds to cannabinoid receptors. *Biochem. Pharmacol.* **50**, 83–90.

Minokoshi, Y., Kim, Y. B., Peroni, O. D., Fryer, L. G., Muller, C., Carling, D., and Kahn, B. B. (2002). Leptin stimulates fatty-acid oxidation by activating AMP-activated protein kinase. *Nature* **415**, 339–343.

Minokoshi, Y., Alquier, T., Furukawa, N., Kim, Y. B., Lee, A., Xue, B., Mu, J., Foufelle, F., Ferre, P., Birnbaum, M. J., Stuck, B. J., and Kahn, B. B. (2004). AMP-kinase regulates food intake by responding to hormonal and nutrient signals in the hypothalamus. *Nature* **428**, 569–574.

Monteleone, P., Matias, I., Martiadis, V., De Petrocellis, L., Maj, M., and Di Marzo, V. (2005). Blood levels of the endocannabinoid anandamide are increased in anorexia nervosa and in binge-eating disorder, but not in bulimia nervosa. *Neuropsychopharmacology* **30**, 1216–1221.

Muccioli, G., Pons, N., Ghe, C., Catapano, F., Granata, R., and Ghigo, E. (2004). Ghrelin and des-acyl ghrelin both inhibit isoproterenol-induced lipolysis in rat adipocytes via a non-type 1a growth hormone secretagogue receptor. *Eur. J. Pharmacol.* **498**, 27–35.

Mulligan, J. D., Gonzalez, A. A., Stewart, A. M., Carey, H. V., and Saupe, K. W. (2007). Upregulation of AMPK during cold exposure occurs via distinct mechanism in brown and white adipose tissue. *J. Physiol.* **580**(2), 677–684.

Murakami, N., Hayashida, T., Kuroiwa, T., Nakahara, K., Ida, T., Mondal, M. S., Nakazato, M., Kojima, M., and Kangawa, K. (2002). Role for central ghrelin in food intake and secretion profile of stomach ghrelin in rats. *J. Endocrinol.* **174**, 283–288.

Murata, M., Okimura, Y., Iida, K., Matsumoto, M., Sowa, H., Kaji, H., Kojima, M., Kangawa, K., and Chihara, K. (2002a). Ghrelin modulates the downstream molecules of insulin signaling in hepatoma cells. *J. Biol. Chem.* **277**, 5667–5674.

Murata, M., Okimura, Y., Iida, K., Matsumoto, M., Sowa, H., Kaji, H., Kojima, M., Kangawa, K., and Chihara, K. (2002b). Ghrelin modulates the downstream molecules of insulin signaling in hepatoma cells. *J. Biol. Chem.* **277**, 5667–5674.

Nagaya, N., Kojima, M., Uematsu, M., Yamagishi, M., Hosoda, H., Oya, H., Hayashi, Y., and Kangawa, K. (2001). Hemodynamic and hormonal effects of human ghrelin in healthy volunteers. *Am. J. Physiol. Regul. Integr. Comp. Physiol.* **280**, R1483–R1487.

Nakazato, M., Murakami, N., Date, Y., Kojima, M., Matsuo, H., Kangawa, K., and Matsukura, S. (2001). A role for ghrelin in the central regulation of feeding. *Nature* **409**, 194–198.

Oka, S., Tsuchie, A., Tokumura, A., Muramatsu, M., Suhara, Y., Takayama, H., Waku, K., and Sugiura, T. (2003). Ether-linked analogue of 2-arachidonoylglycerol (noladin ether) was not detected in the brains of various mammalian species. *J. Neurochem.* **85**, 1374–1381.

Okada, K., Ishii, S., Minami, S., Sugihara, H., Shibasaki, T., and Wakabayashi, I. (1996). Intracerebroventricular administration of the growth hormone-releasing peptide KP-102 increases food intake in free-feeding rats. *Endocrinology* **137**, 5155–5158.

Okumura, H., Nagaya, N., Enomoto, M., Nakagawa, E., Oya, H., and Kangawa, K. (2002). Vasodilatory effect of ghrelin, an endogenous peptide from the stomach. *J. Cardiovasc. Pharmacol.* **39,** 779–783.

Osei-Hyiaman, D., Depetrillo, M., Pacher, P., Liu, J., Radaeva, S., Batkai, S., Harvey-White, J., Mackie, K., Offertaler, L., Wang, L., and Kunos, G. (2005). Endocannabinoid activation at hepatic CB1 receptors stimulates fatty acid synthesis and contributes to diet-induced obesity. *J. Clin. Invest.* **115,** 1298–1305.

Otto, B., Cuntz, U., Fruehauf, E., Wawarta, R., Folwaczny, C., Riepl, R. L., Heiman, M. L., Lehnert, P., Fichter, M., and Tschop, M. (2001). Weight gain decreases elevated plasma ghrelin concentrations of patients with anorexia nervosa. *Eur. J. Endocrinol.* **145,** 669–673.

Overduin, J., Frayo, R. S., Grill, H. J., Kaplan, J. M., and Cummings, D. E. (2005). Role of the duodenum and macronutrient type in ghrelin regulation. *Endocrinology* **146,** 845–850.

Pacher, P., Batkai, S., and Kunos, G. (2006). The endocannabinoid system as an emerging target of pharmacotherapy. *Pharmacol. Rev.* **58,** 389–462.

Pemberton, C. J., Tokola, H., Bagi, Z., Koller, A., Pontinen, J., Ola, A., Vuolteenaho, O., Szokodi, I., and Ruskoaho, H. (2004). Ghrelin induces vasoconstriction in the rat coronary vasculature without altering cardiac peptide secretion. *Am. J. Physiol. Heart Circ. Physiol.* **287,** H1522–H1529.

Penicaud, L., Leloup, C., Fioramonti, X., Lorsignol, A., and Benani, A. (2006). Brain glucose sensing: A subtle mechanism. *Curr. Opin. Clin. Nutr. Metab. Care* **9,** 458–462.

Perreault, M., Istrate, N., Wang, L., Nichols, A. J., Tozzo, E., and Stricker-Krongrad, A. (2004). Resistance to the orexigenic effect of ghrelin in dietary-induced obesity in mice: Reversal upon weight loss. *Int. J. Obes. Relat. Metab. Disord.* **28,** 879–885.

Pettersson, I., Muccioli, G., Granata, R., Deghenghi, R., Ghigo, E., Ohlsson, C., and Isgaard, J. (2002). Natural (ghrelin) and synthetic (hexarelin) GH secretagogues stimulate H9c2 cardiomyocyte cell proliferation. *J. Endocrinol.* **175,** 201–209.

Pi-Sunyer, F. X., Aronne, L. J., Heshmati, H. M., Devin, J., and Rosenstock, J. (2006). Effect of rimonabant, a cannabinoid-1 receptor blocker, on weight and cardiometabolic risk factors in overweight or obese patients: RIO-North America: A randomized controlled trial. *JAMA* **295,** 761–775.

Poykko, S. M., Kellokoski, E., Ukkola, O., Kauma, H., Paivansalo, M., Kesaniemi, Y. A., and Horkko, S. (2006). Plasma ghrelin concentrations are positively associated with carotid artery atherosclerosis in males. *J. Intern. Med.* **260,** 43–52.

Purnell, J. Q., Weigle, D. S., Breen, P. A., and Cummings, D. E. (2003). Ghrelin levels correlate with insulin levels, insulin resistance, and high-density lipoprotein cholesterol, but not with gender, menopausal status, or cortisol levels in humans. *J. Clin. Endocrinol. Metab.* **88,** 5747–5752.

Ravussin, E., Tschop, M., Morales, S., Bouchard, C., and Heiman, M. L. (2001). Plasma ghrelin concentration and energy balance: Overfeeding and negative energy balance studies in twins. *J. Clin. Endocrinol. Metab.* **86,** 4547–4551.

Russell, R. R., III, Li, J., Coven, D. L., Pypaert, M., Zechner, C., Palmeri, M., Giordano, F. J., Mu, J., Birnbaum, M. J., and Young, L. H. (2004). AMP-activated protein kinase mediates ischemic glucose uptake and prevents postischemic cardiac dysfunction, apoptosis, and injury. *J. Clin. Invest.* **114,** 495–503.

Sanchis-Segura, C., Cline, B. H., Marsicano, G., Lutz, B., and Spanagel, R. (2004). Reduced sensitivity to reward in CB1 knockout mice. *Psychopharmacology (Berl.)* **176,** 223–232.

Scheen, A. J., Finer, N., Hollander, P., Jensen, M. D., and Van Gaal, L. F. (2006). Efficacy and tolerability of rimonabant in overweight or obese patients with type 2 diabetes: A randomised controlled study. *Lancet* **368,** 1660–1672.

Schutte, A. E., Huisman, H. W., Schutte, R., van Rooyen, J. M., Malan, L., and Malan, N. T. (2007). Aging influences the level and functions of fasting plasma ghrelin levels: The POWIRS-Study. *Regul. Pept.* **139,** 65–71.

Shibasaki, T., Yamauchi, N., Takeuchi, K., Ishii, S., Sugihara, H., and Wakabayashi, I. (1998). The growth hormone secretagogue KP-102-induced stimulation of food intake is modified by fasting, restraint stress, and somatostatin in rats. *Neurosci. Lett.* **255,** 9–12.

Shiiya, T., Nakazato, M., Mizuta, M., Date, Y., Mondal, M. S., Tanaka, M., Nozoe, S., Hosoda, H., Kangawa, K., and Matsukura, S. (2002). Plasma ghrelin levels in lean and obese humans and the effect of glucose on ghrelin secretion. *J. Clin. Endocrinol. Metab.* **87,** 240–244.

Shuto, Y., Shibasaki, T., Otagiri, A., Kuriyama, H., Ohata, H., Tamura, H., Kamegai, J., Sugihara, H., Oikawa, S., and Wakabayashi, I. (2002). Hypothalamic growth hormone secretagogue receptor regulates growth hormone secretion, feeding, and adiposity. *J. Clin. Invest.* **109,** 1429–1436.

Simiand, J., Keane, M., Keane, P. E., and Soubrie, P. (1998). SR 141716, a CB1 cannabinoid receptor antagonist, selectively reduces sweet food intake in marmoset. *Behav. Pharmacol.* **9,** 179–181.

Sipe, J. C., Waalen, J., Gerber, A., and Beutler, E. (2005). Overweight and obesity associated with a missense polymorphism in fatty acid amide hydrolase (FAAH). *Int. J. Obes. (Lond.)* **29,** 755–759.

Steffens, S., and Mach, F. (2006). Cannabinoid receptors in atherosclerosis. *Curr. Opin. Lipidol.* **17,** 519–526.

Steffens, S., Veillard, N. R., Arnaud, C., Pelli, G., Burger, F., Staub, C., Karsak, M., Zimmer, A., Frossard, J. L., and Mach, F. (2005). Low dose oral cannabinoid therapy reduces progression of atherosclerosis in mice. *Nature* **434,** 782–786.

Sturm, K., MacIntosh, C. G., Parker, B. A., Wishart, J., Horowitz, M., and Chapman, I. M. (2003). Appetite, food intake, and plasma concentrations of cholecystokinin, ghrelin, and other gastrointestinal hormones in undernourished older women and well-nourished young and older women. *J. Clin. Endocrinol. Metab.* **88,** 3747–3755.

Sugiura, T., Kondo, S., Sukagawa, A., Nakane, S., Shinoda, A., Itoh, K., Yamashita, A., and Waku, K. (1995). 2-Arachidonoylglycerol: A possible endogenous cannabinoid receptor ligand in brain. *Biochem. Biophys. Res. Commun.* **215,** 89–97.

Sun, Y., Ahmed, S., and Smith, R. G. (2003). Deletion of ghrelin impairs neither growth nor appetite. *Mol. Cell. Biol.* **23,** 7973–7981.

Sun, Y., Wang, P., Zheng, H., and Smith, R. G. (2004). Ghrelin stimulation of growth hormone release and appetite is mediated through the growth hormone secretagogue receptor. *Proc. Natl. Acad. Sci. USA* **101,** 4679–4684.

Teff, K. L., Elliott, S. S., Tschop, M., Kieffer, T. J., Rader, D., Heiman, M., Townsend, R. R., Keim, N. L., D'Alessio, D., and Havel, P. J. (2004). Dietary fructose reduces circulating insulin and leptin, attenuates postprandial suppression of ghrelin, and increases triglycerides in women. *J. Clin. Endocrinol. Metab.* **89,** 2963–2972.

Thompson, N. M., Gill, D. A., Davies, R., Loveridge, N., Houston, P. A., Robinson, I. C., and Wells, T. (2004). Ghrelin and des-octanoyl ghrelin promote adipogenesis directly *in vivo* by a mechanism independent of the type 1a growth hormone secretagogue receptor. *Endocrinology* **145,** 234–242.

Torsello, A., Luoni, M., Schweiger, F., Grilli, R., Guidi, M., Bresciani, E., Deghenghi, R., Muller, E. E., and Locatelli, V. (1998). Novel hexarelin analogs stimulate feeding in the rat through a mechanism not involving growth hormone release. *Eur. J. Pharmacol.* **360,** 123–129.

Tschop, M., Smiley, D. L., and Heiman, M. L. (2000). Ghrelin induces adiposity in rodents. *Nature* **407,** 908–913.

Tschop, M., Wawarta, R., Riepl, R. L., Friedrich, S., Bidlingmaier, M., Landgraf, R., and Folwaczny, C. (2001a). Post-prandial decrease of circulating human ghrelin levels. *J. Endocrinol. Invest.* **24,** RC19–RC21.

Tschop, M., Weyer, C., Tataranni, P. A., Devanarayan, V., Ravussin, E., and Heiman, M. L. (2001b). Circulating ghrelin levels are decreased in human obesity. *Diabetes* **50,** 707–709.

Tucci, S. A., Rogers, E. K., Korbonits, M., and Kirkham, T. C. (2004). The cannabinoid CB1 receptor antagonist SR141716 blocks the orexigenic effects of intrahypothalamic ghrelin. *Br. J. Pharmacol.* **143,** 520–523.

Tups, A., Helwig, M., Khorooshi, R. M., Archer, Z. A., Klingenspor, M., and Mercer, J. G. (2004). Circulating ghrelin levels and central ghrelin receptor expression are elevated in response to food deprivation in a seasonal mammal *(Phodopus sungorus). J. Neuroendocrinol.* **16,** 922–928.

Underdown, N. J., Hiley, C. R., and Ford, W. R. (2005). Anandamide reduces infarct size in rat isolated hearts subjected to ischaemia-reperfusion by a novel cannabinoid mechan-ism. *Br. J. Pharmacol.* **146,** 809–816.

Van Gaal, L. F., Rissanen, A. M., Scheen, A. J., Ziegler, O., and Rossner, S. (2005). Effects of the cannabinoid-1 receptor blocker rimonabant on weight reduction and cardiovascular risk factors in overweight patients: 1-year experience from the RIO-Europe study. *Lancet* **365,** 1389–1397.

Van Sickle, M. D., Duncan, M., Kingsley, P. J., Mouihate, A., Urbani, P., Mackie, K., Stella, N., Makriyannis, A., Piomelli, D., Davison, J. S., Marnett, L. J., Di Marzo, V., et al. (2005). Identification and functional characterization of brainstem cannabinoid CB2 receptors. *Science* **310,** 329–332.

Vavvas, D., Apazidis, A., Saha, A. K., Gamble, J., Patel, A., Kemp, B. E., Witters, L. A., and Ruderman, N. B. (1997). Contraction-induced changes in acetyl-CoA carboxylase and 5′-AMP-activated kinase in skeletal muscle. *J. Biol. Chem.* **272,** 13255–13261.

Walsh, D., Nelson, K. A., and Mahmoud, F. A. (2003). Established and potential therapeutic applications of cannabinoids in oncology. *Support Care Cancer* **11,** 137–143.

Weekers, F., Van Herck, E., Isgaard, J., and Van den, B. G. (2000). Pretreatment with growth hormone-releasing peptide-2 directly protects against the diastolic dysfunction of myocardial stunning in an isolated, blood-perfused rabbit heart model. *Endocrinology* **141,** 3993–3999.

Weigle, D. S., Cummings, D. E., Newby, P. D., Breen, P. A., Frayo, R. S., Matthys, C. C., Callahan, H. S., and Purnell, J. Q. (2003). Roles of leptin and ghrelin in the loss of body weight caused by a low fat, high carbohydrate diet. *J. Clin. Endocrinol. Metab.* **88,** 1577–1586.

Wiley, K. E., and Davenport, A. P. (2002). Comparison of vasodilators in human internal mammary artery: Ghrelin is a potent physiological antagonist of endothelin-1. *Br. J. Pharmacol.* **136,** 1146–1152.

Williams, C. M., and Kirkham, T. C. (2002). Observational analysis of feeding induced by Delta9-THC and anandamide. *Physiol. Behav.* **76,** 241–250.

Williams, D. L., Grill, H. J., Cummings, D. E., and Kaplan, J. M. (2006). Overfeeding-induced weight gain suppresses plasma ghrelin levels in rats. *J. Endocrinol. Invest.* **29,** 863–868.

Winder, W. W., and Hardie, D. G. (1996). Inactivation of acetyl-CoA carboxylase and activation of AMP-activated protein kinase in muscle during exercise. *Am. J. Physiol.* **270,** E299–E304.

Witczak, C. A., Fujii, N., Hirshman, M. F., and Goodyear, L. J. (2007). CaMKK{alpha} regulates skeletal muscle glucose uptake independent of AMPK and Akt activation. *Diabetes* **56**(5), 1403–1409.

Woods, A., Dickerson, K., Heath, R., Hong, S. P., Momcilovic, M., Johnstone, S. R., Carlson, M., and Carling, D. (2005). C(Ca2+)/calmodulin-dependent protein kinase kinase-beta acts upstream of AMP-activated protein kinase in mammalian cells. *Cell Metab.* **2,** 21–33.

Wortley, K. E., Anderson, K. D., Garcia, K., Murray, J. D., Malinova, L., Liu, R., Moncrieffe, M., Thabet, K., Cox, H. J., Yancopoulos, G. D., Wiegand, S. J., and Sleeman, M. W. (2004). Genetic deletion of ghrelin does not decrease food intake but influences metabolic fuel preference. *Proc. Natl. Acad. Sci. USA* **101,** 8227–8232.

Wortley, K. E., del Rincon, J. P., Murray, J. D., Garcia, K., Iida, K., Thorner, M. O., and Sleeman, M. W. (2005). Absence of ghrelin protects against early-onset obesity. *J. Clin. Invest.* **115,** 3573–3578.

Wren, A. M., Small, C. J., Ward, H. L., Murphy, K. G., Dakin, C. L., Taheri, S., Kennedy, A. R., Roberts, G. H., Morgan, D. G., Ghatei, M. A., and Bloom, S. R. (2000). The novel hypothalamic peptide ghrelin stimulates food intake and growth hormone secretion. *Endocrinology* **141,** 4325–4328.

Wren, A. M., Small, C. J., Thomas, E. L., Abbott, C. R., Ghatei, M. A., Bell, J. D., and Bloom, S. R. (2004). Continuous subcutaneous administration of ghrelin results in accumulation of adipose tissue, independent of hyperphagiaor bodyweight gain. *Endocrine Abstracts Spring,* OC35 (Abstract).

Yin, W., Mu, J., and Birnbaum, M. J. (2003). Role of AMP-activated protein kinase in cyclic AMP-dependent lipolysis In 3T3-L1 adipocytes. *J. Biol. Chem.* **278,** 43074–43080.

Zhou, G., Myers, R., Li, Y., Chen, Y., Shen, X., Fenyk-Melody, J., Wu, M., Ventre, J., Doebber, T., Fujii, N., Musi, N., Hirshman, M. F., *et al.* (2001). Role of AMP-activated protein kinase in mechanism of metformin action. *J. Clin. Invest.* **108,** 1167–1174.

Zong, H., Ren, J. M., Young, L. H., Pypaert, M., Mu, J., Birnbaum, M. J., and Shulman, G. I. (2002). AMP kinase is required for mitochondrial biogenesis in skeletal muscle in response to chronic energy deprivation. *Proc. Natl. Acad. Sci. USA* **99,** 15983–15987.

CHAPTER SEVEN

Ghrelin and Feedback Systems

Katsunori Nonogaki*

Contents

I. Introduction	150
II. Regulation of Ghrelin Secretion	151
A. Starvation and feeding	151
B. Efferent vagus nerve	152
C. Body weight	152
D. Hormones and peptides	153
E. Brain neurotransmitters	157
F. Gastric bypass	158
III. Afferent Pathways of Ghrelin from the Stomach to the Hypothalamus	158
IV. Hyperphagia, Obesity, and Des-acyl Ghrelin	159
V. Hypothalamic Gene Expression and Plasma Des-acyl Ghrelin	161
VI. Conclusion	161
Acknowledgments	162
References	162

Abstract

Ghrelin is produced primarily in the stomach in response to hunger, and circulates in the blood. Plasma ghrelin levels increase during fasting and decrease after ingesting glucose and lipid, but not protein. The efferent vagus nerve contributes to the fasting-induced increase in ghrelin secretion. Ghrelin secreted by the stomach stimulates the afferent vagus nerve and promotes food intake. Ghrelin also stimulates pituitary gland secretion of growth hormone (GH) via the afferent vagus nerve. GH inhibits stomach ghrelin secretion. These findings indicate that the vagal circuit between the central nervous system and stomach has a crucial role in regulating plasma ghrelin levels. Moreover, body mass index modulates plasma ghrelin levels. In a lean state and anorexia nervosa, plasma ghrelin levels are increased, whereas in obesity, except in Prader–Willi syndrome, plasma ghrelin levels are decreased and the feeding- and sleeping-induced decline in plasma ghrelin levels is disrupted. There are two forms of ghrelin: active

* Center of Excellence, Division of Molecular Metabolism and Diabetes, Tohoku University Graduate School of Medicine, Miyagi 980-8575, Japan

n-octanoyl-modified ghrelin and des-acyl ghrelin. Fasting increases both ghrelin types compared with the fed state. Hyperphagia and obesity are likely to decrease plasma des-acyl ghrelin, but not n-octanoyl-modified ghrelin levels. Hypothalamic serum and glucocorticoid-inducible kinase-1 and serotonin 5-HT2C/1B receptor gene expression levels are likely to be proportional to plasma des-acyl ghrelin levels during fasting, whereas they are likely to be inversely proportional to plasma desacyl ghrelin levels in an increased energy storage state such as obesity. Thus, a dysfunction of the ghrelin feedback systems might contribute to the pathophysiology of obesity and eating disorders. © 2008 Elsevier Inc.

I. Introduction

Although ghrelin is distributed in a variety of tissues, including the pancreas, brain, kidney, testis, and placenta, it is produced primarily in the stomach in response to hunger and starvation, and circulates in the blood, serving as a peripheral signal to the central nervous system (CNS) to stimulate feeding (Ariyasu *et al.*, 2001; Barreiro *et al.*, 2002a,b; Date *et al.*, 2002a; Gnanapavan *et al.*, 2002; Kim *et al.*, 2003; Kojima *et al.*, 1999; Korbonits *et al.*, 2001a,b; Lu *et al.*, 2002; Prado *et al.*, 2004). In the stomach, ghrelin-containing cells are more abundant in the fundus than in the pylorus (Date *et al.*, 2000; Tomasetto *et al.*, 2001; Tschöp *et al.*, 2001a). *In situ* hybridization and immunohistochemical analyses indicate that ghrelin-containing cells are a distinct endocrine cell type found in the mucosal layer of the stomach (Date *et al.*, 2000; Rindi *et al.*, 2002). Although four types of endocrine cells have been identified in the oxyntic mucosa with the following relative abundance: enterochromaffin-like, D, enterochromaffin, and X/A-like cells, the X/A-like cells are filled with ghrelin (Date *et al.*, 2000; Dornonville de la Cour *et al.*, 2001; Yabuki *et al.*, 2004). Ghrelin in the secretory granules of the X/A-like cells is acyl-modified. There are also ghrelin-immunoreactive cells in the duodenum, jejunum, ileum, and colon (Date *et al.*, 2000; Hosoda *et al.*, 2000; Sakata *et al.*, 2002). In the intestine, the ghrelin concentration gradually decreases from the duodenum to the colon. As in the stomach, the main

Table 1 Regulators of ghrelin secretion

Stimulators of ghrelin secretion
  Fasting, lean states, the parasympathetic nervous system, testosterone, growth hormone releasing hormone
Inhibitors of ghrelin secretion
  Feeding (glucose, lipids), obesity, insulin, growth hormone, growth hormone releasing peptide 6, 5-HT, LPS, interleukin-1, leptin, somatostatin, urocortin-1, thyroid hormone, cholecystokinin, Peptide YY, gastric bypass, melatonin

molecular forms of intestinal ghrelin are n-octanoyl ghrelin and des-acyl ghrelin (Date *et al.*, 2000).

A variety of substances in addition to feeding state can modulate plasma ghrelin levels (Table 1). In this chapter, we provide recent evidences and new concept in the regulation of ghrelin secretion by the stomach and feedback systems between the stomach and the CNS *in vivo*.

II. REGULATION OF GHRELIN SECRETION

A. Starvation and feeding

Plasma ghrelin levels increase during fasting and decrease after feeding. Plasma ghrelin levels increase immediately before each meal and fall to minimum levels within 1 h after eating (Ariyasu *et al.*, 2001; Cummings *et al.*, 2001; Faulconbridge *et al.*, 2003; Shiiya *et al.*, 2002; Tschöp *et al.*, 2001b). Oral or intravenous glucose administration decreases plasma ghrelin levels (Greenman *et al.*, 2004; McCowen *et al.*, 2002; Shiiya *et al.*, 2002). The preprandial increase in ghrelin levels occurs in humans that initiate meals voluntarily without any time- or food-related cues (Cummings *et al.*, 2002a). Indeed, ghrelin levels and hunger scores are positively correlated (Cummings *et al.*, 2004). Furthermore, the postprandial suppression of plasma ghrelin levels is proportional to the ingested caloric load (Callahan *et al.*, 2004). These findings suggest that ghrelin is an initiation signal for meal consumption.

Because gastric distension by water intake does not change plasma ghrelin levels, physical expansion of the stomach does not induce ghrelin release (Asakawa *et al.*, 2001; Cummings *et al.*, 2001; Shiiya *et al.*, 2002; Tschöp *et al.*, 2001a). Plasma ghrelin levels decrease after a high-fat meal and oral lipid load (Erdmann *et al.*, 2003; Greenman *et al.*, 2004), whereas they do not change after an oral protein load (Greenman *et al.*, 2004). These findings indicate that plasma ghrelin levels increase during fasting, whereas they decrease after ingesting glucose and lipid, but not protein.

Ghrelin levels do not change after intragastric saline or glucose infusion in rats, if gastric emptying is prevented using a pyloric cuff (Williams *et al.*, 2003a). Ghrelin levels decrease if intragastric glucose is administered with normal gastric emptying, whereas saline has no effect. These data suggest that the meal-related ghrelin decline does not involve gastric distension or gastric chemosensing, but requires a postgastric factor.

Prolonged fasting for 3 days does not change ghrelin levels compared to the baseline state (Chan *et al.*, 2004). A high-fat diet decreases plasma ghrelin levels, whereas a low-protein diet increases plasma ghrelin levels in rats (Beck *et al.*, 2002; Lee *et al.*, 2002). In humans, a high-carbohydrate diet causes a larger drop in ghrelin levels than a high-fat diet (Monteleone *et al.*, 2003). In humans, a physiologic dose of orally administered essential amino acids or a protein meal does not cause a drop in ghrelin levels

(Greenman *et al.*, 2004; Knerr *et al.*, 2003). Ghrelin levels increase during sleep in humans, and this increase is blunted in obese subjects or by sleep deprivation (Dzaja *et al.*, 2004; Yildiz *et al.*, 2004).

B. Efferent vagus nerve

The contribution of the vagus nerve to the regulation of plasma ghrelin levels was recently evaluated. Vagotomy affects neither baseline ghrelin levels nor the suppression of ghrelin levels by nutrient load (Williams *et al.*, 2003b). The food deprivation-induced plasma ghrelin level increase, however, is completely prevented by subdiaphragmatic vagotomy and is substantially reduced by atropine treatment (Williams *et al.*, 2003b). These findings indicate that the efferent vagus nerve contributes to the increase in ghrelin secretion during fasting, whereas it does not contribute to the suppression of ghrelin levels after feeding (Fig. 1).

C. Body weight

Plasma ghrelin levels are low in obese people, other than those with Prader–Willi syndrome, and high in lean people (Bellone *et al.*, 2002; Cummings *et al.*, 2002b; DelParigi *et al.*, 2002; Hansen *et al.*, 2002; Haqq *et al.*, 2003a,b; Rosicka *et al.*, 2003; Shiiya *et al.*, 2002; Tschöp *et al.*, 2001b). Despite their obesity with hyperleptinemia, subjects with Prader–Willi syndrome have elevated plasma ghrelin levels. Plasma ghrelin levels are highly increased in patients with anorexia nervosa and return to normal levels on weight gain and recovery from the disease (Ariyasu *et al.*, 2001; Cuntz *et al.*, 2002;

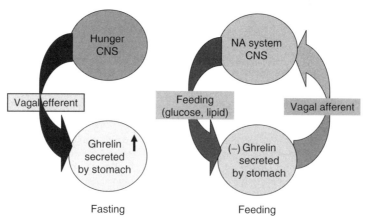

Figure 1 Fasting and feeding-induced ghrelin regulation. The efferent vagus nerve contributes to stomach ghrelin secretion induced by fasting, whereas the afferent vagus nerve contributes to convey orexigenic ghrelin signals from the stomach to the CNS. Hypothalamic noradrenergic systems may contribute to promote feeding induced by peripheral ghrelin. CNS, central nervous system; NA, noradrenergic; (−), inhibition.

Otto *et al.*, 2001; Ravussin *et al.*, 2001; Soriano-Guillen *et al.*, 2004; Tanaka *et al.*, 2003). Ghrelin levels in the cerebrospinal fluid are negatively correlated with body mass index (Tritos *et al.*, 2003). Ghrelin levels in obese subjects increase when they lose weight (Hansen *et al.*, 2002). Plasma ghrelin levels are lower in patients with insulin resistance, such as those with polycystic ovarian syndrome or type 2 diabetes mellitus (Pagotto *et al.*, 2002; Pöykkö *et al.*, 2003; Schöfl *et al.*, 2002). Patients with gastric bypass lose weight, and their plasma ghrelin levels decrease (Cummings *et al.*, 2002a; Geloneze *et al.*, 2003; Leonetti *et al.*, 2003). Changes in plasma ghrelin levels associated with food intake are diminished in these patients. Plasma ghrelin levels also decrease in patients with short bowel syndrome (Krsek *et al.*, 2003).

D. Hormones and peptides

1. Growth hormone

Exogenous growth hormone (GH) decreases stomach ghrelin mRNA expression and plasma ghrelin levels, but does not affect stomach ghrelin stores (Qi *et al.*, 2003). Intravenous and intracerebroventricular ghrelin injections stimulate pituitary gland GH release in rats and healthy humans (Date *et al.*, 2002b; Hataya *et al.*, 2001; Takaya *et al.*, 2000). Chemical and surgical blockade of the afferent gastric vagus nerve attenuates pituitary gland GH secretion induced by the peripheral administration of ghrelin (Date *et al.*, 2002). These findings suggest that there is an inhibitory feedback loop between pituitary GH and stomach ghrelin production, and that the afferent vagus nerve contributes to ghrelin-induced GH secretion (Fig. 2A).

2. Pancreatic hormones

a. Insulin Insulin inhibits ghrelin levels (Lucidi *et al.*, 2002; Saad *et al.*, 2002). Ghrelin levels are lowered by a euglycemic hyperinsulinemic clamp, suggesting that insulin responsiveness is a major regulator of ghrelin level (McCowen *et al.*, 2002). In humans, ghrelin levels decrease during graded hyperinsulinemic euglycemic clamp conditions (Anderwald *et al.*, 2003). Hyperinsulinemic clamping suppresses ghrelin levels in the euglycemic state as well as in hyper- and hypoglycemic states, despite a rise in GH levels during the hypoglycemic state (Flanagan *et al.*, 2003; Schaller *et al.*, 2003). Other studies, however, showed that insulin does not decrease circulating ghrelin levels in human (Caixas *et al.*, 2002).

In subjects with type 2 diabetes, a 5-g glucose bolus did not change insulin levels in the first 10 min, whereas ghrelin levels dropped to a similar extent as those in control subjects, suggesting that this effect is insulin independent (Briatore *et al.*, 2003). In another human study, however, hyperglycemia did not change ghrelin levels (Schaller *et al.*, 2003). Patients with type 1 diabetes show no change in ghrelin levels after food intake (Holdstock *et al.*, 2003).

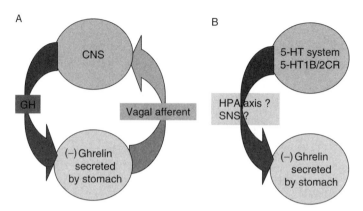

Figure 2 (A) GH-induced ghrelin regulation. The vagus nerve contributes to stimulate ghrelin and GH secretion. GH inhibits ghrelin secretion in the stomach: GH, growth hormone; CNS, central nervous system; (−), inhibition. (B) A possible feedback system between brain 5-HT systems and ghrelin secretion. Fasting increases hypothalamic 5-HT2C and 5-HT1B receptor gene expression. Pharmacologic activation of the 5-HT2C and 5-HT1B receptors inhibits the increases in plasma active ghrelin levels during fasting. 5-HT1B/2CR, 5-HT1B receptor and 5-HT2C receptor; HPA axis, hypothalamic-pituitary-adrenal axis; SNS, sympathetic nervous system.

b. Somatostatin Exogenous treatment with somatostatin (SST) and its analogues, such as octreotide, suppress plasma ghrelin levels (Barkan *et al.*, 2003; Haqq *et al.*, 2003a; Norrelund *et al.*, 2002; Tan *et al.*, 2004). Endogenous SST inhibits basal ghrelin gene expression in a tissue-specific manner and independently and directly inhibits pituitary ACTH synthesis and release. Thus, endogenous SST exerts an inhibitory effect on ghrelin synthesis and the adrenal axis through independent pathways (Luque, 2006). Endogenous SST, however, does not significantly contribute to fasting-induced increases in ghrelin (Luque, 2006).

Plasma ghrelin levels are increased in SST knockout mice (Sst−/−) compared with wild-type mice (Luque, 2006). Consistent with elevations in plasma ghrelin levels, Sst−/− mice have an increase in stomach ghrelin mRNA levels, whereas hypothalamic and pituitary expression of ghrelin is not altered (Luque, 2006). Despite the increase in total ghrelin levels, circulating levels of active n-octanoyl ghrelin are not altered in Sst−/− mice (Luque, 2006).

c. Glucagon Glucagon significantly decreases plasma ghrelin levels in healthy subjects (Arafat *et al.*, 2005). The inhibitory effect of glucagon on ghrelin could not be explained by changes in plasma glucose or insulin levels (Arafat *et al.*, 2005, 2006), and glucagon does not have direct inhibitory effects on ghrelin-producing cells in the stomach (Kamegai *et al.*, 2004). In subjects with hypothalamic-pituitary lesion and at least one pituitary hormone, such as GH, ACTH, and TSH, deficiency, the inhibitory effect of glucagon on ghrelin was abolished. These findings suggest that glucagon acts centrally to induce a reduction of plasma ghrelin levels in healthy subjects, and the hypothalamic-pituitary axis plays an essential role in the suppression of ghrelin induced by peripheral glucagon administration (Arafat et al., 2006).

3. Leptin

Gastric ghrelin mRNA and plasma ghrelin levels are increased in ob/ob mice, which are deficient in leptin, whereas they are decreased in db/db mice, which are deficient in the leptin receptor (Ariyasu et al., 2002; Toshinai et al., 2001; Fig. 3). Leptin administration decreases gastric ghrelin gene expression in ob/ob mice. In humans with leptin deficiency, ghrelin levels are low, appropriate to their body mass index (Haqq et al., 2003b). Leptin transgene expression in the rat hypothalamus increases circulating ghrelin levels (Bagnasco et al., 2002). Some studies, however, report opposite effects of leptin (Asakawa et al., 2003). Leptin administration in physiologic and pharmacologic doses to healthy volunteers does not affect ghrelin levels over several hours to a few days (Chan et al., 2004). These findings suggest that plasma leptin levels are inversely proportional to plasma ghrelin levels, although exogenous administration of leptin does not regulate ghrelin levels independently of changes in adiposity (Fig. 3).

4. Thyroid hormone

Plasma ghrelin levels are reduced in hyperthyroidism and are normalized by medical antithyroid treatment that leads to body weight gain (Riis et al., 2003). Hyperinsulinemia suppresses ghrelin regardless of thyroid status.

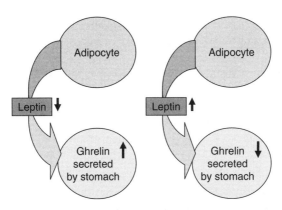

Figure 3 Endogenous leptin and ghrelin regulation. Plasma ghrelin levels are high in obese mice and human with leptin deficiency, whereas plasma ghrelin levels are low in obese mice and human with leptin receptor deficiency and high leptin levels other than subjects with Prader–Willi syndrome.

Ghrelin is not a primary stimulator of appetite and food intake in hyperthyroidism (Riis et al., 2003), and the mechanisms underlying the suppressive effect of hyperthyroidism on ghrelin secretion remain unclear. Changes in body weight in hyperthyroidism might contribute to alterations in ghrelin levels.

5. Cholecystokinin

Peripherally administered cholecystokinin (CCK) induces a decrease in meal size in rats and humans (Barkan et al., 2003). This peptide, released from the proximal small intestine, functions as a postprandial satiety signal (Basa et al., 2003; Batterham et al., 2002, 2003a,b). Although ghrelin and CCK have opposite effects on feeding, ghrelin exhibits characteristics similar to CCK in the short-term regulation of feeding. Both ghrelin and CCK, after release from the gastrointestinal tract, transmit starvation and satiety signals to the brain through the growth hormone secretagogue receptor 1a (GHS-R1a) and CCK type A receptors, respectively, located in the vagal capsaicin-sensitive afferents (Beck et al., 2002; Bedendi et al., 2003; Bednarek et al., 2000; Bellone et al., 2002; Bennett et al., 1997). Thus, vagal afferent fibers represent a major target of these peripheral feeding regulators, ghrelin and CCK.

Ghrelin administration after CCK treatment does not induce feeding; CCK administration after ghrelin treatment does not reduce feeding. In normal subjects, plasma ghrelin levels rise before meal onset and decline 30 min after feeding. Plasma CCK levels in lean subjects fed a solid meal peak around 60 min after eating (Date et al., 2005).

6. Urocortin-1

Urocortin-1, a member of the corticotropin-releasing factor family, affects the pituitary-adrenal axis, the cardiovascular system, circulating neurohormones, and renal function, and suppresses appetite (Davis et al., 2004). Urocortin-1 also suppresses plasma ghrelin levels while increasing plasma adrenocorticotropic hormone, cortisol, and atrial natriuretic peptide secretion in healthy male volunteers (Davis et al., 2004). Ghrelin increases glucocorticoids via a central effect on corticotropin-releasing hormone secretion and n-octanoyl ghrelin is the form of ghrelin that activates the GHS-R1a and modulates corticotropin-releasing hormone-related neuronal activity.

7. Others

Acute subcutaneous injection of GH-releasing peptide 6 decreases ghrelin levels in normal rats (Tschöp et al., 2002). Male patients with hypogonadism resulting in testosterone deficiency have lower ghrelin values, which increase after testosterone replacement (Pagotto et al., 2003). Melatonin treatment inhibits plasma ghrelin levels in rats (Mustonen et al., 2001).

E. Brain neurotransmitters

1. Interleukin-1 and prostaglandins

Intraperitoneally injected lipopolysaccharides (LPS) decrease fasting plasma ghrelin levels 3 h postinjection and the values return to preinjection levels at 24 h in rats (Wang et al., 2006). Interleukin-1 (IL-1) receptor antagonists and indomethacin prevent the first 3-h LPS-induced decline in ghrelin levels. IL-1β suppresses plasma ghrelin levels, whereas urocortin-1 does not (Wang et al., 2006). Ghrelin injected intravenously prevents an LPS-induced reduction in gastric emptying and food intake (Wang et al., 2006). These findings indicate that the IL-1 and prostaglandin pathways are part of the early mechanisms by which LPS suppress fasting plasma ghrelin levels, and that exogenous ghrelin can normalize LPS-induced alterations in digestive function. There might be a negative feedback system between LPS, IL-1, and ghrelin in the regulation of feeding via prostaglandins. The mechanisms by which central prostaglandins alter plasma ghrelin levels remain to be resolved. IL-1 and prostaglandin-induced increases in the hypothalamic-pituitary axis and sympathetic outflow might contribute to suppress ghrelin secretion.

2. Serotonin

Brain serotonin (5-hydroxytryptamine; 5-HT) systems contribute to regulate eating behavior and energy homeostasis. Serotonin 5-HT2C and 5-HT1B receptors mediate the anorexic effects of 5-HT drugs, such as fenfluramine, which stimulates 5-HT release and inhibits 5-HT reuptake, and m-chlorophenylpiperazine (mCPP), a 5-HT2C receptor agonist. Fasting for 24 h increases the expression of hypothalamic 5-HT2C and 5-HT1B receptor genes in association with increases in plasma active ghrelin levels compared with the fed state in mice (Nonogaki et al., 2006a). Treatment with mCPP or fenfluramine significantly inhibits the increases in plasma active ghrelin levels (Nonogaki et al., 2006a). mCPP or fenfluramine significantly increases the expression of hypothalamic proopiomelanocortin and cocaine- and amphetamine-regulated transcript genes, while having no effect on the expression of hypothalamic neuropeptide Y, agouti-related protein, or ghrelin genes (Nonogaki et al., 2006a). These results suggest that there is a negative feedback loop between the 5-HT systems and plasma active ghrelin levels in energy homeostasis in mice (Fig. 2B).

3. Thyroid stimulating hormone-releasing hormone

Brainstem thyroid stimulating hormone-releasing hormone (TRH) is synthesized in the caudal raphe nuclei of the brainstem and spinal vagal and sympathetic motor neurons. Food intake in rats fasted overnight is significantly reduced by intracisternal-injected TRH antibody and peripherally administered atropine (Ao et al., 2006). Brainstem TRH might contribute to regulate food intake through vagus nerve-mediated stimulation of ghrelin secretion (Ao *et al.*, 2006).

F. Gastric bypass

Total ghrelin secretion following gastric bypass surgery is reduced by up to 77% compared with normal-weight control groups and up to 72% compared with weight-matched obese groups (Cummings *et al.*, 2002a; Geloneze *et al.*, 2003; Leonetti *et al.*, 2003). Furthermore, the normal meal-related fluctuations and diurnal rhythm of ghrelin levels are absent in these patients. The mechanism for decreasing plasma ghrelin levels in patients with gastric bypass is not known. One hypothesis is that direct contact between the gastric mucosa and food is important for the production and secretion of ghrelin (Adams *et al.*, 1996).

III. AFFERENT PATHWAYS OF GHRELIN FROM THE STOMACH TO THE HYPOTHALAMUS

Peripherally injected ghrelin stimulates hypothalamic neurons (Hewson and Dickson, 2000; Ruter *et al.*, 2003; Wang *et al.*, 2002) as well as food intake (Wren *et al.*, 2001). In general, peptides injected peripherally do not pass the blood–brain barrier (BBB). Indeed, the rate at which peripheral ghrelin passes the BBB is very low, although ghrelin can cross the BBB in a complex manner (Banks *et al.*, 2002). Peripheral ghrelin might have to activate the appropriate hypothalamic regions via an indirect pathway.

The existence of ghrelin receptors on vagal afferent neurons in the rat nodose ganglion suggests that ghrelin signals from the stomach are transmitted to the brain via the vagus nerve (Date *et al.*, 2006; Date *et al.*, 2002b; Sakata *et al.*, 2003; Zhang *et al.*, 2004a). Moreover, the finding that intracerebroventricular administration of ghrelin induces c-Fos in the dorsomotor nucleus of the vagus and stimulates gastric acid secretion indicates that ghrelin activates the vagus system (Date *et al.*, 2001).

In contrast, vagotomy inhibits the ability of ghrelin to stimulate food intake and GH release in mice and rats (Andrews and Sanger, 2002; Bellone *et al.*, 2002; Date *et al.*, 2002b). A similar effect is observed when capsaicin, a specific afferent neurotoxin, is applied to vagus nerve terminals to induce sensory denervation. Basal levels of ghrelin are not affected by vagotomy, whereas fasting-induced increases in plasma ghrelin are completely abolished by subdiaphragmatic vagotomy or atropine treatment (Asakawa *et al.*, 2001; Williams *et al.*, 2003b). Blockade of the vagal afferent pathway abolishes ghrelin-induced feeding. These findings indicate that the vagal afferent pathway is important for conveying orexigenic ghrelin signals from the stomach to the brain (Fig. 1).

Peripheral ghrelin signaling, which travels to the nucleus tractus solitarius at least in part via the vagus nerve, increases noradrenaline in the arcuate nucleus of the hypothalamus, thereby stimulating feeding, at least partially through α_1- and β_2-noradrenergic receptors (Date *et al.*, 2006). In addition, bilateral midbrain transections rostral to the nucleus tractus solitarius, or toxin-induced loss of neurons in the hindbrain that express dopamine β-hydroxylase (a noradrenaline synthetic enzyme), abolish ghrelin-induced feeding (Date *et al.*, 2006). These findings indicate that the noradrenergic system is likely to require for the central control of feeding behavior induced by peripherally administered ghrelin (Fig. 1).

IV. Hyperphagia, Obesity, and Des-acyl Ghrelin

There are two forms of ghrelin, an active n-octanoyl-modified ghrelin with bioactivity and nonactive des-acyl ghrelin without bioactivity (Ariyasu *et al.*, 2002; Hosoda *et al.*, 2000; Kojima and Kangawa, 2005). Because the acylation of ghrelin is assumed to be essential for its actions, des-acyl ghrelin, which lacks the fatty acid modification of ghrelin, was assumed to be devoid of biologic effects (Kojima and Kangawa, 2005). Active ghrelin is an endogenous ligand for the GHS-R1a, whereas des-acyl ghrelin is GHS-R1a independent (Kojima and Kangawa, 2005). It is therefore unlikely that des-acyl ghrelin competes with active ghrelin for binding sites.

Several recent studies, however, indicate that des-acyl ghrelin has biologic effects on proliferation, survival, and metabolism of cardiomyocytes, endothelial cells, adipocytes, myocytes, and myelocytes, and on food intake (Asakawa *et al.*, 2005; Baldanzi *et al.*, 2004; Bedendi *et al.*, 2003; Cassoni *et al.*, 2004; Chen *et al.*, 2005; Kojima and Kangawa, 2005; Thompson *et al.*, 2004; Toshinai *et al.*, 2006; Zhang *et al.*, 2004b), although the effects of des-acyl ghrelin remain controversial. Toshinai *et al.* (2006) reported that central des-acyl ghrelin increases food intake, while peripheral des-acyl ghrelin has no effect on food intake in rats, or C57BL6J and ddy mice. Asakawa *et al.* (2005) and Chen *et al.* (2005) reported that either central or peripheral administration of des-acyl ghrelin suppresses food intake in ddy mice and rats.

Both plasma active and des-acyl ghrelin levels in association with hypothalamic SGK-1, 5-HT2C, and 5-HT1B receptor gene expression are increased during fasting and decreased after feeding (Nonogaki *et al.*, 2006a,b; Fig. 4A). A^y mice have dominant alleles at the agouti locus (A), which produces ectopic expression of the agouti peptide, an antagonist of the hypothalamic MC4 and MC3 receptors (Fan *et al.*, 1997; Lu *et al.*, 1994).

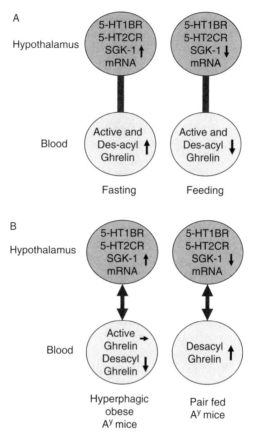

Figure 4 (A) Alterations of hypothalamic 5-HT2C/1B receptor and SGK-1 gene expression in a normal state. (B) Alterations of hypothalamic 5-HT2C/1B receptor and SGK-1 gene expression in chronic hyperphagia and obesity.

Although obese A^y mice have lower levels of plasma ghrelin than wild-type mice (Martin et al., 2006), the decreases appear to be due to decreased plasma des-acyl ghrelin levels (Nonogaki et al., 2006c). Plasma des-acyl ghrelin levels, but not active ghrelin levels, are decreased in non-obese A^y mice that consume more food (Nonogaki et al., 2006b,c). The low plasma des-acyl ghrelin levels are observed not only in the fed state, but also in a 24-h fasted state of non-obese A^y mice (Nonogaki et al., 2006c). Moreover, decreases in plasma des-acyl ghrelin levels occur in 8-week-old obese A^y mice. Thus, decreases in plasma des-acyl ghrelin, but not active ghrelin, levels precede the development of obesity in A^y mice. In addition, restricted feeding reverses the decrease in plasma des-acyl ghrelin levels in non-obese A^y mice and increases plasma des-acyl ghrelin levels in obese A^y mice. These findings suggest that chronic hyperphagia induced by the agouti peptide, rather than the direct effects of the agouti peptide, contribute to the decreases in des-acyl ghrelin levels, leading to an increased ratio of active/des-acyl ghrelin (Nonogaki et al., 2006c).

V. Hypothalamic Gene Expression and Plasma Des-acyl Ghrelin

In the hypothalamus, ghrelin neurons contact the cell bodies and dendrites of neuropeptide Y, agouti-related peptide and proopiomelanocortin neurons in the arcuate nucleus (Cowley et al., 2003) and orexin neurons in the lateral hypothalamus (Toshinai et al., 2003, 2006). A recent study, however, demonstrated that alterations in the expression of hypothalamic neuropeptide Y, agouti-related peptide, proopiomelanocortin, cocaine- and amphetamine-regulated transcript, and orexin genes are not proportional to the alterations of plasma des-acyl ghrelin and the ratio of active and des-acyl ghrelin (Nonogaki et al., 2006b,c). On the other hand, hypothalamic 5-HT2C and 5-HT1B receptor gene expression increases in association with decreases in plasma des-acyl ghrelin in hyperphagic A^y mice, whereas it decreased in association with increases in plasma des-acyl ghrelin levels in A^y mice under restricted feeding state (Nonogaki et al., 2006c) (Fig. 4B).

Serum- and glucocorticoid-induced protein kinase (SGK) is a serine/threonine-specific protein kinase that is transcriptionally regulated by serum, glucorticoids, and mineral corticoids. Fasting or obesity with hyperphagia increase hypothalamic SGK-1 gene expression. Hypothalamic SGK-1 mRNA levels are proportional to daily food intake and body weight in age-matched C57BL6J, KK, and KKA^y mice (Nonogaki et al., 2006b). Plasma des-acyl ghrelin, but not active ghrelin, levels are inversely proportional to daily food intake and body weight among these animals. The increases in hypothalamic SGK-1 gene expression in these animals are not accompanied by increases in plasma corticosterone levels. Under conditions of increased energy usage, such as fasting, hypothalamic SGK-1 gene expression and plasma des-acyl ghrelin levels are positively correlated, while during conditions of increased energy storage, they are negatively correlated (Nonogaki et al., 2006b; Fig. 4A and B).

VI. Conclusion

Ghrelin is produced primarily in the stomach in response to hunger and circulates in the blood. There are at least three feedback systems between the CNS and ghrelin secretion by the stomach. First, plasma ghrelin levels are increased during fasting, whereas they are decreased after the ingestion of glucose and lipid, but not protein. Second, ghrelin stimulates GH secretion by the pituitary gland, whereas GH inhibits ghrelin secretion by the stomach. The efferent vagus nerve contributes to the increase in ghrelin secretion during fasting, whereas the afferent vagus nerve contributes to feeding and GH secretion. The vagal nerve circuits between the CNS and the stomach have a crucial role in the regulation of ghrelin secretion. Third, fasting increases hypothalamic 5-HT2C and 5-HT1B receptor gene expression. Pharmacologic activation of the 5-HT2C and 5-HT1B receptors inhibits the increases in plasma active ghrelin levels during fasting. Pathways other than feeding and GH, such as the hypothalamic-pituitary-adrenal axis and sympathetic outflow to the stomach, might contribute to the inhibitory mechanisms of the 5-HT system. People with obesity due to cause other than Prader–Willi syndrome and leptin deficiency are likely to have decreased plasma ghrelin levels and do not exhibit the decreases in plasma ghrelin levels induced by feeding and sleeping.

Ghrelin has two forms, active n-octanoyl-modified ghrelin and des-acyl ghrelin. Fasting increases both types of ghrelin compared with the fed state. Chronic hyperphagia and obesity are likely to decrease plasma des-acyl ghrelin levels, but not n-octanoyl-modified ghrelin levels. Hypothalamic SGK-1 and 5-HT2C/1B receptor gene expression are likely to be proportional to plasma des-acyl ghrelin levels during fasting, whereas they are likely to be inversely proportional to plasma des-acyl ghrelin levels during conditions of increased energy storage such as obesity. Thus, a dysfunction of the ghrelin feedback systems might contribute to the pathophysiology of obesity and eating disorders.

ACKNOWLEDGMENTS

We thank Ms. Kana Nozue for her assistance, and Dr. Karin Mesches and Dr. Michael Mesches for their critical reading. This work was supported by a Grant in-Aid for Scientific Research (C2) and Human Science Research (KH21016), Takeda Research Foundation.

REFERENCES

Adams, E. F., Buchfelder, M., Lei, T., Huang, B., Hamacher, C., Derwahl, M., Bowers, C. Y., and Fahlbusch, R. (1996). In vitro responses of GH-secreting tumours with and without gsp oncogenes to octreotide, GHRH and growth hormone-releasing peptide. In "Pituitary Adenomas: From Basic Research to Diagnosis and Therapy" (K. von Werder and R. Fahlbusch, Eds.), pp. 43–47. Amsterdam, Elsevier.

Anderwald, C., Brabant, G., Bernroider, E., Horn, R., Brehm, A., Waldhausl, W., and Roden, M. (2003). Insulin-dependent modulation of plasma ghrelin and leptin concentrations is less pronounced in type 2 diabetic patients. *Diabetes* **52,** 1792–1798.

Andrews, P. L., and Sanger, G. J. (2002). Abdominal vagal afferent neurones: An important target for the treatment of gastrointestinal dysfunction. *Curr. Opin. Pharmacol.* **2**, 650–656.

Ao, Y., Go, V. L., Toy, N., Li, T., Wang, Y., Song, M. K., Reeve, J. R., Jr., Liu, Y., and Yang, H. (2006). Brainstem thyrotropin-releasing hormone regulates food intake through vagal-dependent cholinergic stimulation of ghrelin secretion. *Endocrinology* **147**(12), 6004–6012.

Arafat, A. M., Otto, B., Rochlitz, H., Tschop, M., Bahr, V., Mohlig, M., Diederich, S., Spranger, J., and Pfeiffer, A. F. (2005). Glucagon inhibits ghrelin secretion in humans. *Eur. J. Endocrinol.* **153**, 397–402.

Arafat, A. M., Perschel, F. H., Otto, B., Weickert, M. O., Rochlitz, H., Schöfl, C., Spranger, J., Möhlig, M., and Pfeiffe, A. F. H. (2006). Glucagon suppression of ghrelin secretion is exerted at hypothalamus-pituitary level. *J. Clin. Endocrinol. Metab.* **91**, 3528–3533.

Ariyasu, H., Takaya, K., Tagami, T., Ogawa, Y., Hosoda, K., Akamizu, T., Suda, M., Koh, T., Natsui, K., Toyooka, S., Shirakami, G., Usui, T., *et al.* (2001). Stomach is a major source of circulating ghrelin, and feeding state determines plasma ghrelin-like immunoreactivity levels in humans. *J. Clin. Endocrinol. Metab.* **86**, 4753–4758.

Ariyasu, H., Takaya, K., Hosoda, H., Iwakura, H., Ebihara, K., Mori, K., Ogawa, Y., Hosoda, K., Akamizu, T., Kojima, M., Kangawa, K., and Nakao, K. (2002). Delayed short-term secretory regulation of ghrelin in obese animals: Evidenced by a specific RIA for the active form of ghrelin. *Endocrinology* **143**, 3341–3350.

Asakawa, A., Inui, A., Kaga, T., Yuzuriha, H., Nagata, T., Ueno, N., Makino, S., Fujimiya, M., Niijima, A., Fujino, M. A., and Kasuga, M. (2001). Ghrelin is an appetite-stimulatory signal from stomach with structural resemblance to motilin. *Gastroenterology* **120**, 337–345.

Asakawa, A., Inui, A., Kaga, T., Katsuura, G., Fujimiya, M., Fujino, A., and Kasuga, M. (2003). Antagonism of ghrelin receptor reduces food intake and body weight gain in mice. *Gut* **52**, 947–952.

Asakawa, A., Inui, A., Fujimiya, M., Sakamaki, R., Shinfuku, N., Ueta, Y., Meguid, M. M., and Kasuga, M. (2005). Stomach regulates energy balance via acylated ghrelin and des-acyl ghrelin. *Gut* **54**, 18–24.

Bagnasco, M., Dube, M. G., Kalra, P. S., and Kalra, S. P. (2002). Evidence for the existence of distinct central appetite, energy expenditure, and ghrelin stimulation pathways as revealed by hypothalamic site-specific leptin gene therapy. *Endocrinology* **143**, 4409–4421.

Baldanzi, G., Filigheddu, N., Cutrupi, S., Catapano, S. F., Bonissoni, S., Fubini, A., Malan, D., Baj, G., Granata, R., Broglio, F., Papotti, M., Surico, N., *et al.* (2004). Ghrelin and des-acyl ghrelin inhibit cell death in cardiomyocytes and endothelial cells through ERK1/2 and PI 3-kinase/AKT. *J. Cell Biol.* **159**, 1029–1037.

Banks, W. A., Tschöp, M., Robinson, S. M., and Heiman, M. L. (2002). Extent and direction of ghrelin transport across the blood–brain barrier is determined by its unique primary structure. *J. Pharmacol. Exp. Ther.* **302**, 822–827.

Barkan, A. L., Dimaraki, E. V., Jessup, S. K., Symons, K. V., Ermolenko, M., and Jaffe, C. A. (2003). Ghrelin secretion in humans is sexually dimorphic, suppressed by somatostatin, and not affected by the ambient growth hormone levels. *J. Clin. Endocrinol. Metab.* **88**, 2180–2184.

Barreiro, M. L., Gaytan, F., Caminos, J. E., Pinilla, L., Casanueva, F. F., Aguilar, E., Dieguez, C., and Tena-Sempere, M. (2002a). Cellular location and hormonal regulation of ghrelin expression in rat testis. *Biol. Reprod.* **67**, 1768–1776.

Barreiro, M. L., Pinilla, L., Aguilar, E., and Tena-Sempere, M. (2002b). Expression and homologous regulation of GH secretagogue receptor mRNA in rat adrenal gland. *Eur. J. Endocrinol.* **147**, 677–688.

Basa, N. R., Wang, L., Arteaga, J. R., Heber, D., Livingston, E. H., and Tache, Y. (2003). Bacterial lipopolysaccharide shifts fasted plasma ghrelin to postprandial levels in rats. *Neurosci. Lett.* **343**, 25–28.
Batterham, R. L., Cowley, M. A., Small, C. J., Herzog, H., Cohen, M. A., Dakin, C. L., Wren, A. M., Brynes, A. E., Low, M. J., Ghatei, M. A., Cone, R. D., and Bloom, S. R. (2002). Gut hormone PYY(3–36) physiologically inhibits food intake. *Nature* **418**, 650–654.
Batterham, R. L., Cohen, M. A., Ellis, S. M., LeRoux, C. W., Withers, D. J., Frost, G. S., Ghatei, M. A., and Bloom, S. R. (2003a). Inhibition of food intake in obese subjects by peptide YY3–36. *N. Engl. J. Med.* **349**, 941–948.
Batterham, R. L., LeRoux, C. W., Cohen, M. A., Park, A. J., Ellis, S. M., Patterson, M., Frost, G. S., Ghatei, M. A., and Bloom, S. R. (2003b). Pancreatic polypeptide reduces appetite and food intake in humans. *J. Clin. Endocrinol. Metab.* **88**, 3989–3992.
Beck, B., Musse, N., and Stricker-Krongrad, A. (2002). Ghrelin, macronutrient intake and dietary preferences in Long-Evans rats. *Biochem. Biophys. Res. Commun.* **292**, 1031–1035.
Bedendi, I., Alloatti, G., Marcantoni, A., Malan, D., Catapano, F., Ghe, C., Deghenghi, R., Ghigo, E., and Muccioli, G. (2003). Cardiac effects of ghrelin and its endogenous derivatives des-octanoyl ghrelin and des-Gln(14)-ghrelin. *Eur. J. Pharmacol.* **476**, 87–95.
Bednarek, M. A., Feighner, S. D., Pong, S. S., McKee, K. K., Silva, M., Warren, V. A., Howard, A. D., Van der Ploeg, L. H. T., and Heck, J. V. (2000). Structure and function studies on the new growth hormone releasing peptide, ghrelin: Minimal sequence of ghrelin necessary for activation of growth hormone secretagogue receptor 1a. *J. Med. Chem.* **43**, 4370–4376.
Bellone, S., Rapa, A., Vivenza, D., Castellino, N., Petri, A., Bellone, J., Me, E., Broglio, F., Prodam, F., Ghigo, E., and Bona, G. (2002). Circulating ghrelin levels as function of gender, pubertal status and adiposity in childhood. *J. Endocrinol. Invest.* **25**, 13–15.
Bennett, P. A., Thomas, G. B., Howard, A. D., Feighner, S. D., Van der Ploeg, L. H. T., Smith, R. G., and Robinson, I. C. A. F. (1997). Hypothalamic growth hormone secretagogue-receptor (GHS-R) expression is regulated by growth hormone in the rat. *Endocrinology* **138**, 4552–4557.
Briatore, L., Andraghetti, G., and Cordera, R. (2003). Acute plasma glucose increase, but not early insulin response, regulates plasma ghrelin. *Eur. J. Endocrinol.* **149**, 403–406.
Caixas, A., Bashore, C., Nash, W., Pi-Sunyer, F., and Laferrere, B. (2002). Insulin, unlike food intake, does not suppress ghrelin in human subjects. *J. Clin. Endocrinol. Metab.* **87**, 1902–1906.
Callahan, H. S., Cummings, D. E., Pepe, M. S., Breen, P. A., Matthys, C. C., and Weigle, D. S. (2004). Postprandial suppression of plasma ghrelin level is proportional to ingested caloric load but does not predict intermeal interval in humans. *J. Clin. Endocrinol. Metab.* **89**, 1319–1324.
Cassoni, P., Ghe, C., Marrocco, T., Tarabra, E., Allia, E., Catapano, F., Deghenghi, R., Ghigo, E., Papotti, M., and Muccioli, G. (2004). Expression of ghrelin and biological activity of specific receptors for ghrelin and des-acyl ghrelin in human prostate neoplasms and related cell lines. *Eur. J. Endocrinol.* **150**, 173–184.
Chan, J. L., Bullen, J., Lee, J. H., Yiannakouris, N., and Mantzoros, C. S. (2004). Ghrelin levels are not regulated by recombinant leptin administration and/or three days of fasting in healthy subjects. *J. Clin. Endocrinol. Metab.* **89**, 335–343.
Chen, C. Y., Inui, A., Asakawa, A., Fujino, K., Kato, I., Chen, C. C., Ueno, N., and Fujimiya, M. (2005). Des-acyl ghrelin acts by CRF type 2 receptors to disrupt fasted stomach motility in conscious rats. *Gastroenterology* **129**, 8–25.

Cowley, M. A., Smith, R. G., Diano, S., Tschöp, M., Pronchuk, N., Grove, K. L., Strasburger, C. J., Bidlingmaier, M., Esterman, M., Heiman, M. L., Garcia-Segura, L. M., Nillni, E. A., *et al.* (2003). The distribution and mechanism of action of ghrelin in the CNS demonstrates a novel hypothalamic circuit regulating energy homeostasis. *Neuron* **37,** 649–661.

Cummings, D. E., Purnell, J. Q., Frayo, R. S., Schmidova, K., Wisse, B. E., and Weigle, D. S. (2001). A preprandial rise in plasma ghrelin levels suggests a role in meal initiation in humans. *Diabetes* **50,** 1714–1719.

Cummings, D. E., Weigle, D. S., Frayo, R. S., Breen, P. A., Ma, M. K., Dellinger, E. P., and Purnell, J. Q. (2002a). Plasma ghrelin levels after diet-induced weight loss or gastric bypass surgery. *N. Engl. J. Med.* **346,** 1623–1630.

Cummings, D. E., Clement, K., Purnell, J. Q., Vaisse, C., Foster, K. E., Frayo, R. S., Schwartz, M. W., Basdevant, A., and Weigle, D. S. (2002b). Elevated plasma ghrelin levels in Prader Willi syndrome. *Nat. Med.* **8,** 643–644.

Cummings, D. E., Frayo, R. S., Marmonier, C., Aubert, R., and Chapelot, D. (2004). Plasma ghrelin levels and hunger scores in humans initiating meals voluntarily without time- and food-related cues. *Am. J. Physiol. Endocrinol. Metab.* **287,** E297–E304.

Cuntz, U., Fruhauf, E., Wawarta, R., Tschöp, M., Folwaczny, C., Riepl, R., Lehnert, P., Fichter, M., and Otto, B. (2002). A role for the novel weight-regulating hormone ghrelin in anorexia nervosa. *Am. Clin. Lab.* **21,** 22–23.

Date, Y., Kojima, M., Hosoda, H., Sawaguchi, A., Mondal, M. S., Suganuma, T., Matsukura, S., Kangawa, K., and Nakazato, M. (2000). Ghrelin, a novel growth hormone-releasing acylated peptide, is synthesized in a distinct endocrine cell type in the gastrointestinal tracts of rats and humans. *Endocrinology* **141,** 4255–4261.

Date, Y., Nakazato, M., Murakami, N., Kojima, M., Kangawa, K., and Matsukura, S. (2001). Ghrelin acts in the central nervous system to stimulate gastric acid secretion. *Biochem. Biophys. Res. Commun.* **280,** 904–907.

Date, Y., Nakazato, M., Hashiguchi, S., Dezaki, K., Mondal, M. S., Hosoda, H., Kojima, M., Kangawa, K., Arima, T., Matsuo, H., Yada, T., and Matsukura, S. (2002a). Ghrelin is present in pancreatic alpha-cells of humans and rats and stimulates insulin secretion. *Diabetes* **51,** 124–129.

Date, Y., Murakami, N., Toshinai, K., Matsukura, S., Niijima, A., Matsuo, H., Kangawa, K., and Nakazato, M. (2002b). The role of the gastric afferent vagal nerve in ghrelin-induced feeding and growth hormone secretion in rats. *Gastroenterology* **123,** 1120–1128.

Date, Y., Toshinai, K., Koda, S., Miyazato, M., Shimbara, T., Tsuruta, T., Niijima, A., Kangawa, K., and Nakazato, M. (2005). Peripheral interaction of ghrelin with cholecystokinin on feeding regulation. *Endocrinology* **146,** 3518–3525.

Date, Y., Shimbara, T., Koda, S., Toshinai, K., Ida, T., Murakami, N., Miyazato, M., Kokame, K., Ishizuka, Y., Ishida, Y., Kageyama, H., Shioda, S., *et al.* (2006). Peripheral ghrelin transmits orexigenic signals through the noradrenergic pathway from the hindbrain to the hypothalamus. *Cell Metab.* **4,** 323–331.

Davis, M. E., Pemberton, C. J., Yandle, T. G., Lainchbury, J. G., Rademaker, M. T., Nicholls, M. G., Frampton, C. M., and Richards, A. M. (2004). Urocortin-1 infusion in normal humans. *J. Clin. Endocrinol. Metab.* **89,** 1402–1409.

DelParigi, A., Tschöp, M., Heiman, M. L., Salbe, A. D., Vozarova, B., Sell, S. M., Bunt, J. C., and Tataranni, P. A. (2002). High circulating ghrelin: A potential cause for hyperphagia and obesity in Prader-Willi syndrome. *J. Clin. Endocrinol. Metab.* **87,** 5461–5464.

Dornonville de la Cour, C., Bjorkqvist, M., Sandvik, A. K., Bakke, I., Zhao, C. M., Chen, D., and Hakanson, R. (2001). A-like cells in the rat stomach contain ghrelin and do not operate under gastrin control. *Regul. Pept.* **99,** 141–150.

Dzaja, A., Dalal, M. A., Himmerich, H., Uhr, M., Pollmacher, T., and Schuld, A. (2004). Sleep enhances nocturnal plasma ghrelin levels in healthy subjects. *Am. J. Physiol. Endocrinol. Metab.* **286,** E963–E967.
Erdmann, J., Lippl, F., and Schusdziarra, V. (2003). Differential effect of protein and fat on plasma ghrelin levels in man. *Regul. Pept.* **116,** 101–107.
Fan, W., Boston, B. A., Kesterson, R. A., Hruby, V. J., and Cone, R. D. (1997). Role of melanocortinergic neurons in feeding and the agouti obesity syndrome. *Nature* **385,** 165–168.
Faulconbridge, L. F., Cummings, D. E., Kaplan, J. M., and Grill, H. J. (2003). Hyperphagic effects of brainstem ghrelin administration. *Diabetes* **52,** 2260–2265.
Flanagan, D. E., Evans, M. L., Monsod, T. P., Rife, F., Heptulla, R. A., Tamborlane, W. V., and Sherwin, R. S. (2003). The influence of insulin on circulating ghrelin. *Am. J. Physiol. Endocrinol. Metab.* **284,** E313–E316.
Geloneze, B., Tambascia, M. A., Pilla, V. F., Geloneze, S. R., Repetto, E. M., and Pareja, J. C. (2003). Ghrelin: A gut-brain hormone: Effect of gastric bypass surgery. *Obes. Surg.* **13,** 17–22.
Gnanapavan, S., Kola, B., Bustin, S. A., Morris, D. G., McGee, P., Fairclough, P., Bhattacharya, S., Carpenter, R., Grossman, A. B., and Korbonits, M. (2002). The tissue distribution of the mRNA of ghrelin and subtypes of its receptor, GHS-R, in humans. *J. Clin. Endocrinol. Metab.* **87,** 2988.
Greenman, Y., Golani, N., Gilad, S., Yaron, M., Limor, R., and Stern, N. (2004). Ghrelin secretion is modulated in a nutrient- and gender-specific manner. *Clin. Endocrinol.* **60,** 382–388.
Hansen, T. K., Dall, R., Hosoda, H., Kojima, M., Kangawa, K., Christiansen, J. S., and Jorgensen, J. O. (2002). Weight loss increases circulating levels of ghrelin in human obesity. *Clin. Endocrinol.* **56,** 203–206.
Haqq, A. M., Stadler, D. D., Rosenfeld, R. G., Pratt, K. L., Weigle, D. S., Frayo, R. S., LaFranchi, S. H., Cummings, D. E., and Purnell, J. Q. (2003a). Circulating ghrelin levels are suppressed by meals and octreotide therapy in children with Prader-Willi syndrome. *J. Clin. Endocrinol. Metab.* **88,** 3573–3576.
Haqq, A. M., Farooqi, I. S., O'Rahilly, S., Stadler, D. D., Rosenfeld, R. G., Pratt, K. L., LaFranchi, S. H., and Purnell, J. Q. (2003b). Serum ghrelin levels are inversely correlated with body mass index, age, and insulin concentrations in normal children and are markedly increased in Prader-Willi syndrome. *J. Clin. Endocrinol. Metab.* **88,** 174–178.
Hataya, Y., Akamizu, T., Takaya, K., Kanamoto, N., Ariyasu, H., Saijo, M., Moriyama, K., Shimatsu, A., Kojima, M., Kangawa, K., and Nakao, K. (2001). A low dose of ghrelin stimulates growth hormone (GH) release synergistically with GH-releasing hormone in human. *J. Clin. Endocrinol. Metab.* **86,** 4552–4555.
Hewson, A. K., and Dickson, S. L. (2000). Systemic administration of ghrelin induces Fos and Egr-1 proteins in the hypothalamic arcuate nucleus of fasted and fed rats. *J. Neuroendocrinol.* **12,** 1047–1049.
Holdstock, C., Ludvigsson, J., and Karlsson, F. A. (2003). Abnormal ghrelin secretion in new onset childhood Type 1 diabetes. *Diabetologia* **47,** 150–151.
Hosoda, H., Kojima, M., Matsuo, H., and Kangawa, K. (2000). Ghrelin and des-acyl ghrelin: Two major forms of rat ghrelin peptide in gastrointestinal tissue. *Biochem. Biophys. Res. Commun.* **279,** 909–913.
Kamegai, J., Tamura, H., Shimizu, T., Ishii, S., Sugihara, H., and Oikawa, S. (2004). Effects of insulin, leptin, and glucagon on ghrelin secretion from isolated perfused rat stomach. *Regul. Pept.* **119,** 77–81.
Kim, M. S., Yoon, C. Y., Park, K. H., Shin, C. S., Park, K. S., Kim, S. Y., Cho, B. Y., and Lee, H. K. (2003). Changes in ghrelin and ghrelin receptor expression according to feeding status. *Neuroreport* **14,** 1317–1320.

Knerr, I., Groschl, M., Rascher, W., and Rauh, M. (2003). Endocrine effects of food intake: Insulin, ghrelin, and leptin responses to a single bolus of essential amino acids in humans. *Ann. Nutr. Metab.* **47,** 312–318.

Kojima, M., and Kangawa, K. (2005). Ghrelin: Structure and function. *Physiol. Rev.* **85,** 495–522.

Kojima, M., Hosoda, H., Date, Y., Nakazato, M., Matsuo, H., and Kangawa, K. (1999). Ghrelin is a growth-hormone-releasing acylated peptide from stomach. *Nature* **402,** 656–660.

Korbonits, M., Bustin, S. A., Kojima, M., Jordan, S., Adams, E. F., Lowe, D. G., Kangawa, K., and Grossman, A. B. (2001a). The expression of the growth hormone secretagogue receptor ligand ghrelin in normal and abnormal human pituitary and other neuroendocrine tumors. *J. Clin. Endocrinol. Metab.* **86,** 881–887.

Korbonits, M., Kojima, M., Kangawa, K., and Grossman, A. B. (2001b). Presence of ghrelin in normal and adenomatous human pituitary. *Endocrine* **14,** 101–104.

Krsek, M., Rosicka, M., Papezova, H., Krizova, J., Kotrlikova, E., Haluz'k, M., Justova,V., Lacinova, Z., and Jarkovska, Z. (2003). Plasma ghrelin levels and malnutrition: A comparison of two etiologies. *Eat Weight Disord.* **8,** 207–211.

Lee, H. M., Wang, G., Englander, E. W., Kojima, M., and Greeley, J. G., Jr. (2002). Ghrelin, a new gastrointestinal endocrine peptide that stimulates insulin secretion: Anteric distribution, ontogeny, influence of endocrine, and dietary manipulations. *Endocrinology* **143,** 185–190.

Leonetti, F., Silecchia, G., Iacobellis, G., Ribaudo, M. C., Zappaterreno, A., Tiberti, C., Iannucci, C. V., Perrotta, N., Bacci, V., Basso, M. S., Basso, N., and Di Mario, U. (2003). Different plasma ghrelin levels after laparoscopic gastric bypass and adjustable gastric banding in morbid obese subjects. *J. Clin. Endocrinol. Metab.* **88,** 4227–4231.

Lu, D., Willard, D., Patel, I. R., Kadwell, S., Overton, L., Kost, T., Luther, M., Chen, W., Woychik, R. P., Wilkison, W. O., and Cone, R. D. (1994). Agouti protein is an antagonist of the melanocyte-stimulating-hormone receptor. *Nature* **371,** 799–802.

Lu, S., Guan, J. L., Wang, Q. P., Uehara, K., Yamada, S., Goto, N., Date, Y., Nakazato, M., Kojima, M., Kangawa, K., and Shioda, S. (2002). Immunocytochemical observation of ghrelin-containing neurons in the rat arcuate nucleus. *Neurosci. Lett.* **321,** 157–160.

Lucidi, P., Murdolo, G., DiLoreto, C., De Cicco, A., Parlanti, N., Fanelli, C., Santeusanio, F., Bolli, G. B., and De Feo, P. (2002). Ghrelin is not necessary for adequate hormonal counterregulation of insulin-induced hypoglycemia. *Diabetes* **51,** 2911–2914.

Luque, R. M. (2006). Evidence that endogenous SSt inhibits ACTH and ghrelin expression by independent pathways. *Am. J. Physiol. Endocrinol. Metab.* **291,** E395–E403.

Martin, N. M., Houston, P. A., Patterson, M., Sajedi, A., Carmignac, D. F., Ghatei, M. A., Bloom, S. R., and Small, C. J. (2006). Abnormalities of the somatotrophic axis in the obese agouti mouse. *Int. J. Obes.* **30,** 430–438.

McCowen, K. C., Maykel, J. A., Bistrian, B. R., and Ling, P. R. (2002). Circulating ghrelin concentrations are lowered by intravenous glucose or hyperinsulinemic euglycemic conditions in rodents. *J. Endocrinol.* **175,** R7–R11.

Monteleone, P., Bencivenga, R., Longobardi, N., Serritella, C., and Maj, M. (2003). Differential responses of circulating ghrelin to high-fat or high-carbohydrate meal in healthy women. *J. Clin. Endocrinol. Metab.* **88,** 5510–5514.

Mustonen, A. M., Nieminen, P., and Hyvarinen, H. (2001). Preliminary evidence that pharmacologic melatonin treatment decreases rat ghrelin levels. *Endocrine* **16,** 43–46.

Nonogaki, K., Ohashi-Nozue, K., and Oka, Y. (2006a). A negative feedback system between brain serotonin systems and plasma active ghrelin levels in mice. *Biochem. Biophys. Res. Commun.* **341,** 703–707.

Nonogaki, K., Nozue, K., and Oka, Y. (2006b). Induction of hypothalamic serum- and glucocorticoid-induced protein kinase-1 gene expression and its relation to plasma

des-acyl ghrelin in energy homeostasis in mice. *Biochem. Biophys. Res. Commun.* **344**, 696–699.
Nonogaki, K., Nozue, K., and Oka, Y. (2006c). Hyperphagia alters expression of hypothalamic 5-HT2C and 5-HT1B receptor genes and plasma des-acyl ghrelin levels in A[y] mice. *Endocrinology* **147**, 5893–5900.
Norrelund, H., Hansen, T. K., Orskov, H., Hosoda, H., Kojima, M., Kangawa, K., Weeke, J., Moller, N., Christiansen, J. S., and Jorgensen, J. O. (2002). Ghrelin immunoreactivity in human plasma is suppressed by somatostatin. *Clin. Endocrinol.* **57**, 539–546.
Otto, B., Cuntz, U., Fruehauf, E., Wawarta, R., Folwaczny, C., Riepl, R. L., Heiman, M. L., Lehnert, P., Fichter, M., and Tschöp, M. (2001). Weight gain decreases elevated plasma ghrelin concentrations of patients with anorexia nervosa. *Eur. J. Endocrinol.* **145**, 669–673.
Pagotto, U., Gambineri, A., Vicennati, V., Heiman, M. L., Tschöp, M., and Pasquali, R. (2002). Plasma ghrelin, obesity, and the polycystic ovary syndrome: Correlation with insulin resistance and androgen levels. *J. Clin. Endocrinol. Metab.* **87**, 5625–5629.
Pagotto, U., Gambineri, A., Pelusi, C., Genghini, S., Cacciari, M., Otto, B., Castaneda, T., Tschöp, M., and Pasquali, R. (2003). Testosterone replacement therapy restores normal ghrelin in hypogonadal men. *J. Clin. Endocrinol. Metab.* **88**, 4139–4143.
Pöykkö, S. M., Kellokoski, E., Horkko, S., Kauma, H., Kesaniemi, Y. A., and Ukkola, O. (2003). Low plasma ghrelin is associated with insulin resistance, hypertension, and the prevalence of type 2 diabetes. *Diabetes* **52**, 2546–2553.
Prado, C. L., Pugh-Bernard, A. E., Elghazi, L., Sosa-Pineda, B., and Sussel, L. (2004). Ghrelin cells replace insulin-producing βcells in two mouse models of pancreas development. *Proc. Natl. Acad. Sci. USA* **101**, 2924–2929.
Qi, X., Reed, J., Englander, E. W., Chandrashekar, V., Bartke, A., and Greeley, G. H., Jr. (2003). Evidence that growth hormone exerts a feedback effect on stomach ghrelin production and secretion. *Exp. Biol. Med.* **228**, 1028–1032.
Ravussin, E., Tschöp, M., Morales, S., Bouchard, C., and Heiman, M. L. (2001). Plasma ghrelin concentration and energy balance: Overfeeding and negative energy balance studies in twins. *J. Clin. Endocrinol. Metab.* **86**, 4547–4551.
Riis, A. L., Hansen, T. K., Moller, N., Weeke, J., and Jorgensen, J. O. (2003). Hyperthyroidism is associated with suppressed circulating ghrelin levels. *J. Clin. Endocrinol. Metab.* **88**, 853–857.
Rindi, G., Necchi, V., Savio, A., Torsello, A., Zoli, M., Locatelli, V., Raimondo, F., Cocchi, D., and Solcia, E. (2002). Characterisation of gastric ghrelin cells in man and other mammals: Studies in adult and fetal tissues. *Histochem. Cell Biol.* **117**, 511–519.
Rosicka, M., Krsek, M., Matoulek, M., Jarkovska, Z., Marek, J., Justova, V., and Lacinova, Z. (2003). Serum ghrelin levels in obese patients: The relationship to serum leptin levels and soluble leptin receptors levels. *Physiol. Res.* **52**, 61–66.
Ruter, J., Kobelt, P., Tebbe, J. J., Avsar, Y., Veh, R., Wang, L., Klapp, B. F., Wiedenmann, B., Tache, Y., and Monnikes, H. (2003). Intraperitoneal injection of ghrelin induces Fos expression in the paraventricular nucleus of the hypothalamus in rats. *Brain Res.* **991**, 26–33.
Saad, M. F., Bernaba, B., Hwu, C. M., Jinagouda, S., Fahmi, S., Kogosov, E., and Boyadjian, R. (2002). Insulin regulates plasma ghrelin concentration. *J. Clin. Endocrinol. Metab.* **87**, 3997–4000.
Sakata, I., Nakamura, K., Yamazaki, M., Matsubara, M., Hayashi, Y., Kangawa, K., and Sakai, T. (2002). Ghrelin-producing cells exist as two types of cells, closed- and opened-type cells, in the rat gastrointestinal tract. *Peptides* **23**, 531–536.

Sakata, I., Yamazaki, M., Inoue, K., Hayashi, Y., Kangawa, K., and Sakai, T. (2003). Growth hormone secretagogue receptor expression in the cells of the stomach-projected afferent nerve in the rat nodose ganglion. *Neurosci. Lett.* **342,** 183–186.

Schaller, G., Schmidt, A., Pleiner, J., Woloszczuk, W., Wolzt, M., and Luger, A. (2003). Plasma ghrelin concentrations are not regulated by glucose or insulin: A double-blind, placebo-controlled crossover clamp study. *Diabetes* **52,** 16–20.

Schöfl, C., Horn, R., Schill, T., Schlosser, H. W., Muller, M. J., and Brabant, G. (2002). Circulating ghrelin levels in patients with polycystic ovary syndrome. *J. Clin. Endocrinol. Metab.* **87,** 4607–4610.

Shiiya, T., Nakazato, M., Mizuta, M., Date, Y., Mondal, M. S., Tanaka, M., Nozoe, S., Hosoda, H., Kangawa, K., and Matsukura, S. (2002). Plasma ghrelin levels in lean and obese humans and the effect of glucose on ghrelin secretion. *J. Clin. Endocrinol. Metab.* **87,** 240–244.

Soriano-Guillen, L., Barrios, V., Campos-Barros, A., and Argente, J. (2004). Ghrelin levels in obesity and anorexia nervosa: Effect of weight reduction or recuperation. *J. Pediatr.* **144,** 36–42.

Takaya, K., Ariyasu, H., Kanamoto, N., Iwakura, H., Yoshimoto, A., Harada, M., Mori, K., Komatsu, Y., Usui, T., Shimatsu, A., Ogawa, Y., Hosoda, K., *et al.* (2000). Ghrelin strongly stimulates growth hormone release in humans. *J. Clin. Endocrinol. Metab.* **85,** 4908–4911.

Tan, T. M., Vanderpump, M., Khoo, B., Patterson, M., Ghatei, M. A., and Goldstone, A. P. (2004). Somatostatin infusion lowers plasma ghrelin without reducing appetite in adults with Prader-Willi syndrome. *J. Clin. Endocrinol. Metab.* **89,** 4162–4165.

Tanaka, M., Naruo, T., Yasuhara, D., Tatebe, Y., Nagai, N., Shiiya, T., Nakazato, M., Matsukura, S., and Nozoe, S. (2003). Fasting plasma ghrelin levels in subtypes of anorexia nervosa. *Psychoneuroendocrinology* **28,** 829–835.

Thompson, N. M., Gill, D. A., Davies, R., Loveridge, N., Houston, P. A., Robinson, I. C., and Wells, T. (2004). Ghrelin and des-octanoyl ghrelin promote adipogenesis directly in vivo by a mechanism independent of the type 1a growth hormone secretagogue receptor. *Endocrinology* **145,** 234–242.

Toshinai, K., Mondal, M. S., Nakazato, M., Date, Y., Murakami, N., Kojima, M., Kangawa, K., and Matsukura, S. (2001). Upregulation of ghrelin expression in the stomach upon fasting, insulin-induced hypoglycemia, and leptin administration. *Biochem. Biophys. Res. Commun.* **281,** 1220–1225.

Toshinai, K., Date, Y., Murakami, N., Shimada, M., Mondal, M. S., Shimbara, T., Guan, L. L., Wand, Q. P., Funahashi, H., Sakurai, T., Shioda, S., Matsukura, S., *et al.* (2003). Ghrelin-induced food intake is mediated via the orexin pathway. *Endocrinology* **144,** 1506–1512.

Toshinai, K., Yamaguchi, H., Sun, Y., Smith, R. G., Yamanaka, A., Sakurai, T., Date, Y., Mondal, M. S., Shimbara, T., Kawagoe, T., Murakami, N., Miyazato, M., *et al.* (2006). Des-acyl ghrelin induces food intake by a mechanism independent of the growth hormone secretagogue receptor. *Endocrinology* **147,** 2306–2314.

Tomasetto, C., Wendling, C., Rio, M. C., and Poitras, P. (2001). Identification of cDNA encoding motilin related peptide/ghrelin precursor from dog fundus. *Peptides* **22,** 2055–2059.

Tritos, N. A., Kokkinos, A., Lampadariou, E., Alexiou, E., Katsilambros, N., and Maratos-Flier, E. (2003). Cerebrospinal fluid ghrelin is negatively associated with body mass index. *J. Clin. Endocrinol. Metab.* **88,** 2943–2946.

Tschöp, M., Wawarta, R., Riepl, R. L., Friedrich, S., Bidlingmaier, M., Landgraf, R., and Folwaczny, C. (2001a). Post-prandial decrease of circulating human ghrelin levels. *J. Endocrinol. Invest.* **24,** 19–21.

Tschöp, M., Weyer, C., Tataranni, P. A., Devanarayan, V., Ravussin, E., and Heiman, M. L. (2001b). Circulating ghrelin levels are decreased in human obesity. *Diabetes* **50,** 707–709.

Tschöp, M., Statnick, M. A., Suter, T. M., and Heiman, M. L. (2002). GH-releasing peptide-2 increases fat mass in mice lacking NPY: indication for a crucial mediating role of hypothalamic agouti-related protein. *Endocrinology* **143,** 558–568.

Wang, L., Basa, N. R., Shaikh, A., Luckey, A., Heber, D., St-Pierre, D. H., and Tache, Y. (2006). LPS inhibits fasted plasma ghrelin levels in rats: Role of IL-1 and PGs and functional implications. *Am. J. Physiol. Gastrointest. Liver Physiol.* **291,** G611–G620.

Wang, L., Saint-Pierre, D. H., and Tache, Y. (2002). Peripheral ghrelin selectively increases Fos expression in neuropeptide Y-synthesizing neurons in mouse hypothalamic arcuate nucleus. *Neurosci. Lett.* **325,** 47–51.

Williams, D. L., Cummings, D. E., Grill, H. J., and Kaplan, J. M. (2003a). Meal-related ghrelin suppression requires postgastric feedback. *Endocrinology* **144,** 2765–2767.

Williams, D. L., Grill, H. J., Cummings, D. E., and Kaplan, J. M. (2003b). Vagotomy dissociates short- and long-term controls of circulating ghrelin. *Endocrinology* **144,** 5184–5187.

Wren, A. M., Seal, L. J., Cohen, M. A., Brynes, A. E., Frost, G. S., Murphy, K. G., Dhillo, W. S., Ghatei, M. A., and Bloom, S. R. (2001). Ghrelin enhances appetite and increases food intake in humans. *J. Clin. Endocrinol. Metab.* **86,** 5992–5995.

Yabuki, A., Ojima, T., Kojima, M., Nishi, Y., Mifune, H., Matsumoto, M., Kamimura, R., Masuyama, T., and Suzuki, S. (2004). Characterization and species differences in gastric ghrelin cells from mice, rats and hamsters. *J. Anat.* **205,** 239–246.

Yildiz, B. O., Suchard, M. A., Wong, M. L., McCann, S. M., and Licinio, J. (2004). Alterations in the dynamics of circulating ghrelin, adiponectin, and leptin in human obesity. *Proc. Natl. Acad. Sci. USA* **101,** 4531–4536.

Zhang, W., Lin, T. R., Hu, Y., Fan, Y., Zhao, L., Stuenkel, E. L., and Mulholland, M. W. (2004a). Ghrelin stimulates neurogenesis in the dorsal motor nucleus of the vagus. *J. Physiol.* **559,** 729–737.

Zhang, W., Zhao, L., Lin, T. R., Chai, B., Fan, Y., Gantz, I., and Mulholland, M. W. (2004b). Inhibition of adipogenesis by ghrelin. *Mol. Biol. Cell* **15,** 2484–2491.

CHAPTER EIGHT

Ghrelin Gene-Related Peptides Modulate Rat White Adiposity

Andrés Giovambattista,* Rolf C. Gaillard,[†] and Eduardo Spinedi*,[†]

Contents

I. Introduction	172
II. Effects of Ghrelin on Rat Retroperitoneal Adipocyte Endocrine Functions	176
A. Animals	176
B. RP adipocyte isolation	176
C. Set up of the adipocyte culture system for evaluation of the leptin-releasing activity of ghrelin-related substances	177
D. Adipocyte RNA isolation and analyses	177
E. Leptin measurement	178
F. RP adipocyte endocrine function	179
III. Desacyl Ghrelin as a Potential Physiological Modulator of Adiposity	183
A. Isolation and differentiation of RP preadipocytes	184
B. Determination of glucose concentration in culture medium	185
C. RT-PCR analysis	186
IV. Discussion and Remarks	193
Acknowledgments	199
References	199

Abstract

It is known that ghrelin and des-N-octanoyl (desacyl) ghrelin modulate food intake and adipogenesis *in vivo*. However, desacyl ghrelin represents the majority of ghrelin forms found in the circulation. The present study explored whether ghrelin gene-derived compounds could modulate, *in vitro*, adipocyte endocrine function and preadipocyte differentiation. Retroperitoneal (RP) adipocytes were cultured in the absence or presence of either ghrelin or desacyl ghrelin and in combination with either inhibitors of protein synthesis, insulin,

---

* Neuroendocrine Unit, Multidisciplinary Institute on Cell Biology (CONICET-CICPBA), 1900 La Plata, Argentina
[†] Division of Endocrinology, Diabetology and Metabolism, University Hospital (CHUV), CH 1011 Lausanne, Switzerland

dexamethasone (DXM), or GHSR1a antagonist. The results indicate that both ghrelin forms possess a direct leptin-releasing activity (LRA) on RP adipocytes and significantly enhanced adipocyte *ob* mRNA expression. These activities were related and unrelated to the activation of GHSR1a after coincubation with ghrelin and desacyl ghrelin, respectively. Moreover, desacyl ghrelin facilitated RP preadipocyte differentiation. Desacyl ghrelin enhanced cell lipid content, and PPARγ2, and LPL mRNAs expression. The LRAs developed by different substances tested followed a rank order: ghrelin > desacyl ghrelin = insulin ≥ DXM. Additionally, desacyl ghrelin was able to enhance medium glucose consumption by mature adipocytes in culture. These data strongly support that adipogenesis and adipocyte function are processes directly and positively modulated by ghrelin gene-derived peptides, thus further indicating that, besides their effects on food intake, ghrelin gene-derived peptides could play an important role on adiposity for maintaining homeostasis. © 2008 Elsevier Inc.

I. INTRODUCTION

Adipose tissue cells (adipocytes) are highly specialized cells that play a very important role in energy metabolism and homeostasis. Although considered as passive cells for a very long time, and besides their known more for their role of storing energy (in the form of triglycerides), it has been shown that these cells regulate homeostasis via endocrine, paracrine, and autocrine mechanisms (Mohamed-Ali *et al.*, 1998). These functions thus influence metabolic activity at the adipocyte level itself as well as several other tissues. Many, although not all, of the substances produced by adipocytes are shown in Fig. 1. Although the most important substance secreted by adipocytes is fatty acids, adipokines is the accepted term for multiple proteins and factors secreted by adipocytes (Frühbeck *et al.*, 2001; Rajala and Scherer, 2003; Trayhurn and Beattie, 2001; Trayhurn and Wood, 2004). These substances are characterized as having a high diversity in both structure and function. Among them, we can rapidly mention cytokines/chemokines, acute-phase proteins of the inflammatory process, and long (still open) miscellaneous substances. However, adipocyte leptin has played a revolutionary role in relation to the control of energy balance of the organism. Thus, adipocytes became highly specialized endocrine cells. The adipogenic process is a complex one; however, it could be simply resumed as follows: Cells termed preadipocytes, once arrived to growth arrest, can differentiate by the action of appropriate combination of mitogenic and adipogenic signals. Then, cells continue subsequent steps of differentiation, leading to the acquisition of morphological and biochemical characteristics of mature, highly specialized, adipose tissue cells.

Among many other ways, energy homeostasis is, at least in part, under the control of substances derived from the stomach (Asakawa *et al.*, 2005).

Ghrelin Peptides and Adipose Tissue Function 173

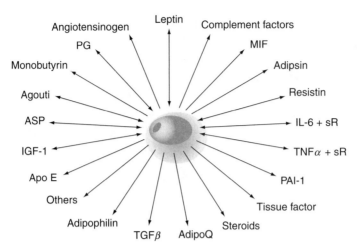

Figure 1 The adipocyte as a source of adipokines and factors, and the autocrine regulation of adipose tissue cell function.

In fact, the stomach-derived peptide hormone named ghrelin (Kojima et al., 1999) is mainly produced in this organ by the X/A-like cells within the oxyntic glands of the gastric fundus mucosa (Sakata et al., 2002). However, although in minor amount, it is expressed in several tissues: placenta (Gualillo et al., 2001a; testis (Tena-Sempere et al., 2002); ovary (Caminos et al., 2003b); kidney (Mori et al., 2000); pituitary (Caminos et al., 2003a); small intestine (Date et al., 2000); pancreas (Volante et al., 2002a); immune cells (Hattori et al., 2001); brain (Lu et al., 2002); and lung (Volante et al., 2002b). Ghrelin is a 28-amino acid peptide with an acyl side chain, n-octanoic acid at serine 3, and this structure (Fig. 2) is essential for developing its biological activity. Two distinct growth hormone (GH) secretagogue receptor (GHSR) cDNAs have been isolated (Howard et al., 1996), the type 1a and 1b. A highly conserved gene in the human, chimpanzee, rat, and mouse encodes the GHSR. This gene is located at 3q26.2, and spans ~4.3 kb (Petersen et al., 2001). Ghrelin binds to GHSR1a (Kojima et al., 2001), a receptor consisting of 366 amino acids with seven-transmembrane regions. This receptor also specifically binds peptidyl (hexarelin) and nonpeptidyl (spiroindoline MK-0677) compounds (Smith et al., 1997). The GHSR1a is G-protein–coupled receptor (Gnanapavan et al., 2002; Guan et al., 1997), which is different from that of GH releasing hormone (GHRH). The GHSR1b form is a nonspliced, nonfunctional receptor mRNA variant, consisting of 289 amino acids with five-transmembrane regions, the nucleotide sequence 1–265 being identical to type 1a. While type 1a activation leads ionic Ca release through the stimulation of subunit Gα11 subunit, type 1b intracellular signaling seems to be due to activation of phospholipase C. Other ghrelin-related peptides have been identified; for example, des-Gln14-ghrelin,

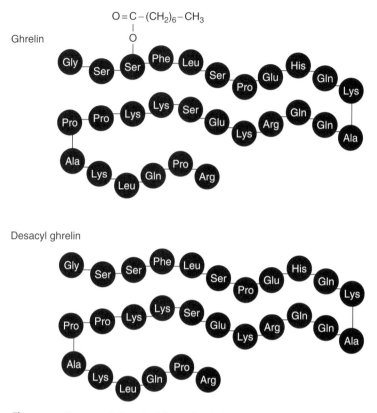

Figure 2 Structural chemical formula of ghrelin gene-derived peptides.

a 27-amino acid peptide, also acylated on serine 3, but a deletion of glutamine in position 14 characterizes this structure. Des-Gln14-ghrelin is generated by alternative splicing of the ghrelin gene (Hosoda et al., 2000a; Kojima et al., 1999) and, because of the presence of the acyl side chain, has similar ghrelin biological activities. Interestingly, a nonacylated form of ghrelin (desacyl ghrelin) can be found in both stomach and blood (Hosoda et al., 2000b). Although both ghrelin and desacyl ghrelin are present in the peripheral circulation, desacyl ghrelin represents the major circulating form of ghrelin-related peptides in mammals. It happens because the fatty acid attached to the serine 3 is highly unstable (Hosoda et al., 2004). In fact, physiological ghrelin plasma concentration (in the fmol/ml range) in humans represents ∼6% from total ghrelin-related forms in circulation (Yoshimoto et al., 2002), thus resulting desacyl ghrelin as the principal responsible for such difference.

Ghrelin strongly stimulates food intake, which leads to weight gain, and also reduces fat depot utilization after administration in rodents (Broglio et al., 2003) and humans (Tschop et al., 2000; Wren et al., 2000, 2001).

In rodents, ghrelin expression is upregulated by fasting, hypoglycemia, or leptin administration (Toshinai et al., 2001), and also after chronic undernutrition (Gualillo et al., 2002). Thus clearly indicating that negative energy balance condition of the organism triggers ghrelin expression and secretion. Conversely, caloric intake or chronically positive balance condition suppresses ghrelin expression and release. There is evidence that no gender-dependent difference accounts for ghrelin levels in both humans (Otto et al., 2001) and rats (Gualillo et al., 2001b).

Regarding the mechanism of action of ghrelin in food intake stimulation, it is accepted that central or peripheral ghrelin administration results in an enhancement in the hypothalamic expression of mRNA for agouti-related protein (AgRP) and neuropeptide Y (NPY) (Nakazato et al., 2001). However, because ghrelin analogs are able to stimulate food intake in NPY null mice, a major role for AgRP in hypothalamic ghrelin effect on food intake is more accepted (Tung et al., 2001). Mechanisms opposed to these have been proposed and these relate to the adipocyte-derived anorectic signal, leptin (Horvath et al., 2001; Kamegai et al., 2001; Toshinai et al., 2003; Tschop et al., 2002). In spite of the fact that the major source of ghrelin-related peptides is located in the periphery rather than in the central nervous system (CNS), Banks et al. (2002) found an important physiological implication regarding the passing of ghrelin across the blood–brain barrier (BBB) in mice. They described that there exists a bidirectional passage of this peptide, and that while ghrelin is easily delivered from the CNS to the periphery, due to a saturable process; conversely, desacyl ghrelin rapidly enters from the circulation into the brain by a nonsaturable diffusion process. This finding simply accords with the fact that desacyl ghrelin is the most abundant ghrelin-related form found in the circulation. However, up to now, only GHSR has been described to be present in the arcuate nucleus (AN) of hypothalamus (Mano-Otagiri et al., 2006), mainly on GHRH and NPY neurons (Mano-Otagiri and Shibasaki, 2006), a nonfunctional receptor for desacyl ghrelin. Moreover, it has been suggested that hypothalamic effects of desacyl ghrelin on food intake, although controversial, take place even in GHSR-deficient rats (Toshinai et al., 2006).

Finally, regarding some effects of ghrelin-related gene products on adiposity, there are data from the literature indicating that rats lacking hypothalamic GHSR display not only decreased GH secretion and food intake, but also adiposity (Shuto et al., 2002). It is known that GHSR expression is enhanced during adipogenesis in rats (Choi et al., 2003); and that ghrelin administration in rodents induces weight gain, a process due to, at least in part, to decreased fat utilization (Kalra et al., 1999). Few studies have provided data sustaining ghrelin adipogenic activity (Choi et al., 2003; Thompson et al., 2004). Moreover, only one report supports for a direct adipogenic effect of ghrelin in rats (Choi et al., 2003). While ghrelin effects are exerted through GHSR1a (Matsumoto et al., 2001), conversely, des-N-octanoyl (desacyl) ghrelin seems to act after the activation of a receptor different from GHSR1a (Thompson *et al.*, 2004; Toshinai *et al.*, 2006). However, both ghrelin-related forms share common effects on different tissues, for example, inhibiting cell death of cardiomyocytes and endothelial cells (Baldanzi *et al.*, 2002), and they develop similar vasodilator potency (Kleinz *et al.*, 2006). Conversely, other reports indicate that inverse effects of these two ghrelin-related compounds operate for regulation of energy balance (Asakawa *et al.*, 2005; Chen *et al.*, 2005b). It has been suggested that desacyl ghrelin can either stimulate (Nonogaki *et al.*, 2006; Toshinai *et al.*, 2006) or inhibit (Chen *et al.*, 2005a) food intake in rats.

Therefore, we explored the hypothesis that a direct effect of peripheral ghrelin gene-derived compounds (ghrelin and desacyl ghrelin) on adiposity could play a physiological role in the maintenance of homeostasis.

II. Effects of Ghrelin on Rat Retroperitoneal Adipocyte Endocrine Functions

A. Animals

Adult (300–330 g BW) male Fischer 344N rats were used in our experiments. Animals were maintained in plastic cages (five rats per cage), in a temperature (21–23 °C)- and light (lights on 0700–1900 h)-controlled room. Food and water was available *ad libitum*. Rats were killed by decapitation following NIH guidelines for care and use of experimental animals on experimental day at 0900 h. Then, retroperitoneal (RP) fat pads, aseptically dissected, were transferred into sterile Petri dishes and weighed, and immediately after 10 ml of sterile Krebs–Ringer–MOPS medium was added (Sigma Chem. Co., St. Louis, MO; Krebs–Ringer:double distilled H_2O:MOPS = 1:3:1; pH 7.4) medium was added.

B. RP adipocyte isolation

RP adipocytes were obtained by previously described methodology (Perello *et al.*, 2003a), although with minor modifications. Preweighed RP fat pads were received in sterile plastic tubes, containing 4 ml/g fat of Krebs–Ringer–MOPS medium containing 1% BSA (w/v), antibiotics, and 1 mg/ml Collagenase type 1 (Sigma; pH 7.4). Tissues were then incubated for 40 min (at 37 °C) with gentle shaking. Thereafter, fat suspension was filtered through a nylon cloth and centrifuged (30 s at 400 rpm) at room temperature. Infranatant was then aspirated, and adipocytes were washed (three consecutive times) with 10 ml of fresh sterile Krebs–Ringer–MOPS–BSA medium, followed by centrifugation. Washed cells were then diluted with an appropriate volume of Dulbecco's modified Eagle's medium

(DMEM) (Sigma) [supplemented with 1 % BSA (w/v) (Sigma), 1% FCS (v/v), and antibiotics, pH 7.4 (culture medium)]; the volume used was that necessary to render in ~250,000 adipocytes/ml of medium.

C. Set up of the adipocyte culture system for evaluation of the leptin-releasing activity of ghrelin-related substances

After isolation, adipocytes were seeded (800 μl of suspension per well) onto 24-well plates; then, 200 μl of the following test solutions were added: culture medium either alone (basal) or containing insulin (0.01–10 nM; Novo Nordisk Pharma AG, Switzerland) (Piermaria et al., 2003), dexamethasone phosphate (1–100 nM; Sidus Lab., Argentina) (Piermaria et al., 2003), ghrelin (0.001–1 nM; Phoenix Pharma., Inc., Belmont, CA), desacyl ghrelin (0.01–10 nM; Phoenix), cycloheximide (35.5 μM; Sigma, previously dissolved in 0.01% ethanol) (Glasow et al., 2001), actinomycin D (0.5 μM; Sigma) (Glasow et al., 2001), ghrelin antagonist (Holst et al., 2003) [10–1000 nM; ghrelin-A (D-Arg^1, D-Phe^5, D-Trp^{7,9}, Leu^{11})-substance P; Phoenix] or, when appropriate, different combinations of these substances. After addition of different solutions, RP adipocytes were cultured (at 37 °C) for different time periods (0.5–48 h) in a sterile atmosphere of 95% air–5% CO_2 (85–87% humidity). At the end of incubation, media were aspirated and kept frozen (−20 °C) for further measurement of leptin concentrations. Remaining adipocytes were used for RNA extraction.

D. Adipocyte RNA isolation and analyses

Total RNA was isolated from remaining adipocytes after 24 h culture, by a modification of the single-step, acid guanidinium isothiocyanate–phenol–chloroform extraction method described by Chomczynski and Sacchi (1987) (Trizol; Invitrogen, Life Tech.; catalog number 15596-026). The yield and quality of extracted RNA were assessed by 260/280 nm optical density ratio and electrophoresis in denaturing conditions on 2% agarose gel.

1. RT-PCR

For RT-PCR analysis, 1 μg of total RNA was incubated with 0.2 mM dNTPs, 1 mM $MgSO_4$, 1 M *ob* primers, 0.25 M β-actin primers, 0.1 U/μl AMV reverse transcriptase (5 U/μl), and 0.1 U/μl Tfl DNA polymerase (5 U/μl); final volume: 25 μl. Amplifications were done in a thermal cycler (Perkin–Elmer) in the following conditions: 48 °C—45 min for reverse transcription step (one cycle); 94 °C—2 min for AMV reverse transcriptase inactivation and RNA/cDNA/primer denaturing (one cycle); 94 °C—30 s for denaturing; 55 °C—1 min for annealing; 68 °C—2 min for extension (40 cycles); 68 °C—7 min for final extension (one cycle), and 4 °C for soak

(Promega Access RT-PCR System No. A1250). Primers were designed for a high homology region of the *ob* gene: (F) 5'-CCC ATT CTG AGT TTG TTC A-3', and (R) 5'-GCA TTC AGG GCT AAG GTC-3' (300 bp) (GenBank accession number: U48849). In this semiquantitative technique, the second set of primers was specific for the β-actin gene, with the following sequences: (F) 5'-TTG TAA CAA ACT GGG ACG ATA TGG-3', and (R) 5'-GAT CTT GAT CTT CAT GGT GCT AGG-3' (764 bp) (GenBank accession number: NM031144). Controls without reverse transcriptase were systematically performed to detect cDNA contamination. The amplified products were analyzed on 2% agarose gel and visualized by ethidium bromide UV transillumination in a Digital Imaging System (Kodak Digital Science, Electrophoresis Documentation and Analysis 120 System).

2. Northern blot

For Northern blot analysis, 10-μg samples of total RNA per lane were run in 1% formaldehyde–agarose denaturing gels and transferred to Hybond-N membranes (Amersham, Arlington Heights, IL) in 20× sodium chloride-sodium citrate (SSC) (3 M NaCl and 0.3 M sodium citrate·2H$_2$O) solution. Denaturing gels were run at 75 V for 4 h and washed twice with 10× SSC before setting up transfer in 20× SSC. *ob* and β-actin cDNAs were obtained by RT-PCR as described earlier. Probes were labeled to high specific activity with (α-^{32}P)dCTP (Amersham) by nick translation. Hybridizations were done overnight at 65 °C. Membranes were washed twice with 2× SSC + 0.1% sodium dodecyl sulfate (SDS) for 15 min at room temperature and once with 0.1× SSC + 0.1% SDS for 15 min at 55 °C. Membranes were then exposed to Kodak BioMax MR film for 12 h at −80 °C with dual intensifying screen and analyzed in a Digital Imaging System (Kodak Digital Science).

E. Leptin measurement

Medium leptin concentrations were determined by a specific radioimmunoassay developed in our laboratory (Giovambattista *et al.*, 2000). Briefly, synthetic murine leptin (PrePro Tech, Inc.) was used for both labeled peptide, standards, and for the development of antileptin serum. Leptin was radioiodinated by the chloramine-T method and purified in a Sephacryl S-300 (Sigma) column equilibrated with sodium phosphate (0.05 M)–BSA (2 g/liter)–sodium azide (10 mg/liter) solution (pH 7.4). The antileptin serum was raised in rabbit using murine leptin coupled to BSA. The detection range of the standard curve was 0.2–25 ng/ml. The assay displayed 2% and 0% crossreactivity with human leptin and mouse/rat anterior pituitary hormones, respectively. The intra- and interassay coefficients of variation were 5–8% and 10–13%, respectively.

F. RP adipocyte endocrine function

1. Time course of the insulin-stimulated leptin output by isolated RP adipocytes

In order to assess the optimal conditions of our adipocyte system in terms of leptin output, cells were cultured in the absence (baseline) or presence of 1 nM insulin for different periods of times: 0.5, 12, 24, and 48 h. The results are included in Fig. 3. As depicted in Fig. 3 (upper panel), spontaneous leptin output by RP adipocytes in culture enhanced in a time-dependent fashion, being maximal at culture time 48 h. When insulin (1 nM) was tested, adipocytes also developed a time-dependent response in terms of leptin secretion into the medium, and these adipocyte responses to insulin were significantly ($p < 0.05$) higher than the respective baseline, regardless of the time examined. When adipocyte responses to insulin stimulation were expressed as net leptin released into the medium (total release minus the baseline), the data indicated that adipocyte responses were developed in a concentration-related fashion (Fig. 3, lower panel).

2. Acute effects of insulin, dexamethasone, and ghrelin on leptin secretion by RP adipocytes

Because an endocrine function is characterized by a relatively fast secretory response, it became important to test whether ghrelin could affect preformed pool of adipocyte leptin in a short time period of incubation. For this purpose, mature RP adipocytes were cultured, in metabolic conditions, for 0.5 h in the absence (baseline) or presence of several concentrations of ghrelin. The data in Table 1 show that the preformed pool of adipocyte leptin is low, as indicated by the spontaneous adipokine output (baseline). Interestingly, ghrelin was able to significantly ($p < 0.05$) enhance RP adipocyte leptin release when 1 nM or higher concentration was employed. For comparison purposes, the leptin-releasing activity (LRA) of 1 nM insulin and 50 nM dexamethasone (DXM) are also included. These data clearly indicate that two well-characterized leptin secretagogues (insulin and DXM) as well as ghrelin are able to acutely stimulate a significant amount of leptin release by incubated RP adipocytes.

3. Evaluation of the leptin-releasing activity of different leptin secretagogues by RP adipocytes in 24-h culture

The following experiments were run in order to assess whether changes in physiological ghrelin concentrations, such as those resembling an *in vivo* 24-h fasting period, could play an important role in maintaining homeostasis. For this purpose, RP adipocytes were cultured for 24 h in the absence or presence of graded concentrations of ghrelin. The results presented in Fig. 4 indicate that RP adipocytes in culture were able to enhance leptin secretion into the medium in a concentration-dependent fashion. Similarly, insulin and DXM also induced concentration-related adipocyte responses but the

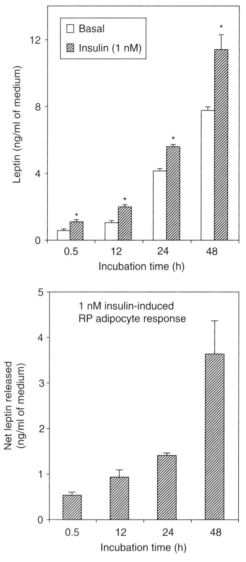

Figure 3 Spontaneous (basal) and 1 nM insulin-stimulated leptin secretion into the medium by isolated retroperitoneal (RP) adipocytes in culture for different periods of times (upper panel). Time-dependent net leptin released (total secretion minus the respective baseline) by dispersed white RP adipocytes in culture with 1 nM insulin (lower panel). Values are the mean ± SEM of four independent experiments ($n = 5$ wells per condition in each experiment). *$p < 0.05$ versus basal values.

Table 1 Medium leptin concentration 0.5 h after incubation of mature RP adipocytes with medium alone (baseline) or containing either ghrelin, insulin, or dexamethasone (DXM)

	Medium leptin (ng/ml)
Baseline	0.58 ± 0.05
Ghrelin (nM)	
0.01	0.66 ± 0.07
0.1	$1.19 \pm 0.08^*$
1	$1.31 \pm 0.10^*$
10	$1.48 \pm 0.16^*$
Insulin (nM)	
1	$1.11 \pm 0.09^*$
DXM	
50	$1.37 \pm 0.13^*$

* $p < 0.05$ versus baseline values.
Data are the mean $\pm$ SEM ($n = 4$ experiments), with 5 wells per condition per experiment.

responses were weaker than that of ghrelin (Table 2). A comparison of the LRA activities of ghrelin, insulin, and DXM, when the effect of each substance was evaluated at concentration 1 nM, is presented in Table 3.

4. Effects of the combination of physiological ghrelin concentration with insulin or dexamethasone on RP adipocyte endocrine function

Because morning physiological ghrelin circulating concentrations, in *ad libitum* eating rats, are in the 0.1 nM range (Hosoda *et al.*, 2000b; Yoshimoto *et al.*, 2002), additional experiments were performed. These explored the aim regarding the behavior of RP adipocytes when cultured, for 24 h, in the presence of ghrelin (0.1 nM) in combination with other known specific leptin secretagogue, such as insulin (1 nM) or DXM (10 nM). Interestingly, the LRA of ghrelin combined with either stimulus resulted in the addition of individual effects (Fig. 5).

5. Effects of adipocyte protein synthesis and GHSR1a blockage on the leptin-releasing activity of ghrelin

These experiments were performed in 24-h cultured RP adipocytes in the presence of culture medium alone or containing ghrelin (0.1 nM). Parallel experiments were run in adipocytes exposed to medium containing a transcription inhibitor (0.5 μM actinomycin D) or a ribosomal protein synthesis inhibitor (35.5 μM cycloheximide) or a GHR-1a antagonist (ghrelin-A; 10–1000 nM) either in the absence or presence of ghrelin (0.1 nM). The results indicated that (Fig. 6): (a) while 0.5 μM actinomycin D and 1000 nM ghrelin-A did not modify basal leptin output, conversely, spontaneous leptin

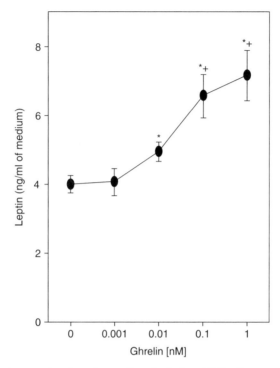

Figure 4 Leptin secretion into the medium by isolated RP adipocytes in culture, for 24 h, in the presence of medium alone (0 nM ghrelin) or containing graded concentrations of ghrelin. Values are the mean ± SEM of—three to four independent experiments ($n = 5$–6 wells per condition in each experiment). *$p < 0.05$ versus basal values. $^+ p < 0.05$ versus 0.01 nM ghrelin values.

release was significantly ($p < 0.05$) diminished by cyclo-heximide; and (b) the LRA of 0.1 nM ghrelin was fully abrogated by either actinomycin, cycloheximide, or ghrelin-A. Moreover, a lower concentration of ghrelin-A (100 nM) although did significantly reduce the LRA of 0.1 nM ghrelin, such effect was only partial (data not shown). These results further indicate that active RP adipocyte RNA and protein synthesis, and GHR-1a are involved in the ghrelin effect on adipokine production.

6. Effect of GHSR1a antagonist on ghrelin modulation of *ob* gene function

These experiments were performed on RP adipocytes obtained 24 h after culture medium alone (basal) or containing 0.1 nM ghrelin either alone or combined with 1000 nM ghrelin-A. *ob* mRNA expression was analyzed by either RT-PCR (Fig. 7, upper panels) or Northern blot (Fig. 7, lower panels), in the culture conditions mentioned earlier. However, adipocytes obtained after culture with ghrelin-A (1000 nM) were also investigated.

Table 2 Leptin-releasing activity on RP adipocytes 24-h cultured without (baseline) or with graded concentrations of either insulin or DXM

	Medium leptin (ng/ml)
Baseline	3.99 ± 0.17
Insulin (nM)	
0.01	3.89 ± 0.27
0.1	4.25 ± 0.32
1	6.35 ± 0.51*
10	6.93 ± 0.59*
DXM (nM)	
1	5.35 ± 0.23*
10	5.78 ± 0.38*
50	6.55 ± 0.49*,+
100	7.99 ± 0.74*,+

* $p < 0.05$ versus baseline values.
+ $p < 0.05$ versus DXM 1 nM values.
Data are the mean ± SEM ($n = 5$ experiments), with 6 wells per condition per experiment.

Table 3 Leptin-releasing activity (LRA), on RP adipocytes in culture (for 24 h), of 1 nM insulin and 1 nM DXM in comparison with that developed by 1 nM ghrelin

Secretagogue (1 nM)	LRA
Ghrelin	1.00
Insulin	0.87
DXM	0.74

The results of analyses indicated that basal *ob* mRNA expression was enhanced ($p < 0.05$ versus basal values) in adipocytes cultured for 24 h in the presence of 0.1 nM ghrelin. The stimulation induced by 0.1 nM ghrelin was completely abolished when 1000 nM ghrelin-A was present in the culture medium, although the ghrelin-A itself did not modify *ob* mRNA expression.

III. Desacyl Ghrelin as a Potential Physiological Modulator of Adiposity

As previously mentioned, desacyl ghrelin is the ghrelin-related form more abundant in the peripheral circulation. Then, we attempted to investigate whether this nonoctanoyl ghrelin form could be implicated as a potential modulator of the adipogenic process.

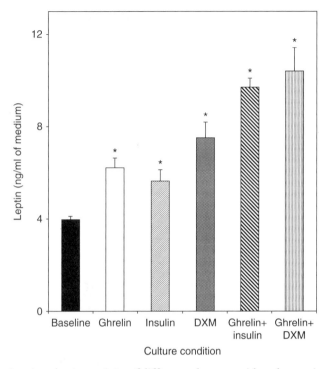

Figure 5 Leptin-releasing activity of different substances, either alone or in combination, on RP adipocytes in 24 h of culture. Substances/combinations tested were: medium alone (baseline), ghrelin (0.1 nM), insulin (1 nM), dexamethasone (DXM; 10 nM), 0.1 nM ghrelin plus 1 nM insulin and 0.1 nM ghrelin plus 10 nM DXM. Values are the mean ± SEM of three to four independent experiments ($n = 6$ wells per condition in each experiment). *$p < 0.05$ versus baseline values.

A. Isolation and differentiation of RP preadipocytes

Isolated fibroblastic preadipocytes were prepared following previously reported methods (Giovambattista et al., 2006; Mitchell et al., 1997). Briefly, after dissection of RP fat pads (from adult male Sprague–Dawley rats), they were weighed and transferred into sterile tubes containing (4 ml/g of fat) Krebs–Ringer–MOPS medium (Sigma), with 1 % BSA (w/v), antibiotics, 1 mg/ml Collagenase type 1 (Sigma; pH 7.4), and incubated for 40 min at 37 °C with gentle shaking. Then, the suspension was filtered through a nylon cloth and centrifuged (30 s at 400 rpm) at room temperature. The pellet containing the stroma-vascular fraction was resuspended and washed (two to three times) with DMEM (Sigma) containing 10% FCS (Gibco BRL), 100 IU/ml penicillin (Gibco BRL), and 100 μg/ml streptomycin (Gibco BRL). After the final wash, cells were resuspended in an appropriate volume of medium in order to obtain $\sim 2 \times 10^4$ cells/0.5 ml of medium. This volume was plated onto 24-well plates and cultured for 48–72 h at

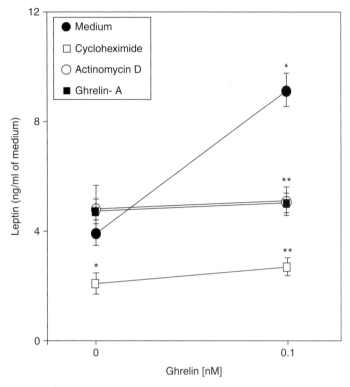

Figure 6 Effect of cycloheximide (35.5 μM), actinomycin D (0.5 μM), or GHSR1a antagonist (ghrelin-A; 1000 nM) on spontaneous (medium: 0 nM ghrelin) and 0.1 nM ghrelin-induced leptin secretion by 24 h-cultured RP adipocytes. Values are the mean ± SEM of three to four independent experiments (n = 5–6 wells per condition in each experiment). *p < 0.05 versus medium 0 nM ghrelin values. **p < 0.05 versus medium 0.1 nM ghrelin values.

37 °C in a 5% CO_2 atmosphere. Thereafter, differentiation was induced by the addition of DMEM medium [containing 10% FCS, 100 IU/ml penicillin, 100 μg/ml streptomycin, 125 μM indomethacin (Montpellier Lab., Argentina)] and 5 μg/ml insulin (Sidus Lab.) (now defined as differentiation medium; DM) (Knight *et al.*, 1987). DM was replaced by fresh DM every other day, and up to day 8 of culture.

B. Determination of glucose concentration in culture medium

Media glucose concentrations were evaluated by a previously reported (Moreno *et al.*, 2006) enzymatic-colorimetric assay purchased from Weiner Lab. (Argentina).

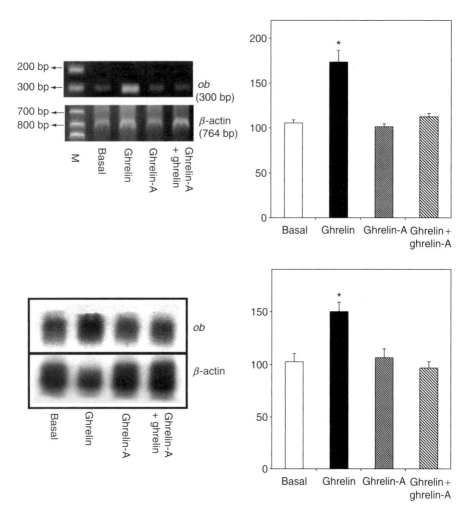

Figure 7 Representative RT-PCR (upper panel) and Northern blot (lower panel) detections of *ob* and *β*-actin mRNAs in 24-h cultured isolated RP adipocytes. Cells were cultured with medium alone (basal) or containing 0.1 nM ghrelin, 1000 nM ghrelin antagonist (ghrelin-A) or 0.1 nM ghrelin plus 1000 nM ghrelin-A (M: molecular marker, 100-bp ladder). Semiquantitative analyses, expressed in arbitrary units, are also shown in respective right-side panels. Values are the mean ± SEM of three different experiments ($n = 5$ wells per condition in each experiment). *$p < 0.05$ versus basal values.

C. RT-PCR analysis

Total RNA was isolated as mentioned in Section II.D.1. Primers were designed for a high homology region of the PPARγ2 gene: (F) 5′-TGG GTG AAA CTC TGG GAG AT-3′, and (R) 5′-CCA TAG TGG AAG

CCT GAT GC-3′ (453 bp) (GenBank accession number: Y12882); LPL gene: (F) 5′-AGG ACC CCT GAA GAC AC-3′, and (R) 5′-GGC ACC CAA CTC TCA TA-3′ (149 bp) (GenBank accession number: NM012598); and *ob* gene: (F) 5′-CCC ATT CTG AGT TTG TTC A-3′, and (R) 5′-GCA TTC AGG GCT AAG GTC-3′ (300 bp) (GenBank accession number: U48849). In this semiquantitative technique, the second set of primers was specific for the β-actin gene, having the following sequences: (F) 5′-TTG TAA CAA ACT GGG ACG ATA TGG-3′, and (R) 5′-GAT CTT GAT CTT CAT GGT GCT AGG-3′ (764 bp) (GenBank accession number: NM031144). Controls without reverse transcriptase were systematically performed to detect cDNA contamination. The amplified products were analyzed on 2% agarose gel and visualized by ethidium bromide UV transillumination in a Digital Imaging System (Kodak).

1. Effect of desacyl ghrelin on adipogenesis

As depicted earlier, after 48–72 h in culture, cells were incubated with DM either alone or with desacyl ghrelin (0.1–10 nM; Phoenix). On culture days 4, 6, and 8, media were removed and kept frozen ($-80\ °C$) until the measurement of leptin concentrations. Remaining cells were washed with PBS twice, fixed in 3.7% formaldehyde for 1 h and then stained with 0.6% (w/v) Oil Red O solution (60% isopropanol, 40% water) for 2 h at room temperature. Cells were then washed with water to remove the unbound dye. Stained Oil Red O was eluted with isopropanol and quantified by measuring the optical absorbance at 510 nm (Fujimoto *et al.*, 2004). Additional cells incubated with DM alone or containing 1 nM desacyl ghrelin were harvested, on culture day 8, for total RNA extraction and further evaluation of PPARγ2 and lipoproteinlipase (LPL) mRNA expression. These results indicate that DM-induced adipogenesis was developed in a time-related fashion. Interestingly, 1 nM desacyl ghrelin (but not lower concentrations, data not shown) added into the DM facilitated preadipocyte differentiation induced by incubation with DM (Frost and Lane, 1985). In fact, cell lipid content was significantly ($p < 0.05$ versus DM alone values) increased by 1 nM desacyl ghrelin, regardless of the culture day examined (Fig. 8). Cells incubated with DM alone already released a significant amount of leptin on day 2 of culture, an effect remaining until culture day 8 (Fig. 8). Interestingly, addition of 1 nM or higher desacyl ghrelin was able to significantly enhance ($p < 0.05$ versus DM alone values) leptin secretion into the medium from day 6 of differentiation (Fig. 9). Moreover, harvested cells, on day 8 of culture, were examined for evaluation of other differentiation markers; it was found that 1 nM desacyl ghrelin significantly ($p < 0.05$ versus DM only values) enhanced mRNA expression of both LPL and PPARγ2 (Fig. 10).

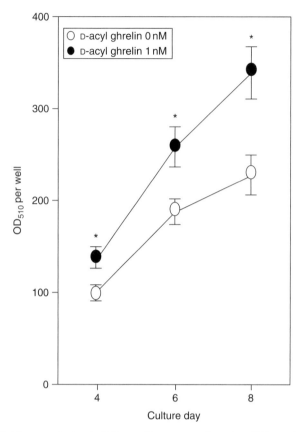

Figure 8 Lipid accumulation in RP preadipocytes undergoing differentiation (on different days) induced by medium either alone (desacyl ghrelin: D-acyl ghrelin 0 nM) or containing 1 nM D-acyl ghrelin. Values are the mean ± SEM of four different experiments ($n = 6$ wells per condition per experiment). $*p < 0.05$ versus D-acyl ghrelin 0 nM values on similar culture day.

2. Effect of desacyl ghrelin on, *in vitro* differentiated, RP adipocyte function

These studies were performed on RP adipocytes, obtained after *in vitro* differentiation (cultured RP preadipocytes for 8 days in the presence of DM only). After removing the DM, it was replaced by incubation medium (IM) either alone or containing graded concentrations of desacyl ghrelin (0.1–10 nM). Then RP adipocytes were incubated for different periods of time (6–24 h). At the end of each culture time, medium was removed and leptin concentration was then evaluated. The results indicate (Fig. 11) that spontaneous leptin secretion was already detected 6 h after culture. Moreover, basal leptin output increased with the incubation time. When adipocytes were exposed to desacyl ghrelin, this ghrelin form was able to enhance

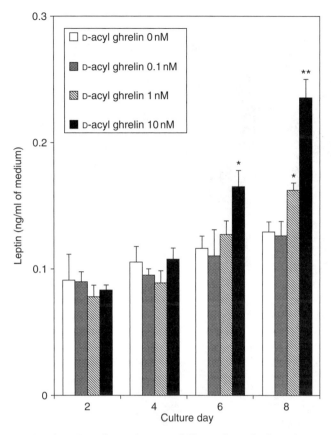

Figure 9 Leptin release into the medium, on different days of culture, by preadipocytes undergoing differentiation in the presence of medium alone (D-acyl ghrelin 0 nM) or containing graded (0.1–10 nM) concentrations of D-acyl ghrelin. Values are the mean ± SEM of four different experiments ($n = 6$ wells per condition per experiment). * $p < 0.05$ versus D-acyl ghrelin 0 nM values on similar day. **$p < 0.05$ versus D-acyl ghrelin 1 nM values on similar day.

leptin secretion above baseline, although this occurred in relation to both the concentration of the compound and the culture time. In fact, only 1 and 10 nM were able to exert this effect and only when adipocytes were exposed to desacyl ghrelin for 12 and 24 h. Interestingly, 12-h incubation of adipocytes with 1 nM desacyl ghrelin resulted in significant ($p < 0.05$) enhancement of *ob* mRNA expression over values obtained for adipocytes incubated with IM only for a similar time period (Table 4). We also compared the LRA of different leptin secretagogues on RP adipocytes after *in vitro* complete differentiation. Table 5 shows the results of RP adipocytes, incubated 12 h, in the absence (baseline) or presence of either desacyl ghrelin (0.1–10 nM), ghrelin (0.01–1 nM), or insulin (1–100 nM).

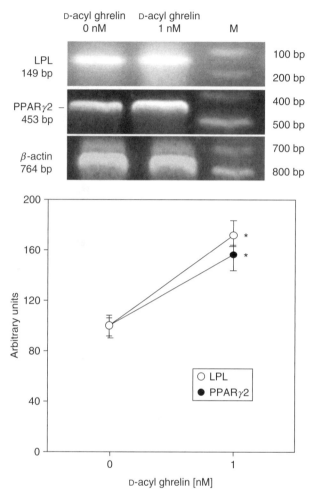

Figure 10 Representative RT-PCR detection of RP adipocyte PPARγ2, LPL, and β-actin mRNAs expression (upper panel). Fully differentiated adipocytes were obtained after 8-day incubation of preadipocytes with either differentiation medium alone (D-acyl ghrelin 0 nM) or containing 1 nM D-acyl ghrelin (M: molecular marker, 100-bp ladder). Semiquantitative analyses, expressed in arbitrary units, are also depicted (lower panel). Values are the mean ± SEM of three different experiments ($n = 5$ wells per condition in each experiment). *$p < 0.05$ versus respective D-acyl ghrelin 0 nM values.

All substances tested developed concentration-related adipocyte responses. Being ghrelin more effective (on molar basis) than both desacyl ghrelin and insulin to secrete leptin into the culture medium, the leptin-releasing activities of these compounds (when tested at 1 nM) were: ghrelin > desacyl ghrelin = insulin. The glucose medium consumption (Fig. 12) by 12-h cultured RP adipocytes in the presence of desacyl ghrelin (1 or 10 nM) was

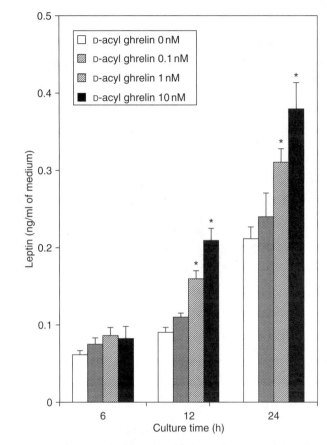

Figure 11 Time- and concentration-related leptin-releasing activity of D-acyl ghrelin on RP adipocytes in culture (24 h). Data are the mean ± SEM ($n = 4$ experiments, with 5 wells per condition in each experiment). *$p < 0.05$ versus D-acyl ghrelin 0 nM values on similar culture time.

significantly ($p < 0.05$) greater than values obtained in basal conditions (secretagogue 0 nM). Moreover, the effect of desacyl ghrelin and insulin on medium glucose consumption was equivalent.

3. Effects of RNA and protein synthesis inhibitors, and GHSR1a antagonist on desacyl ghrelin-induced *ob* production

RP adipocytes (obtained after *in vitro* differentiation) were 12-h cultured without (medium only) or with 1 nM desacyl ghrelin alone or in combination with either 0.5 μM actinomycin D or 35.5 μM cycloheximide, respectively. These results are shown in Fig. 13. As depicted, whereas incubation with 0.5 μM actinomycin D had no significant effect on spontaneous

Table 4 Adipocyte *ob* mRNA expression, relative to β-actin, after 12 h of culture with incubation medium (IM) either alone (baseline) or containing 1 nM desacyl ghrelin

	Arbitrary units
Baseline	101 ± 11
Desacyl Ghrelin (1 nM)	$152 \pm 13^*$

* $p < 0.05$ versus baseline values.
Desacyl ghrelin. Data, expressed in arbitrary units, are the mean $\pm$ SEM ($n = 3$ experiments, with 6 wells per condition in each experiment).

Table 5 Medium leptin concentration after 12 h incubation of RP adipocytes with medium alone (baseline) or containing graded concentrations (in nM) of either ghrelin, desacyl ghrelin or insulin

	Medium leptin (ng/ml/12 h)
Baseline	0.092 ± 0.008
Ghrelin	
0.01	$0.156 \pm 0.007^*$
0.1	$0.191 \pm 0.009^*$
1	$0.205 \pm 0.012^{**}$
Desacyl ghrelin	
0.1	$0.144 \pm 0.007^*$
1	$0.163 \pm 0.019^*$
10	$0.215 \pm 0.011^{**}$
Insulin	
1	$0.137 \pm 0.012^*$
10	$0.215 \pm 0.010^{**}$
100	$0.238 \pm 0.018^{**}$

* $p < 0.05$ versus baseline values.
** $p < 0.05$ versus baseline and values obtained with the respective lowest concentration.
Data are the mean $\pm$ SEM; $n = 4$ experiments, with 5 wells per condition per experiment.

(incubation with medium alone) adipocyte leptin output, conversely, this parameter was significantly ($p < 0.05$) reduced by incubation with 35.5 μM cycloheximide. As it can also be seen in Fig. 13, 1 nM desacyl ghrelin-elicited leptin output was fully prevented when adipocytes were coincubated with 0.5 μM actinomycin as well as with 35.5 μM cycloheximide. These data provide evidence that RNA and protein synthesis are mediating the leptin-releasing activity of desacyl ghrelin. Finally, to determine whether desacyl ghrelin (1 nM)-stimulated leptin secretion by cultured adipocytes could be mediated by GHSR1a, 12-h cultured adipocytes were exposed to GHSR1a-A (10–1000 nM) either alone or in combination

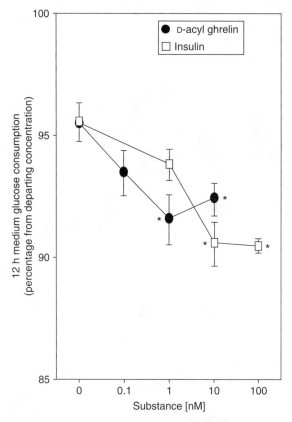

Figure 12 Medium glucose consumption by cultured (12 h) adipocytes with medium alone (0 nM substance) or containing several concentrations of different test substances (D-acyl ghrelin or insulin). Values are the mean ± SEM ($n = 4$ experiments, with 5 wells per condition per experiment). *$p < 0.05$ versus 0 nM substance values.

with 1 nM desacyl ghrelin. The results indicate (Fig. 13) that lower concentrations (data not shown) or even 1000 nM GHSR1a-A did not modify either spontaneous leptin output or 1 nM desacyl ghrelin-elicited leptin output into the culture medium arguing for a noninvolvement of GHSR1a in this desacyl ghrelin effect on RP adipocytes.

IV. Discussion and Remarks

The present data strongly support that both ghrelin and desacyl ghrelin stimulate white adipose tissue endocrine function. Moreover, we demonstrated that desacyl ghrelin, the most common ghrelin gene-derived form

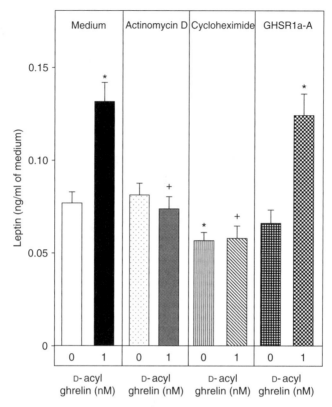

Figure 13 Effects of actinomycin D (0.5 μM), cycloheximide (35.5 μM), and GHSR1a antagonist (GHSR1a-A; 1000 nM) on spontaneous (D-acyl ghrelin 0 nM) and 1 nM D-acyl ghrelin-stimulated leptin release by 12-h cultured adipocytes (desacyl ghrelin: D-acyl ghrelin). Date are the mean ± SEM of three independent experiments ($n = 6$ wells per condition in each experiment). $^*p < 0.05$ versus medium only values (D-acyl ghrelin 0 nM). $^+p < 0.05$ versus medium 1 nM D-acyl ghrelin values.

present in peripheral circulation also facilitated rat white RP preadipocyte differentiation. Although these compounds share some similar activities at the adipose tissue level, their actions seem to take place through different receptor-mediated mechanisms. While ghrelin effects were fully abolished by a specific GHSR1a antagonist, desacyl ghrelin effect was not modified by the blockage of this adipocyte receptor. This study reinforces the concept that ghrelin-related substances regulate both energy balance (Gualillo et al., 2002; Kojima et al., 1999; Nonogaki et al., 2006) and adipogenesis (Choi et al., 2003; Thompson et al., 2004; Tschop et al., 2000).

In fact, ghrelin stimulated rat white RP adipocyte *ob* production, mRNA expression, and secretion through a GHSR1a-mediated mechanism. Ghrelin is a natural endogenous ligand for GHSR1a; it stimulates GH release from somatotropes (Kojima et al., 1999) and food intake (Kalra et al., 2003).

In vitro studies claim for an adipogenic, PPAR-γ2-mediated effect of ghrelin on rat adiposity (Choi *et al.*, 2003), and because ghrelin inhibits isoproterenol-induced lipolysis (Muccioli *et al.*, 2004; Thompson *et al.*, 2004), a positive modulatory effect of ghrelin on adiposity could be assumed. However, controversy is derived from a relatively recent report claiming for an inhibitory effect of ghrelin on preadipocyte differentiation and proliferation (Zhang *et al.*, 2004). Thus, in our *in vitro* study, we clarified that the adipocyte receptor is involved in the mediation of ghrelin stimulatory effect on leptin production. Indeed, we were able to demonstrate that physiological ghrelin concentration (e.g., 0.1 nM) stimulated rat RP adipocyte *ob* gene expression and protein secretion. Surprisingly, we observed that the leptin-releasing activity exerted by ghrelin at the adipocyte level occurred acutely. Ghrelin was able to significantly increase RP adipocyte leptin secretion after 30 min in culture. The literature indicates that the ghrelin activity on *in vivo* adipogenesis could be observed only after a long time period of ghrelin exposition (Thompson *et al.*, 2004). Thus, our data strongly argues for an important leptin-releasing effect of ghrelin, in addition to its effect on newly synthesized leptin. However, because of the conditions characterizing some of our studies, it could be reasonable to expect that after 24 h culture of adipocytes with ghrelin, an overall increase in leptin secretion could be due to its effect on adipogenesis (Choi *et al.*, 2003; Thompson *et al.*, 2004). This direct effect of ghrelin on rat RP adipocytes could be a result of physiological relevance. It is recognized that stomach-secreted ghrelin is peripherally converted to desacyl ghrelin (Hosoda *et al.*, 2000b), thus its rapid leptin-releasing activity on adipocytes could be playing a relevant physiological role for maintaining homeostasis. In fact, circulating ghrelin and leptin converge for hypothalamic regulation of food intake (Horvath *et al.*, 2001; Inui, 2001). Moreover, during negative balance conditions of the individual (e.g., fasting) while leptin levels fall (Ahima *et al.*, 1996), conversely, those of ghrelin concentrations increase (Kalra *et al.*, 2003). But after food intake initiation, a reactive response takes place, and is characterized by an increment and a diminution in the circulating levels of leptin and ghrelin, respectively (Kalra *et al.*, 2003). As we and other authors found (Bornstein *et al.*, 2000), preformed leptin secretion is very low, thus only a long time period of increased ghrelin in the circulation could result in the triggering of *ob* protein synthesis and, in turn, release into the peripheral circulation shortly after meal initiation (Kalra *et al.*, 1999). Fasting is a condition also characterized by, among others, high peripheral glucocorticoid levels (Ahima *et al.*, 1996); this substance also could play a cooperative role with ghrelin forms to further enhance adipocyte leptin production (Chautard *et al.*, 1999; Spinedi and Gaillard, 1998) and thus cooperating for the maintenance of the homeostasis.

We found a relatively greater leptin-releasing activity of ghrelin than those developed by insulin and DXM. Moreover, we also observed that ghrelin effect could be additional to the one induced by these stimuli, although the mechanisms of action are different (Bradley et al., 2001). In fact, insulin and DXM are substances well recognized as enhancers of leptin production (Bradley et al., 2001; Hardie et al., 1996). Thus reinforcing the idea that altogether substances could cooperate for the control of the organism's energy balance, both by specific individual pleiotropic effects and also by enhancing adipocyte endocrine function.

Regarding the desacylated ghrelin form, desacyl ghrelin, there are only few data in the literature describing an activity on rat adipose tissue function. As mentioned, desacyl ghrelin represents the major ghrelin gene-related form found in the circulation (Hosoda et al., 2004). In contrast to ghrelin effect, desacyl ghrelin has been shown not to stimulate food intake when peripherally administered (Toshinai et al., 2004), and also that it lacks GHSR1a binding capability (Bednarek et al., 2000; Torsello et al., 2002). In one study, an *in vivo* effect of desacyl ghrelin on adiposity has been reported (Thompson et al., 2004). Now, we have demonstrated that desacyl ghrelin possesses an important positive *in vitro* effect on the adipogenic process (Norman et al., 2003), and also on adipokine production. In fact, desacyl ghrelin was found to stimulate cell lipid accumulation and mRNA expression of both PPARγ2 and LPL, all well-recognized markers of the adipogenic process (Ailhaud et al., 1992; Lowell, 1999; Mitchell et al., 1997). In addition, its adipogenic effect was accompanied by an enhancement in leptin secretion. Moreover, on mature RP adipocytes (obtained by *in vitro* differentiation), desacyl ghrelin enhanced both *ob* mRNA expression and secretion. As already mentioned for ghrelin (Chen et al., 2005b; Muccioli et al., 2004), desacyl ghrelin (Muccioli et al., 2004) has the property of inhibiting lipolysis induced by isoproterenol. This common effect of both ghrelin-related peptides clearly contributes to their adipogenic activities. Although the ghrelin stimulatory effect on rat white adipose tissue leptin production is due to the activation of GHSR1a, conversely, the respective of desacyl ghrelin seems to be due to a distinct mechanism. Nevertheless, both peptides share some similar properties on adiposity. Thus strongly supporting a relevant peripheral role of circulating ghrelin forms in the overall adipogenic process.

We have demonstrated that ghrelin- and desacyl ghrelin-induced leptin secretion by adipocytes is a process involving both adipocyte RNA and protein synthesis. However, while ghrelin actions were mediated by GHSR1a, conversely, desacyl ghrelin effects seem to be unrelated to that receptor. Actinomycin D prevented the leptin-releasing activity of both ghrelin gene-related peptides, without changes in spontaneous *ob* protein secretion. We also observed that cycloheximide-exposed adipocytes displayed decreased spontaneous leptin secretion and lack of any leptin secretion when stimulated with either ghrelin or desacyl ghrelin. Thus, as reported for other substances (Bradley and Cheatham, 1999), our data strongly support that the ghrelin- and desacyl ghrelin-elicited leptin production by rat white adipocytes involves transcriptional and translational process of protein synthesis. Important to denote is that high concentrations of GHSR1a-A were needed to abolish the ghrelin effect on adipocytes; the antagonistic action of this compound was claimed as less potent than ghrelin itself (Holst et al., 2003). Although ghrelin-A was reported to be more potent as a ghrelin inverse agonist (Holst et al., 2003), this action seems not to interfere in our *in vitro* adipocyte system. In fact, we found that this product did not modify spontaneous adipocyte *ob* mRNA expression or leptin output. However, whether ghrelin effect is mediated by other adipocyte receptor, as occurred with its adipogenic effect, remains to be determined (Thompson et al., 2004). Important to stress is that the ghrelin antagonist employed in our studies was able to fully abolish the leptin-releasing activity of ghrelin. Then, it is reasonable to expect that the direct leptin-releasing activity of desacyl ghrelin should be unrelated to any ghrelin effect due to *in vitro* acylation of desacyl ghrelin by adipocyte acylation-stimulating protein (Cianflone et al., 2003). Important to denote is that the leptin-releasing activity, on RP adipocytes, developed by different substances tested follows a rank order: ghrelin $>$ desacyl ghrelin $=$ insulin $\geq$ DXM.

Interestingly, we also observed that glucose uptake from the IM induced by desacyl ghrelin was equivalent to that obtained after incubation with insulin; thus supporting, as reported for ghrelin (Kim et al., 2004; Patel et al., 2006), that desacyl ghrelin could also be implicated in adipocyte regulation of glucose homeostasis. Moreover, the leptin-releasing activity of desacyl ghrelin, similarly to that occurred for ghrelin (Giovambattista et al., 2006), was additive to that of insulin. Because several effects of desacyl ghrelin have taken place at 1 nM concentration, our data argue in favor for physiological (Hosoda et al., 2000b; Nishi et al., 2005) effects of desacyl ghrelin in mammals. However, it cannot be ruled out that increases in leptin output could also be partially due to desacyl ghrelin-induced adipogenesis.

The relevance of the effects of these ghrelin gene-related substances on rat white adipose tissue function could have even other implications. Because the amount of adipocyte preformed leptin contributing to the circulating pool of the adipokine is low (Bornstein et al., 2000; Giovambattista et al., 2006), a prolonged period of high circulating levels of ghrelin gene-related forms could be an important factor for the induction of *ob* production. Negative balance energy conditions are also characterized by increased circulating levels of glucocorticoid (Ahima et al., 1996); then ghrelin forms (Choi et al., 2003; Giovambattista et al., 2006; Matsumoto et al., 2001) and glucocorticoid (Chautard et al., 1999; Giovambattista et al., 2006) could cooperate to enhance adipocyte leptin production, a step preceding its release shortly after food intake initiation (Giovambattista et al., 2000). Fasting-induced changes in peripheral ghrelin compounds, leptin, and insulin levels take place similarly in wild-type and GHSR null mice (Sun et al., 2004); thus, it has been suggested that ghrelin could regulate circulating leptin via a GHSR-unrelated mechanism, for example, the autonomic nervous system activity (Asakawa *et al.*, 2001) and gender factors (Greenman *et al.*, 2004).

Besides physiological actions of ghrelin-related peptides, a question still remains open: could ghrelin forms be involved in any physiopathological condition affecting adiposity? Several reports sustain that different genetic obese models are characterized, among others, by enhanced production of ghrelin (Asakawa *et al.*, 2001; Beck *et al.*, 2003; Haqq *et al.*, 2003; Kanda *et al.*, 2003; Meyer *et al.*, 2004). Moreover, circulating ghrelin concentrations in obese humans did not change after meal (English *et al.*, 2002) or body mass reduction (Morpurgo *et al.*, 2003), an opposite effect to that developed by lean subjects. Leptin administration in lean mice enhances ghrelin expression in the stomach (Toshinai *et al.*, 2001), and ghrelin stimulates pancreatic insulin secretion (Date *et al.*, 2002; Lee *et al.*, 2002). Observations, which agree for the sustaining of a gastric–adipoinsular axis (Kieffer and Habener, 2000) interaction further suggest that ghrelin gene-related compounds could play a relevant role in obesity. Thus, enhanced circulating levels of glucocorticoid (Perello *et al.*, 2003b) and ghrelin forms (Asakawa *et al.*, 2001; Beck *et al.*, 2003) could cooperate for maintaining/exacerbating obese phenotypes.

This chapter, through the data presented, provides evidence for a potent direct stimulatory effect of ghrelin and desacyl ghrelin on rat white adipocyte *ob* gene expression and leptin release. Moreover, it was observed that the most common ghrelin form found in circulation, desacyl ghrelin, resulted in an important facilitatory factor of the adipogenic process, and a clear regulator of adipocyte glucose metabolism. It was also found that the adipocyte receptor mediating ghrelin-induced leptin production is GHSR1a, different from the one previously claimed for its adipogenic activity (Thompson *et al.*, 2004). Moreover, the desacyl ghrelin effect on adipose tissue function seems to also be different from GHSR1a; however, further research is necessary to characterize the receptor involved in desacyl ghrelin activities. Additionally, it could be of relevance to study the potential effect(s) of another ghrelin gene-related compound, obestatin, on adipose tissue function (Zhang *et al.*, 2005). In contrast to ghrelin, this peptide has been shown to cooperate in the overall homeostatic process due to its intrinsic anorexigenic activity (Zhang *et al.*, 2005).

Because the ghrelin gene-derived peptides studied are both inducers of adipogenesis and stimulate leptin production, ghrelin-related compounds could have a relevant physiological role in the restoration of homeostasis after changes in energetic balance (Dallman *et al.*, 1999; McEwen and Wingfield, 2003; Thompson *et al.*, 2004). However, we should not discard the possible importance of ghrelin for the development of metabolic disorders (Bagnasco *et al.*, 2002; Nakazato *et al.*, 2001; Tschop *et al.*, 2000).

ACKNOWLEDGMENTS

The present work was supported by grants from FONCyT (PICT 05-13634), FNSR (3200BO-105657/1), CONICET (PIP 6176), and Fondation pour la Recherche en Endocrinologie (2005/2006).

REFERENCES

Ahima, R. S., Prabakaran, D., Mantzoros, C., Qu, D., Lowell, B., Maratos-Flier, E., and Flier, J. S. (1996). Role of leptin in the neuroendocrine response to fasting. *Nature* **382**, 250–252.
Ailhaud, G., Grimaldi, P., and Negrel, R. (1992). Cellular and molecular aspects of adipose tissue development. *Annu. Rev. Nutr.* **12**, 207–233.
Asakawa, A., Inui, A., Kaga, T., Yuzuriha, H., Nagata, T., Ueno, N., Makino, S., Fujimiya, M., Niijima, A., Fujino, M. A., and Kasuga, M. (2001). Ghrelin is an appetite-stimulatory signal from stomach with structural resemblance to motilin. *Gastroenterology* **120**, 337–345.
Asakawa, A., Inui, A., Fujimiya, M., Sakamaki, R., Shinfuku, N., Ueta, Y., and Kasuga, M. (2005). Stomach regulates energy balance via acylated ghrelin and desacyl ghrelin. *Gut* **54**, 18–24.
Bagnasco, M., Dube, M. G., Kalra, P. S., and Kalra, S. P. (2002). Evidence for the existence of distinct central appetite, energy expenditure, and ghrelin stimulation pathways as revealed by hypothalamic site-specific leptin gene therapy. *Endocrinology* **143**, 4409–4421.
Baldanzi, G., Filigheddu, N., Cutrupi, S., Catapano, F., Bonissoni, S., Fubini, A., Malan, D., Baj, G., Granata, R., Broglio, F., Papotti, M., Surico, N., *et al.* (2002). Ghrelin and des-acyl ghrelin inhibit cell death in cardiomyocytes and endothelial cells through ERK1/2 and PI 3-kinase/AKT. *J. Cell Biol.* **159**, 1029–1037.
Banks, W. A., Tschop, M., Robinson, S. M., and Heiman, M. L. (2002). Extent and direction of ghrelin transport across the blood-brain barrier is determined by its unique primary structure. *J. Pharmacol. Exp. Ther.* **309**, 822–827.
Beck, B., Richy, S., and Stricker-Krongrad, A. (2003). Ghrelin and body weight regulation in the obese Zucker rat in relation to feeding state and dark/light cycle. *Exp. Biol. Med.* **228**, 1124–1131.
Bednarek, M. A., Feighner, S. D., Pong, S. S., McKee, K. K., Hreniuk, D. L., Silva, M. B., Warren, V. A., Howard, A. D., Van Der Ploeg, L. H., and Heck, J. V. (2000). Structure-function studies on the new growth hormone-releasing peptide, ghrelin: Minimal sequence of ghrelin necessary for activation of growth hormone secretagogue receptor 1a. *J. Med. Chem.* **43**, 4370–4376.
Bornstein, S. R., Abu-Asab, M., Glasow, A., Path, G., Hauner, H., Tsokos, M., Chrousos, G. P., and Scherbaum, W. A. (2000). Immunohistochemical and ultrastructural localization of leptin and leptin receptor in human white adipose tissue and differentiating human adipose cells in primary culture. *Diabetes* **49**, 532–538.
Bradley, R. L., and Cheatham, B. (1999). Regulation of *ob* gene expression and leptin secretion by insulin and dexamethasone in rat adipocytes. *Diabetes* **48**, 272–278.
Bradley, R. L., Cleveland, K. A., and Cheatham, B. (2001). The adipocyte as a secretory organ: Mechanisms of vesicle transport and secretory pathways. *Recent Prog. Horm. Res.* **56**, 329–358.
Broglio, F., Gottero, C., Arvat, E., and Gigho, E. (2003). Endocrine and non-endocrine actions of ghrelin. *Horm. Res.* **59**, 109–117.

Caminos, J. E., Nogueiras, R., Blanco, M., Seona, L. M., Bravo, S., Alvarez, C. V., Garcia-Caballero, T., Casanueva, F. F., and Dieguez, C. (2003a). Cellular distribution and regulation of ghrelin messenger ribonucleic acid in the rat pituitary gland. *Endocrinology* **144,** 5089–5097.

Caminos, J. E., Tena-Sempere, M., Gaytan, F., Sánchez-Criado, J. E., Barreiro, M. L., Nogueiras, R., Casanueva, F. F., Dieguez, C., Aguilar, E., and Dieguez, C. (2003b). Expresión of ghrelin in the cyclic and pregnant rat ovary. *Endocrinology* **144,** 1594–1602.

Chautard, T., Spinedi, E., Voirol, M., Pralong, F. P., and Gaillard, R. C. (1999). Role of glucocorticoids in the response of the hypothalamo-corticotrope, immune and adipose systems to repeated endotoxin administration. *Neuroendocrinology* **69,** 360–369.

Chen, C. Y., Chao, Y., Chang, F. Y., Chien, E. J., Lee, S. D., and Doong, M. L. (2005a). Intracisternal des-acyl ghrelin inhibits food intake and non-nutrient gastric emptying in conscious rats. *Int. J. Mol. Med.* **16,** 695–699.

Chen, C. Y., Inui, A., Asakawa, A., Fujino, K., Kato, I., Chen, C. C., Ueno, N., and Fujimiya, M. (2005b). Des-acyl ghrelin acts by CRF type 2 receptors to disrupt fasted stomach motility in conscious rats. *Gastroenterology* **129,** 8–25.

Choi, K., Roh, S. G., Hong, Y. H., Shrestha, Y. B., Hishikawa, D., Chen, C., Kojima, M., Kangawa, K., and Sasaki, S. (2003). The role of ghrelin and growth hormone secretagogues receptor on rat adipogenesis. *Endocrinology* **144,** 754–759.

Chomczynski, P., and Sacchi, N. (1987). Single-step method of RNA isolation by acid guanidinium thiocyanate-phenol-chloroform extraction. *Anal. Biochem.* **162,** 156–159.

Cianflone, K., Xia, Z., and Chen, L. Y. (2003). Critical review of acylation-stimulating protein physiology in humans and rodents. *Biochim. Biophys. Acta* **1609,** 127–143.

Dallman, M. F., Akana, S. F., Bhatnagar, S., Bell, M. E., Choi, S., Chu, A., Horsley, C., Levin, N., Meijer, O., Soriano, L. R., Strack, A. M., and Viau, V. (1999). Starvation: Early signals, sensors, and sequelae. *Endocrinology* **140,** 4015–4023.

Date, Y., Kojima, M., Hosoda, H., Sawaguchi, A., Mondal, M. S., Suganuma, T., Matsukura, S., Kangawa, K., and Nakazato, M. (2000). Ghrelin, a novel growth hormone-releasing acylated peptide, is synthesized in a distinct endocrine cell type in the gastrointestinal tracts of rats and humans. *Endocrinology* **141,** 4255–4261.

Date, Y., Nakazato, M., Hashiguchi, S., Dezaki, K., Mondal, M. S., Hosoda, H., Kojima, M., Kangawa, K., Arima, T., Matsuo, H., Yada, T., and Matsukura, S. (2002). Ghrelin is present in pancreatic alpha-cells of humans and rats and stimulates insulin secretion. *Diabetes* **51,** 124–129.

English, P. J., Ghatei, M. A., Malik, I. A., Bloom, S. R., and Wilding, J. P. (2002). Food fails to suppress ghrelin levels in obese humans. *J. Clin. Endocrinol. Metab.* **87,** 2984–2987.

Frost, S. C., and Lane, M. D. (1985). Evidence for the involvement of vicinal sulfhydryl groups in insulin-activated hexose transport by 3T3-L1 adipocytes. *J. Biol. Chem.* **260,** 2646–2652.

Frühbeck, G., Gomez-Ambrosi, J., Muruzabal, F. J., and Burell, M. A. (2001). The adipocyte: A model for integration of endocrine and metabolic signaling in energy metabolism regulation. *Am. J. Physiol.* **280,** E827–E847.

Fujimoto, M., Masuzaki, H., Tanaka, T., Yasue, S., Tomita, T., Okazawa, K., Fujikura, J., Chusho, H., Ebihara, K., Hayashi, T., Hosoda, K., and Nakao, K. (2004). An angiotensin II AT1 receptor antagonist, telmisartan augments glucose uptake and GLUT4 protein expression in 3T3-L1 adipocytes. *FEBS Lett.* **576,** 492–497.

Giovambattista, A., Chisari, A. N., Gaillard, R. C., and Spinedi, E. (2000). Food intake-induced leptin secretion modulates hypothalamo-pituitary-adrenal axis response and hypothalamic Ob-Rb expression to insulin administration. *Neuroendocrinology* **72,** 341–349.

Giovambattista, A., Piermaria, J., Suescun, M. O., Calandra, R. S., Gaillard, R. C., and Spinedi, E. (2006). Direct effect of ghrelin on leptin production by cultured rat white adipocytes. *Obesity* **14,** 19–27.

Glasow, A., Kiess, W., Anderegg, U., Berthold, A., Bottner, A., and Kratzsch, J. (2001). Expression of leptin (Ob) and leptin receptor (Ob-R) in human fibroblasts: Regulation of leptin secretion by insulin. *J. Clin. Endocrinol. Metab.* **86,** 4472–4479.

Gnanapavan, S., Kola, B., Bustin, S. A., Morris, D. G., McGee, P., Fairclough, P., Bhattacharya, S., Carpenter, R., Grossman, A. B., and Korbonits, M. (2002). The tissue distribution of the mRNA of ghrelin and subtypes of its receptor, GHS-R, in humans. *J. Clin. Endocrinol. Metab.* **87,** 2988–2991.

Greenman, Y., Golani, N., Gilad, S., Yaron, M., Limor, R., and Stern, N. (2004). Ghrelin secretion is modulated in a nutrient- and gender-specific manner. *Clin. Endocrinol. (Oxf)* **60,** 382–388.

Gualillo, O., Caminos, J., Blanco, M., Garcia-Caballero, T., Kojima, M., Kangawa, K., Dieguez, C., and Casanueva, F. F. (2001a). Ghrelin, a novel placental-derived hormone. *Endocrinology* **142,** 788–794.

Gualillo, O., Caminos, J. E., Kojima, M., Kangawa, K., Arvat, E., Ghigo, E., Casanueva, F. F., and Dieguez, C. (2001b). Gender and gonads influences on ghrelin mRNA levels in rat stomach. *Eur. J. Endocrinol.* **144,** 687–690.

Gualillo, O., Caminos, J. E., Nogueiras, R., Seoane, L. M., Arvat, E., Ghigo, E., Casanueva, F. F., and Dieguez, C. (2002). Effect of food restriction on ghrelin in normal-cycling female rats and in pregnancy. *Obes. Res.* **10,** 682–687.

Guan, X. M., Yu, H., Palyha, O. G., McKee, K. K., Feighner, S. D., Sirnathsinghji, D. J., Smith, R. G., Van der Ploeg, L. H., and Howard, A. D. (1997). Distribution of mRNA encoding the growth hormone secretagogue receptor in brain and peripheral tissues. *Brain Res. Mol. Brain Res.* **48,** 23–29.

Hardie, L. J., Guilhot, N., and Trayhurn, P. (1996). Regulation of leptin production in cultured mature white adipocytes. *Horm. Metab. Res.* **28,** 685–689.

Haqq, A. M., Farooqi, I. S., O'Rahilly, S., Stadler, D. D., Rosenfeld, R. G., Pratt, K. L., LaFranchi, S. H., and Purnell, J. Q. (2003). Serum ghrelin levels are inversely correlated with body mass index, age, and insulin concentrations in normal children and are markedly increased in Prader-Willi syndrome. *J. Clin. Endocrinol. Metab.* **88,** 174–178.

Hattori, N., Saito, T., Yagyu, T., Jiang, B. H., Kitawa, K., and Inagaki, C. (2001). GH, GH receptor, GH secretagogue receptor, and ghrelin expression in human T cells, B cells, and neutrophils. *J. Clin. Endocrinol. Metab.* **86,** 4284–4291.

Holst, B., Cygankiewicz, A., Jensen, T. H., Ankersen, M., and Schwartz, T. W. (2003). High constitutive signaling of the ghrelin receptor-identification of a potent inverse agonist. *Mol. Endocrinol.* **17,** 2201–2210.

Horvath, T. L., Diano, S., Sotonyi, P., Heiman, M., and Tschop, M. (2001). Minireview: Ghrelin and the regulation of energy balance—a hypothalamic perspective. *Endocrinology* **142,** 4163–4169.

Hosoda, H., Kojima, M., Matsuo, H., and Kangawa, K. (2000a). Purification and characterization of rat des-Gln14-ghrelin, a second endogenous ligand for the growth hormone secretagogue receptor. *J. Biol. Chem.* **275,** 21995–22000.

Hosoda, H., Kojima, M., Matsuo, H., and Kangawa, K. (2000b). Ghrelin and des-acyl ghrelin: Two major forms of rat ghrelin peptide in gastrointestinal tissue. *Biochem. Biophys. Res. Commun.* **279,** 909–913.

Hosoda, H., Doi, K., Nagaya, N., Okumura, H., Nakagawa, E., Enomoto, M., Ono, F., and Kangawa, K. (2004). Optimum collection and storage conditions for ghrelin measurements: Octanoyl modification of ghrelin is rapidly hydrolyzed to desacyl ghrelin in blood samples. *Clin. Chem.* **50,** 1077–1080.

Howard, A. D., Feighner, S. D., Cully, D. F., Arena, J. P., Liberator, P. A., Rosenblum, C. I., Hamelin, M., Hreniuk, D. L., Palyha, O. C., Anderson, J., Paress, P. S., Diaz, C., *et al.* (1996). A receptor in pituitary and hypothalamus that functions in growth hormone release. *Science* **273,** 974–977.

Inui, A. (2001). Ghrelin: An orexigenic and somatotrophic signal from the stomach. *Nat. Rev. Neurosci.* **2,** 551–560.

Kalra, S. P., and Kalra, P. S. (2003). Neuropeptide Y: A physiological orexigen modulated by the feedback action of ghrelin and leptin. *Endocrine* **22,** 49–56.

Kalra, S. P., Dube, M. G., Pu, S., Xu, B., Horvath, T. L., and Kalra, P. S. (1999). Interacting appetite-regulating pathways in the hypothalamic regulation of body weight. *Endocr. Rev.* **20,** 68–100.

Kamegai, J., Tamura, H., Shimizu, T., Ishii, S., Sugihara, H., and Wakababayashi, I. (2001). Chronic central infusion of ghrelin increases hypothalamic neuropeptide Y and agouti-related protein mRNA levels and body weight in rats. *Diabetes* **50,** 2438–2443.

Kanda, T., Takahashi, T., Itoh, T., Kusaka, K., Yamakawa, J., Kudo, S., Takeda, T., Tsugawa, H., and Takekoshi, N. (2003). Upregulation of cardiac ghrelin mRNA in leptin-deficient and leptin receptor-deficient mice with viral myocarditis. *J. Int. Med. Res.* **31,** 503–508.

Kieffer, T. J., and Habener, J. F. (2000). The adipoinsular axis: Effects of leptin on pancreatic beta-cells. *Am. J. Physiol. Endocrinol. Metab.* **278,** E1–E14.

Kim, M. S., Yoon, C. Y., Jang, P. G., Park, Y. J., Shin, C. S., Ryu, J. W., Pak, Y. K., Park, J. Y., Lee, K. U., Kim, S. Y., Lee, H. K., Kim, Y. B., *et al.* (2004). The mitogenic and antiapoptotic actions of ghrelin in 3T3-L1 adipocytes. *Mol. Endocrinol.* **18,** 2291–2301.

Kleinz, M. J., Maguire, J. J., Skepper, J. N., and Davenport, A. P. (2006). Functional and immunocytochemical evidence for a role of ghrelin and des-octanoyl ghrelin in the regulation of vascular tone in man. *Cardiovasc. Res.* **69,** 227–235.

Knight, D. M., Chapman, A. B., Navre, M., Drinkwater, L., Bruno, J. J., and Ringold, G. M. (1987). Requirements for triggering of adipocyte differentiation by glucocorticoids and indomethacin. *Mol. Endocrinol.* **1,** 36–43.

Kojima, M., Hosoda, H., Date, Y., Nakazato, M., Matsuo, H., and Kangawa, K. (1999). Ghrelin is a growth-hormone-releasing acylated peptide from stomach. *Nature* **402,** 656–660.

Kojima, M., Hosoda, H., Matsuo, H., and Kangawa, K. (2001). Ghrelin: Discovery of the natural endogenous ligand for the growth hormone secretagogue receptor. *Trends Endocrinol. Metab.* **12,** 118–122.

Lee, H. M., Wang, G., Englander, E. W., Kojima, M., and Greeley, G. H., Jr. (2002). Ghrelin, a new gastrointestinal endocrine peptide that stimulates insulin secretion: Enteric distribution, ontogeny, influence of endocrine, and dietary manipulations. *Endocrinology* **143,** 185–190.

Lowell, B. B. (1999). PPARgamma: An essential regulator of adipogenesis and modulator of fat cell function. *Cell* **99,** 239–242.

Lu, S., Guan, J. L., Wang, Q. P., Uehara, K., Yamada, S., Goto, N., Date, Y., Nakazato, M., Kojima, M., Kangawa, K., and Shioda, S. (2002). Immunocytochemical observation of ghrelin-containing neurons in the rat arcuate nucleus. *Neurosci. Lett.* **321,** 157–160.

Mano-Otagiri, A., and Shibasaki, T. (2006). Expression of growth hormone secretagogue receptor on growth hormone-releasing hormone neurons and neuropeptide Y neurons in the arcuate nucleus of rat hypothalamus. *J. Nippon Med. Sch.* **73,** 176–177.

Mano-Otagiri, A., Nemoto, T., Sekino, A., Yamauchi, N., Shuto, Y., Sugihara, H., Oikawa, S., and Shibasaki, T. (2006). Growth hormone-releasing hormone (GR) neurons in the arcuate nuecleus (Arc) of the hypothalamus are decreased in transgenic rats whose expression of ghrelin receptor is attenuated: Evidence of ghrelin receptor is involved in the up-regulation of GHRH expression in the Arc. *Endocrinology* **147,** 4093–4103.

Matsumoto, M., Hosoda, H., Kitajima, Y., Morozumi, N., Minamitake, Y., Tanaka, S., Matsuo, H., Kojima, M., Hayashi, Y., and Kangawa, K. (2001). Structure-activity relationship of ghrelin: Pharmacological study of ghrelin peptides. *Biochem. Biophys. Res. Commun.* **287,** 142–146.

McEwen, B. S., and Wingfield, J. C. (2003). The concept of allostasis in biology and biomedicine. *Horm. Behav.* **43,** 2–15.

Meyer, C. W., Korthaus, D., Jagla, W., Cornali, E., Grosse, J., Fuchs, H., Klingenspor, M., Roemheld, S., Tschop, M., Heldmaier, G., De Angelis, M. H., and Nehls, M. (2004). A novel missense mutation in the mouse growth hormone gene causes semidominant dwarfism, hyperghrelinemia, and obesity. *Endocrinology* **145,** 2531–2541.

Mitchell, S. E., Rees, W. D., Hardie, L. J., Hoggard, N., Tadayyon, M., Arch, J. R., and Trayhurn, P. (1997). *ob* gene expression and secretion of leptin following differentiation of rat preadipocytes to adipocytes in primary culture. *Biochem. Biophys. Res. Commun.* **230,** 360–364.

Mohamed-Ali, V., Pinkey, J. H., and Coppack, S. W. (1998). Adipose tissue as an endocrine and paracrine organ. *Int. J. Obes.* **22,** 1145–1158.

Moreno, G., Perello, M., Camihort, G., Luna, G., Console, G., Gaillard, R. C., and Spinedi, E. (2006). Impact of transient correction of increased adrenocortical activity in hypothalamo-damaged, hyperadipose female rats. *Int. J. Obes. (Lond.)* **30,** 73–82.

Mori, K., Yoshimoto, A., Takaya, K., Hosoda, K., Ariyasu, H., Yahata, K., Mukuyama, M., Sugawara, A., Hosoda, H., Kojima, M., Nangawa, K., and Nakao, K. (2000). Kidney produces a novel acylated peptide, ghrelin. *FEBS Lett.* **486,** 213–216.

Morpurgo, P. S., Resnik, M., Agosti, F., Cappiello, V., Sartorio, A., and Spada, A. (2003). Ghrelin secretion in severely obese subjects before and after a 3-week integrated body mass reduction program. *J. Endocrinol. Invest.* **26,** 723–727.

Muccioli, G., Pons, N., Ghe, C., Catapano, F., Granata, R., and Ghigo, E. (2004). Ghrelin and des-acyl ghrelin both inhibit isoproterenol-induced lipolysis in rat adipocytes via a non-type 1a growth hormone secretagogue receptor. *Eur. J. Pharmacol.* **498,** 27–35.

Nakazato, M., Murakami, N., Date, Y., Kojima, M., Matsuo, H., Kangawa, K., and Matsura, S. (2001). A role for ghrelin in the central regulation of feeding. *Nature* **409,** 194–198.

Nishi, Y., Hiejima, H., Mifune, H., Sato, T., Kangawa, K., and Kojima, M. (2005). Developmental changes in the pattern of ghrelin's acyl modification and the levels of acyl-modified ghrelins in murine stomach. *Endocrinology* **146,** 2709–2715.

Nonogaki, K., Ohashi-Nozue, K., and Oka, Y. (2006). Induction of hypothalamic serum- and glucocorticoid-induced protein kinase-1 gene expression and its relation to plasma des-acyl ghrelin in energy homeostasis in mice. *Biochem. Biophys. Res. Commun.* **344,** 696–699.

Norman, D., Isidori, A. M., Frajese, V., Caprio, M., Chew, S. L., Grossman, A. B., Clark, A. J., Michael Besser, G., and Fabbri, A. (2003). ACTH and α-MSH inhibit leptin expression and secretion in 3T3-L1 adipocytes: Model for a central-peripheral melanocortin-leptin pathway. *Mol. Cell. Endocrinol.* **200,** 99–109.

Otto, B., Cuntz, U., Fruehauf, E., Wawarta, R., Folwaczny, C., Riepl, R. L., Heiman, M. L., Lehnert, P., Fichter, M., and Tschop, M. (2001). Weight gain decreases plasma ghrelin concentrations of patients with anorexia nervosa. *Eur. J. Endocrinol.* **145,** 669–673.

Patel, A. D., Stanley, S. A., Murphy, K. G., Frost, C. G., Gardiner, J. V., Kent, A. S., White, N. E., Ghatei, M. A., and Bloom, S. R. (2006). Ghrelin stimulates insulin-induced glucose uptake in adipocytes. *Regul. Pept.* **134,** 17–22.

Perello, M., Castrogiovanni, D., Moreno, G., Gaillard, R. C., and Spinedi, E. (2003a). Neonatal hypothalamic androgenization in the female rat induces changes in peripheral insulin sensitivity and adiposity function at adulthood. *Neuro Endocrinol. Lett.* **24,** 241–248.

Perello, M., Gaillard, R. C., Chisari, A., and Spinedi, E. (2003b). Adrenal enucleation in MSG-damaged hyperleptinemic male rats transiently restores adrenal sensitivity to leptin. *Neuroendocrinology* **78**, 176–184.

Petersen, S., Rasch, A. C., Penshorn, M., Beil, F. U., and Schulte, H. M. (2001). Genomic structure and transcriptional regulation of human growth hormone secretagogue receptor. *Endocrinology* **142**, 2649–2659.

Piermaria, J., Console, G., Perello, M., Moreno, G., Gaillard, R. C., and Spinedi, E. (2003). Impact of estradiol on parametrial adipose tissue function: Evidence for establishment of a new set point of leptin sensitivity in control of energy metabolism in female rat. *Endocrine* **20**, 239–245.

Rajala, M. W., and Scherer, P. E. (2003). Minireview: The adipocyte—at the crossroads of energy homeostasis, inflammation, and atherosclerosis. *Endocrinology* **144**, 3765–3773.

Sakata, I., Nakamura, K., Yamazaki, M., Matsubara, M., Hayashi, Y., Kangawa, K., and Sakai, T. (2002). Ghrelin-producing cells exist as two types of cells, closed- and opened-type cells, in the rat gastrointestinal tract. *Peptides* **23**, 531–536.

Shuto, Y., Shibasaki, T., Otagiri, A., Kuriyama, H., Ohata, H., Tamura, H., Kamegai, J., Sugihara, H., Oikawa, S., and Wakabayashi, I. (2002). Hypothalamic growth hormone secretagogue receptor regulates growth hormone secretion, feeding, and adiposity. *J. Clin. Invest.* **109**, 1429–1436.

Smith, R. G., Van der Ploeg, L. H., Howard, A. D., Feighner, S. D., Cheng, K., Hickey, G. J., Wyvratt, M. J., Jr., Fisher, M. H., Nargund, R. P., and Patchett, A. A. (1997). Peptidomimetic regulation of growth hormone secretion. *Endocr. Rev.* **18**, 621–645.

Spinedi, E., and Gaillard, R. C. (1998). A regulatory loop between the hypothalamo-pituitary-adrenal (HPA) axis and circulating leptin: A physiological role of ACTH. *Endocrinology* **139**, 4016–4020.

Sun, Y., Wang, P., Zheng, H., and Smith, R. G. (2004). Ghrelin stimulation of growth hormone release and appetite is mediated through the growth hormone secretagogue receptor. *Proc. Natl. Acad. Sci. USA* **101**, 4679–4684.

Tena-Sempere, M., Barreiro, M. L., Gonzalez, L. C., Gaytan, F., Zhang, F. P., Caminos, J. E., Pinilla, L., Casanueva, F. F., Dieguez, C., and Aguilar, E. (2002). Novel expression and functional role of ghrelin in rat testis. *Endocrinology* **143**, 717–725.

Thompson, N. M., Gill, D. A., Davies, R., Loveridge, N., Houston, P. A., Robinson, I. C., and Wells, T. (2004). Ghrelin and des-octanoyl ghrelin promote adipogenesis directly *in vivo* by a mechanism independent of the type 1a growth hormone secretagogue receptor. *Endocrinology* **145**, 234–242.

Torsello, A., Ghe, C., Bresciani, E., Catapano, F., Ghigo, E., Deghenghi, R., Locatelli, V., and Muccioli, G. (2002). Short ghrelin peptides neither displace ghrelin binding *in vitro* nor stimulate GH release *in vivo*. *Endocrinology* **143**, 1968–1971.

Toshinai, K., Mondal, M. S., Nazazato, M., Date, Y., Murakami, N., Kojima, M., Kangawa, K., and Matsura, S. (2001). Upregulation of ghrelin expression in the stomach upon fasting insulin-induced hypoglycemia, and leptin administration. *Biochem. Biophys. Res. Commun.* **281**, 1220–1225.

Toshinai, K., Date, Y., Murakami, N., Shimada, M., Mondal, M. S., Shimbara, T., Guan, J. L., Wang, Q. P., Funahashi, H., Sakurai, T., Shioda, S., Matsukura, K., *et al.* (2003). Ghrelin-induced food intake is mediated via the orexin pathway. *Endocrinology* **144**, 1506–1512.

Toshinai, K., Yamaguchi, H., Sun, Y., Smith, R. G., Yamanaka, A., Sakurai, T., Date, Y., Mondal, M. S., Shimbara, T., Kagawoe, T., Murakami, N., Miyazato, M., *et al.* (2006). Des-acyl ghrelin induces food intake by a mechanism independent of growth hormone secretagogue receptor. *Endocrinology* **147**, 2306–2314.

Trayhurn, P., and Beattie, J. H. (2001). Physiological role of adipose tissue: White adipose tissue as an endocrine and secretory organ. *Proc. Nutr. Soc.* **60,** 329–339.

Trayhurn, P., and Wood, I. S. (2004). Adipokines: Inflammation and the pleiotropic role of white adipose tissue. *Br. J. Nutr.* **92,** 347–355.

Tschop, M., Smiley, D. L., and Heiman, M. L. (2000). Ghrelin induces adiposity in rodents. *Nature* **407,** 908–913.

Tschop, M., Statnick, M. A., Suter, T. M., and Heiman, M. L. (2002). GH-releasing peptide-2 increases fat mass in mice lacking NPY: Indication for a crucial mediating role of hypothalamic agouti-related protein. *Endocrinology* **143,** 558–568.

Tung, Y. C., Hewson, A. K., and Dickson, S. L. (2001). Actions of leptin on growth hormone secretagogue-responsive neurons in the rat hypothalamic arcuate nucleus recorded *in vitro*. *J. Neuroendocrinol.* **13,** 209–215.

Volante, M., Allia, E., Gugliotta, P., Funaro, A., Broglio, F., Deghenghi, R., Muccioli, G., Gigho, E., and Papotti, M. (2002a). Expression of ghrelin and of the GH secretagogue receptor by pancreatic islet cells and related endocrine tumors. *J. Clin. Endocrinol. Metab.* **87,** 1300–1308.

Volante, M., Fulcheri, E., Allia, E., Cerrato, M., Pucci, A., and Papotti, M. (2002b). Ghrelin expression in fetal, infant, and adult human lung. *J. Histochem. Cytochem.* **50,** 1013–1021.

Wren, A. M., Seal, L. J., Cohen, M. A., Brynes, A. E., Frost, G. S., Murphy, K. G., Dhillo, W. S., Ghatei, M. A., and Bloom, S. R. (2001). Ghrelin enhances appetite and increases food intake in humans. *J. Clin. Endocrinol. Metab.* **86,** 5992–5995.

Wren, A. M., Small, C. J., Ward, H. L., Murphy, K. G., Dakin, C. L., Taheri, S., Kennedy, A. R., Roberts, G. H., Morgan, D. G., Ghatei, M. A., and Bloom, S. R. (2000). The novel hypothalamic peptide ghrelin stimulates food intake and growth hormone secretion. *Endocrinology* **141,** 4325–4328.

Yoshimoto, A., Mori, K., Sugawara, A., Mukoyama, M., Yahata, K., Takayoshi, S., Takaya, K., Hosoda, H., Kojima, M., Kangawa, K., and Nakao, K. (2002). Plasma ghrelin and desacyl ghrelin concentrations in renal failure. *J. Am. Soc. Nephrol.* **13,** 2748–2752.

Zhang, W., Zhao, L., Lin, T. R., Chai, B., Fan, Y., Gantz, I., and Mulholland, M. W. (2004). Inhibition of adipogenesis by ghrelin. *Mol. Biol. Cell* **15,** 2482–2491.

Zhang, J. V., Ren, P.-G., Avsian-Kretchmer, O., Luo, C.-W., Rauch, R., Klein, C., and Hsueh, A. J. W. (2005). Obestatin, a peptide encoded by the ghrelin gene, opposes ghrelin's effects on food intake. *Science* **310,** 996–999.

CHAPTER NINE

Cardiac, Skeletal, and Smooth Muscle Regulation by Ghrelin

Adelino F. Leite-Moreira,* Amândio Rocha-Sousa,* *and* Tiago Henriques-Coelho*

Contents

I. Introduction	208
II. Structure and Distribution of Ghrelin	210
A. Structure	210
B. Distribution	210
C. Receptors	214
III. Contractile Effects of Ghrelin	217
A. Myocardium	217
B. Smooth muscle	223
C. Skeletal muscle	227
Acknowledgments	230
References	230

Abstract

Ghrelin, mainly secreted from gastric mucosa, is the endogenous ligand for the growth hormone secretagogue receptor and induces a potent release of growth hormone. Ghrelin is widely expressed in different tissues and therefore has both endocrine and paracrine/autocrine effects. In this chapter, we summarize: (1) structure and distribution of ghrelin and its receptors; (2) myocardial effects of ghrelin, describing its acute and chronic actions on cardiac function; (3) ghrelin effects on smooth muscle, namely vascular smooth muscle, intraocular and gastrointestinal smooth muscle; and (4) skeletal actions of ghrelin. Ghrelin has a potent vasodilator effect, thereby reducing cardiac afterload and increasing cardiac output. In models of heart failure and myocardial ischemia, ghrelin administration has beneficial effects. At smooth muscle, ghrelin modulates vascular tone, increases gut transit, and relaxes iris muscles. In the skeletal muscle, ghrelin regulates resting membrane potential. In conclusion,

---

* Department of Physiology, Faculty of Medicine, University of Porto, Alameda Professor Hernâni Monteiro, 4200-319 Porto, Portugal

there are increasing evidences that ghrelin is a peptide with paracrine actions that can modulate cardiac, smooth, and skeletal muscle functions. © 2008 Elsevier Inc.

I. INTRODUCTION

Growth hormone secretagogues (GHSs) are small synthetic molecules that act through a specific G-protein–coupled receptor called growth hormone secretagogue receptor (GHSR). This was an orphan receptor (i.e., had no known natural ligand) until 1999, when Kojima *et al.* (1999) succeeded in a series of brilliant experiments in identifying the endogenous ligand for the GHSR, which they named ghrelin, from "ghre" the Indo-European root for "to grow" in reference to its ability to stimulate GH release. Ghrelin is a 28-amino acid peptide with a serine 3 (Ser^3) N-octanoic acid modification that is secreted by the X/A-like cells of the gastric fundus (Date *et al.*, 2000) that circulates in the bloodstream and acts directly on the pituitary to release GH. As the endogenous ligand for GHSR, ghrelin induces a GH release in a dose-dependent manner (Kojima *et al.*, 1999). Intravenous injection of ghrelin induces a GH peak after 5–15 min, which returns to basal levels 1 h later (Arvat *et al.*, 2000; Kojima *et al.*, 1999; Takaya *et al.*, 2000).

The surprising discovery that the principal site of ghrelin synthesis is the stomach and not the hypothalamus raised the hypothesis that ghrelin could be an endocrine link between stomach, hypothalamus and pituitary. Tschop *et al.* (2000) demonstrated that daily administration of ghrelin caused weight gain and increased food intake, suggesting an involvement in regulation of energy balance. Plasma ghrelin concentration is increased in fasting conditions and reduced after feeding (Cummings *et al.*, 2001; Tschop *et al.*, 2001a), suggesting that ghrelin may act as an initiation signal for food intake and may be controlled by some nutritional factors (Peeters, 2005). The importance of ghrelin in body weight regulation was strengthened with the observation that lean subjects have higher ghrelin levels than obese subjects (Tschop *et al.*, 2001b). In fact, ghrelin is the most powerful stimulator of appetite of all known peptides that act, at least in part by activating the genes encoding agouti-related peptide and neuropeptide Y (NPY), two potent appetite stimulators, in the arcuate nucleus (Kamegai *et al.*, 2000; Nakazato *et al.*, 2001).

Besides being the endogenous GHS with orexigenic activity, ghrelin has multiple other actions (Lago *et al.*, 2005). Ghrelin affects several other systems like gastrointestinal (Masuda *et al.*, 2000; Peeters, 2005; Trudel *et al.*, 2002), cardiovascular (Cao *et al.*, 2006; Enomoto *et al.*, 2003), pulmonary (Santos *et al.*, 2006; Volante *et al.*, 2002a), reproductive (Gualillo *et al.*, 2001; Tanaka *et al.*, 2003), and central nervous system (Lin *et al.*, 2004; Matsumura *et al.*, 2002), among others (Fig. 1). However, ghrelin knockout mice did not present growth or feeding problems, suggesting that alternative

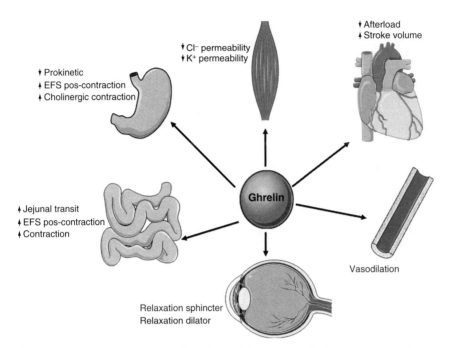

Figure 1 Global view of contractile effects of ghrelin. Ghrelin has described action on the cardiac muscle, vascular smooth muscle, ocular smooth muscle, gastrointestinal smooth muscle, and skeletal muscle. (See Color Insert.)

pathways can compensate many of the known effects of ghrelin (Sun et al., 2003).

In the cardiovascular system, ghrelin seems to have important roles at hemodynamic, contractile and vascular levels. Afterload reduction and cardiac output elevation, without increasing heart rate, both in normal subjects and in patients with dilated cardiomyopathy, are the major hemodynamic effects of ghrelin (Marleau et al., 2006; Nagaya and Kangawa, 2003; Nagaya et al., 2001a, 2004). The most important vascular effect of ghrelin is vasodilatation that seems to be endothelial and GH independent, suggesting an action at smooth muscle level (Henriques-Coelho et al., 2004; Kleinz et al., 2006; Shimizu et al., 2003; Wiley and Davenport, 2002). *In vitro* studies revealed that ghrelin exerts a direct negative inotropic effect mediated by a receptor different from GHSR1a (Bedendi et al., 2003; Soares et al., 2006). Binding sites for GHS have also been found in human skeletal muscle, suggesting that specific receptors can mediate biological activities in this tissue (Papotti et al., 2000). These studies highlight the role of ghrelin in muscle contraction and other studies demonstrated that ghrelin has important actions in gastrointestinal (Depoortere et al., 2006) and ocular (Rocha-Sousa et al., 2006) smooth muscles. This chapter reviews the effects of ghrelin on cardiac, smooth and skeletal muscle.

II. Structure and Distribution of Ghrelin

A. Structure

Ghrelin is a 28-amino acid peptide in which the Ser^3 is N-octanoylated (Kojima *et al.*, 1999). This modification by an acyl group is essential for ghrelin's bioactivity, and there is evidence that the first 4 or 5 residues (Gly-Ser-Ser(n-octanoyl)-Phe-Leu) that include this acyl group are sufficient for calcium mobilization *in vitro* (Bednarek *et al.*, 2000). Human ghrelin gene is located on chromosome 3 (3p25-26) and encodes for a 117-amino acid-long pre-proghrelin (Kojima *et al.*, 1999). A ghrelin variant resulting from alternative splicing was identified as des-Gln14-ghrelin (Hosoda *et al.*, 2000a). It has 27 amino acids with N-octanoic acid modification and the same activity potency. Several minor forms of ghrelin were described with acyl chains of 10 or 11 carbon atoms or peptide chains with the missing last amino acid (arginine) at position 28 (Hosoda *et al.*, 2000a, 2003). However, all these variants are only present in low amounts, indicating that the major active form of ghrelin is a 28-amino acid peptide with octanoylated Ser3.

Interestingly, Hosoda *et al.* (2000b) identified a nonacylated form of ghrelin, the desacyl ghrelin that exists at significant levels in both stomach and blood. In blood, desacyl-ghrelin circulates in amounts far greater than acylated ghrelin, does not replace ghrelin from its hypothalamic and pituitary binding sites, and shows no GH-releasing and other endocrine activities. Increasing number of studies has been reported that desacyl-ghrelin has some biological effects, including the modulation of cell proliferation and, to a small extent, adipogenesis (Cassoni *et al.*, 2004; Nanzer *et al.*, 2004; Thompson *et al.*, 2004). Ghrelin circulates in plasma binded to high-density lipoproteins (HDLs) that contain a potent esterase, paraoxonase, which may be involved in deacylation of acyl-modified ghrelin (Beaumont *et al.*, 2003). Therefore, desacyl-ghrelin may represent either a precursor of acyl-modified ghrelin or the product of its deacylation.

In conclusion, two major forms of ghrelin are found in tissues and plasma: n-octanoyl-modified and desacyl-ghrelin. The normal ghrelin concentration of plasma samples in humans is 10–20 fmol/ml for n-octanoyl ghrelin and 100–150 fmol/ml for total ghrelin, including both acyl-modified and desacyl ghrelin (Ariyasu *et al.*, 2001; Kojima and Kangawa, 2005).

B. Distribution

Ghrelin was first isolated from rat stomach where it was localized, to round and electron-dense granules, in the X/A-like neuroendocrine cells that account for 20% of the endocrine cell population in adult oxyntic glands of gastric fundus (Date *et al.*, 2000). A small number of immunopositive cells for ghrelin are also present along small and large intestines, and its concentration decreases from the duodenum to the colon (Date et al., 2000; Hosoda et al., 2000b).

The pancreas is another ghrelin-producing organ, but the cell type that produces ghrelin is still controversial (Date et al., 2002; Gnanapavan et al., 2002; Volante et al., 2002b). Ghrelin mRNA is expressed in the kidney that can be an important site for clearance and degradation of ghrelin (Gnanapavan et al., 2002; Mori et al., 2000). Lung is also an important source of ghrelin, having been demonstrated to be the organ where the highest concentration of ghrelin protein occurs (Ghelardoni et al., 2006). In the central nervous system, ghrelin has been found in the hypothalamic arcuate nucleus, a region that controls appetite and in neurons localized adjacent to the third ventricle that controls food intake (Cowley et al., 2003; Lu et al., 2002). Ghrelin has also been found in pituitary where it may act in an autocrine/paracrine way (Korbonits et al., 2001; Lin et al., 2004). Other organs have been shown to express ghrelin as well as testis (Barreiro et al., 2002), ovary (Caminos et al., 2003), and placenta (Gualillo et al., 2001; Table 1).

A recent study by Ghelardoni et al. compared systematically ghrelin gene expression and ghrelin protein concentration in several tissues. They found that ghrelin gene was expressed in stomach, small intestine, brain, cerebellum, pituitary, heart, pancreas, salivary gland, adrenal, ovary, and testis, whereas protein was detected in stomach, small intestine, brain, cerebellum, pituitary, lung, skeletal muscle, pancreas, salivary gland, adrenal, ovary, and testis, demonstrating that gene and protein expression were dissociated (Ghelardoni et al., 2006; Table 1). Nevertheless, the physiological significance of the widespread tissue distribution of ghrelin remains to be determined.

During fetal development, ghrelin appears as a new hormone important for organ development and growth. In contrast to adult, fetal stomach is not the major source of ghrelin because its gastric concentration is very low and only increases after birth (Hayashida et al., 2002). There is increasing evidence that pancreas and lung are alternative sources of ghrelin in the fetus (Rindi et al., 2002). Pancreas begins to express ghrelin at midgestation and its mRNA levels are six to seven times greater than in the fetal stomach (Chanoine and Wong, 2004). With regard to lung development, ghrelin is intensively expressed during pseudoglandular stage, indicating that the lung is an additional important source of circulating ghrelin (Santos et al., 2006; Volante et al., 2002a). Thus, onset of both pancreatic and pulmonary expression of ghrelin precedes that of gastric ghrelin. Recent evidences revealed the importance of maternal ghrelin to fetal development. Nakahara and collaborators demonstrated that a single ghrelin injection into the mother increased circulating fetal ghrelin within 5 min after injection, suggesting that maternal ghrelin transits easily to the fetal circulation. There are also evidences that treatment of mothers with ghrelin can increase lung and body weight at birth (Nakahara et al., 2006; Santos et al., 2006). However, the precise role of ghrelin in islet development or in lung branching remains to be establish.

Table 1 Ghrelin and its receptors in human tissues

	Ghrelin				Ghrelin receptors				
References	Gnanapavan et al., 2002	Kleinz et al., 2006	Ghelardoni et al., 2006	Ghelardoni et al., 2006	GHSR Papotti et al., 2000	Takeshita et al., 2006	GHSR1a Kleinz et al., 2006	GHSR1a Gnanapavan et al., 2002	GHSR1b Gnanapavan et al., 2002
Study	mRNA[a]	Binding studies	mRNA	Protein	Binding studies[a]	mRNA binding studies	Binding studies	GHSR1a[a]	GHSR1b[a]
Method	Standard and real-time RT-PCR	Immunocyto-chemistry	Standard RT-PCR	Enzyme immunoassay	Radioreceptor assay ([^{125}I] Tyr-Ala-hexarelin)	Standard RT-PCR immunocyto-chemistry	Immunocyto-chemistry	Classical and real-time RT-PCR	Classical and real-time RT-PCR
Tissues	Fundus	Endothelium lining;	Stomach	Stomach	Myocardium	Mucosae and muscular layer:	Endothelium lining;	Pituitary	Skin
	Jejunum	Small intramyo-cardial vessels	Small intestine	Small intestine	Adrenal gland	Esophagus	Small intramio-cardial;	Thyroid	Myocardium
	Duodenum	Small coronary arteries	Brain	Brain	Testis	Stomach	Pulmonary; Renal and adrenal vessels	Pancreas	Pituitary
	Antrum	Artrial chamber	Cerebelum	Cerebellum	Aortic smooth muscle	Duodenum	Endothelial cells lining	Spleen	Thyroid
	Lung	Small pulmonary vessels	Pituitary	Pituitary	Aortic endothelium	Jejunum	Saphenous vein; Mammary and coronary artery	Myocardium	Pancreas
	Pancreas	Small intrarenal vessels	Myocardium	Lung	Coronary	Ileum	Atrial; Ventricular cardiomyo-cites	Adrenal	Ileum

Vein	Coronary arteries	Pancreas	Skeletal muscle	Carotid	Colon	Vascular smooth muscle	Right colon
Gall bladder	Saphenous vein	Salivary gland	Pancreas	Lung			Liver
Lymph node	Mamary artery	Adrenal	Salivary gland	Ovary			Breast
Esophagus		Ovary	Adrenal	Liver			Spleen
Left colon		Testis	Ovary	Skeletal muscle			Duodenum
Buccal			Testis	Kidney			Placenta
Pituitary				Pituitary gland			Lung
Breast				Thyroid gland			Adrenal
Kidney				Adipose tissue			Buccal
Ovary				Vena cava			Fundus
Prostate				Uterus			Lymph node
Right colon				Skin			Gall bladder
Ileum				Lymph node			Atrium
Liver							Lymphocytes
Spleen							Kidney
Fallopian tube							Antrum
Lymphocytes							Bladder
Testis							Prostate
Fat							Fallopian tube
Placenta							Vein
Adrenal							Left colon
Muscle							Muscle
Bladder							Ovary
Atrium							Testis
Thyroid							Fat
Myocardium							Esophagus
Skin							Jejunum

[a] Decreasing order.
RT-PCR, reverse transcription and polymerase chain reaction.

C. Receptors

The GHSR1 was identified in 1996 by expression cloning (Howard et al., 1996) and is part of a G-protein–coupled receptors' superfamily that contains motilin, neuromedin U, and neurotensin receptors (Guan et al., 1997; Howard et al., 2000; Vincent et al., 1999). The ghrelin receptor is well conserved across vertebrate species, which suggests that ghrelin and its receptor have important physiological functions (Palyha et al., 2000). The gene of the human GHSR1 is located on chromosome 3q26.2. Two different splice forms of the human GHSR are known: GHSR1a and GHSR1b. GHSR1a is a typical G-protein–coupled seven-transmembrane receptor with binding and functional properties of the ghrelin receptor. GHSR1b is produced by alternative splicing and does not have known biological activity.

Regarding GHSR distribution, there are some conflicting data, which might be explained by the use of different techniques that differ widely in specificity and sensitivity (Korbonits et al., 2004). Papotti et al. (2000), using a radioreceptor assay, detected the presence of GHS receptors in a wide range of tissues: myocardium (highest binding activity), adrenal, gonads, arteries, lung, liver, skeletal muscle, kidney, pituitary, thyroid, adipose tissue, veins, uterus, skin, and lymphonode. Gnanapavan et al. (2002), using classical and real-time reverse transcription and polymerase chain reaction, described GHSR1a mRNA expression in pituitary, thyroid gland, pancreas, spleen, myocardium, and adrenal gland, whereas unspliced nonfunctional-type 1b GHSR mRNA was found in all studied tissues (Table 2). These controversial findings may suggest the existence of a different receptors subtype, other than GHSR1a. The significance of the widespread expression of the nonfunctional GHSR1b is still unknown.

GHSR1a desensitization and internalization represents an important physiological mechanism that modulates receptor responsiveness and acts as a filter for intracellular signaling. Therefore, after ghrelin binding to GHSR1a, the ligand–receptor complex is internalized via clathrin-coated pits and GHSR1a can be either recycled back to the plasma membrane after dissociation in early endosomes or degraded within lysosomes (Camina et al., 2004; Fig. 2).

Signal transduction pathways of GHSR are well defined in somatotroph cells (Cunha and Mayo, 2002; Malagon et al., 2003) and involves three second-messenger pathways: cAMP-dependent protein kinase (AMPK), protein kinase C (PKC), and inositol(1,4,5)-triphosphate (IP_3). Ligand binding to the GHSR activates phospholipase C (PLC), hydrolyzes phosphatidylinositol-4,5-biphosphate, and originates diacylglycerol (DAG) and IP_3. The first elevation of calcium levels is transient and is a result of the binding of IP_3 to the IP_3 receptor on the endoplasmic reticulum. Second elevation of intracellular levels of calcium is persistent and is a result of PKC activation by DAG. PKC inhibits the potassium channel through tyrosine phosphorylation and

Table 2 Muscle localization of ghrelin and its receptors

	Ghrelin		GHSR1a	GHSR1b	GHSRu	CD36	References
	mRNA	Protein					
Myocardium	✓	–	✓	✓	✓		Ghelardoni et al., 2006; Gnanapavan et al., 2002; Papotti et al., 2000; Katugampola et al., 2001.
Skeletal	✓	✓	✓(?)	✓			Ghelardoni et al., 2006; Gnanapavan et al., 2002
Smooth							
Eye	✓						Rocha-Sousa et al., 2006
Vascular							Ghelardoni et al., 2006; Katugampola et al., 2001; Kleinz et al., 2006
Muscle			✓	✓			
Endothelium	✓	✓			✓(?)	✓	
Gut	✓	✓	✓				Ghelardoni et al., 2006; Takeshita et al., 2006

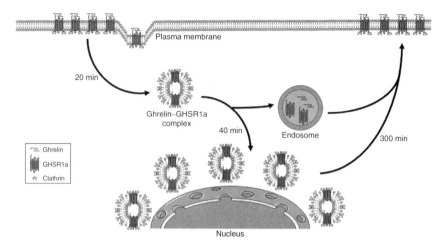

Figure 2 Internalization pathway of GHSR1a. Under unstimulated conditions, GHSR1a is mainly expressed at the plasma membrane. After 20 min exposure to ghrelin, the ghrelin–GHSR1a complex progressively disappears by endocytosis via clathrin-coated pits. The complex accumulates in the perinuclear region after 60 min. GHSR1a shows a slow recycling and its membrane levels return to normal after 360 min (Camina et al., 2004). (See Color Insert.)

induces depolarization that causes the opening of voltage-dependent L-type calcium channels present in the cell membrane (Malagon et al., 2003). The overture of these channels is also activated by the facilitator effect of ghrelin on Na^+-influx through ion channels (Malagon et al., 2003). Ghrelin, by a PLC- and PKC-dependent pathway, activates the 44- and 42-kDa extracellular signal-regulated protein kinases (ERK1/2) and the transcriptional factor Elk1 (Mousseaux et al., 2006). Ghrelin modulates cAMP cellular levels acting on AMPK, having a stimulatory effect on AMPK in the heart and in the hypothalamus and an inhibitory effect in the liver and adipose tissue (Kola et al., 2005) Ghrelin induces vasodilatation and GH secretion via nitric oxide (NO)/cGMP-signaling pathway (Rodriguez-Pacheco et al., 2005; Shimizu et al., 2003). It was reported that binding of ghrelin to GHSR1a in differentiated adipocytes increases the expression of peroxisome proliferator-activated receptor gama 2 (PPARγ2), a nuclear receptor that is a key regulator of transcriptional pathways in adipogenesis (Choi et al., 2003).

Increasing evidences point to the existence of a novel unidentified ghrelin receptor. The activation of cyclooxygenase by nonacylated, acylated, and truncated forms of ghrelin indicates that these ligands bind to a putative receptor (GHSRu) with an unknown structure (Bedendi et al., 2003; Cao et al., 2006). This putative receptor might mediate the effects of both acylated and nonacylated forms of ghrelin on the survival of H9c2 cardiomyocytes because the binding occurs through activation of ERK1/2 and phosphatidylinositol 3-OHkinase (PI-3K) (Baldanzi et al., 2002). Further studies need to be performed in order to clarify the effects of ghrelin independent from GHSR1a.

III. CONTRACTILE EFFECTS OF GHRELIN

A. Myocardium

1. Ghrelin and its receptors

The axis growth hormone/insulin-like growth factor-I (GH/IGF-1) has important myocardial effects by acting on specific cardiac receptors (Colao et al., 2001). Synthetic peptidyl and nonpeptidyl GHS, besides the strong stimulation of GH secretion by activation of GHSR1a, also have direct GH-independent cardiac actions (Benso et al., 2004; Muccioli et al., 2000). Hexarelin, a peptidyl GHS, improve heart postischemic recovery in GH-deficient rats (Locatelli et al., 1999) and increase left ventricular ejection in patients with severe GH deficiency (Bisi et al., 1999; Imazio et al., 2002). GHSRs were detected mainly in the myocardium by radioreceptor assay with [^{125}I]Tyr-Ala-hexarelin (Papotti et al., 2000). After the discovery of ghrelin, a lot of interest was dedicated to explore its cardiovascular effects. Since then, a lot of findings were obtained with regard to the distribution, production, and physiological effects of ghrelin in the myocardium and in the vasculature (Isgaard and Johansson, 2005; Tables 1 and 2).

Katugampola et al. (2001) analyzed ghrelin-binding sites in cardiovascular system using autoradiographical localization with [^{125}I-His9]-ghrelin. They demonstrated that in the myocardium, right atria have higher density of receptors than the left ventricle, whereas in the vasculature, aorta and pulmonary artery have the highest density of receptors in comparison with the saphenous vein or the coronary artery. Furthermore, receptor density was modified by vascular diseases, and there was an upregulation of ghrelin in vessels with advanced intimal thickening. Kleinz et al. (2006) using standard immunocytochemistry and confocal microscopy showed that ghrelin receptor is present in human cardiomyocytes, vascular smooth muscle cells, and endothelial cells and that ghrelin is localized to intracellular vesicles in endothelial cells of human arteries and veins. Iglesias et al. (2004) demonstrated that ghrelin is synthesized and secreted by cardiomyocytes and is involved in the protection of these cells from apoptosis through paracrine/autocrine effects (Tables 1 and 2).

At the moment, it is possible to state that the myocardium expresses four GHS receptors: GHSR1a, GHSR1b (nonfunctional), GHSRu (unknown), and CD36. Quantitative studies using real-time PCR demonstrated that the myocardium is the fifth and the second tissue in human body that express more GHSR1a and GHSR1b mRNA, respectively (Gnanapavan et al., 2002).

The physiological function of GHSR1b subtype is still unknown. GHSRu was proposed after the discovery of the antiapoptotic effect of ghrelin on cardiomyocytes and endothelial cells that do not express GHSR1a but recognizes both nonacylated and acylated ghrelin (Baldanzi et al., 2002). CD36 is a multifunctional B-type scavenger receptor specifically expressed in the adipose tissue, platelets, monocytes/macrophages, denditric cells, microvascular endothelium, and myocardium (Febbraio et al., 2001). Bodart et al. (2002) demonstrated that activation of CD36 by hexarelin induces an increase in coronary pressure in isolated perfused hearts, suggesting that CD36 may mediate the coronary vasospasm seen in hypercholesterolemia and atherosclerosis.

2. Acute effects

A single intravenous injection of ghrelin in healthy humans decreases mean arterial pressure, without altering mean pulmonary arterial pressure, and increases cardiac index and stroke volume, without changing hear rate (Nagaya et al., 2001a). On the other hand, a subcutaneous injection of ghrelin in healthy volunteers increases left ventricular (LV) ejection fraction and stroke volume, and decreases LV end-systolic volume, without changing mean arterial pressure or catecholamine levels (Enomoto et al., 2003). These results suggest that intravenous injection of ghrelin decreases LV afterload and increases cardiac output, whereas subcutaneous ghrelin may increase LV contractility. The decrease in mean arterial pressure seems to be, at least in part, mediated by its action on the central nervous system in the nucleus of the solitary tract (Matsumura et al., 2002). In the coronary circulation, ghrelin exerts a vasoconstrictor effect that is dependent on Ca^{2+} and PKC (Pemberton et al., 2003). Regarding the *in vitro* effects of ghrelin on myocardial contractility, Bedendi et al. (2003) demonstrated that ghrelin, des-Gln14-ghrelin, and desoctanoyl ghrelin show similar negative inotropic effects on papillary muscles that seem to be independent of GH release but dependent on the release of cyclooxygenase metabolites from endothelial cells. Moreover, ghrelin seems to have also a negative lusitropic characterized by a slower rate and an earlier onset of myocardial relaxation that is modulated by prostaglandins and NO (Soares et al., 2006; Figs. 3 and 4; Table 3).

3. Chronic effects

Chronic subcutaneous administration of ghrelin improved cardiac performance in rats with heart failure (HF) as indicated by an increase in cardiac output, stroke volume, LV maximum dP/dt and LV fractional shortening (Nagaya et al., 2001b). Additionally, ghrelin increased diastolic thickness of the noninfarcted posterior wall and attenuated the development of LV remodeling and cardiac cachexia (Nagaya et al., 2001b). In patients with HF, repeated intravenous administration of ghrelin improved LV function,

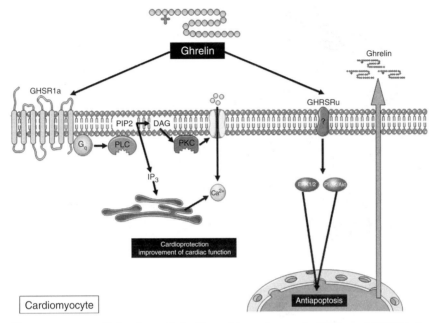

Figure 3 Myocardial actions of ghrelin and its pathways. Activation of GHSR1a by ghrelin stimulates a G-protein that activates PLC-signaling pathway producing IP_3 and DAG. Both IP_3 and DAG lead to an elevation of myocardial Ca^{2+} levels via stimulation of Ca^{2+} influx through the voltage-gated Ca^{2+} channel and Ca^{2+} release from sarcoplasmic reticulum. A putative receptor, GHSRu, with unknown structure mediates the effects of ghrelin on the survival of myocytes through extracellular signal-regulated kinase 1/2 (ERK1/2) and PI-3K activation (Baldanzi et al., 2002). Ghrelin is also synthesized and secreted by cardiomyocytes, probably mediating paracrine/autocrine effects (Iglesias et al., 2004). (See Color Insert.)

exercise capacity (increased peak workload and peak oxygen consumption during exercise) and muscle wasting (increased muscle strength and lean body mass) (Nagaya et al., 2004; Table 3).

Ghrelin has a protective effect against isoproterenol-induced myocardial injury, and its plasmatic and myocardial levels were increased in this model. Ghrelin administration has beneficial hemodynamic effects, ameliorated cardiomegaly, attenuated myocardial lipid peroxidation injury, and relieved cardiac fibrosis in rats with isoproterenol-induced myocardial injury (Chang et al., 2004b; Li et al., 2006; Table 3).

The effects of ghrelin in the monocrotaline model of pulmonary hypertension were studied by our group (Henriques-Coelho et al., 2004). We demonstrated that subcutaneous administration of ghrelin results in a reduction of right ventricular (RV) peak systolic pressure, RV diastolic disturbances, RV hypertrophy, pulmonary vascular remodeling, and in amelioration of LV dysfunction induced by monocrotaline (Henriques-Coelho et al., 2004). Animals with pulmonary hypertension and RV hypertrophy presented an

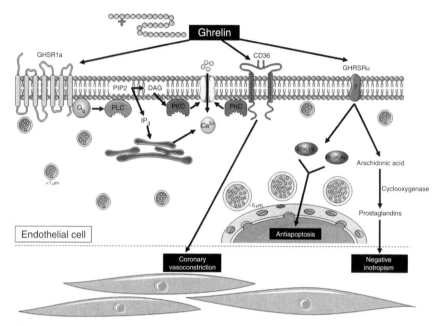

Figure 4 Endothelial actions of ghrelin and its pathways. Ghrelin can be detected in the cytoplasm of endothelial cells in two subcellular compartments: Vesicle-like structures (diameter <1 μm) within the cytoplasm and structures of ~5 μm diameter localized closed to the cell nucleus (Kleinz et al., 2006). Activation of GHSR1a induces an increase in cytosolic Ca^{2+} levels. Ghrelin binding to GHSRu inhibits cell death in endothelial cells through extracellular-signal-regulated kinase 1/2 (ERK1/2) and PI-3K/Akt (Baldanzi et al., 2002). Ghrelins have a negative inotropic effect mediated by cyclooxygenase metabolites produced by endothelial cells by interaction with GHSRu (Bedendi et al., 2003). CD36 is a new ghrelin receptor that mediates coronary vasoconstriction that seems to involve PKC and a calcium channel (Bodart et al., 2002). (See Color Insert.)

increase of more than 20 times of ghrelin mRNA in RV myocardium, suggesting that ghrelin expression in the myocardium can be modulated by load (Henriques-Coelho et al., 2004; Table 3).

4. Clinical implications

Ghrelin exerts important cardioprotective effects in experimental HF since its administration improves cardiac structure and function, reduces infarct size, and attenuates the development of cardiac cachexia (Chang et al., 2004a; Fazio et al., 2003; Frascarelli et al., 2003; Nagaya et al., 2001b). In humans, ghrelin decreases systemic vascular resistance, increases cardiac output and ameliorates muscle wasting in atients with chronic HF (Nagaya et al., 2004). Thus, ghrelin may be a new therapeutic agent for the treatment of severe chronic HF, improving not only cardiac function but also cardiac cachexia, an independent risk factor in patients with end-stage chronic HF (Anker et al., 1997).

Table 3 Contractile actions of ghrelin

Type	Action	Cellular mediation
Striate muscle		
Skeletal muscle	Reduction of the chloride conductivity	G-coupled protein/Ca^{2+} increase/PKC/closing Cl^- or K^+ channels. Dependent of GHSR1a (?)
	Reduction of the potassium condutivity	
Cardiac muscle	Antiapoptotic effect in cardiomyocytes	Independent of GHSR1a
	Negative inotropic and lusitropic effects *in vitro*	Independent of GHSR1a
	Reduces afterload and mean arterial pressure	By its potent vasodilator effect
	Improves cardiac contractility in pathological cardiac conditions	
Smooth muscle		
Vascular smooth muscle	Relaxing the human mammary artery	Independent of GHSR1a
	Human systemic hypotensive effect	Release of endothelial NO
	Improving endothelial dysfunction	
Intraocular smooth muscles	Relaxing effect of the constrictor of the eye pupillae	Release of prostaglandins
	Relaxing effect of the dilator of the eye puppillae	Independent of GHSR1a
		Dependent of the GHSR1a

(*continued*)

Table 3 (*continued*)

Type	Action	Cellular mediation
Gastrointestinal system	Prokinectic activity in stomach	Vagus-dependent effect
	Increases the intestinal transit	Vagus-dependent effect
	Gastric antrum and jejunum induces fasted motor activity in feeded state animals	
	Increasing the frequency of migrating myoelectric complex	
	Increases the electric field stimulation pos-contraction (cholinergic and non cholinergic) in rat stomach strips	GHSR1a-independent pathway
	GHRP-6 increases the cholinergic contraction	GHSR1a-independent pathway
	Potentiates the cholinergic contraction masked by the nitrogenic nerves in mice fundic strips ghrelin	Facilitate the release takycinin through the release of substance P in the vagal terminals
	Elicited a concentration-dependent contraction of jejunal strips rats intestine	

Ghrelin plasmatic levels appear to be associated with other cardiovascular diseases, namely hypertension (Poykko et al., 2003), atherosclerosis (Pöykko et al., 2006) and metabolic syndrome (Ukkola et al., 2006). A correlation seems to exist between serum ghrelin and RV cardiovascular indexes in clinically healthy obese men. Tritos et al. (2004) demonstrated an independent association between serum ghrelin levels and height-adjusted RV mass, RV end-diastolic and end-systolic volumes, RV ejection fraction, whereas no significant association was found between serum ghrelin and indexes of LV structure or function (Tritos et al., 2004). These findings indicate a close interaction between the endocrine and cardiovascular systems in obesity.

Baessler et al. (2006) studied genetic variants within the GHSR and its association with parameters of left ventricular mass (LVM) and geometry, and found that common variants in the GHSR region are associated with parameters of LVM and geometry independent of blood pressure and body mass in the general population and, thus, may be involved in the pathogenesis of LVH (Baessler et al., 2006).

B. Smooth muscle

1. Vascular smooth muscle

a. Ghrelin and its receptor In the vascular smooth muscle, the actions of ghrelin and other somatosecretagogues (GHS) include the modulation of contraction, cell proliferation, and arterial pressure control. Papotti et al. (2000) described, for the first time, the existence of hexarelin-binding sites in the vascular smooth muscle and endothelium (689–1725 fmol/mg of protein), which were even higher than those found in pituitary gland (Tables 1 and 2).

Ghrelin was then identified in the endothelial membrane of small intramyocardial blood vessels, small coronary arteries, saphenous vein, coronary arteries and internal mammary artery (Kleinz et al., 2006). In the same study, ghrelin receptors were detected in small intramyocardial vessels, endothelial cells and in the tunica media of saphenous vein, internal mammary artery and coronary artery. Intracellulary, ghrelin was identified in the endoplasmatic reticulum of endothelial cells, while its receptor was mainly present in the surface of the vascular smooth muscle, with weak presence in the endothelium (Kleinz et al., 2006; Tables 1 and 2).

The density of GHR in human coronary artery is so high that it is comparable to that of AT2 receptor (Katugampola and Davenport, 2003).

b. Acute effects Ghrelin has a relaxing effect in human mammary artery precontracted by endothelin-1. This effect was similar in potency and magnitude of the response to that caused by adrenomedullin (Wiley and Davenport, 2002). Also this effect is very slow (10–20 min to plateau), which is 2.5–5 times slower than the response to the NO donor DEO/NO in the same artery (Wiley and Davenport, 2001). Ghrelin and desoctanoyl-ghrelin show comparable endothelium-independent vasodilator effect and efficacy (Kleinz et al., 2006). In man, an intravenous bolus of ghrelin produces a sustained decrease of the mean arterial pressure (Nagaya et al., 2001b). In healthy volunteers, the potent hypotensive action of ghrelin is long lasting without concomitant change in the heart rate, total plasma proteins and plasma osmolality (Nagaya et al., 2001b). This hypotensive effect is also present in rats being abolished in the presence of blockers of small (apamin), intermediate and large conductance (carybdotoxin) K_{Ca} channels. That outcome was increased when NOS was chronically blocked by L-NAME. So, this acute hypotensive effect may result from activation of the vascular/endothelial K_{Ca} channels (Shinde et al., 2005), which is consistent with a previous report which demonstrated that hexarelin, a synthetic GHS, evoked an increase of K^+ conductance in pituitary somatotrophs (Herrington and Hille, 1994). The subcellular pathway of that effect could be the connection between ghrelin and vascular K^+ channels (endothelial or vascular) which, via coupling to a subunit of the G-protein, promote the release of EDHF and vasodilation of resistance arteries (Vequaud and Thorin, 2001; Fig. 5). The effect of ghrelin, together with its high circulating levels (100 pmol/liter), hypotensive effect and localization of GHSR receptors in the vascular system, implies that ghrelin plays a role in the regulation of this system (Table 3).

c. Chronic effects Ghrelin improved endothelial dysfunction and increased eNOS expression through a GH-independent mechanism in GH-deficient rat aortic arteries. This effect is mainly dependent on the endothelial release of NO and independent of the endothelial prostaglandins (Shimizu et al., 2003).

In the coronary artery media and intima layers of patients with coronary artery disease, ghrelin receptor's density is significantly increased (with no changes in affinity), when compared to those without disease (Katugampola et al., 2002). Likewise, in a model of vascular calcium accumulation induced by excess of vitamin D_3 and nicotine, the plasma levels of ghrelin and aortic expression of its mRNA were significantly reduced. After treatment with ghrelin, aorta calcification ameliorated with a decrease in calcium contents, calcium deposition, ALP activity (alkaline phosphatase activity), inhibition of aorta calcium deposition, inhibition of nodular structures, and increase of OPN (osteopontin) formation. In calcified vascular smooth muscle cells, treatment with ghrelin decreased calcium deposition, ALP activity, and calcium overload and increased OPN mRNA expression. So, ghrelin was downregulated in plasma of animals with calcified arteries. Its exogenous administration could effectively antagonize vascular calcification through a

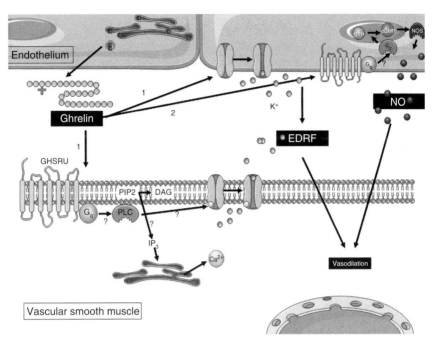

Figure 5 Smooth muscle effects of ghrelin and its pathways. In vascular smooth muscle, ghrelin provides vasodilation through two GHSR1a-independent pathways. Ghrelin stimulates an unknown receptor (GHSRu) which through a G-protein–coupled pathway will close the Ca^{2+}-dependent K^+ channels of the endothelium or vascular smooth muscle and release the endothelial-derived relaxing factor (EDRF) which will relax the vascular smooth muscle (pathway 1) (Shinde et al., 2005). Another pathway for the relaxation is the stimulation of a G-protein–coupled system and activation of the guanidyl cyclase (Gc). This will increase the production of cGMP with consequent induction of the NO synthase promoting the NO-mediated vasodilation (pathway 2) (Shimizu et al., 2003). (See Color Insert.)

downregulation of endothelin expression, whose levels and mRNA were increased in this condition (Li et al., 2005; Table 3).

2. Intraocular smooth muscles

a. Ghrelin and its receptor The constrictor and the dilator of the pupillae are muscles mainly innervated by the cholinergic and the adrenergic autonomic system, respectively. These are the systems primarily implied in the contraction of those muscles. In the last decade, several substances were identified to have the ability of relaxing both muscles (Barilan et al., 2003; Geyer et al., 1998; Pianka et al., 2000; Yamaji et al., 2003; Yousufzai et al., 1999). In a study, mRNA of ghrelin was identified in the rat iris posterior epithelium and in the cilliary epithelium, while ghrelin was shown to relax both muscles as well (Rocha-Sousa et al., 2006; Table 2).

b. Effects Ghrelin was recently shown to relax the iris sphincter and dilator muscles of the rabbit and rat. This effect has a rapid onset and is mediated by the release of prostaglandins (Rocha-Sousa *et al.*, 2006). This rapid onset is distinct from what is observed in vascular smooth muscle which is very slow (Wiley and Davenport, 2002).

The role of the GHSR1a in this effect is different in the sphincter and dilator muscle. In the iris sphincter, it is independent of the GHSR1a, while in the dilator it is dependent of that receptor. The relaxing effects of ghrelin, at least in the sphincter iris muscle, do not seem to be species dependent as similar effects were observed in rabbits and rats (Rocha-Sousa *et al.*, 2006; Table 3).

3. Gastrointestinal smooth muscles

a. Ghrelin and its receptor In this different type of smooth muscle, ghrelin exerts different effects depending on segment of the gastrointestinal tract being analyzed. The stomach is essential organ for ghrelin production and was the first identified location (Kojima *et al.*, 1999). The production and actions of ghrelin are widespread through all the gut. Papotti *et al.* (2000) found residual binding sites for hexarelin in the stomach and colon, with less binding sites than these found in the myocardium, vessels and skeletal muscle. Ghrelin mRNA was identified in the fundus, antrum, jejunum, duodenum and esophagus in decreasing order of concentration (Gnanapavan *et al.*, 2002). Also Ghelardoni *et al.* found ghrelin mRNA and protein in the stomach and small intestine. In the stomach, the mRNA level was maximal, while its protein level was quite below that measured in the lung. So, in the gut, protein expression is dissociated from gene expression (Ghelardoni *et al.*, 2006; Tables 1 and 2).

The GHSR1a was identified in the enteric nervous system of the human (and rat) stomach and large intestine by imunohystochemistry (Dass *et al.*, 2003) and RT-PCR (Depoortere *et al.*, 2005). Furthermore, ghrelin was localized to the myenteric plexus of guinea pig small intestine, mainly in the cholinergic neurons (Xu *et al.*, 2005). In mice, the effect of ghrelin on the gut is mediated by GHSR1a with the particularity that its dose response curve is bell-shaped (Kitazawa *et al.*, 2005). The bell-shaped curve is explained by a possible susceptibility to a rapid desensitization of the ghrelin receptor or by the existence of high- and low-affinity receptors (Depoortere *et al.*, 2005; Tables 1 and 2).

The ghrelin receptor was found to be expressed in the human esophagus, stomach, duodenum, jejunum, ileum, and colon, scattered throughout the mucosa and muscular layers (Takeshita *et al.*, 2006; Tables 1 and 2).

b. Actions Ghrelin has an important prokinetic action in the stomach, and increases the intestinal transit (Tack *et al.*, 2005, 2006; Trudel *et al.*, 2002). In rodents, ghrelin was found to accelerate gastric emptying, stimulate interdigestive motility, via a mechanism prevented by vagatomy and to enhance small bowel transit (Asakawa et al., 2001; Masuda et al., 2000). Ghrelin, given intravenous or intracerebroventricular in feeding animals, induced fasted motor activity in the duodenum and increased motor activity in the gastric antrum (Fujino et al., 2003) and jejunum (Edholm et al., 2004). The effect of ghrelin in the stomach is mediated by GHSR1a, being the central effect dependent of the vagus nerve and of the NPY brain neurons. Peripherally administered ghrelin, which acts directly on the ghrelin receptor located on the stomach and duodenum, induced fasted motor activity in vagally denervated animals. The effect of ghrelin to induce fasted motor activity is inhibited by the decrease of gastric pH (Fujino et al., 2003). In the rat jejunum, ghrelin increases the frequency of the migrating myoelectric complex with a weak antagonistic action of D-Lys3-GHRP-6 (a GHSR1a antagonist) (Edholm et al., 2004; Table 3).

In isolated rat stomach smooth muscle (antrum and fundus strips), ghrelin tends to increase the small cholinergic-mediated contraction, evoked during electric field stimulation (EFS) and increases the EFS-evoked after contraction, mediated by the combination of cholinergic and noncholinergic excitatory activity (Dass et al., 2003; Depoortere et al., 2005). In rabbit stomach, GHRP-6 (a GHS) potentiates the cholinergic contraction in a way independent of GHSR1a. This observation was confirmed by the weak activity of a GHSR agonist in the presence of a motilin receptor antagonist, and that the GHSR1a antagonist potentiates the EFS contraction as ghrelin does (Depoortere et al., 2003; Table 3).

In mice fundic strips, ghrelin potentiates the cholinergic contraction masked by the nitrogenic nerves (Kitazawa et al., 2005). That potentiation is also mediated by the ability of ghrelin to facilitate the tachykinin-dependent response to EFS through the release of substance P from the vagus terminals (Bassil et al., 2006). All these effects are obtained in conditions of EFS, while alone ghrelin does not have any effect on basal tension of the muscular strips. Surprisingly, the GHSR1a inhibitor D-Lys3-GHRP-6 is itself a potent contractor of the muscular strips through the stimulation of the $5HT_{2B}$ receptor (Depoortere et al., 2006). In rats' intestine, ghrelin elicited a concentration-dependent contraction of jejunal strips (Edholm et al., 2004). This effect is absent in mouse, rat, or human colon (Bassil et al., 2005; Dass et al., 2003; Table 3).

C. Skeletal muscle

1. Ghrelin and its receptors

The effects of ghrelin and other somatosegretagogues (GHS) on the contractile function of skeletal muscles are related to the regulation of the resting membrane potential and the conductivity to chloride, potassium, and sodium ions. Binding sites for the GHS, namely hexarelin, have been found in human skeletal muscle, suggesting that a specific receptor can mediate its biological activity in this muscle (Papotti *et al.*, 2000). Other studies found ghrelin mRNA, but not GHSR1a, in the skeletal muscle (Gnanapavan *et al.*, 2002; Tables 1 and 2). These findings suggest that the effects of ghrelin in this muscle, like in the heart and in the iris sphincter, are dependent of a different type of GHS receptor (Bedendi *et al.*, 2003; Rocha-Sousa *et al.*, 2006; Soares *et al.*, 2006).

2. Acute effects

The importance of the GHS in the skeletal muscle function is related with the possibility to restore the membrane potential abnormally changed by age. In aging animals, the mechanical threshold of the skeletal muscles is shifted toward more negative voltage (De Luca *et al.*, 1992; Pierno *et al.*, 1998) explained by the reduction of the resting conductivity to chloride (Cl^-) due to downregulation of the Cl^- channels (Pierno *et al.*, 1999). In these animals, the conductivity to ATP-sensitive potassium channels is modified, which is associated to the redox potential modification (Tricarico and Camerino, 1994). Also, the activity of calcium-activated potassium channels are enhanced in advanced ages (Tricarico *et al.*, 1997a), with an increase of the macroscopic potassium conductance (De Luca *et al.*, 1994; Tricarico *et al.*, 1997b). Finally, the skeletal muscle sodium channels have a reduction in the single-channel conductivity, in some muscle fibers, with a greater availability of extrajunctional sarcolemma channels. Age phenotype fibers also have a reduction of the velocity of activated and inactivated sodium currents (Desaphy *et al.*, 1998). In conclusion, the aged rats have an increase in the membrane excitability with consequent reduction of the stimulus needed to initiate an action potential.

Some authors postulate that these changes could be related to the reduction of GH secretion by the pituitary. To test this hypothesis some authors tried to supplement aged rats with GH and GHS, to verify their effects on the membrane potential and sodium, chloride, and potassium conductivity. Desaphy *et al.* found that GHS, namely hexarelin, has the capacity to restore the firing capacity of fast-twitch muscle fibers as do GH in aged animals (Desaphy *et al.*, 1998). The administration of GH in aged animals significantly ameliorated the lower Cl^- conductivity and the reduction of muscular mechanical threshold. This reverse of Cl^- conductivity was shown to be resulted from the okadaic acid-sensitive phosphatase that counteracts the abnormally elevated PKC activity responsible for the inhibition of Cl^- channels (De Luca *et al.*, 1997). Also the recovery of the K^+ conductivity promoted by the GH/IGF-1 axis is dependent on the normalization of the intracellular Ca^{2+} level that is affected in this process (Tricarico *et al.*, 1997b).

Pierno *et al.* (2003) postulated that ghrelin might possess the same effect as GH in reversing age effects in the skeletal muscle. However, contrary to

GH, ghrelin reduces the chloride channels conductivity of the membrane for Cl⁻ and K⁺. This reduction is mediated by the stimulation of a specific G-protein–coupled receptor with activation of PLC, which increases intracellular Ca^{2+}, through the release of IP_3-dependent stores. The high cytosolic Ca^{2+} induces persistent PKC stimulation, which closes the Cl⁻ and K⁺ channels and decreases the chloride and potassium conductivity (Pierno et al., 2003). The effect is totally suppressed by D-Lys3 GHRP-6; however for the author, this effect was independent of GHSR1a (Pierno et al., 2003). The role of the GHSR1a in this effect was not clear at that time because until recently (Ghelardoni et al., 2006) the mRNA of the receptor was not identified in the skeletal muscle (Gnanapavan et al., 2002; Fig. 6; Table 3).

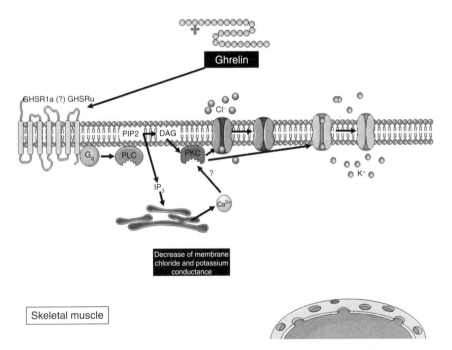

Figure 6 Ghrelin's skeletal muscle effects and its subcellular pathways. Ghrelin stimulates a membrane receptor (possibly other than GHSR1a), coupled to a G-protein–coupled system. That system activates a PLC-signaling pathway producing IP_3 and DAG. Both IP_3 and DAG produce persistent increase of the Ca^{2+} levels which will stimulate the PKC. PKC produces a phosphorylation of the Cl⁻ and K⁺ channels with decrease in chloride and potassium membrane conductivity (Pierno et al., 2003). (See Color Insert.)

3. Chronic effects

In contrast to GH, the chronic treatment with GHS fails to ameliorate the low gCl and high gK typical of aged animals, and so they are not a therapeutic option for restoring the muscular activity in aged animal. This is explained by the direct effect of GHS in the skeletal muscle receptor, which may counteract the effects mediated by the GHS-released GH (Pierno et al., 2003).

ACKNOWLEDGMENTS

Supported by Portuguese grants from FCT (POCTI/SAU-FCT/60803/2004 and POLI/SAU-MMO/61547/2004), through Unidade I&D Cardiovascular (51/94-FCT). We sincerely thank to Servier Medical Art for the helpful image database that was used for construction of this paper images. There are no financial or other relations that could lead to a conflict of interest.

REFERENCES

Anker, S. D., Chua, T. P., Ponikowski, P., Harrington, D., Swan, J. W., Kox, W. J., Poole-Wilson, P. A., and Coats, A. J. (1997). Hormonal changes and catabolic/anabolic imbalance in chronic heart failure and their importance in cardiac cachexia. *Circulation* **96,** 526–534.

Ariyasu, H., Takaya, K., Tagami, T., Ogawa, Y., Hosoda, K., Akamizu, T., Suda, M., Koh, T., Natsui, K., Toyooka, S., Shirakami, G., Usui, T., et al. (2001). Stomach is a major source of circulating ghrelin, and feeding state determines plasma ghrelin-like immunoreactivity levels in humans. *J. Clin. Endocrinol. Metab.* **86,** 4753–4758.

Arvat, E., Di Vito, L., Broglio, F., Papotti, M., Muccioli, G., Dieguez, C., Casanueva, F. F., Deghenghi, R., Camanni, F., and Ghigo, E. (2000). Preliminary evidence that ghrelin, the natural GH secretagogue (GHS)-receptor ligand, strongly stimulates GH secretion in humans. *J. Endocrinol. Invest.* **23,** 493–495.

Asakawa, A., Inui, A., Kaga, T., Yuzuriha, H., Nagata, T., Fujimiya, M., Katsuura, G., Makino, S., Fujino, M. A., and Kasuga, M. (2001). A role of ghrelin in neuroendocrine and behavioral responses to stress in mice. *Neuroendocrinology* **74,** 143–147.

Baessler, A., Kwitek, A. E., Fischer, M., Koehler, M., Reinhard, W., Erdmann, J., Riegger, G., Doering, A., Schunkert, H., and Hengstenberg, C. (2006). Association of the ghrelin receptor gene region with left ventricular hypertrophy in the general population: Results of the MONICA/KORA augsburg echocardiographic substudy. *Hypertension* **47,** 920–927.

Baldanzi, G., Filigheddu, N., Cutrupi, S., Catapano, F., Bonissoni, S., Fubini, A., Malan, D., Baj, G., Granata, R., Broglio, F., Papotti, M., Surico, N., et al. (2002). Ghrelin and des-acyl ghrelin inhibit cell death in cardiomyocytes and endothelial cells through ERK1/2 and PI 3-kinase/AKT. *J. Cell Biol.* **159,** 1029–1037.

Barilan, A., Nachman-Rubistein, R., Oron, Y., and Geyer, O. (2003). Muscarinic blockers potentiate β adrenergic relaxation of bovine iris sphincter. *Graefes Arch. Clin. Exp. Ophthalmol.* **241,** 226–231.

Barreiro, M. L., Gaytan, F., Caminos, J. E., Pinilla, L., Casanueva, F. F., Aguilar, E., Dieguez, C., and Tena-Sempere, M. (2002). Cellular location and hormonal regulation of ghrelin expression in rat testis. *Biol. Reprod.* **67,** 1768–1776.

Bassil, A. K., Dass, N. B., Murray, C. D., Muir, A., and Sanger, G. J. (2005). Prokineticin-2, motilin, ghrelin and metoclopramide: Prokinetic utility in mouse stomach and colon. *Eur. J. Pharmacol.* **524**, 138–144.

Bassil, A. K., Dass, N. B., and Sanger, G. J. (2006). The prokinetic-like activity of ghrelin in rat isolated stomach is mediated via cholinergic and tachykininergic motor neurones. *Eur. J. Pharmacol.* **544**, 146–152.

Beaumont, N. J., Skinner, V. O., Tan, T. M., Ramesh, B. S., Byrne, D. J., MacColl, G. S., Keen, J. N., Bouloux, P. M., Mikhailidis, D. P., Bruckdorfer, K. R., Vanderpump, M. P., and Srai, K. S. (2003). Ghrelin can bind to a species of high density lipoprotein associated with paraoxonase. *J. Biol. Chem.* **278**, 8877–8880.

Bedendi, I., Alloatti, G., Marcantoni, A., Malan, D., Catapano, F., Ghe, C., Deghenghi, R., Ghigo, E., and Muccioli, G. (2003). Cardiac effects of ghrelin and its endogenous derivatives des-octanoyl ghrelin and des-Gln14-ghrelin. *Eur. J. Pharmacol.* **476**, 87–95.

Bednarek, M. A., Feighner, S. D., Pong, S. S., McKee, K. K., Hreniuk, D. L., Silva, M. V., Warren, V. A., Howard, A. D., Van Der Ploeg, L. H., and Heck, J. V. (2000). Structure-function studies on the new growth hormone-releasing peptide, ghrelin: Minimal sequence of ghrelin necessary for activation of growth hormone secretagogue receptor 1a. *J. Med. Chem.* **16**, 4370–4376.

Benso, A., Broglio, F., Marafetti, L., Lucatello, B., Seardo, M. A., Granata, R., Martina, V., Papotti, M., Muccioli, G., and Ghigo, E. (2004). Ghrelin and synthetic growth hormone secretagogues are cardioactive molecules with identities and differences. *Semin. Vasc. Med.* **4**, 107–114.

Bisi, G., Podio, V., Valetto, M. R., Broglio, F., Bertuccio, G., Del Rio, G., Arvat, E., Boghen, M. F., Deghenghi, R., Muccioli, G., Ong, H., and Ghigo, E. (1999). Acute cardiovascular and hormonal effects of GH and hexarelin, a synthetic GH-releasing peptide, in humans. *J. Endocrinol. Invest.* **22**, 266–272.

Bodart, V., Febbraio, M., Demers, A., McNicoll, N., Pohankova, P., Perreault, A., Sejlitz, T., Escher, E., Silverstein, R. L., Lamontagne, D., and Ong, H. (2002). CD36 mediates the cardiovascular action of growth hormone-releasing peptides in the heart. *Circ. Res.* **90**, 844–849.

Camina, J. P., Carreira, M. C., El Messari, S., Llorens-Cortes, C., Smith, R. G., and Casanueva, F. F. (2004). Desensitization and endocytosis mechanisms of ghrelin-activated growth hormone secretagogue receptor 1a. *Endocrinology* **145**, 930–940.

Caminos, J. E., Tena-Sempere, M., Gaytan, F., Sanchez-Criado, J. E., Barreiro, M. L., Nogueiras, R., Casanueva, F. F., Aguilar, E., and Dieguez, C. (2003). Expression of ghrelin in the cyclic and pregnant rat ovary. *Endocrinology* **144**, 1594–1602.

Cao, J. M., Ong, H., and Chen, C. (2006). Effects of ghrelin and synthetic GH secretagogues on the cardiovascular system. *Trends Endocrinol. Metab.* **17**, 13–18.

Cassoni, P., Ghe, C., Marrocco, T., Tarabra, E., Allia, E., Catapano, F., Deghenghi, R., Ghigo, E., Papotti, M., and Muccioli, G. (2004). Expression of ghrelin and biological activity of specific receptors for ghrelin and des-acyl ghrelin in human prostate neoplasms and related cell lines. *Eur. J. Endocrinol.* **150**, 173–184.

Chanoine, J. P., and Wong, A. C. (2004). Ghrelin gene expression is markedly higher in fetal pancreas compared with fetal stomach: Effect of maternal fasting. *Endocrinology* **145**, 3813–3820.

Chang, L., Ren, Y., Liu, X., Li, W. G., Yang, J., Geng, B., Weintraub, N. L., and Tang, C. (2004a). Protective effects of ghrelin on ischemia/reperfusion injury in the isolated rat heart. *J. Cardiovasc. Pharmacol.* **43**, 165–170.

Chang, L., Zhao, J., Li, G. Z., Geng, B., Pan, C. S., Qi, Y. F., and Tang, C. S. (2004b). Ghrelin protects myocardium from isoproterenol-induced injury in rats. *Acta Pharmacol. Sin.* **25**, 1131–1137.

Choi, K., Roh, S. G., Hong, Y. H., Shrestha, Y. B., Hishikawa, D., Chen, C., Kojima, M., Kangawa, K., and Sasaki, S. (2003). The role of ghrelin and growth hormone secretagogues receptor on rat adipogenesis. *Endocrinology* **144**, 754–759.

Colao, A. M., Marzullo, P., Di Somma, C., and Lombardi, G. (2001). Growth hormone and the heart. *Clin. Endocrinol.* **54**, 137–154.

Cowley, M. A., Smith, R. G., Diano, S., Tschop, M., Pronchuk, N., Grove, K. L., Strasburger, C. J., Bidlingmaier, M., Esterman, M., Heiman, M. L., Garcia-Segura, L. M., Nillni, E. A., *et al.* (2003). The distribution and mechanism of action of ghrelin in the CNS demonstrates a novel hypothalamic circuit regulating energy homeostasis. *Neuron* **37**, 649–661.

Cummings, D. E., Purnell, J. Q., Frayo, R. S., Schmidova, K., Wisse, B. E., and Weigle, D. S. (2001). A preprandial rise in plasma ghrelin levels suggests a role in meal initiation in humans. *Diabetes* **50**, 1714–1719.

Cunha, S. R., and Mayo, K. E. (2002). Ghrelin and growth hormone (GH) secretagogues potentiate GH-releasing hormone (GHRH)-induced cyclic adenosine 30,50-monophosphate production in cells expressing transfected GHRH and GH secretagogue receptors. *Endocrinology* **143**, 4570–4582.

Dass, N. B., Munonyara, M., Bassil, A. K., Hervieu, G. J., Osbourne, S., Corcoran, S., Morgan, M., and Sanger, G. J. (2003). Growth hormone secretagogue receptors in rat and human gastrointestinal tract and the effects of ghrelin. *Neuroscience* **120**, 443–453.

Date, Y., Kojima, M., Hosoda, H., Sawaguchi, A., Mondal, M. S., Suganuma, T., Matsukura, S., Kangawa, K., and Nakazato, M. (2000). Ghrelin, a novel growth hormone-releasing acylated peptide, is synthesized in a distinct endocrine cell type in the gastrointestinal tracts of rats and humans. *Endocrinology* **141**, 4255–4261.

Date, Y., Nakazato, M., Hashiguchi, S., Dezaki, K., Mondal, M. S., Hosoda, H., Kojima, M., Kangawa, K., Arima, T., Matsuo, H., Yada, T., and Matsukura, S. (2002). Ghrelin is present in pancreatic alpha-cells of humans and rats and stimulates insulin secretion. *Diabetes* **51**, 124–129.

De Luca, A., Pierno, S., Huxtable, R. J., Falli, P., Franconi, F., Giotti, A., and Camerino, D. C. (1992). Effects of taurine depletion on membrane electrical properties of rat skeletal muscle. *Adv. Exp. Med. Biol.* **315**, 199–205.

De Luca, A., Pierno, S., and Conte Camerino, D. (1994). Pharmacological interventions for the changes of chloride channel conductance of aging rat skeletal muscle. *Ann. N. Y. Acad. Sci.* **717**, 180–188.

De Luca, A., Pierno, S., Cocchi, D., and Conte Camerino, D. (1997). Effects of chronic growth hormone treatment in aged rats on the biophysical and pharmacological properties of skeletal muscle chloride channels. *Br. J. Pharmacol.* **121**, 369–374.

Depoortere, I., Thijs, T., Thielemans, L., Robberecht, P., and Peeters, T. L. (2003). Interaction of the growth hormone-releasing peptides ghrelin and growth hormone-releasing peptide-6 with the motilin receptor in the rabbit gastric antrum. *J. Pharmacol. Exp. Ther.* **305**, 660–667.

Depoortere, I., De Winter, B., Thijs, T., De Man, J., Pelckmans, P., and Peeters, T. (2005). Comparison of the gastroprokinetic effects of ghrelin, GHRP-6 and motilin in rats *in vivo* and *in vitro*. *Eur. J. Pharmacol.* **515**, 160–168.

Depoortere, I., Thijs, T., and Peeters, T. (2006). The contractile effect of the ghrelin receptor antagonist, D-Lys3-GHRP-6, in rat fundic strips is mediated through 5-HT receptors. *Eur. J. Pharmacol.* **537**, 160–165.

Desaphy, J. F., De Luca, A., Pierno, S., Imbrici, P., and Camerino, D. C. (1998). Partial recovery of skeletal muscle sodium channel properties in aged rats chronically treated with growth hormone or the GH-secretagogue hexarelin. *J. Pharmacol. Exp. Ther.* **286**, 903–912.

Edholm, T., Levin, F., Hellstrom, P. M., and Schmidt, P. T. (2004). Ghrelin stimulates motility in the small intestine of rats through intrinsic cholinergic neurons. *Regul. Pept.* **121,** 25–30.

Enomoto, M., Nagaya, N., Uematsu, M., Okumura, H., Nakagawa, E., Ono, F., Hosoda, H., Oya, H., Kojima, M., Kanmatsuse, K., and Kangawa, K. (2003). Cardiovascular and hormonal effects of subcutaneous administration of ghrelin, a novel growth hormone-releasing peptide in healthy humans. *Clin. Sci. (Lond.)* **105,** 431–435.

Fazio, S., Sabatini, D., Capaldo, B., Vigorito, C., Giordano, A., Guida, R., Pardo, F., Frascarelli, S., Ghelardoni, S., Ronca-Testoni, S., and Zucchi, R. (2003). Effect of ghrelin and synthetic growth hormone secretagogues in normal and ischemic rat heart. *Basic Res. Cardiol.* **98,** 401–405.

Febbraio, M., Hajjar, D. P., and Silverstein, R. L. (2001). CD36: A class B scavenger receptor involved in angiogenesis, atherosclerosis, inflammation, and lipid metabolism. *J. Clin. Invest.* **108,** 785–791.

Frascarelli, S., Ghelardoni, S., Ronca-Testoni, S., and Zucchi, R. (2003). Effect of ghrelin and synthetic growth hormone secretagogues in normal and ischemic rat heart. *Basic Res. Cardiol.* **98,** 401–405.

Fujino, K., Inui, A., Asakawa, A., Kihara, N., Fujimura, M., and Fujimiya, M. (2003). Ghrelin induces fasted motor activity of the gastrointestinal tract in conscious fed rats. *J. Physiol.* **550,** 227–240.

Geyer, O., Bar-ilan, A., Nachman, R., Lazar, M., and Oron, Y. (1998). Beta3-adrenergic relaxation of bovine iris sphincter muscle. *FEBS Lett.* **429,** 356–358.

Ghelardoni, S., Carnicelli, V., Frascarelli, S., Ronca-Testoni, S., and Zucchi, R. (2006). Ghrelin tissue distribution: Comparison between gene and protein expression. *J. Endocrinol. Invest.* **29,** 115–121.

Gnanapavan, S., Kola, B., Bustin, S. A., Morris, D. G., McGee, P., Fairclough, P., Bhattacharya, S., Carpenter, R., Grossman, A. B., and Korbonits, M. (2002). The tissue distribution of the mRNA of ghrelin and subtypes of its receptor, GHS-R, in humans. *J. Clin. Endocrinol. Metab.* **87,** 2988–2991.

Gualillo, O., Caminos, J., Blanco, M., Garcia-Caballero, T., Kojima, M., Kangawa, K., Dieguez, C., and Casanueva, F. (2001). Ghrelin, a novel placental-derived hormone. *Endocrinology* **142,** 788–794.

Guan, X.-M., Yu, H., Palyha, O. C., McKee, K. K., Feighner, S. D., Sirinathsinghji, D. J. S., Smith, R. G., Van der Ploeg, L. H. T., and Howard, A. D. (1997). Distribution of mRNA encoding the growth hormone secretagogue receptor in brain and peripheral tissues. *Mol. Brain Res.* **48,** 23–29.

Hayashida, T., Nakahara, K., Mondal, M. S., Date, Y., Nakazato, M., Kojima, M., Kangawa, K., and Murakami, N. (2002). Ghrelin in neonatal rats: Distribution in stomach and its possible role. *J. Endocrinol.* **173,** 239–245.

Henriques-Coelho, T., Correia-Pinto, J., Roncon-Albuquerque, R., Jr., Baptista, M. J., Lourenco, A. P., Oliveira, S. M., *et al.* (2004). Endogenous production of ghrelin and beneficial effects of its exogenous administration in monocrotaline-induced pulmonary hypertension. *Am. J. Physiol. Heart Circ. Physiol.* **287,** H2885–H2890.

Herrington, J., and Hille, B. (1994). Growth hormone-releasing hexapeptide elevates intracellular calcium in rat somatotropes by two mechanisms. *Endocrinology* **135,** 1100–1108.

Hosoda, H., Kojima, M., Matsuo, H., and Kangawa, K. (2000a). Purification and characterization of rat des-Gln14-Ghrelin, a second endogenous ligand for the growth hormone secretagogue receptor. *J. Biol. Chem.* **275,** 21995–22000.

Hosoda, H., Kojima, M., Matsuo, H., and Kangawa, K. (2000b). Ghrelin and des-acyl ghrelin: Two major forms of rat ghrelin peptide in gastrointestinal tissue. *Biochem. Biophys. Res. Commun.* **279,** 909–913.

Hosoda, H., Kojima, M., Mizushima, T., Shimizu, S., and Kangawa, K. (2003). Structural divergence of human ghrelin. Identification of multiple ghrelin-derived molecules produced by post-translational processing. *J. Biol. Chem.* **278,** 64–70.

Howard, A. D., Feighner, S. D., Cully, D. F., Arena, J. P., Liberator, P. A., Rosenblum, C. I., Hamelin, M., Hreniuk, D. L., Palyha, O. C., Anderson, J., Paress, P. S., Diaz, C., *et al.* (1996). A receptor in pituitary and hypothalamus that functions in growth hormone release. *Science* **273,** 974–977.

Howard, A. D., Wang, R., Pong, S. S., Mellin, T. N., Strack, A., Guan, X. M., Zeng, Z., Williams, D. L., Jr., Feighner, S. D., Nunes, C. N., Murphy, B., Stair, J. N., *et al.* (2000). Identification of receptors for neuromedin U and its role in feeding. *Nature* **406,** 70–74.

Iglesias, M. J., Pineiro, R., Blanco, M., Gallego, R., Dieguez, C., Gualillo, O., Gonzalez-Juanatey, J. R., and Lago, F. (2004). Growth hormone releasing peptide (ghrelin) is synthesized and secreted by cardiomyocytes. *Cardiovasc. Res.* **62,** 481–488.

Imazio, M., Bobbio, M., Broglio, F., Benso, A., Podio, V., Valetto, M. R., Bisi, G., Ghigo, E., and Trevi, G. P. (2002). GH-independent cardiotropic activities of hexarelin in patients with severe left ventricular dysfunction due to dilated and ischemic cardiomyopathy. *Eur. J. Heart Fail.* **4,** 185–191.

Isgaard, J., and Johansson, I. (2005). Ghrelin and GHS on cardiovascular applications/functions. *J. Endocrinol. Invest.* **28,** 838–842.

Kamegai, J., Tamura, H., Shimizu, T., Ishii, S., Sugihara, H., and Wakabayashi, I. (2000). Central effect of ghrelin, an endogenous growth hormone secretagogue, on hypothalamic peptide gene expression. *Endocrinology* **141,** 4797–4800.

Katugampola, S., and Davenport, A. (2003). Emerging roles for orphan G-protein-coupled receptors in the cardiovascular system. *Trends Pharmacol. Sci.* **24**(1), 30–35.

Katugampola, S., Pallikaros, Z., and Davenport, A. P. (2001). [125I-His9]-ghrelin, a novel radioligand for localizing GHS orphan receptors in human and rat tissue; up regulation of receptors with atherosclerosis. *Br. J. Pharmacol.* **134,** 143–149.

Katugampola, S. D., Kuc, R. E., Maguire, J. J., and Davenport, A. P. (2002). G-protein-coupled receptors in human atherosclerosis: Comparison of vasoconstrictors (endothelin and thromboxane) with recently de-orphanized (urotensin-II, apelin and ghrelin) receptors. *Clin. Sci. (Lond.)* **103,** 171S–175S.

Kitazawa, T., De Smet, B., Verbeke, K., Depoortere, I., and Peeters, T. L. (2005). Gastric motor effects of peptide and non-peptide ghrelin agonists in mice *in vivo* and *in vitro*. *Gut* **54,** 1078–1084.

Kleinz, M. J., Maguire, J. J., Skepper, J. N., and Davenport, A. P. (2006). Functional and immunocytochemical evidence for a role of ghrelin and des-octanoyl ghrelin in the regulation of vascular tone in man. *Cardiovasc. Res.* **69,** 227–235.

Kojima, M., and Kangawa, K. (2005). Ghrelin: Structure and function. *Physiol. Rev.* **85,** 495–522.

Kojima, M., Hosoda, H., Date, Y., Nakazato, M., Matsuo, H., and Kangawa, K. (1999). Ghrelin is a growth-hormone releasing acylated peptide from stomach. *Nature* **402,** 656–660.

Kola, B., Hubina, E., Tucci, S. A., Kirkham, T. C., Garcia, E. A., Mitchell, S. E., Williams, L. M., Hawley, S. A., Hardie, D. G., Grossman, A. B., and Korbonits, M. (2005). Cannabinoids and ghrelin have both central and peripheral metabolic and cardiac effects via AMP-activated protein kinase. *J. Biol. Chem.* **280,** 25196–25201.

Korbonits, M., Kojima, M., Kangawa, K., and Grossman, A. B. (2001). Presence of ghrelin in normal and adenomatous human pituitary. *Endocrine* **14,** 101–104.

Korbonits, M., Goldstone, A. P., Gueorguiev, M., and Grossman, A. B. (2004). Ghrelin– a hormone with multiple functions. *Front Neuroendocrinol.* **25,** 27–68.

Lago, F., Gonzalez-Juanatey, J. R., Casanueva, F. F., Gomez-Reino, J., Dieguez, C., and Gualillo, O. (2005). Ghrelin, the same peptide for different functions: Player or bystander? *Vitam. Horm.* **71**, 405–432.
Li, G. Z., Jiang, W., Zhao, J., Pan, C. S., Cao, J., Tang, C. S., and Chang, L. (2005). Ghrelin blunted vascular calcification *in vivo* and *in vitro* in rats. *Regul. Pept.* **129**, 167–176.
Li, L., Zhang, L. K., Pang, Y. Z., Pan, C. S., Qi, Y. F., Chen, L., Wang, X., Tang, C. S., and Zhang, J. (2006). Cardioprotective effects of ghrelin and des-octanoyl ghrelin on myocardial injury induced by isoproterenol in rats. *Acta Pharmacol. Sin.* **27**, 527–535.
Lin, Y., Matsumura, K., Fukuhara, M., Kagiyama, S., Fujii, K., and Iida, M. (2004). Ghrelin acts at the nucleus of the solitary tract to decrease arterial pressure in rats. *Hypertension* **43**, 1–6.
Locatelli, V., Rossoni, G., Schweiger, F., Torsello, A., De Gennaro Colonna, V., Bernareggi, M., Deghenghi, R., Muller, E. E., and Berti, F. (1999). Growth hormone-independent cardioprotective effects of hexarelin in the rat. *Endocrinology* **140**, 4024–4031.
Lu, S., Guan, J. L., Wang, Q. P., Uehara, K., Yamada, S., Goto, N., Date, Y., Nakazato, M., Kojima, M., Kangawa, K., and Shioda, S. (2002). Immunocytochemical observation of ghrelin-containing neurons in the rat arcuate nucleus. *Neurosci. Lett.* **321**, 157–160.
Malagon, M. M., Luque, R. M., Ruiz-Guerrero, E., Rodriguez-Pacheco, F., Garcia-Navarro, S., Casanueva, F. F., Gracia-Navarro, F., and Castano, J. P. (2003). Intracellular signaling mechanisms mediating ghrelin-stimulated growth hormone release in somatotropes. *Endocrinology* **144**, 5372–5380.
Marleau, S., Mulumba, M., Lamontagne, D., and Ong, H. (2006). Cardiac and peripheral actions of growth hormone and its releasing peptides: Relevance for the treatment of cardiomyopathies. *Cardiovasc. Res.* **69**, 26–35.
Masuda, Y., Tanaka, T., Inomata, N., Ohnuma, N., Tanaka, S., Itoh, Z., Hosoda, H., Kojima, M., and Kangawa, K. (2000). Ghrelin stimulates gastric acid secretion and motility in rats. *Biochem. Biophys. Res. Commun.* **276**, 905–908.
Matsumura, K., Tsuchihashi, T., Fujii, K., Abe, I., and Iida, M. (2002). Central ghrelin modulates sympathetic activity in conscious rabbits. *Hypertension* **40**, 694–699.
Mori, K., Yoshimoto, A., Takaya, K., Hosoda, K., Ariyasu, H., Yahata, K., Mukoyama, M., Sugawara, A., Hosoda, H., Kojima, M., Kangawa, K., and Nakao, K. (2000). Kidney produces a novel acylated peptide, ghrelin. *FEBS Lett.* **486**, 213–216.
Mousseaux, D., Le Gallic, L., Ryan, J., Oiry, C., Gagne, D., Fehrentz, J. A., Galleyrand, J. C., and Martinez, J. (2006). Regulation of ERK1/2 activity by ghrelin-activated growth hormone secretagogue receptor 1A involves a PLC/PKCvarepsilon pathway. *Br. J. Pharmacol.* **148**, 350–365.
Muccioli, G., Broglio, F., Valetto, M. R., Ghe, C., Catapano, F., Graziani, A., Papotti, M., Bisi, G., Deghenghi, R., and Ghigo, E. (2000). Growth hormone-releasing peptides and the cardiovascular system. *Ann. Endocrinol. (Paris)* **61**, 27–31.
Nagaya, N., and Kangawa, K. (2003). Ghrelin improves left ventricular dysfunction and cardiac cachexia in heart failure. *Curr. Opin. Pharmacol.* **3**, 146–151.
Nagaya, N., Kojima, M., Uematsu, M., Yamagishi, M., Hosoda, H., Oya, H., Hayashi, Y., and Kangawa, K. (2001a). Hemodynamic and hormonal effects of human ghrelin in healthy volunteers. *Am. J. Physiol. Regul. Integr. Comp. Physiol.* **280**, R1483–R1487.
Nagaya, N., Uematsu, M., Kojima, M., Ikeda, Y., Yoshihara, F., Shimizu, W., Hosoda, H., Hirota, Y., Ishida, H., Mori, H., and Kangawa, K. (2001b). Chronic administration of ghrelin improves left ventricular dysfunction and attenuates development of cardiac cachexia in rats with heart failure. *Circulation* **104**, 1430–1435.
Nagaya, N., Uematsu, M., Kojima, M., Ikeda, Y., Yoshihara, F., Shimizu, W., Hosoda, H., Hirota, Y., Ishida, H., Mori, H., and Kangawa, K. (2004). Effects of ghrelin

administration on left ventricular function, exercise capacity and muscle wasting in patients with chronic heart failure. *Circulation* **110,** 3674–3679.

Nakahara, K., Nakagawa, M., Baba, Y., Sato, M., Toshinai, K., Date, Y., Nakazato, M., Kojima, M., Miyazato, M., Kaiya, H., Hosoda, H., Kangawa, K., *et al.* (2006). Maternal ghrelin plays an important role in rat fetal development during pregnancy. *Endocrinology* **147,** 1333–1342.

Nakazato, M., Murakami, N., Date, Y., Kojima, M., Matsuo, H., Kangawa, K., and Matsukura, S. (2001). A role for ghrelin in the central regulation of feeding. *Nature* **409**(37), 194–198.

Nanzer, A. M., Khalaf, S., Mozid, A. M., Fowkes, R. C., Patel, M. V., Burrin, J. M., Grossman, A. B., and Korbonits, M. (2004). Ghrelin exerts a proliferative effect on a rat pituitary somatotroph cell line via the mitogen-activated protein kinase pathway. *Eur. J. Endocrinol.* **151,** 233–240.

Palyha, O. C., Feighner, S. D., Tan, C. P., McKee, K. K., Hreniuk, D. L., Gao, Y. D., Schleim, K. D., Yang, L., Morriello, G. J., Nargund, R., Patchett, A. A., Howard, A. D., *et al.* (2000). Ligand activation domain of human orphan growth hormone (GH) secretagogue receptor (GHS-R) conserved from Pufferfish to humans. *Mol. Endocrinol.* **14,** 160–169.

Papotti, M., Ghe, C., Cassoni, P., Catapano, F., Deghenghi, R., Ghigo, E., and Muccioli, G. (2000). Growth hormone secretagogue binding sites in pheripheral human tissues. *J. Clin. Endocrinol. Metab.* **85,** 3803–3807.

Peeters, T. L. (2005). Ghrelin: A new player in the control of gastrointestinal functions. *Gut* **54,** 1638–1649.

Pemberton, C., Wimalasena, P., Yandle, T., Soule, S., and Richards, M. (2003). C-terminal pro-ghrelin peptides are present in the human circulation. *Biochem. Biophys. Res. Commun.* **310,** 567–573.

Pianka, P., Oron, Y., Lazar, M., and Geyer, O. (2000). Nonadrenergic, noncholinergic relaxation of bovine iris sphincter: Role of endogenous nitric oxide. *Invest. Ophthalmol. Vis. Sci.* **41,** 880–886.

Pierno, S., De Luca, A., Camerino, C., Huxtable, R. J., and Camerino, D. C. (1998). Chronic administration of taurine to aged rats improves the electrical and contractile properties of skeletal muscle fibers. *J. Pharmacol. Exp. Ther.* **286,** 1183–1190.

Pierno, S., De Luca, A., Beck, C. L., George, A. L., Jr., and Conte Camerino, D. (1999). Aging-associated down-regulation of ClC-1 expression in skeletal muscle: Phenotypic-independent relation to the decrease of chloride conductance. *FEBS Lett.* **449,** 12–16.

Pierno, S., De Luca, A., Desaphy, J. F., Fraysse, B., Liantonio, A., Didonna, M. P., Lograno, M., Cocchi, D., Smith, R. G., and Camerino, D. C. (2003). Growth hormone secretagogues modulate the electrical and contractile properties of rat skeletal muscle through a ghrelin-specific receptor. *Br. J. Pharmacol.* **139,** 575–584.

Poykko, S. M., Kellokoski, E., Horkko, S., Kauma, H., Kesaniemi, Y. A., and Ukkola, O. (2003). Low plasma ghrelin is associated with insulin resistance, hypertension, and the prevalence of type 2 diabetes. *Diabetes* **52,** 2546–2553.

Pöykko, S. M., Kellokoski, E., Ukkola, O., Kauma, H., Paivansalo, M., Kesaniemi, Y. A., and Horkko, S. (2006). Plasma ghrelin concentrations are positively associated with carotid artery atherosclerosis in males. *J. Intern. Med.* **260,** 43–52.

Rindi, G., Savio, A., Torsello, A., Zoli, M., Locatelli, V., Cocchi, D., Paolotti, D., and Solcia, E. (2002). Ghrelin expression in gut endocrine growths. *Histochem. Cell Biol.* **117,** 521–525.

Rocha-Sousa, A., Saraiva, J., Henriques-Coelho, T., Falcão-Reis, F., Correia-Pinto, J., and Leite-Moreira, A. F. (2006). "Ghrelin as a novel locally produced relaxing peptide of the iris sphincter and dilator muscles." *Exp. Eye Res.* **83,** 1179–1187.

Rodriguez-Pacheco, F., Luque, R. M., Garcia-Navarro, S., Gracia-Navarro, F., Castano, J. P., and Malagon, M. M. (2005). Ghrelin induces growth hormone (GH) secretion via nitric oxide (NO)/cGMP signaling. *Ann. N. Y. Acad. Sci.* **1040,** 452–453.

Santos, M., Bastos, P., Gonzaga, S., Roriz, J. M., Baptista, M. J., Nogueira-Silva, C., Melo-Rocha, G., Henriques-Coelho, T., Roncon-Albuquerque, R., Jr., Leite-Moreira, A. F., De Krijger, R. R., Tibboel, D., *et al.* (2006). Ghrelin expression in human and rat fetal lungs and the effect of ghrelin administration in nitrofen-induced congenital diaphragmatic hernia. *Pediatr. Res.* **59,** 531–537.

Shimizu, Y., Nagaya, N., Teranishi, Y., Imazu, M., Yamamoto, H., Shokawa, T., Kangawa, K., Kohno, N., and Yoshizumi, M. (2003). Ghrelin improves endothelial dysfunction through growth hormone-independent mechanisms in rats. *Biochem. Biophys. Res. Comun.* **310,** 830–835.

Shinde, U. A., Desai, K. M., Yu, C., and Gopalakrishnan, V. (2005). Nitric oxide synthase inhibition exaggerates the hypotensive response to ghrelin: Role of calcium-activated potassium channels. *J. Hypertens.* **23,** 713–715.

Soares, J. B., Rocha-Sousa, A., Castro-Chaves, P., Henriques-Coelho, T., and Leite-Moreira, A. F. (2006). Inotropic and lusitropic effects of ghrelin and their modulation by the endocardial endothelium, NO, prostaglandins, GHS-R1a and K(Ca) channels. *Peptides* **27,** 1616–1623.

Sun, Y., Ahmed, S., and Smith, R. G. (2003). Deletion of ghrelin impairs neither growth nor appetite. *Mol. Cell Biol.* **23,** 7973–7981.

Tack, J., Depoortere, I., Bisschops, R., Verbeke, K., Janssens, J., and Peeters, T. (2005). Influence of ghrelin on gastric emptying and meal-related symptoms in idiopathic gastroparesis. *Aliment. Pharmacol. Ther.* **22,** 847–853.

Tack, J., Depoortere, I., Bisschops, R., Delporte, C., Coulie, B., Meulemans, A., Janssens, J., and Peeters, T. (2006). Influence of ghrelin on interdigestive gastrointestinal motility in humans. *Gut* **55,** 327–333.

Takaya, K., Ariyasu, H., Kanamoto, N., Iwakura, H., Yoshimoto, A., Harada, M., Mori, K., Komatsu, Y., Usui, T., Shimatsu, A., Ogawa, Y., Hosoda, K., *et al.* (2000). Ghrelin strongly stimulates growth hormone (GH) release in humans. *J. Clin. Endocrinol. Metab.* **85,** 4908–4911.

Takeshita, E., Matsuura, B., Dong, M., Miller, L. J., Matsui, H., and Onji, M. (2006). Molecular characterization and distribution of motilin family receptors in the human gastrointestinal tract. *J. Gastroenterol.* **41,** 223–230.

Tanaka, K., Minoura, H., Isobe, T., Yonaha, H., Kawato, H., Wang, D. F., Yoshida, T., Kojima, M., Kangawa, K., and Toyoda, N. (2003). Ghrelin is involved in the decidualization of human endometrial stromal cells. *J. Clin. Endocrinol. Metab.* **88,** 2335–2340.

Thompson, N. M., Gill, D. A., Davies, R., Loveridge, N., Houston, P. A., Robinson, I. C., and Wells, T. (2004). Ghrelin and des-octanoyl ghrelin promote adipogenesis directly *in vivo* by a mechanism independent of the type 1a growth hormone secretagogue receptor. *Endocrinology* **145,** 234–242.

Tricarico, D., and Camerino, D. C. (1994). ATP-sensitive K+ channels of skeletal muscle fibers from young adult and aged rats: Possible involvement of thiol-dependent redox mechanisms in the age-related modifications of their biophysical and pharmacological properties. *Mol. Pharmacol.* **46,** 754–761.

Tricarico, D., Petruzzi, R., and Camerino, D. C. (1997a). Changes of the biophysical properties of calcium-activated potassium channels of rat skeletal muscle fibres during aging. *Pflugers Arch.* **434,** 822–829.

Tricarico, D., Mallamaci, R., Barbieri, M., and Conte Camerino, D. (1997b). Modulation of ATP-sensitive K+ channel by insulin in rat skeletal muscle fibers. *Biochem. Biophys. Res. Commun.* **232,** 536–539.

Tritos, N. A., Kissinger, K. V., Manning, W. J., and Danias, P. G. (2004). Association between ghrelin and cardiovascular indexes in healthy obese and lean men. *Clin. Endocrinol. (Oxf)* **60,** 60–66.

Trudel, L., Tomasetto, C., Rio, M. C., Bouin, M., Plourde, V., Eberling, P., and Poitras, P. (2002). Ghrelin/motilin-related peptide is a potent prokinetic to reverse gastric postoperative ileus in rat. *Am. J. Physiol. Gastrointest. Liver Physiol.* **282,** G948–G952.

Tschop, M., Smiley, D. L., and Heiman, M. L. (2000). Ghrelin induces adiposity in rodents. *Nature* **407,** 908–913.

Tschop, M., Wawarta, R., Riepl, R. L., Friedrich, S., Bidlingmaier, M., Landgraf, R., and Folwaczny, C. (2001a). Post-prandial decrease of circulating human ghrelin levels. *J. Endocrinol. Invest.* **24,** 19–21.

Tschop, M., Weyer, C., Tataranni, P. A., Devanarayan, V., Ravussin, E., and Heiman, M. L. (2001b). Circulating ghrelin levels are decreased in human obesity. *Diabetes* **50,** 707–709.

Ukkola, O., Poykko, S. M., and Antero Kesaniemi, Y. (2006). Low plasma ghrelin concentration is an indicator of the metabolic syndrome. *Ann. Med.* **38,** 274–279.

Vequaud, P., and Thorin, E. (2001). Endothelial G protein beta-subunits trigger nitric oxide-but not endothelium-derived hyperpolarizing factor-dependent dilation in rabbit resistance arteries. *Circ. Res.* **89,** 648–649.

Vincent, J. P., Mazella, J., and Kitabgi, P. (1999). Neurotensin and neurotensin receptors. *Trends Pharmacol. Sci.* **20,** 302–309.

Volante, M., Fulcheri, E., Allia, E., Cerrato, M., Pucci, A., and Papotti, M. (2002a). Ghrelin expression in fetal, infant and adult human lung. *J. Histochem. Cytochem.* **50,** 1013–1021.

Volante, M., Allia, E., Gugliotta, P., Funaro, A., Broglio, F., Deghenghi, R., Muccioli, G., Ghigo, E., and Papotti, M. (2002b). Expression of ghrelin and of the GH secretagogue receptor by pancreatic islet cells and related endocrine tumors. *J. Clin. Endocrinol. Metab.* **87,** 1300–1308.

Wiley, K. E., and Davenport, A. P. (2001). Nitric oxide-mediated modulation of the endothelin-1 signalling pathway in the human cardiovascular system. *Br. J. Pharmacol.* **132,** 213–220.

Wiley, K. E., and Davenport, A. P. (2002). Comparison of vasodilators in human internal mammary artery: Ghrelin is a potent physiological antagonist of endothelin-1. *Br. J. Pharmacol.* **136,** 1146–1152.

Xu, L., Depoortere, I., Tomasetto, C., Zandecki, M., Tang, M., Timmermans, J. P., and Peeters, T. L. (2005). Evidence for the presence of motilin, ghrelin, and the motilin and ghrelin receptor in neurons of the myenteric plexus. *Regul. Pept.* **124,** 119–125.

Yamaji, K., Yoshitomi, T., and Usui, S. (2003). Effect of somatostatin and galanin on isolated rabbit iris sphincter and dilator muscles. *Exp. Eye Res.* **77,** 609–614.

Yousufzai, S. Y., Ali, N., and Abdel-Latif, A. A. (1999). Effects of adrenomedullin on cyclic AMP formation and on relaxation in iris sphincter smooth muscle. *Invest. Ophthalmol. Vis. Sci.* **40,** 3245–3253.

CHAPTER TEN

GHRELIN AND BONE

Martijn van der Velde,* Patric Delhanty,* Bram van der Eerden,* Aart Jan van der Lely,* *and* Johannes van Leeuwen*

Contents

I. Bone Balance: Resorption and Formation	239
II. Interplay Between the Gastrointestinal System and Bone: The Effect of Gastrectomy/Fundectomy on Bone	240
III. Effects of GH and GHS on Bone Metabolism	242
IV. Correlation Between Ghrelin and Bone Parameters in Clinical Studies	244
V. Effects of Ghrelin on Osteoblastic Cells *In Vitro*	245
VI. Conclusions	249
References	249

Abstract

A consequence of gastrectomy is loss of bone mass. Several mechanisms have been proposed, such as malabsorption of vitamins and minerals. Additionally, a peptide hormone produced in the stomach has been shown to mediate a calcitropic effect on bone. The identity of this peptide has not been elucidated, but ghrelin, produced by A-like cells in the fundus of the stomach, could be a good candidate. Ghrelin stimulates growth hormone (GH) secretion both *in vivo* and *in vitro*, and could by this means have a positive effect on bone. There is also evidence for direct effects of ghrelin on bone. We discuss here the role that ghrelin may play in bone metabolism, based on the most recent literature. © 2008 Elsevier Inc.

I. BONE BALANCE: RESORPTION AND FORMATION

Bone is a multifunctional organ: it protects soft tissues, works as levers for skeletal muscle action, and serves as the body's major store of calcium. In addition, it supports hematopoiesis and houses the brain and spinal cord (Harada and Rodan, 2003). Bone is also a dynamic tissue: it is constantly

---

* Department of Internal Medicine, Erasmus MC, Dr. Molewaterplein 50, 3015 CE, Rotterdam, The Netherlands

being remodeled and can therefore adapt to environmental needs. A full remodeling cycle comprises bone removal, or resorption, by osteoclasts followed by bone formation by osteoblasts, two processes that are tightly coupled. During aging in humans, the equilibrium between bone formation and resorption shifts from net formation during growth, balance in the third through fifth decade of life, to net resorption in old age (Seeman and Delmas, 2006). Osteoporosis is a disease characterized by excessive bone resorption, most prominently observed in women after menopause, but in both men and women there is net bone resorption with increasing age. The major cause is the postmenopausal decline in levels of estrogen, a potent factor in bone maintenance, which results in low bone mass and microarchitectural deterioration of the bone, leading to weaker bones and increased risk for fractures. In Caucasian populations, 50% of women over 50 years are at risk of a fragility fracture later on in life (Sambrook and Cooper, 2006). Current treatment of osteoporosis mainly targets bone resorption. Bisphosphonates and estrogen replacement are used to inhibit this process. Anabolic therapy is needed to simulate bone formation. Currently, the only therapy targeting bone formation is treatment with the anabolic parathyroid hormone (PTH).

II. Interplay Between the Gastrointestinal System and Bone: The Effect of Gastrectomy/Fundectomy on Bone

Maintenance of skeletal integrity is dependent on several organs, for example, the skin, the kidney, the immune system, and the brain. However, at least equally important is the gastrointestinal tract, where minerals [calcium in particular (Hessov *et al.*, 1984)], vitamin D (Jahnsen *et al.*, 2002), vitamin K (Szulc and Meunier, 2001), and nutrients are taken up from the food. It is therefore not surprising that gastrointestinal disease can lead to low bone mineral density (BMD) (Aukee *et al.*, 1975; Bisballe *et al.*, 1991; Blichert-Toft *et al.*, 1979; Inoue *et al.*, 1992; Klein *et al.*, 1987; Mellstrom *et al.*, 1993; Zittel *et al.*, 1997). In addition to malabsorption of the above-mentioned dietary compounds and minerals, increased cytokine production as a direct consequence of inflammatory bowel disease (Hyams *et al.*, 1997) and celiac disease (Sylvester *et al.*, 2002) have negative effects on bone mineral content (BMC), and treatment of these diseases with glucocorticoids accentuates these effects. Gastrectomy causes malabsorption of vitamin D (Bisballe *et al.*, 1991) and calcium (Tovey *et al.*, 1992), leading to osteomalacia, and this in turn can lead to secondary hyperparathyroidism, resulting in osteoporosis (Vestergaard, 2003). However, osteomalacia represents only a small portion of postgastrectomy bone disease (Bernstein and Leslie, 2003; Bisballe *et al.*, 1991; Eddy, 1971; Morgan *et al.*, 1965), and the mechanism remains elusive. Experimental gastrectomy in animals also leads to osteopenia. Lack of gastric acid to utilize dietary calcium was also proposed as a mechanism for osteopenia in these animals, but short-term inhibition of acid secretion did not result in bone loss. Moreover, calcium supplementation did not prevent bone loss after gastrectomy (Persson et al., 1993).

Other mechanisms have been proposed for gastrectomy-related bone loss based on the concept that the stomach is a source of endocrine hormones that impact on bone metabolism. One proposed mechanism is that gastrin, a stomach-derived hormone, enhances the uptake of calcium into bone by stimulating the release of a putative hormone from enterochromaffin-like (ECL)-acid producing cells in the gastric fundus (Hakanson et al., 1990). This putative hormone has been tentatively called "gastrocalcin." Larsson et al. (2001, 2002) showed that gastric fundal extracts, containing ECL cells, evoked a rise in intracellular calcium in human, mouse, and rat osteoblast cell lines. This response was abolished by preincubating the extracts with an exopeptidase, indicating that "gastrocalcin" should be a peptide. They hypothesized that if ghrelin, a circulating peptide hormone derived from the gastric fundus (Date et al., 2000; Kojima et al., 1999), was the putative "gastrocalcin," it should also evoke a rise in intracellular calcium in osteoblasts. It did not, suggesting that ghrelin was not "gastrocalcin." However, ghrelin could just be another factor present in the heterogeneous extract from the stomach, also affecting osteoblasts, and may be via other intracellular pathways (Baldanzi et al., 2002). Moreover, it was shown that A-like cells in the rat stomach produced ghrelin and did not operate under gastrin control, clearly distinguishing ghrelin and "gastrocalcin" (Dornonville et al., 2001). The identity of the latter has not been elucidated to date. A positive role for gastrin in bone metabolism is doubtful because both after gastrectomy (hypogastrinaemia) and fundectomy (compensatory hypergastrinaemia) osteopenia occurred (Cui et al., 2001). Omeprazole, a proton pump inhibitor, inhibits gastric acid release and causes ECL cell hyperplasia and subsequent hypergastrinaemia, much like postfundectomy hypergastrinaemia. Osteopenia developed despite the hypergastrinaemia, and this is not in accordance with the hypothesis that gastrin releases the calcitropic gastrocalcin. A role for ghrelin was proposed, but the number of A-like cells was not changed after omeprazole treatment. Accordingly, ghrelin levels were not altered. These findings contradict the finding of Persson et al. (1993) who did not find osteopenia in rats after omeprazole treatment, but it should be noted that the duration of treatment was much shorter in that study. The mechanism of omeprazole-induced osteopenia remains elusive (Cui et al., 2001). Lehto-Axtelius et al. (2002) showed that fundectomy in rats reduced the number of ECL cells, but also reduced the number of A-like cells and that serum ghrelin progressively decreased with removal of increasing amounts of gastric fundus. Puzio et al. (2005) also showed that fundectomy lowered serum ghrelin levels in rats, and that this was accompanied by decreased mechanical and geometrical properties of the femora in these animals, but there was no effect on bone length or BMD.

Dornonville de la Cour *et al.* (2005) showed that ghrelin treatment in mice partially reversed gastrectomy-induced reduction in body weight, lean body mass, and body fat, but not in bone mass. The majority of the circulating ghrelin is unacylated, while only a small part is acylated. Gastrectomy reduced the levels of unacylated and acylated ghrelin in the circulation. Infusion of acylated ghrelin restored unacylated ghrelin but not acylated ghrelin to levels found in sham-operated mice. Apparently, the infused acylated ghrelin is rapidly deacylated in the serum (De Vriese *et al.*, 2004; Shanado *et al.*, 2004). Therefore, a role of acylated ghrelin in maintenance of BMD cannot be excluded. Acylated, but not unacylated, ghrelin is known to stimulate GH secretion and to increase serum insulin-like growth factor-1 (IGF-1) (Kojima *et al.*, 1999). Dornonville de la Cour *et al.* (2005) found that gastrectomy did not change serum IGF-1, neither did infusion of acylated ghrelin. This is in line with the findings of Sun *et al.* (2003) who reported no differences in serum IGF-1 between wild-type and ghrelin-deficient mice. In conclusion, the reduction in BMD after gastrectomy could possibly be caused by the lack of action of acylated ghrelin, likely via a mechanism distinct from the GH–IGF-1 axis.

III. EFFECTS OF GH AND GHS ON BONE METABOLISM

Before birth, linear bone growth is mainly regulated by autocrine/paracrine IGF-1 and IGF-2, largely independent of GH. After birth, growth is dependent on GH and IGF-1. A pituitary adenoma during childhood or adulthood stimulates growth by increased GH secretion, resulting in pituitary gigantism or acromegaly, respectively. Conversely, severe dwarfism results when GH secretion or action is impaired (van der Eerden *et al.*, 2003). This could be caused by defects in the formation of GH-secreting cells or GH insensitivity, including defects in the GH receptor or IGF-1 deletion (Pfaffle *et al.*, 1993; Rosenfeld *et al.*, 2000; Savage *et al.*, 2001; Wit *et al.*, 1989; Woods and Savage, 1996). IGF-2 is essential for embryonic growth (DeChiara *et al.*, 1991), and IGF-1 has a continuous function throughout development and adulthood (Liu *et al.*, 1993). Salmon and Daughaday (1957) postulated that GH effects are mediated primarily by IGF-1 released from the liver, the so-called "somatomedin hypothesis." Later studies have challenged this view, including one showing that local injection of GH stimulated unilateral tibial bone growth, leaving the contralateral tibia unaffected, so showing a direct, IGF-1-independent effect of GH (Isaksson *et al.*, 1982). Green *et al.* (1985) and Zezulak and Green (1986), who showed independent, but cooperative actions of both GH and IGF-1 on adipocytes, formulated the "dual effector hypothesis," which could be applied to longitudinal bone growth as well (Isaksson *et al.*, 1982). The growth-promoting effects could be mediated by either liver-derived or locally produced IGF-1, in response to GH. By employing knockout models, including a liver-specific IGF-1-deficient mouse and a mouse deficient for both liver-specific IGF-1 and acid-labile subunit, it has become clear that both local and systemic IGF-1 play a role in linear bone growth (Butler et al., 2002; Sjogren et al., 1999; Yakar et al., 1999, 2000, 2002). Besides having important roles in bone development, GH and IGF-1 are also very important in bone remodeling throughout life, stimulating both bone formation and resorption, so increasing bone remodeling (Johannsson et al., 1996; Ohlsson et al., 1998; Olney, 2003).

Alterations in the GH/IGF-1 axis may be involved in the pathogenesis of osteoporosis, and low serum IGF-1 has been detected in osteoporotic elderly (Pun et al., 1990). GH deficiency in children and adults affects skeletal mineralization (Drake et al., 2001; Root and Root, 2002), and their treatment with GH initially increases bone resorption leading to an increased number of remodeling units. Subsequent stimulation of bone formation leads to a net increase in bone mass, but only when the treatment is continued for a prolonged period. This is also called the "biphasic model" of GH action on bone (Gotherstrom et al., 2001; Koranyi et al., 2001; Ohlsson et al., 1998; ter Maaten et al., 1999). GH replacement therapy is successfully used in GH-deficient patients, but data on GH therapy in osteoporotic non-GH-deficient patients is scarce. Moreover, study results are contradictory, demonstrating a positive effect of GH on bone mass in postmenopausal women (Landin-Wilhelmsen et al., 2003) and in men with idiopathic osteoporosis (Gillberg et al., 2002), but in other studies no change or rather a decrease in BMD were reported (Ohlsson et al., 1998). The potential therapeutic benefits turned out to depend on the dose, frequency, and duration of the treatment (Butterfield et al., 1997; Wells and Houston, 2001). It is for this reason that treatments with growth hormone secretagogues (GHSs), which mimic the physiological release of GH, have been postulated to be a more successful therapeutic approach (Cocchi et al., 2005).

GHSs are a class of synthetic compounds that can stimulate GH secretion both *in vivo* and *in vitro* (Locatelli and Torsello, 1997). Both peptidyl and nonpeptidyl GHSs have been examined. Two studies found that treatment with the nonpeptidyl GHS, MK-0677, increased markers of both bone formation and resorption in humans (Murphy et al., 1999; Svensson et al., 1998). This is in line with the previously described biphasic effect of GH: after an initial bone resorption phase creating an increased number of bone remodeling units, bone formation takes place leading to increased bone mass (Ohlsson et al., 1998). Murphy et al. (2001) combined treatment of MK-0677 with alendronate, a compound that abrogates bone turnover. MK-0677 only had additional beneficial effects on the femoral neck, which could be clinically relevant, since osteoporosis-related fractures tend to occur at this site. In animal studies, peptidyl GHSs, for example hexarelin, increase the expression of bone formation markers, but unlike GH or nonpeptidyl GHS seemed to reduce the expression of bone resorption markers in rats (Sibilia *et al.*, 1999) and dogs (Cella *et al.*, 1996). In another study in rats, hexarelin and ipamorelin, another peptidyl GHSs, increased cortical and total bone mass to a similar extent as caused by GH (Svensson *et al.*, 2000). The differences in action between peptidyl and nonpeptidyl GHS suggest that perhaps the peptidyl GHSs have GH-independent effects. Peptidyl GHSs appear to specifically stimulate bone formation, so a direct effect on osteoblasts is very likely.

Ghrelin stimulates GH release both *in vitro* and *in vivo* (Kojima *et al.*, 1999) and this effect is dependent on the GHSR1a (Sun *et al.*, 2004). GH, in turn, stimulates IGF-1 release from the liver. Following intraperitoneal infusion for 4 weeks, ghrelin increased BMD in wild-type rats. More importantly, it also increased BMD in spontaneous dwarf rats, deficient in GH, showing a direct GH-independent effect (Fukushima *et al.*, 2005). The concentrations of ghrelin used here did not evoke a rise in body weight and food intake in another study (Tschop *et al.*, 2000), excluding possible indirect effects on BMD of ghrelin through its orexigenic activity. Remarkably, ghrelin-deficient mice have the same BMD as wild-type mice (Sun *et al.*, 2003). Similarly, mice lacking GHSR have unaltered BMD, despite having slightly lowered serum IGF-1 levels (Sun *et al.*, 2004).

IV. CORRELATION BETWEEN GHRELIN AND BONE PARAMETERS IN CLINICAL STUDIES

Anorexia nervosa is a disorder accompanied by low BMD. Multiple factors contribute to this, including hypogonadism, undernutrition, excessive exercise, low levels of nutritionally dependent IGF-1, acquired GH resistance, and hypercortisolaemia. But anorexic subjects are also characterized by high ghrelin levels (Hotta *et al.*, 2004, 2005; Misra *et al.*, 2004, 2005; Misra and Klibanski, 2006; Tolle *et al.*, 2003). Besides stimulating GH secretion (Arvat *et al.*, 2001; Kojima *et al.*, 2005; Takaya *et al.*, 2000; Tannenbaum *et al.*, 2003; Tassone *et al.*, 2003), ghrelin also stimulates ACTH secretion (Arvat *et al.*, 2001; Tassone *et al.*, 2003; Vulliemoz *et al.*, 2004) and suppresses LH pulsatility (Vulliemoz *et al.*, 2004), possibly leading to increased bone turnover. In combination with the other factors mentioned above, this results in lower bone mass. In contrast, obesity is characterized by low ghrelin levels and high BMD. The only exception is the Prader–Willi syndrome, in which obesity is linked to high ghrelin levels and low BMD, which could be explained by a dysfunctional GH/IGF-1 axis (Hoybye, 2004). On the other hand, Oh *et al.* (2005) found no association between serum-active ghrelin levels and BMD in middle-aged Korean men. Recently, total plasma ghrelin was found to be weakly inversely correlated with markers for bone resorption (after adjustment for age and

BMI) in a cohort of older men, but not women, in the Rancho Bernardo study (Weiss *et al.*, 2006).

Clearly, these observational studies do not imply direct causality between ghrelin and bone metabolism. Besides, hormonal and nutritional statuses are confounding factors since they interact with and modulate ghrelin action. Only *in vitro* studies using isolated bone cells can reveal a direct relationship between ghrelin and bone metabolism and unravel the underlying mechanism.

V. Effects of Ghrelin on Osteoblastic Cells *In Vitro*

Several studies have now shown that ghrelin can either stimulate (Andreis *et al.*, 2003; Belloni *et al.*, 2004; Granata *et al.*, 2006; Jeffery *et al.*, 2002; Kim *et al.*, 2004; Mazzocchi *et al.*, 2004; Pettersson *et al.*, 2002; Sirotkin *et al.*, 2006; Yeh *et al.*, 2005) or inhibit (Baiguera *et al.*, 2004; Cassoni *et al.*, 2001; Volante *et al.*, 2003) proliferation in several distinct cell types. Correspondingly, ghrelin can stimulate (Baiguera *et al.*, 2004; Belloni *et al.*, 2004) or inhibit (Baldanzi *et al.*, 2002; Belloni *et al.*, 2004; Chung *et al.*, 2006; Granata *et al.*, 2006; Iglesias *et al.*, 2004; Kim *et al.*, 2004; Sirotkin *et al.*, 2006) apoptosis in several cell lines. Belloni *et al.* (2004) showed that ghrelin inhibited apoptosis in rat osteoblasts. Four studies will be discussed below that have been performed on ghrelin and osteoblasts by (Fukushima *et al.*, 2005; Kim *et al.*, 2005; Maccarinelli *et al.*, 2005) and our own study (Delhanty *et al.*, 2006). In all studies, ghrelin was effective in the nanomolar range.

Maccarinelli *et al.* (2005) showed for the first time that ghrelin, besides stimulating proliferation, could also promote differentiation in primary fetal rat calvarial osteoblastic cells. Both ghrelin and hexarelin stimulated proliferation in a similar manner. Besides inducing proliferation, ghrelin and hexarelin equipotently stimulated alkaline phosphatase (ALP) activity, a marker of osteoblast differentiation. Another marker of osteoblast activity, osteocalcin (OC) production, was stimulated by both ghrelin and hexarelin. EP-40737, a hexarelin analogue that was found to stimulate GH secretion, but not food intake in rats (Torsello *et al.*, 2000), was also tested. In contrast to ghrelin and hexarelin, this compound inhibited osteoblast proliferation, but changed neither ALP activity nor OC production. Fukushima *et al.* (2005) assessed the effects of ghrelin and GHRP-6 (a hexapeptidyl GHS) on proliferation and differentiation of primary rat calvarial osteoblasts. Both peptides showed a similar stimulation of proliferation. The effects of ghrelin on proliferation were completely abolished by the GHSR1a antagonist [D-Lys3]-GHRP-6, suggesting that they are mediated through GHSR1a. The effects of ghrelin on osteoblastic differentiation were shown by increased expression of two early (Collagen I and ALP) and one late (OC) osteoblast differentiation marker. Ghrelin also stimulated both ALP activity and mineralized nodule formation. There was no effect of ghrelin on the expression of Runx2, an early transcription factor necessary for osteoblast differentiation, but this could be explained by the fact that the cells used were already more committed toward the osteoblast lineage. Kim et al. (2005) showed that ghrelin not only stimulates proliferation but also inhibits apoptosis in the mouse osteoblastic cell line MC3T3-E1. This effect was shared by GHRP-6. Both effects could be blocked by the GHSR1a antagonist [D-Lys3]-GHRP-6, suggesting the involvement of GHSR1a. Treatment of the MC3T3-E1 cells with TNF-α and cyclohexamide induced apoptosis, which could be reversed by ghrelin. Moreover, ghrelin also suppressed caspase-3 activation that occurred during the in vitro differentiation process. Furthermore, ghrelin increased ALP activity and mineralization in these cells in addition to elevating the expression of several osteoblast differentiation markers, like Collagen I, ALP, and OC. Together these data show that ghrelin increases proliferation and differentiation and decreases apoptosis of rat and murine osteoblasts and that these effects are likely mediated through GHSR1a.

The majority of circulating ghrelin is unacylated, and although long believed to be inactive, there is growing evidence that this form of ghrelin has biological effects (Ariyasu et al., 2005; Baldanzi et al., 2002; Bedendi et al., 2003; Broglio et al., 2004; Cassoni et al., 2004; Chen et al., 2005a,b; Gauna et al., 2004, 2005, 2006; Granata et al., 2006; Heijboer et al., 2006; Martini et al., 2006; Matsuda et al., 2006; Muccioli et al., 2004; Nakahara et al., 2006; Sato et al., 2006; Toshinai et al., 2006; Tsubota et al., 2005). No effects of unacylated ghrelin have been reported on osteoblastic cells, apart from our study (Delhanty et al., 2006). We used the well-characterized human osteoblastic cell line SV-HFO (Chiba et al., 1993; Eijken et al., 2005; Janssen et al., 1999; Jansen et al., 2004; van Driel et al., 2006), which can be differentiated into functional osteoblasts in 21 days by addition of glucocorticoids to the culture. Ghrelin as well as unacylated ghrelin increased the proliferation of SV-HFO cells. For both peptides, the response declined during culture, being most effective in the first week and completely absent in the third week. In contrast to the studies on rat osteoblasts, we did not observe any effect of ghrelin and also of unacylated ghrelin on ALP activity or mineralization in SV-HFO cells.

Both Maccarinelli et al. (2005) and Fukushima et al. (2005) showed that fetal rat osteoblasts express the functional receptor for ghrelin, GHSR1a. In addition, the latter study also showed the presence of this receptor in the rat osteoblastic cell line UMR-106. Expression of GHSR1a was confirmed by RT-PCR in MC3T3-E1 mouse osteoblasts, as well as in ROS17/2.8 and UMR-106 rat osteoblasts and in the human osteosarcoma cell lines

MG63 and SaOS2 by Kim et al. (2005). Protein expression of GHSR1a in all these cell lines was confirmed by Western blot. Also, GHSR1a immunoreactivity was found in cell lysates from rat and mouse calvaria and mouse femur. In contrast to the previous studies, which have shown expression of GHSR1a in rat and mouse osteoblastic cells, we have not been able to detect the expression of GHSR1a in human osteoblastic cells or osteoarthritic bone biopsies (Delhanty et al., 2006). This is in line with a study showing a widespread expression of GHSR1a in human tissues, but not in bone (Gnanapavan et al., 2002). On the other hand, we did observe expression of its mRNA splice variant, GHSR1b, in SV-HFO cells and osteoarthritic bone biopsies (Delhanty et al., 2006). The GHSR1b mRNA encodes a truncated form of the ghrelin receptor. The level of mRNA expression of GHSR1b increased with time in the differentiating human osteoblast culture, reaching 400% of the expression in the third week compared to the first week of culture.

Kim et al. (2005) showed that ghrelin-induced phosphorylation of MAPK could be blocked by PD98059, an MAPK inhibitor, suggesting that this is the downstream signal mediating the proliferative effect. Moreover, PD98059 also blocked the GHRP-6-mediated effects on proliferation. Additionally, they showed that ghrelin-stimulated phosphorylation of MAPK promotes osteoblastic differentiation (Kim et al., 2005). However, the role of MAPK on the differentiation of osteoblasts is controversial. Some studies show a stimulatory (Jaiswal et al., 2000; Takeuchi et al., 1997; Xiao et al., 2000, 2002) and other an inhibitory (Chaudhary and Avioli, 2000; Gallea et al., 2001; Higuchi et al., 2002; Kretzschmar et al., 1997) effect of MAPK signaling on osteoblast differentiation. Another report showed that overexpression of ghrelin in preadipocytes inhibited adipocyte differentiation (Zhang et al., 2004). Since the differentiation of mesenchymal progenitors into adipocytes or osteoblasts is reciprocally regulated (Nuttall and Gimble, 2004; Pei and Tontonoz, 2004), it is proposed that ghrelin could play a role in this process, albeit that for osteoblastic differentiation Runx2 is indispensable and that Fukushima et al. (2005) showed that ghrelin did not affect the expression of this transcription factor. However, as noted above, this could be explained by the cells being already more committed toward the osteoblast lineage in the latter study. In our study, the effects on proliferation were shown to be sensitive to U0126, PD98059 (MAPK inhibitors), and Wortmannin (PI-3K inhibitor), suggesting that MAPK and PI-3K pathways are involved in the effects of both peptides on proliferation (Delhanty et al., 2006). Moreover, both acylated and unacylated ghrelin increased levels of phosphorylated MAPK, as shown by a specific ELISA. This is confirmed by other studies showing that ghrelin stimulated proliferation in various cell types through MAPK and PI-3K pathways (Baldanzi et al., 2002; Kim et al., 2004; Nanzer et al., 2004).

Cocchi *et al.* (2005) provided data suggesting that the calvarial osteoblasts used in their study did not express ghrelin and therefore concluded that *in vivo* the physiological effect on these cells would be mediated by circulating, rather than locally produced ghrelin. This is in contrast with a study by Fukushima *et al.* (2005) who showed, by RT-PCR and immunohistochemistry, that calvarial osteoblasts of fetal rat origin expressed ghrelin. We also showed that ghrelin mRNA was present in cells in human osteoblasts and bone biopsies, suggesting an autocrine/paracrine role for ghrelin (Delhanty *et al.*, 2006). This was confirmed by the presence of ghrelin peptide in conditioned medium of cultured human osteoblasts.

Our study with human osteoblasts shows that ghrelin stimulates their proliferation likely via an autocrine/paracrine, non-GHSR1a-mediated mechanism. The presence of a non-GHSR1a-mediated mechanism is also suggested by the effect of unacylated ghrelin, devoid of any binding to GHSR1a, on proliferation. It is intriguing that the increased expression of GHSR1b during the last week of culture coincides with the diminished effects of both ghrelin and unacylated ghrelin on proliferation. GHSR1b is a truncated form of GHSR1a, but according to mutational studies, still contains the residues required for ligand binding (Feighner *et al.*, 1998). GHSR1b could therefore in theory still bind ligand or modify its interaction with the cognate receptor. Considering an antagonistic (dominant-negative) role for GHSR1b on GHSR1a signaling, as suggested by Chan and Cheng (2004), and its expression pattern in time, the diminished effects observed during glucocorticoid-induced differentiation in SV-HFO human osteoblasts might be explained (Delhanty *et al.*, 2006). Potentially, receptor heterodimerization could provide a mechanism through which GHSR1b inhibits GHSR1a signaling. Alternatively, GHSR1b could prevent ghrelin from binding to GHSR1a, or to an alternative receptor. However, Smith *et al.* (2005) showed that GHSR1b is not expressed on cell surfaces, putting into question its role in modulating or affecting signal transduction. Interestingly, we did not observe any effects of ghrelin and unacylated ghrelin on markers of mineralization, unlike other studies performed in osteoblasts of rodent origin (Cocchi *et al.*, 2005; Fukushima *et al.*, 2005; Kim *et al.*, 2005; Maccarinelli *et al.*, 2005). Perhaps there is a species difference in effects of ghrelin on osteoblast differentiation. However, Kim *et al.* (2005) showed the expression of GHSR1a in the human osteoblast-osteosarcoma cell lines SaOS2 and MG63, indicating that a species difference in GHSR1a expression does not seem to be an explanation. On the other hand, a possible effect of ghrelin on the differentiation of these two osteoblast-osteosarcoma cell lines has never been investigated. Considering the absence of GHSR1a in the SV-HFO human osteoblasts, it is tempting to speculate that the effects on osteoblast differentiation are mediated through GHSR1a and that the effects on proliferation are mediated through another receptor. However, this still leaves open the question why Kim *et al.* (2005) and Fukushima *et al.* (2005)

found that the effects of ghrelin on osteoblast proliferation were blocked by [D-Lys3]-GHRP-6, a GHSR1a antagonist. Theoretically, the effect on proliferation could be mediated by a receptor similar to GHSR1a, also sensitive to [D-Lys3]-GHRP-6. Alternatively, there is a species difference in the receptor mediating the effects of ghrelin on the proliferation of osteoblasts. However, recent evidence suggests that [D-Lys3]-GHRP-6 is an agonist of the 5-HT (serotonin) receptors (Depoortere *et al.*, 2006), which are expressed on osteoblasts (Westbroek *et al.*, 2001), so this does not exclude a direct effect of [D-Lys3]-GHRP-6, independent of ghrelin action.

VI. Conclusions

Despite the confusing clinical and *in vivo* animal data, *in vitro* studies unequivocally show positive effects of ghrelin on osteoblasts. Ghrelin has positive effects on the proliferation of rat, mouse, and human osteoblasts and also on the differentiation of mouse and rat osteoblasts. No study has shown an effect of ghrelin on the differentiation of human osteoblasts. There could be a species difference in the effect of ghrelin on osteoblast differentiation. GHSR1a was expressed in rat and mouse osteoblasts, but not in human osteoblasts. Our data on human osteoblasts suggests that another receptor mediates the effect on proliferation, especially since unacylated ghrelin, not capable of activating GHSR1a, clearly stimulated proliferation in these cells. Furthermore, the involvement of an autocrine/paracrine mechanism seems likely since several studies show ghrelin expression in bone.

It is clear that ghrelin has effects on bone. However, more studies are needed to delineate the exact role of acylated and unacylated ghrelin and their mechanism of action in bone. Potentially, the effects of ghrelin are very much dependent on the proliferation and/or differentiation status of the osteoblasts. Also interaction with other, yet unknown, factors in the serum used in the various cell culture studies might play a role. Currently only data on the bone-forming cell, the osteoblast, are available, but data on the effects of ghrelin on osteoclasts are so far absent and are needed to complete the picture.

REFERENCES

Andreis, P. G., Malendowicz, L. K., Trejter, M., Neri, G., Spinazzi, R., Rossi, G. P., and Nussdorfer, G. G. (2003). Ghrelin and growth hormone secretagogue receptor are expressed in the rat adrenal cortex: Evidence that ghrelin stimulates the growth, but not the secretory activity of adrenal cells. *FEBS Lett.* **536,** 173–179.

Ariyasu, H., Takaya, K., Iwakura, H., Hosoda, H., Akamizu, T., Arai, Y., Kangawa, K., and Nakao, K. (2005). Transgenic mice overexpressing des-acyl ghrelin show small phenotype. *Endocrinology* **146,** 355–364.

Arvat, E., Maccario, M., Di Vito, L., Broglio, F., Benso, A., Gottero, C., Papotti, M., Muccioli, G., Dieguez, C., Casanueva, F. F., Deghenghi, R., Camanni, F., *et al.* (2001). Endocrine activities of ghrelin, a natural growth hormone secretagogue (GHS), in humans: Comparison and interactions with hexarelin, a nonnatural peptidyl GHS, and GH-releasing hormone. *J. Clin. Endocrinol. Metab.* **86**, 1169–1174.

Aukee, S., Alhava, E. M., and Karjalainen, P. (1975). Bone mineral after partial gastrectomy II. *Scand. J. Gastroenterol.* **10**, 165–169.

Baiguera, S., Conconi, M. T., Guidolin, D., Mazzocchi, G., Malendowicz, L. K., Parnigotto, P. P., Spinazzi, R., and Nussdorfer, G. G. (2004). Ghrelin inhibits *in vitro* angiogenic activity of rat brain microvascular endothelial cells. *Int. J. Mol. Med.* **14**, 849–854.

Baldanzi, G., Filigheddu, N., Cutrupi, S., Catapano, F., Bonissoni, S., Fubini, A., Malan, D., Baj, G., Granata, R., Broglio, F., Papotti, M., Surico, N., *et al.* (2002). Ghrelin and des-acyl ghrelin inhibit cell death in cardiomyocytes and endothelial cells through ERK1/2 and PI 3-kinase/AKT. *J. Cell Biol.* **159**, 1029–1037.

Bedendi, I., Alloatti, G., Marcantoni, A., Malan, D., Catapano, F., Ghe, C., Deghenghi, R., Ghigo, E., and Muccioli, G. (2003). Cardiac effects of ghrelin and its endogenous derivatives des-octanoyl ghrelin and des-Gln14-ghrelin. *Eur. J. Pharmacol.* **476**, 87–95.

Belloni, A. S., Macchi, C., Rebuffat, P., Conconi, M. T., Malendowicz, L. K., Parnigotto, P. P., and Nussdorfer, G. G. (2004). Effect of ghrelin on the apoptotic deletion rate of different types of cells cultured *in vitro*. *Int. J. Mol. Med.* **14**, 165–167.

Bernstein, C. N., and Leslie, W. D. (2003). The pathophysiology of bone disease in gastrointestinal disease. *Eur. J. Gastroenterol. Hepatol.* **15**, 857–864.

Bisballe, S., Eriksen, E. F., Melsen, F., Mosekilde, L., Sorensen, O. H., and Hessov, I. (1991). Osteopenia and osteomalacia after gastrectomy: Interrelations between biochemical markers of bone remodelling, vitamin D metabolites, and bone histomorphometry. *Gut* **32**, 1303–1307.

Blichert-Toft, M., Beck, A., Christiansen, C., and Transbol, I. (1979). Effects of gastric resection and vagotomy on blood and bone mineral content. *World J. Surg.* **3**, 99–102, 133–135.

Broglio, F., Gottero, C., Prodam, F., Gauna, C., Muccioli, G., Papotti, M., Abribat, T., van der Lely, A. J., and Ghigo, E. (2004). Non-acylated ghrelin counteracts the metabolic but not the neuroendocrine response to acylated ghrelin in humans. *J. Clin. Endocrinol. Metab.* **89**, 3062–3065.

Butler, A. A., Yakar, S., and LeRoith, D. (2002). Insulin-like growth factor-I: Compartmentalization within the somatotropic axis? *News Physiol. Sci.* **17**, 82–85.

Butterfield, G. E., Thompson, J., Rennie, M. J., Marcus, R., Hintz, R. L., and Hoffman, A. R. (1997). Effect of rhGH and rhIGF-I treatment on protein utilization in elderly women. *Am. J. Physiol.* **272**, E94–E99.

Cassoni, P., Papotti, M., Ghe, C., Catapano, F., Sapino, A., Graziani, A., Deghenghi, R., Reissmann, T., Ghigo, E., and Muccioli, G. (2001). Identification, characterization, and biological activity of specific receptors for natural (ghrelin) and synthetic growth hormone secretagogues and analogs in human breast carcinomas and cell lines. *J. Clin. Endocrinol. Metab.* **86**, 1738–1745.

Cassoni, P., Ghe, C., Marrocco, T., Tarabra, E., Allia, E., Catapano, F., Deghenghi, R., Ghigo, E., Papotti, M., and Muccioli, G. (2004). Expression of ghrelin and biological activity of specific receptors for ghrelin and des-acyl ghrelin in human prostate neoplasms and related cell lines. *Eur. J. Endocrinol.* **150**, 173–184.

Cella, S. G., Cerri, C. G., Daniel, S., Sibilia, V., Rigamonti, A., Cattaneo, L., Deghenghi, R., and Muller, E. E. (1996). Sixteen weeks of hexarelin therapy in aged dogs: Effects on the somatotropic axis, muscle morphology, and bone metabolism. *J. Gerontol. A Biol. Sci. Med. Sci.* **51**, B439–B447.

Chan, C. B., and Cheng, C. H. (2004). Identification and functional characterization of two alternatively spliced growth hormone secretagogue recepter transcripts from the pituitary of black seabream *Acanthopagrus Schlegeli*. *Mol. Cell. Endocrinol.* **214**, 81–95.

Chaudhary, L. R., and Avioli, L. V. (2000). Extracellular-signal regulated kinase signaling pathway mediates downregulation of type I procollagen gene expression by FGF-2, PDGF-BB, and okadaic acid in osteoblastic cells. *J. Cell Biochem.* **76**, 354–359.

Chen, C. Y., Chao, Y., Chang, F. Y., Chien, E. J., Lee, S. D., and Doong, M. L. (2005a). Intracisternal des-acyl ghrelin inhibits food intake and non-nutrient gastric emptying in conscious rats*Int. J. Mol. Med.* **16**, 695–699.

Chen, C. Y., Inui, A., Asakawa, A., Fujino, K., Kato, I., Chen, C. C., Ueno, N., and Fujimiya, M. (2005b). Des-acyl ghrelin acts by CRF type 2 receptors to disrupt fasted stomach motility in conscious rats. *Gastroenterology* **129**, 8–25.

Chiba, H., Sawada, N., Ono, T., Ishii, S., and Mori, M. (1993). Establishment and characterization of a simian virus 40-immortalized osteoblastic cell line from normal human bone. *Jpn. J. Cancer Res.* **84**, 290–297.

Chung, H., Kim, E., Hee, L. D., Seo, S., Ju, S., Lee, D., Kim, H., and Park, S. (2006). Ghrelin inhibits apoptosis in hypothalamic neuronal cells during oxygen-glucose deprivation. *Endocrinology* **148**(1), 148–159.

Cocchi, D., Maccarinelli, G., Sibilia, V., Tulipano, G., Torsello, A., Pazzaglia, U. E., Giustina, A., and Netti, C. (2005). GH-releasing peptides and bone. *J. Endocrinol. Invest.* **28**, 11–14.

Cui, G. L., Syversen, U., Zhao, C. M., Chen, D., and Waldum, H. L. (2001). Long-term omeprazole treatment suppresses body weight gain and bone mineralization in young male rats. *Scand. J. Gastroenterol.* **36**, 1011–1015.

Date, Y., Kojima, M., Hosoda, H., Sawaguchi, A., Mondal, M. S., Suganuma, T., Matsukura, S., Kangawa, K., and Nakazato, M. (2000). Ghrelin, a novel growth hormone-releasing acylated peptide, is synthesized in a distinct endocrine cell type in the gastrointestinal tracts of rats and humans. *Endocrinology* **141**, 4255–4261.

DeChiara, T. M., Robertson, E. J., and Efstratiadis, A. (1991). Parental imprinting of the mouse insulin-like growth factor II gene. *Cell* **64**, 849–859.

de la Cour, C. D., Lindqvist, A., Egecioglu, E., Tung, Y. C. L., Surve, V., Ohlsson, C., Jansson, J. O., Erlanson-Albertsson, C., Dickson, S. L., and Hakanson, R. (2005). Ghrelin treatment reverses the reduction in weight gain and body fat in gastrectomised mice. *Gut* **54**, 907–913.

Delhanty, P. J. D., van der Eerden, B. C. J., van der Velde, M., Gauna, C., Pols, H. A. P., Jahr, H., Chiba, H., van der Lely, A. J., and van Leeuwen, J. P. T. M. (2006). Ghrelin and unacylated ghrelin stimulate human osteoblast growth via mitogen-activated protein kinase (MAPK)/phosphoinositide 3-kinase (PI3K) pathways in the absence of GHS-R1a. *J. Endocrinol.* **188**, 37–47.

Depoortere, I., Thijs, T., and Peeters, T. (2006). The contractile effect of the ghrelin receptor antagonist, D-Lys3-GHRP-6, in rat fundic strips is mediated through 5-HT receptors. *Eur. J. Pharmacol.* **537**, 160–165.

De Vriese, C., Gregoire, F., Lema-Kisoka, R., Waelbroeck, M., Robberecht, P., and Delporte, C. (2004). Ghrelin degradation by serum and tissue homogenates: Identification of the cleavage sites. *Endocrinology* **145**, 4997–5005.

Dornonville, de la C., Bjorkqvist, M., Sandvik, A. K., Bakke, I., Zhao, C. M., Chen, D., and Hakanson, R. (2001). A-like cells in the rat stomach contain ghrelin and do not operate under gastrin control. *Regul. Pept.* **99**, 141–150.

Drake, W. M., Howell, S. J., Monson, J. P., and Shalet, S. M. (2001). Optimizing gh therapy in adults and children. *Endocr. Rev.* **22**, 425–450.

Eddy, R. L. (1971). Metabolic bone disease after gastrectomy. *Am. J. Med.* **50**, 442–449.

Eijken, M., Hewison, M., Cooper, M. S., de Jong, F. H., Chiba, H., Stewart, P. M., Uitterlinden, A. G., Pols, H. A., and van Leeuwen, J. P. (2005). 11beta-Hydroxysteroid dehydrogenase expression and glucocorticoid synthesis are directed by a molecular switch during osteoblast differentiation. *Mol. Endocrinol.* **19,** 621–631.

Feighner, S. D., Howard, A. D., Prendergast, K., Palyha, O. C., Hreniuk, D. L., Nargund, R., Underwood, D., Tata, J. R., Dean, D. C., Tan, C. P., McKee, K. K., Woods, J. W., *et al.* (1998). Structural requirements for the activation of the human growth hormone secretagogue receptor by peptide and nonpeptide secretagogues. *Mol. Endocrinol.* **12,** 137–145.

Fukushima, N., Hanada, R., Teranishi, H., Fukue, Y., Tachibana, T., Ishikawa, H., Takeda, S., Takeuchi, Y., Fukumoto, S., Kangawa, K., Nagata, K., and Kojima, M. (2005). Ghrelin directly regulates bone formation. *J. Bone Miner. Res.* **20,** 790–798.

Gallea, S., Lallemand, F., Atfi, A., Rawadi, G., Ramez, V., Spinella-Jaegle, S., Kawai, S., Faucheu, C., Huet, L., Baron, R., and Roman-Roman, S. (2001). Activation of mitogen-activated protein kinase cascades is involved in regulation of bone morphogenetic protein-2-induced osteoblast differentiation in pluripotent C2C12 cells. *Bone* **28,** 491–498.

Gauna, C., Meyler, F. M., Janssen, J. A., Delhanty, P. J., Abribat, T., van Koetsveld, P., Hofland, L. J., Broglio, F., Ghigo, E., and van der Lely, A. J. (2004). Administration of acylated ghrelin reduces insulin sensitivity, whereas the combination of acylated plus unacylated ghrelin strongly improves insulin sensitivity. *J. Clin. Endocrinol. Metab.* **89,** 5035–5042.

Gauna, C., Delhanty, P. J., Hofland, L. J., Janssen, J. A., Broglio, F., Ross, R. J., Ghigo, E., and van der Lely, A. J. (2005). Ghrelin stimulates, whereas des-octanoyl ghrelin inhibits, glucose output by primary hepatocytes. *J. Clin. Endocrinol. Metab.* **90,** 1055–1060.

Gauna, C., Delhanty, P. J., van Aken, M. O., Janssen, J. A., Themmen, A. P., Hofland, L. J., Culler, M., Broglio, F., Ghigo, E., and van der Lely, A. J. (2006). Unacylated ghrelin is active on the INS-1E rat insulinoma cell line independently of the growth hormone secretagogue receptor type 1a and the corticotropin releasing factor 2 receptor. *Mol. Cell Endocrinol.* **251,** 103–111.

Gillberg, P., Mallmin, H., Petren-Mallmin, M., Ljunghall, S., and Nilsson, A. G. (2002). Two years of treatment with recombinant human growth hormone increases bone mineral density in men with idiopathic osteoporosis. *J. Clin. Endocrinol. Metab.* **87,** 4900–4906.

Gnanapavan, S., Kola, B., Bustin, S. A., Morris, D. G., McGee, P., Fairclough, P., Bhattacharya, S., Carpenter, R., Grossman, A. B., and Korbonits, M. (2002). The tissue distribution of the mRNA of ghrelin and subtypes of its receptor, GHS-R, in humans. *J. Clin. Endocrinol. Metab.* **87,** 2988.

Gotherstrom, G., Svensson, J., Koranyi, J., Alpsten, M., Bosaeus, I., Bengtsson, B., and Johannsson, G. (2001). A prospective study of 5 years of GH replacement therapy in GH-deficient adults: Sustained effects on body composition, bone mass, and metabolic indices. *J. Clin. Endocrinol. Metab.* **86,** 4657–4665.

Granata, R., Settanni, F., Biancone, L., Trovato, L., Nano, R., Bertuzzi, F., Destefanis, S., Annunziata, M., Martinetti, M., Catapano, F., Ghe, C., Isgaard, J., *et al.* (2006). Acylated and unacylated ghrelin promote proliferation and inhibit apoptosis of pancreatic beta cells and human islets. involvement of cAMP/PKA, ERK1/2 and PI3K/AKT signaling. *Endocrinology* **148**(2), 512–529.

Green, H., Morikawa, M., and Nixon, T. (1985). A dual effector theory of growth-hormone action. *Differentiation* **29,** 195–198.

Hakanson, R., Persson, P., Axelson, J., Johnell, O., and Sundler, F. (1990). Evidence that gastrin enhances 45Ca uptake into bone through release of a gastric hormone. *Regul. Pept.* **28,** 107–118.

Harada, S., and Rodan, G. A. (2003). Control of osteoblast function and regulation of bone mass. *Nature* **423,** 349–355.

Heijboer, A. C., van den Hoek, A. M., Parlevliet, E. T., Havekes, L. M., Romijn, J. A., Pijl, H., and Corssmit, E. P. (2006). Ghrelin differentially affects hepatic and peripheral insulin sensitivity in mice. *Diabetologia* **49,** 732–738.

Hessov, I., Mosekilde, L., Melsen, F., Fasth, S., Hulten, L., Lund, B., Lund, B., and Sorensen, O. H. (1984). Osteopenia with normal vitamin D metabolites after small-bowel resection for Crohn's disease. *Scand. J. Gastroenterol.* **19,** 691–696.

Higuchi, C., Myoui, A., Hashimoto, N., Kuriyama, K., Yoshioka, K., Yoshikawa, H., and Itoh, K. (2002). Continuous inhibition of MAPK signaling promotes the early osteoblastic differentiation and mineralization of the extracellular matrix. *J. Bone Miner. Res.* **17,** 1785–1794.

Hotta, M., Ohwada, R., Katakami, H., Shibasaki, T., Hizuka, N., and Takano, K. (2004). Plasma levels of intact and degraded ghrelin and their responses to glucose infusion in anorexia nervosa. *J. Clin. Endocrinol. Metab.* **89,** 5707–5712.

Hoybye, C. (2004). Endocrine and metabolic aspects of adult Prader-Willi syndrome with special emphasis on the effect of growth hormone treatment. *Growth Horm. IGF Res.* **14,** 1–15.

Hyams, J. S., Wyzga, N., Kreutzer, D. L., Justinich, C. J., and Gronowicz, G. A. (1997). Alterations in bone metabolism in children with inflammatory bowel disease: An *in vitro* study. *J. Pediatr. Gastroenterol. Nutr.* **24,** 289–295.

Iglesias, M. J., Pineiro, R., Blanco, M., Gallego, R., Dieguez, C., Gualillo, O., Gonzalez-Juanatey, J. R., and Lago, F. (2004). Growth hormone releasing peptide (ghrelin) is synthesized and secreted by cardiomyocytes. *Cardiovasc. Res.* **62,** 481–488.

Inoue, K., Shiomi, K., Higashide, S., Kan, N., Nio, Y., Tobe, T., Shigeno, C., Konishi, J., Okumura, H., Yamamuro, T., and Fukunaga, M. (1992). Metabolic bone disease following gastrectomy: Assessment by dual energy X-ray absorptiometry. *Br. J. Surg.* **79,** 321–324.

Isaksson, O. G., Jansson, J. O., and Gause, I. A. (1982). Growth hormone stimulates longitudinal bone growth directly. *Science* **216,** 1237–1239.

Jahnsen, J., Falch, J. A., Mowinckel, P., and Aadland, E. (2002). Vitamin D status, parathyroid hormone and bone mineral density in patients with inflammatory bowel disease. *Scand. J. Gastroenterol.* **37,** 192–199.

Jaiswal, R. K., Jaiswal, N., Bruder, S. P., Mbalaviele, G., Marshak, D. R., and Pittenger, M. F. (2000). Adult human mesenchymal stem cell differentiation to the osteogenic or adipogenic lineage is regulated by mitogen-activated protein kinase. *J. Biol. Chem.* **275,** 9645–9652.

Jansen, J. H., Weyts, F. A., Westbroek, I., Jahr, H., Chiba, H., Pols, H. A., Verhaar, J. A., van Leeuwen, J. P., and Weinans, H. (2004). Stretch-induced phosphorylation of ERK1/2 depends on differentiation stage of osteoblasts. *J. Cell Biochem.* **93,** 542–551.

Janssen, J. M., Bland, R., Hewison, M., Coughtrie, M. W., Sharp, S., Arts, J., Pols, H. A., and van Leeuwen, J. P. (1999). Estradiol formation by human osteoblasts via multiple pathways: Relation with osteoblast function. *J. Cell Biochem.* **75,** 528–537.

Jeffery, P. L., Herington, A. C., and Chopin, L. K. (2002). Expression and action of the growth hormone releasing peptide ghrelin and its receptor in prostate cancer cell lines. *J. Endocrinol.* **172,** R7–R11.

Johannsson, G., Rosen, T., Bosaeus, I., Sjostrom, L., and Bengtsson, B. A. (1996). Two years of growth hormone (GH) treatment increases bone mineral content and density in hypopituitary patients with adult-onset GH deficiency. *J. Clin. Endocrinol. Metab.* **81,** 2865–2873.

Kim, M. S., Yoon, C. Y., Jang, P. G., Park, Y. J., Shin, C. S., Park, H. S., Ryu, J. W., Pak, Y. K., Park, J. Y., Lee, K. U., Kim, S. Y., Lee, H. K., *et al.* (2004). The mitogenic

and antiapoptotic actions of ghrelin in 3T3-L1 adipocytes. *Mol. Endocrinol.* **18,** 2291–2301.
Kim, S. W., Her, S. J., Park, S. J., Kim, D., Park, K. S., Lee, H. Y., Han, B. H., Kim, M. S., Shin, C. S., and Kim, S. Y. (2005). Ghrelin stimulates proliferation and differentiation and inhibits apoptosis in osteoblastic MC3T3-E1 cells. *Bone* **37,** 359–369.
Klein, K. B., Orwoll, E. S., Lieberman, D. A., Meier, D. E., McClung, M. R., and Parfitt, A. M. (1987). Metabolic bone disease in asymptomatic men after partial gastrectomy with Billroth II anastomosis. *Gastroenterology* **92,** 608–616.
Kojima, M., Hosoda, H., Date, Y., Nakazato, M., Matsuo, H., and Kangawa, K. (1999). Ghrelin is a growth-hormone-releasing acylated peptide from stomach. *Nature* **402,** 656–660.
Kojima, S., Nakahara, T., Nagai, N., Muranaga, T., Tanaka, M., Yasuhara, D., Masuda, A., Date, Y., Ueno, H., Nakazato, M., and Naruo, T. (2005). Altered ghrelin and peptide YY responses to meals in bulimia nervosa. *Clin. Endocrinol. (Oxf.)* **62,** 74–78.
Koranyi, J., Svensson, J., Gotherstrom, G., Sunnerhagen, K. S., Bengtsson, B., and Johannsson, G. (2001). Baseline characteristics and the effects of five years of GH replacement therapy in adults with GH deficiency of childhood or adulthood onset: A comparative, prospective study. *J. Clin. Endocrinol. Metab.* **86,** 4693–4699.
Kretzschmar, M., Doody, J., and Massague, J. (1997). Opposing BMP and EGF signalling pathways converge on the TGF-beta family mediator Smad1. *Nature* **389,** 618–622.
Landin-Wilhelmsen, K., Nilsson, A., Bosaeus, I., and Bengtsson, B. A. (2003). Growth hormone increases bone mineral content in postmenopausal osteoporosis: A randomized placebo-controlled trial. *J. Bone Miner. Res.* **18,** 393–405.
Larsson, B., Gritli-Linde, A., Norlen, P., Lindstrom, E., Hakanson, R., and Linde, A. (2001). Extracts of ECL-cell granules/vesicles and of isolated ECL cells from rat oxyntic mucosa evoke a Ca2+ second messenger response in osteoblastic cells. *Regul. Pept.* **97,** 153–161.
Larsson, B., Norlen, P., Lindstrom, E., Zhao, D. W., Hakanson, R., and Linde, A. (2002). Effects of ECL cell extracts and granule/vesicle-enriched fractions from rat oxyntic mucosa on cAMP and IP3 in rat osteoblast-like cells. *Regul. Pept.* **106,** 13–18.
Lehto-Axtelius, D., Chen, D., Surve, V. V., and Hakanson, R. (2002). Post-gastrectomy osteopenia in the rat. *Scand. J. Gastroenterol.* **37,** 437–443.
Liu, J. P., Baker, J., Perkins, A. S., Robertson, E. J., and Efstratiadis, A. (1993). Mice carrying null mutations of the genes encoding insulin-like growth factor I (Igf-1) and type 1 IGF receptor (Igf1r). *Cell* **75,** 59–72.
Locatelli, V., and Torsello, A. (1997). Growth hormone secretagogues: Focus on the growth hormone-releasing peptides. *Pharmacol. Res.* **36,** 415–423.
Maccarinelli, G., Sibilia, V., Torsello, A., Raimondo, F., Pitto, M., Giustina, A., Netti, C., and Cocchi, D. (2005). Ghrelin regulates proliferation and differentiation of osteoblastic cells. *J. Endocrinol.* **184,** 249–256.
Martini, A. C., Fernandez-Fernandez, R., Tovar, S., Navarro, V. M., Vigo, E., Vazquez, M. J., Davies, J. S., Thompson, N. M., Aguilar, E., Pinilla, L., Wells, T., Dieguez, C., *et al.* (2006). Comparative analysis of the effects of ghrelin and unacylated ghrelin on luteinizing hormone secretion in male rats. *Endocrinology* **147,** 2374–2382.
Matsuda, K., Miura, T., Kaiya, H., Maruyama, K., Shimakura, S., Uchiyama, M., Kangawa, K., and Shioda, S. (2006). Regulation of food intake by acyl and des-acyl ghrelins in the goldfish. *Peptides* **27,** 2321–2325.
Mazzocchi, G., Neri, G., Rucinski, M., Rebuffat, P., Spinazzi, R., Malendowicz, L. K., and Nussdorfer, G. G. (2004). Ghrelin enhances the growth of cultured human adrenal zona glomerulosa cells by exerting MAPK-mediated proliferogenic and antiapoptotic effects. *Peptides* **25,** 1269–1277.

Mellstrom, D., Johansson, C., Johnell, O., Lindstedt, G., Lundberg, P. A., Obrant, K., Schoon, I. M., Toss, G., and Ytterberg, B. O. (1993). Osteoporosis, metabolic aberrations, and increased risk for vertebral fractures after partial gastrectomy. *Calcif. Tissue Int.* **53,** 370–377.

Misra, M., and Klibanski, A. (2006). Anorexia nervosa and osteoporosis. *Rev. Endocr. Metab Disord.* **7,** 91–99.

Misra, M., Miller, K. K., Herzog, D. B., Ramaswamy, K., Aggarwal, A., Almazan, C., Neubauer, G., Breu, J., and Klibanski, A. (2004). Growth hormone and ghrelin responses to an oral glucose load in adolescent girls with anorexia nervosa and controls. *J. Clin. Endocrinol. Metab.* **89,** 1605–1612.

Misra, M., Miller, K. K., Kuo, K., Griffin, K., Stewart, V., Hunter, E., Herzog, D. B., and Klibanski, A. (2005). Secretory dynamics of ghrelin in adolescent girls with anorexia nervosa and healthy adolescents. *Am. J. Physiol. Endocrinol. Metab.* **289,** E347–E356.

Morgan, D. B., Paterson, C. R., Woods, C. G., Pulvertaft, C. N., and Fourman, P. (1965). Osteomalacia after gastrectomy. A response to very small doses of vitamin D. *Lancet* **2,** 1089–1091.

Muccioli, G., Pons, N., Ghe, C., Catapano, F., Granata, R., and Ghigo, E. (2004). Ghrelin and des-acyl ghrelin both inhibit isoproterenol-induced lipolysis in rat adipocytes via a non-type 1a growth hormone secretagogue receptor. *Eur. J. Pharmacol.* **498,** 27–35.

Murphy, M. G., Bach, M. A., Plotkin, D., Bolognese, J., Ng, J., Krupa, D., Cerchio, K., and Gertz, B. J. (1999). Oral administration of the growth hormone secretagogue MK-677 increases markers of bone turnover in healthy and functionally impaired elderly adults. The MK-677 Study group. *J. Bone Miner. Res.* **14,** 1182–1188.

Murphy, M. G., Weiss, S., McClung, M., Schnitzer, T., Cerchio, K., Connor, J., Krupa, D., and Gertz, B. J. (2001). Effect of alendronate and MK-677 (a growth hormone secretagogue), individually and in combination, on markers of bone turnover and bone mineral density in postmenopausal osteoporotic women. *J. Clin. Endocrinol. Metab.* **86,** 1116–1125.

Nakahara, K., Nakagawa, M., Baba, Y., Sato, M., Toshinai, K., Date, Y., Nakazato, M., Kojima, M., Miyazato, M., Kaiya, H., Hosoda, H., Kangawa, K., *et al.* (2006). Maternal ghrelin plays an important role in rat fetal development during pregnancy. *Endocrinology* **147,** 1333–1342.

Nanzer, A. M., Khalaf, S., Mozid, A. M., Fowkes, R. C., Patel, M. V., Burrin, J. M., Grossman, A. B., and Korbonits, M. (2004). Ghrelin exerts a proliferative effect on a rat pituitary somatotroph cell line via the mitogen-activated protein kinase pathway. *Eur. J. Endocrinol.* **151,** 233–240.

Nuttall, M. E., and Gimble, J. M. (2004). Controlling the balance between osteoblastogenesis and adipogenesis and the consequent therapeutic implications. *Curr. Opin. Pharmacol.* **4,** 290–294.

Oh, K. W., Lee, W. Y., Rhee, E. J., Baek, K. H., Yoon, K. H., Kang, M. I., Yun, E. J., Park, C. Y., Ihm, S. H., Choi, M. G., Yoo, H. J., and Park, S. W. (2005). The relationship between serum resistin, leptin, adiponectin, ghrelin levels and bone mineral density in middle-aged men. *Clin. Endocrinol.* **63,** 131–138.

Ohlsson, C., Bengtsson, B. A., Isaksson, O. G., Andreassen, T. T., and Slootweg, M. C. (1998). Growth hormone and bone. *Endocr. Rev.* **19,** 55–79.

Olney, R. C. (2003). Regulation of bone mass by growth hormone. *Med. Pediatr. Oncol.* **41,** 228–234.

Pei, L., and Tontonoz, P. (2004). Fat's loss is bone's gain. *J. Clin. Invest.* **113,** 805–806.

Persson, P., Gagnemo-Persson, R., Chen, D., Axelson, J., Nylander, A. G., Johnell, O., and Hakanson, R. (1993). Gastrectomy causes bone loss in the rat: Is lack of gastric acid responsible? *Scand. J. Gastroenterol.* **28,** 301–306.

Pettersson, I., Muccioli, G., Granata, R., Deghenghi, R., Ghigo, E., Ohlsson, C., and Isgaard, J. (2002). Natural (ghrelin) and synthetic (hexarelin) GH secretagogues stimulate H9c2 cardiomyocyte cell proliferation. *J. Endocrinol.* **175,** 201–209.

Pfaffle, R. W., Parks, J. S., Brown, M. R., and Heimann, G. (1993). Pit-1 and pituitary function. *J. Pediatr. Endocrinol.* **6,** 229–233.

Pun, K. K., Lau, P., Wong, F. H., Cheng, C. L., Pun, W. K., Chow, S. P., and Leong, J. C. (1990). 25-Hydroxycholecalciferol and insulin-like growth factor I are determinants of serum concentration of osteocalcin in elderly subjects with and without spinal fractures. *Bone* **11,** 397–400.

Puzio, I., Kapica, M., Filip, R., Bienko, M., and Radzki, R. P. (2005). Fundectomy evokes elevated gastrin and lowered ghrelin serum levels accompanied by decrease in geometrical and mechanical properties of femora in rats. *Bull. Vet. Inst. Pulawy* **49,** 69–73.

Root, A. W., and Root, M. J. (2002). Clinical pharmacology of human growth hormone and its secretagogues. *Curr. Drug Targets Immune. Endocr. Metabol. Disord.* **2,** 27–52.

Rosenfeld, R., Allen, D. B., MacGillivray, M. H., Alter, C., Saenger, P., Anhalt, H., Hintz, R., and Katz, H. P. (2000). Growth hormone use in pediatric growth hormone deficiency and other pediatric growth disorders. *Am. J. Manag. Care* **6,** S805–S816.

Salmon, W. D., Jr., and Daughaday, W. H. (1957). A hormonally controlled serum factor which stimulates sulfate incorporation by cartilage *in vitro*. *J. Lab. Clin. Med.* **49,** 825–836.

Sambrook, P., and Cooper, C. (2006). Osteoporosis. *Lancet* **367,** 2010–2018.

Sato, M., Nakahara, K., Goto, S., Kaiya, H., Miyazato, M., Date, Y., Nakazato, M., Kangawa, K., and Murakami, N. (2006). Effects of ghrelin and des-acyl ghrelin on neurogenesis of the rat fetal spinal cord. *Biochem. Biophys. Res. Commun.* **350,** 598–603.

Savage, M. O., Burren, C. P., Blair, J. C., Woods, K. A., Metherell, L., Clark, A. J., and Camacho-Hubner, C. (2001). Growth hormone insensitivity: Pathophysiology, diagnosis, clinical variation and future perspectives. *Horm. Res.* **55**(Suppl. 2), 32–35.

Seeman, E., and Delmas, P. D. (2006). Bone quality–the material and structural basis of bone strength and fragility. *N. Engl. J. Med.* **354,** 2250–2261.

Shanado, Y., Kometani, M., Uchiyama, H., Koizumi, S., and Teno, N. (2004). Lysophospholipase I identified as a ghrelin deacylation enzyme in rat stomach. *Biochem. Biophys. Res. Commun.* **325,** 1487–1494.

Sibilia, V., Cocchi, D., Pagani, F., Lattuada, N., Moro, G. L., Pecile, A., Rubinacci, A., Muller, E. E., and Netti, C. (1999). Hexarelin, a growth hormone - releasing peptide, counteracts bone loss in gonadectomized male rats. *Growth Horm. IGF Res.* **9,** 219–227.

Sirotkin, A. V., Grossmann, R., Maria-Peon, M. T., Roa, J., Tena-Sempere, M., and Klein, S. (2006). Novel expression and functional role of ghrelin in chicken ovary. *Mol. Cell Endocrinol.* **257–258,** 15–25.

Sjogren, K., Liu, J. L., Blad, K., Skrtic, S., Vidal, O., Wallenius, V., LeRoith, D., Tornell, J., Isaksson, O. G., Jansson, J. O., and Ohlsson, C. (1999). Liver-derived insulin-like growth factor I (IGF-I) is the principal source of IGF-I in blood but is not required for postnatal body growth in mice. *Proc. Natl. Acad. Sci. USA* **96,** 7088–7092.

Smith, R. G., Jiang, H., and Sun, Y. (2005). Developments in ghrelin biology and potential clinical relevance. *Trends Endocrinol. Metab.* **16,** 436–442.

Sun, Y., Wang, P., Zheng, H., and Smith, R. G. (2004). Ghrelin stimulation of growth hormone release and appetite is mediated through the growth hormone secretagogue receptor. *Proc. Natl. Acad. Sci. USA* **101,** 4679–4684.

Sun, Y. X., Ahmed, S., and Smith, R. G. (2003). Deletion of ghrelin impairs neither growth nor appetite. *Mol. Cell. Biol.* **23,** 7973–7981.

Svensson, J., Ohlsson, C., Jansson, J. O., Murphy, G., Wyss, D., Krupa, D., Cerchio, K., Polvino, W., Gertz, B., Baylink, D., Mohan, S., and Bengtsson, B. A. (1998).

Treatment with the oral growth hormone secretagogue MK-677 increases markers of bone formation and bone resorption in obese young males. *J. Bone Miner. Res.* **13,** 1158–1166.

Svensson, J., Lall, S., Dickson, S. L., Bengtsson, B. A., Romer, J., Ahnfelt-Ronne, I., Ohlsson, C., and Jansson, J. O. (2000). The GH secretagogues ipamorelin and GH-releasing peptide-6 increase bone mineral content in adult female rats. *J. Endocrinol.* **165,** 569–577.

Sylvester, F. A., Wyzga, N., Hyams, J. S., and Gronowicz, G. A. (2002). Effect of Crohn's disease on bone metabolism *in vitro*: A role for interleukin-6. *J. Bone Miner. Res.* **17,** 695–702.

Szulc, P., and Meunier, P. J. (2001). Is vitamin K deficiency a risk factor for osteoporosis in Crohn's disease? *Lancet* **357,** 1995–1996.

Takaya, K., Ariyasu, H., Kanamoto, N., Iwakura, H., Yoshimoto, A., Harada, M., Mori, K., Komatsu, Y., Usui, T., Shimatsu, A., Ogawa, Y., Hosoda, K., *et al.* (2000). Ghrelin strongly stimulates growth hormone release in humans. *J. Clin. Endocrinol. Metab.* **85,** 4908–4911.

Takeuchi, Y., Suzawa, M., Kikuchi, T., Nishida, E., Fujita, T., and Matsumoto, T. (1997). Differentiation and transforming growth factor-beta receptor down-regulation by collagen-alpha2beta1 integrin interaction is mediated by focal adhesion kinase and its downstream signals in murine osteoblastic cells. *J. Biol. Chem.* **272,** 29309–29316.

Tannenbaum, G. S., Epelbaum, J., and Bowers, C. Y. (2003). Interrelationship between the novel peptide ghrelin and somatostatin/growth hormone-releasing hormone in regulation of pulsatile growth hormone secretion. *Endocrinology* **144,** 967–974.

Tassone, F., Broglio, F., Destefanis, S., Rovere, S., Benso, A., Gottero, C., Prodam, F., Rossetto, R., Gauna, C., van der Lely, A. J., Ghigo, E., and Maccario, M. (2003). Neuroendocrine and metabolic effects of acute ghrelin administration in human obesity. *J. Clin. Endocrinol. Metab.* **88,** 5478–5483.

ter Maaten, J. C., de Boer, H., Kamp, O., Stuurman, L., and van der Veen, E. A. (1999). Long-term effects of growth hormone (GH) replacement in men with childhood-onset GH deficiency. *J. Clin. Endocrinol. Metab.* **84,** 2373–2380.

Tolle, V., Kadem, M., Bluet-Pajot, M. T., Frere, D., Foulon, C., Bossu, C., Dardennes, R., Mounier, C., Zizzari, P., Lang, F., Epelbaum, J., and Estour, B. (2003). Balance in ghrelin and leptin plasma levels in anorexia nervosa patients and constitutionally thin women. *J. Clin. Endocrinol. Metab.* **88,** 109–116.

Torsello, A., Locatelli, V., Melis, M. R., Succu, S., Spano, M. S., Deghenghi, R., Muller, E. E., and Argiolas, A. (2000). Differential orexigenic effects of hexarelin and its analogs in the rat hypothalamus: Indication for multiple growth hormone secretagogue receptor subtypes. *Neuroendocrinology* **72,** 327–332.

Toshinai, K., Yamaguchi, H., Sun, Y., Smith, R. G., Yamanaka, A., Sakurai, T., Date, Y., Mondal, M. S., Shimbara, T., Kawagoe, T., Murakami, N., Miyazato, M., *et al.* (2006). Des-acyl ghrelin induces food intake by a mechanism independent of the growth hormone secretagogue receptor. *Endocrinology* **147,** 2306–2314.

Tovey, F. I., Hall, M. L., Ell, P. J., and Hobsley, M. (1992). A review of postgastrectomy bone disease. *J. Gastroenterol. Hepatol.* **7,** 639–645.

Tschop, M., Smiley, D. L., and Heiman, M. L. (2000). Ghrelin induces adiposity in rodents. *Nature* **407,** 908–913.

Tsubota, Y., Owada-Makabe, K., Yukawa, K., and Maeda, M. (2005). Hypotensive effect of des-acyl ghrelin at nucleus tractus solitarii of rat. *Neuroreport* **16,** 163–166.

van der Eerden, B. C., Karperien, M., and Wit, J. M. (2003). Systemic and local regulation of the growth plate. *Endocr. Rev.* **24,** 782–801.

van Driel, M., Koedam, M., Buurman, C. J., Roelse, M., Weyts, F., Chiba, H., Uitterlinden, A. G., Pols, H. A., and van Leeuwen, J. P. (2006). Evidence that both

1alpha,25-dihydroxyvitamin D3 and 24-hydroxylated D3 enhance human osteoblast differentiation and mineralization. *J. Cell. Biochem.* **99**, 922–935.

Vestergaard, P. (2003). Bone loss associated with gastrointestinal disease: Prevalence and pathogenesis. *Eur. J. Gastroenterol. Hepatol.* **15**, 851–856.

Volante, M., Allia, E., Fulcheri, E., Cassoni, P., Ghigo, E., Muccioli, G., and Papotti, M. (2003). Ghrelin in fetal thyroid and follicular tumors and cell lines: Expression and effects on tumor growth. *Am. J. Pathol.* **162**, 645–654.

Vulliemoz, N. R., Xiao, E., Xia-Zhang, L., Germond, M., Rivier, J., and Ferin, M. (2004). Decrease in luteinizing hormone pulse frequency during a five-hour peripheral ghrelin infusion in the ovariectomized rhesus monkey. *J. Clin. Endocrinol. Metab.* **89**, 5718–5723.

Weiss, L. A., Langenberg, C., and Barrett-Connor, E. (2006). Ghrelin and bone: Is there an association in older adults?: The Rancho Bernardo study. *J. Bone Miner. Res.* **21**, 752–757.

Wells, T., and Houston, P. A. (2001). Skeletal growth acceleration with growth hormone secretagogues in transgenic growth retarded rats: Pattern-dependent effects and mechanisms of desensitization. *J. Neuroendocrinol.* **13**, 496–504.

Westbroek, I., van der, P. A., de Rooij, K. E., Klein-Nulend, J., and Nijweide, P. J. (2001). Expression of serotonin receptors in bone. *J. Biol. Chem.* **276**, 28961–28968.

Wit, J. M., Drayer, N. M., Jansen, M., Walenkamp, M. J., Hackeng, W. H., Thijssen, J. H., and Van den Brande, J. L. (1989). Total deficiency of growth hormone and prolactin, and partial deficiency of thyroid stimulating hormone in two Dutch families: A new variant of hereditary pituitary deficiency. *Horm. Res.* **32**, 170–177.

Woods, K. A., and Savage, M. O. (1996). Laron syndrome: Typical and atypical forms. *Baillieres Clin. Endocrinol. Metab.* **10**, 371–387.

Xiao, G., Jiang, D., Thomas, P., Benson, M. D., Guan, K., Karsenty, G., and Franceschi, R. T. (2000). MAPK pathways activate and phosphorylate the osteoblast-specific transcription factor, Cbfa1. *J. Biol. Chem.* **275**, 4453–4459.

Xiao, G., Gopalakrishnan, R., Jiang, D., Reith, E., Benson, M. D., and Franceschi, R. T. (2002). Bone morphogenetic proteins, extracellular matrix, and mitogen-activated protein kinase signaling pathways are required for osteoblast-specific gene expression and differentiation in MC3T3-E1 cells. *J. Bone Miner. Res.* **17**, 101–110.

Yakar, S., Liu, J. L., Stannard, B., Butler, A., Accili, D., Sauer, B., and LeRoith, D. (1999). Normal growth and development in the absence of hepatic insulin-like growth factor I. *Proc. Natl. Acad. Sci. USA* **96**, 7324–7329.

Yakar, S., Liu, J. L., and Le Roith, D. (2000). The growth hormone/insulin-like growth factor-I system: Implications for organ growth and development. *Pediatr. Nephrol.* **14**, 544–549.

Yakar, S., Rosen, C. J., Beamer, W. G., Ackert-Bicknell, C. L., Wu, Y., Liu, J. L., Ooi, G. T., Setser, J., Frystyk, J., Boisclair, Y. R., and LeRoith, D. (2002). Circulating levels of IGF-1 directly regulate bone growth and density. *J. Clin. Invest.* **110**, 771–781.

Yeh, A. H., Jeffery, P. L., Duncan, R. P., Herington, A. C., and Chopin, L. K. (2005). Ghrelin and a novel preproghrelin isoform are highly expressed in prostate cancer and ghrelin activates mitogen-activated protein kinase in prostate cancer. *Clin. Cancer Res.* **11**, 8295–8303.

Zezulak, K. M., and Green, H. (1986). The generation of insulin-like growth factor-1-sensitive cells by growth hormone action. *Science* **233**, 551–553.

Zhang, W., Zhao, L., Lin, T. R., Chai, B., Fan, Y., Gantz, I., and Mulholland, M. W. (2004). Inhibition of adipogenesis by ghrelin. *Mol. Biol. Cell* **15**, 2484–2491.

Zittel, T. T., Zeeb, B., Maier, G. W., Kaiser, G. W., Zwirner, M., Liebich, H., Starlinger, M., and Becker, H. D. (1997). High prevalence of bone disorders after gastrectomy. *Am. J. Surg.* **174**, 431–438.

CHAPTER ELEVEN

Ghrelin in Pregnancy and Lactation

Jens Fuglsang*

Contents

I. Introduction	260
II. GHSRs in Pregnancy	260
III. Ghrelin in the Fallopian Tubes, Uterus, and Placenta	262
IV. Ghrelin Secretion in Pregnancy	262
A. Diurnal profile: Effect of meals, sleep, and tobacco	264
B. Ghrelin secretion at delivery	264
V. Placenta as a Source of Ghrelin?	265
VI. Physiological Actions of Ghrelin in Pregnancy	266
A. Appetite and adipose tissue metabolism	266
B. Hyperinsulinemia and insulin resistance	268
C. Ghrelin and the GH axis	268
D. Hypertensive disorders: Preeclampsia	271
E. Direct effects of maternal ghrelin levels on the fetus	272
F. Birth	272
G. Other effects	272
VII. Lactation	273
VIII. Fetal Ghrelin	274
A. Fetal ghrelin and fetal and neonatal growth	275
IX. Future Directions for Research	276
X. Summary	276
Acknowledgments	277
References	277

Abstract

Ghrelin and its receptors are found in the reproductive organs and in the placenta, clearly indicating a role for ghrelin in reproduction. Circulating ghrelin levels peak at mid-gestation, then with advancing gestational age declining ghrelin levels are observed. At the same time the maternal organism increases its fat mass, becomes insulin resistant and the growth hormone (GH) axis is dominated by placental growth hormone circulating in concentrations comparable to GH levels

---

* Gynaecological/Obstetrical Research Laboratory, Aarhus University Hospital, Skejby Hospital, DK-8200 Aarhus N, Denmark; and Gynaecological/Obstetrical Department, Aarhus University Hospital, Skejby Hospital, DK-8200 Aarhus N, Denmark

observed in acromegaly. After delivery, normalization of ghrelin levels occurs before the maternal fat mass is restored at prepregnant levels. The physiological course of ghrelin during the three trimesters of human pregnancy is discussed, as are the physiological roles ghrelin may subserve. Regulation of maternal energy intake may be the prevailing effect of ghrelin in pregnancy and lactation, but several other effects of ghrelin may coexist, including local effects. Finally, ghrelin secretion in the fetus is briefly discussed. © 2008 Elsevier Inc.

I. INTRODUCTION

Early after the recognition of ghrelin as an endogenous growth hormone secretagogue (GHS), several lines of evidence suggested a role for ghrelin in reproduction. Ghrelin was found in the reproductive organs and in the placenta, as were GHS receptors (GHSRs). In human pregnancy, accumulation of adipose tissue occurs, insulin resistance develops, and high growth hormone (GH) levels are present. Ghrelin has been coupled to each of these phenomena in nonpregnant subjects. Notwithstanding, only limited research has addressed the physiology and role of ghrelin in human pregnancy. This chapter aims at discussing maternal ghrelin physiology in pregnancy and lactation.

II. GHSRs IN PREGNANCY

Ghrelin is an endogenous ligand of the GHSR (Kojima et al., 1999). The GHSR1a is activated by ghrelin in its acylated (octanoylated) form and is abundant in the hypothalamic-pituitary area (Gnanapavan et al., 2002). The GHSR1b is not activated by GHS, and the physiological role for this receptor is unknown. Other GHSR subtypes probably exist, and other ghrelin isoforms may bind to and activate one or more of these receptors (Broglio et al., 2006; Thompson et al., 2004).

The fallopian tube endothelium has a high immunostaining for GHSR1a (Gaytan et al., 2005), Table 1, though opposed by another study reporting moderate levels of GHSR1b mRNA but undetectable levels of GHSR1a mRNA (Gnanapavan et al., 2002). Under normal circumstances, the fallopian tubes constitute the milieu in which the fertilization of the egg and the first cell divisions of the zygote normally occur and, interestingly, GHSR mRNA has been observed as early as in the morula stage in mice (Kawamura et al., 2003).

In the nonpregnant endometrium and endocervical epithelial lining, no expression of GHSR1a has been detected as determined by immunostaining (Gaytan et al., 2005), contrasting the detection of GHSR mRNA in the endometrium in different phases of the menstrual cycle (Tanaka et al., 2003).

Table 1 Localization of human ghrelin and GHSRs in relation to pregnancy and lactation

Anatomical location	Ghrelin		GHSR		GHS binding
	mRNA	Peptide	mRNA	Peptide	
Fallopian tubes	$+^a$		$+^a$	$+^b$	
Uterine endometrium, nonpregnant	$+^c$	$+^c$	$+^c$	$-^b$	$+^d$
Uterine endometrium, decidualized	$+^c$	$+^c$	$+^c$		
Placenta	$+^{a, c, e, f}$	$+^{c, e}$	$+^{a, c}$		$-/(+)^d$
Breast, nonlactating	$+^{a, f}$		$+^a$		$-/(+)^{d, g}$
Breast milk		$+^{f, h}$			

[a] Gnanapavan et al. (2002).
[b] Gaytan et al. (2005).
[c] Tanaka et al. (2003).
[d] Papotti et al. (2000).
[e] Gualillo et al. (2001).
[f] Kierson et al. (2006).
[g] Cassoni et al. (2001).
[h] Aydin et al. (2006).
GHSR, growth hormone secretagogue receptor. Empty columns indicate that data are missing.

Ghrelin appears to have a definite role in the decidualization of endometrial stromal cells, as GHSR mRNA is found in decidual tissue (Tanaka et al., 2003). In the human placenta, GHSR mRNA is not detected in early pregnancy (Tanaka et al., 2003). In advanced gestation, placental GHSR1b mRNA has been demonstrated, but not GHSR1a (Gnanapavan et al., 2002).

GHSs bind to a high number of peripheral tissues, including endocrine tissues related to reproductive functions. In the testis and the ovary, a higher level of GHS binding is observed as compared to the pituitary. Modest GHS binding did occur in the (nonpregnant) uterus, and almost no binding was observed in term placentas obtained after spontaneous deliveries (Papotti et al., 2000). In this respect it should be remembered that ligand binding studies do not distinguish between different types of receptors, and that the existence of non-GHSR forms has been made plausible (Broglio et al., 2006; Cassoni et al., 2001). Such receptor types are poorly defined in the placenta. Glycoprotein IV, or CD36, appears to mediate effects of GHS (Bodart et al., 2002), and this protein is also expressed in the placenta (Dye et al., 2001).

Furthermore, the placenta may display different patterns of receptor expression in the different trimesters. This would be consistent with the difference in hormone concentrations observed throughout pregnancy, as described below, but this issue awaits further research. Even the process of delivery may change the expression of receptors, consistent with the low binding of labeled hexarelin to placental tissues obtained after spontaneous deliveries (Papotti et al., 2000).

III. Ghrelin in the Fallopian Tubes, Uterus, and Placenta

Ghrelin mRNA is found in the fallopian tube and endometrium during the normal menstrual cycle (Gnanapavan et al., 2002; Tanaka et al., 2003; Table 1). Decidualization induces a tremendous increase in ghrelin mRNA (Tanaka et al., 2003), and this induction appears to depend on the presence of the pregnancy product, as the endometrium in ectopic pregnancies contains far less ghrelin mRNA (Tanaka et al., 2003). *In vitro* experiments have shown that the endometrial ghrelin mRNA production is greatly enhanced by coculture with placental tissue (Tanaka et al., 2003).

In humans, high placental ghrelin mRNA and ghrelin peptide levels are seen in the first trimesters of pregnancy, primarily in the cytotrophoblast (Gualillo et al., 2001), and also the extravillous trophoblast contains ghrelin mRNA (Tanaka et al., 2003). In contrast, placental ghrelin immunostaining was barely detectable after delivery, despite the presence of ghrelin mRNA (Gualillo et al., 2001). Expression studies in rats have demonstrated slightly different results, with no placental ghrelin mRNA in early pregnancy, peaking mRNA levels in mid/late gestation and thereafter declining ghrelin mRNA levels, with a similar pattern of ghrelin protein occurrence (Gualillo et al., 2001).

IV. Ghrelin Secretion in Pregnancy

Only a few studies have reported on the course of circulating ghrelin levels during pregnancy in humans. We reported the longitudinal changes in serum (total) ghrelin levels during pregnancy in 11 pregnant women (Fuglsang et al., 2005; Fig. 1). In this cohort of normal, nondiabetic women, maximum serum values were observed at mid-pregnancy. After this time, serum levels declined to the lowest levels in pregnancy at the end of the third trimester with a 28% decrease on average compared to mid-pregnancy values. Similar observations have been reported in a cross-sectional study measuring acylated ghrelin (Palik et al., 2007). The lowest circulating ghrelin levels are seen in late pregnancy, and may be even lower than in nonpregnant subjects (Makino et al., 2002; Palik et al., 2007).

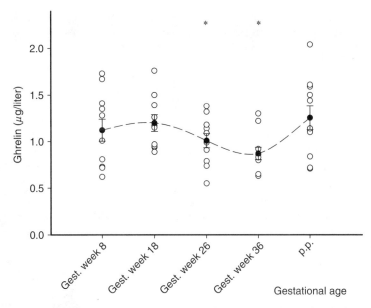

Figure 1 Ghrelin during pregnancy. Open circles correspond to individual measurements, whereas filled circles and bars represent the mean ± SEM ($n = 11$). *, $p < 0.05$ versus gestational week 18. (Repeated measurement ANOVA using the Tukey test for multiple comparisons). (Reproduced with permission Fuglsang et al. (2005), Blackwell Publishing.)

Observations from animal studies differ somewhat from those in humans. In rats, nadir levels of circulating maternal ghrelin were observed at midgestation whereafter levels steadily increased until delivery (Nakahara et al., 2006; Shibata et al., 2004).

No studies have described in detail the relationship between octanoylated and nonoctanoylated ghrelin in either of the three trimesters. Instead, total or acylated ghrelin levels are usually reported. This is of interest, as 70% of circulating ghrelin will be unacylated in nonpregnant subjects, and differential effects of unacylated and acylated ghrelin forms have been described (Broglio et al., 2004). Likewise, the occurrence of ghrelin isoforms (Ghigo et al., 2005) in pregnancy is unknown.

Some lines of evidence suggest that estrogen may influence ghrelin levels in nonpregnant subjects. Serum ghrelin increases after estrogen replacement therapy (Kellokoski et al., 2005), and total ghrelin levels appear higher in females than in males after adjustment for body mass index (BMI) and waist circumference (Monti et al., 2006). However, after acute ghrelin administration, no differences have been observed in GH release between genders in both young and elderly subjects (Broglio et al., 2003a). This finding indicates that at least the pituitary somatotroph response to ghrelin is unrelated to estrogen levels. In pregnancy, estrogen increases to high levels in the last trimester, but in this situation a decline in ghrelin levels is observed (Fuglsang et al., 2005; Palik et al., 2007).

A. Diurnal profile: Effect of meals, sleep, and tobacco

The diurnal profile of serum or plasma ghrelin during pregnancy has not been investigated. In nonpregnant subjects, daytime ghrelin secretion is related to meal ingestion with preprandial rises and postprandial declines in serum levels (Caixas et al., 2002; Cummings et al., 2001; Yildiz et al., 2004). Nighttime ghrelin levels increase in lean subjects, but not in overweight subjects (Cummings et al., 2001; Yildiz et al., 2004). Usually, maternal body weight and adipose tissue weight increase in pregnancy, and this change in body composition may therefore theoretically has an impact on the diurnal ghrelin secretion.

Similarly, no studies have reported human ghrelin levels before and after meal ingestion in pregnancy. However in a study, in pregnant women who had blood samples withdrawn at the onset of at-term labour, circulating ghrelin levels were positively associated with the time since ingestion of the preceding meal (Bouhours-Nouet et al., 2006). In rats, fasting increased ghrelin levels in both pregnant and nonpregnant females but in one setting the relative increase was nearly three times higher in nonpregnant rats (Chanoine and Wong, 2004). Maternal food intake did not affect placental ghrelin gene expression (Chanoine and Wong, 2004). These findings suggest that the diurnal serum profile of ghrelin in advanced pregnancy may be "flat" or have dampened fluctuations compared to nonpregnant subjects, and similar changes in diurnal profile are observed for GHs (Eriksson et al., 1988, 1989).

Smoking may increase ghrelin levels (Bouros et al., 2006; Fagerberg et al., 2003) and it can inflict serious consequences to the fetus; however at delivery, ghrelin levels were similar between smoking and nonsmoking women (Bouhours-Nouet et al., 2006).

B. Ghrelin secretion at delivery

At delivery, only very limited changes in serum ghrelin levels occur. Even though a statistically significant decrease in maternal serum ghrelin is observed within 24–48 h after birth, both the absolute and the relative changes in ghrelin levels are very limited (Fuglsang et al., 2006a). For comparison, insulin levels tend to be restored at nonpregnant levels within 24 h after expulsion of the placenta. The reoccurrence of GH in maternal blood takes place in days, whereas placental growth hormone (PGH) vanishes within hours (Fuglsang et al., 2006a).

V. PLACENTA AS A SOURCE OF GHRELIN?

The major source of circulating ghrelin is thought to be the gastrointestinal tract, especially the stomach (Kojima et al., 1999; Moller et al., 2003). However, both ghrelin mRNA and ghrelin peptide are observed in the human placenta, mainly in the cytotrophoblast (Gualillo et al., 2001; Tanaka et al., 2003). The cytotrophoblast is adjacent to the syncytiotrophoblast, which is the site of several placentally produced hormones, including PGH and human placental lactogen (hPL), and the syncytiotrophoblast is the layer in contact with the maternal circulation. Circulating ghrelin levels appear to peak at the same time as the placenta contains maximum levels of ghrelin mRNA and peptide (Gualillo et al., 2001). Thus, the placenta might be a source of circulating ghrelin in the mother. At least two observations oppose this theory: First, after expulsion of the placenta changes in serum, ghrelin levels are very modest. If the placenta was a significant source of circulating maternal ghrelin in the last trimester, a large decrease in ghrelin levels was expected. Second, from a few observations in twin pregnancies, we reported apparently similar ghrelin levels at term as in singleton pregnancies (Fuglsang et al., 2006a). In twin pregnancies, maternal serum levels of many placental hormones are considerably higher as compared to singleton pregnancies, for example PGH and hPL (Fuglsang et al., 2006a; Spellacy et al., 1978), either due to the increased fetal (energy) demands or simply due to the increased placental size. Additionally, a study in rats found increased levels of placental ghrelin mRNA at mid-gestation when circulating ghrelin was at a minimum. It was concluded that, in rats, placental ghrelin was not related to circulating ghrelin levels (Shibata et al., 2004). Curiously in the same study, gastic ghrelin content was unaltered throughout pregnancy suggesting that neither the stomach was the source for circulating ghrelin (Shibata et al., 2004).

Hence in humans, observations in the third trimester and at delivery argue against a placental contribution to circulating ghrelin levels. Hypothetically, the placenta might instead contribute at mid-gestation when the highest ghrelin mRNA levels and circulating ghrelin levels are observed. Alternatively, autocrine and paracrine mechanisms could be involved. It is of notice that the placenta is of fetal origin and carries the same chromosomal constitution as the fetus, not the mother. Situated at the interface between the maternal and the fetal circulation, placental ghrelin production, therefore, could be involved in several processes of the placenta, controlling, for example, local redistribution of energy between mother and fetus or influencing the syncytiotrophoblastic PGH and hPL production. Expectedly, such effects will not be reflected in circulating maternal ghrelin levels.

Regarding potential autocrine effects in late pregnancy, labeled GHS displayed almost no binding after delivery (Papotti *et al.*, 2000). Assuming that the labor process itself does not affect receptor expression or ghrelin synthesis, which in part seems justified by the nearly unchanged levels of ghrelin before and after birth, low placental ghrelin content and low placental GHS binding capacity coexist in late pregnancy, suggesting that ghrelin is of minor importance for placental metabolism at the time of delivery.

VI. Physiological Actions of Ghrelin in Pregnancy

The role of ghrelin in pregnancy is far from understood and it may subserve several roles. The localization of ghrelin and receptors hereof in the placenta clearly suggests a role for this hormone during pregnancy. Conversely, transgenic ghrelin null mice have been reported to behave and reproduce like wild-type mice (Sun *et al.*, 2003). Hence, ghrelin appears not to be essential for reproduction, at least in mice.

From mid-gestation and forward, increasing insulin resistance develops with respect to glucose and lipid metabolism (Sivan *et al.*, 1999). GH levels increase concomitantly, as does, for example hPL, insulin-like growth factors (IGFs), and leptin (Caufriez *et al.*, 1993; Fuglsang *et al.*, 2005; Mirlesse *et al.*, 1993). Thus, pregnancy is a condition combining aspects of obesity, acromegaly, and diabetes.

A. Appetite and adipose tissue metabolism

Ghrelin administration increases food intake through central mechanisms (Lawrence *et al.*, 2002; Tschop *et al.*, 2000; Wren *et al.*, 2001). Numerous studies have explored the orexigenic effects of ghrelin [for an overview see (Ghigo *et al.*, 2005) or Chapter 6, this volume]; however, its effects on appetite in relation to human pregnancy remain virtually unknown. The well-known preprandial rise in ghrelin levels in nonpregnant subjects may in pregnancy be reflected in the observed association between fasting time and ghrelin levels at delivery (Bouhours-Nouet *et al.*, 2006). On the other hand, fasting ghrelin levels are similar in singleton and twin gestations, despite the obvious difference in maternal weight gain and total weight of the pregnancy product and placenta(s) in twin pregnancies (Fuglsang *et al.*, 2006a). At present, only animal studies give a hint on the potential impact of ghrelin on food intake in pregnancy. In rats, hypothalamic and pituitary ghrelin mRNA expression was fairly constant throughout pregnancy, as was stomach ghrelin content (Shibata *et al.*, 2004). Food restriction resulting in weight loss of dams during pregnancy was accompanied by increased plasma ghrelin levels and increased gastric ghrelin mRNA in late pregnancy (Gualillo et al., 2002). Repeated administration of (acylated) ghrelin in late pregnancy leads to increased maternal food intake, but maternal body weights were not reported (Nakahara et al., 2006). Continuous ghrelin injection in early pregnancy did not change maternal weights (Fernandez-Fernandez et al., 2005).

The central mechanisms involved in appetite regulation are modulated by both ghrelin and leptin (Ghigo et al., 2005; Nakazato et al., 2001). In rats, the satiety effect of leptin decreases during pregnancy; that is, pregnancy induces central leptin resistance (Johnstone and Higuchi, 2001). In human pregnancy, leptin levels increase from the first trimester, and the placenta contributes to the circulating pool of leptin (Henson and Castracane, 2000; Highman et al., 1998; Linnemann et al., 2000). The equilibrium between circulating ghrelin and leptin therefore is shifted toward leptin, at least in terms of total leptin inasmuch the bioactive or free fraction of leptin appears to be unchanged during late pregnancy (Lewandowski et al., 1999). Leptin, though, appears not to affect ghrelin levels (Broglio et al., 2006; Chan et al., 2004). On the other hand, an inverse relationship between serum leptin and acylated ghrelin levels in the third trimester has been reported (Palik et al., 2007).

Intriguingly, circulating ghrelin levels peak at mid-gestation. Maximum fetal growth occurs shortly after this time of pregnancy. Some lines of evidence suggest that maximum maternal fat accretion occurs at mid-gestation until week 30 (Herrera, 2002; Villar et al., 1992) to optimize maternal energy depots before the maximum fetal growth. Lipolysis then predominates in the third trimester (Sivan et al., 1999). Ghrelin levels are decreased in obese nonpregnant subjects (Cappiello et al., 2002; Poykko et al., 2003). The inverse association between BMI and ghrelin levels among nonpregnant individuals (Cappiello et al., 2002; Chan et al., 2004; Fagerberg et al., 2003; Kellokoski et al., 2005; Monti et al., 2006; Poykko et al., 2003) has not been consistently demonstrated in pregnancy, but a significant inverse association was observed to acylated ghrelin (Fuglsang et al., 2005, 2006a; Palik et al., 2007). In pregnancy, the maternal BMI changes as a consequence of the weight gain, but only laborious investigations will discern the weight of the pregnancy product from the increase in maternal adipose tissue, extracellular fluids, blood cells, and so forth. Therefore, the maternal BMI is expected to be a rather rough measure of adiposity. The use of prepregnancy BMI as an estimate of maternal adiposity at term is also encumbered with imprecision. Finally, acylated ghrelin may prove to be a more correct measure when addressing the association between ghrelin and fat mass in pregnancy (Palik et al., 2007), even though ghrelin effects on adipose tissue in animal experiments have been demonstrated to be mediated by both acylated and nonacylated ghrelin (Thompson et al., 2004).

Clearly, the relationship between ghrelin in its various forms and appetite and adipose tissue mass in pregnancy calls for further research.

B. Hyperinsulinemia and insulin resistance

Insulin levels decrease in the first weeks of pregnancy and stabilize until a large rise occurs from approximately week 20 onward. Blood glucose regulation in pregnancy, though, is continuously subjected to the demands of the fetoplacental unit. Hence, throughout the day plasma glucose levels fluctuate with greater amplitudes in pregnant than in nonpregnant women, especially in obesity (Parretti et al., 2001; Phelps et al., 1981; Yogev et al., 2004). Hyperinsulinemia and insulin resistance start to develop in the later half of pregnancy at a time of decreasing ghrelin levels.

An inverse relationship between ghrelin and insulin exists in nonpregnant subjects (Ghigo et al., 2005; Poykko et al., 2003; Saad et al., 2002), and low ghrelin levels are associated with increased insulin resistance and type 2 diabetes (Poykko et al., 2003). In a study in nondiabetic pregnant women, we reported a near-significant inverse association between ghrelin levels and fasting blood glucose and approximations of insulin sensitivity (Fuglsang et al., 2006a; Fig. 2). When excluding the outlier, the association became significant, Fig. 2. In another study, acylated ghrelin was found to be negatively associated with C-peptide levels in normal and gestational diabetic subjects, and a negative association between ghrelin levels and insulin dose in gestational diabetes was observed (Palik et al., 2007).

In nonpregnant individuals, ghrelin decreases insulin levels and leads to hyperglycemia, but does not affect the insulin response to an oral glucose load (Broglio et al., 2003a,b). Interestingly, insulin in turn seems to be able to decrease ghrelin secretion (Flanagan et al., 2003; Saad et al., 2002). These observations fit well with the observations on ghrelin and insulin levels in third trimester pregnancies; however, many other regulatory mechanisms may be involved in the decline in ghrelin levels in late pregnancy, for example effects of increasing body weight of the mother or increasing hormone levels per se. Nevertheless, a role for ghrelin in pancreatic physiology seems likely as GHSR and ghrelin expression has been demonstrated in this organ (Broglio et al., 2006; Gnanapavan et al., 2002; Volante et al., 2002), and ghrelin may even be colocalized with insulin in β-cells (Volante et al., 2002).

Thus, in pregnancy ghrelin may be associated to insulin sensitivity as in nonpregnant subjects but this issue awaits confirmatory findings.

C. Ghrelin and the GH axis

With advancing gestational age, the maternal GH axis shifts from pituitary-derived GH toward a predominance of PGH from the second half of pregnancy, Fig. 3 (Caufriez et al., 1993; Chellakooty et al., 2004; Fuglsang et al., 2005; Mirlesse et al., 1993). A very high-sequence homology exists for the two peptides (Chen et al., 1989) and GH and PGH have the same

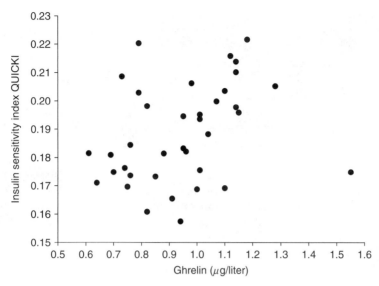

Figure 2 Correlation between fasting ghrelin and insulin sensitivity index$_{QUICKI}$ (ISI$_{QUICKI}$). A trend was observed between ghrelin and ISIs (ISI$_{QUICKI}$: $r = 0.31$, $p = 0.062$; ISI$_{HOMA}$: $r = -0.27$, $p = 0.11$; $n = 37$), with further strengthening of these relationships by exclusion of the outlier (outmost right) (ISI$_{QUICKI}$: $r = 0.44$, $p = 0.007$; ISI$_{HOMA}$: $r = -0.37$, $p = 0.03$; $n = 36$). (Modified with permission Fuglsang et al. (2006), Blackwell Publishing.)

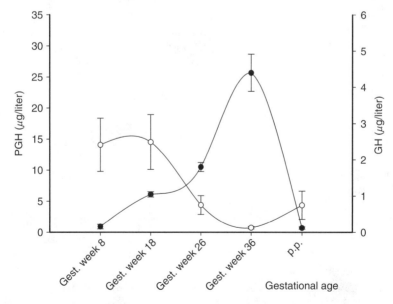

Figure 3 Serum PGH (filled circles) and GH (open circles) during pregnancy. Data are presented as mean ± SEM ($n = 11$). Note the different scaling of the y-axes for PGH and GH. (Reproduced with permission Fuglsang et al. (2005), Blackwell Publishing.)

affinity for the GH receptor (Baumann *et al.*, 1991), but in contrast to GH, PGH is produced in the placenta (Liebhaber *et al.*, 1989; Scippo *et al.*, 1993). Under normal circumstances, pituitary GH is only detectable in trace concentrations after gestational week 20, Fig. 3 (Fuglsang *et al.*, 2005). Third trimester levels of PGH vary considerably between individuals (Chellakooty *et al.*, 2002, 2004), though on average, these PGH levels are comparable to GH levels in acromegaly. In contrast to the diurnal fluctuations in GH levels (Dimaraki *et al.*, 2003; Ho *et al.*, 1988; Veldhuis *et al.*, 1991), PGH appears to be secreted at a rather constant diurnal rate (Eriksson *et al.*, 1988, 1989), though fasting and hypoglycemia may increase PGH levels (Bjorklund *et al.*, 1998; Fuglsang *et al.*, 2006a). Several studies have demonstrated a relationship between third trimester PGH levels and the weight of the newborns (Chellakooty *et al.*, 2004; Coutant *et al.*, 2001; Fuglsang *et al.*, 2003; McIntyre *et al.*, 2000), and PGH may be coupled to fetal growth through lipolytic effects in the maternal organism (Goodman *et al.*, 1991).

Ghrelin levels peak at mid-gestation in contrast to the peak values of PGH observed in the last weeks of pregnancy (Chellakooty *et al.*, 2004; Fuglsang *et al.*, 2005; Mirlesse *et al.*, 1993). Furthermore, only but discrete changes in serum levels of ghrelin are observed immediately after delivery, whereas PGH disappears from the circulation within hours (Fuglsang *et al.*, 2006b). These findings suggest that ghrelin plays no major role in PGH regulation, in accordance with other studies demonstrating a different regulation of PGH as compared to GH. In experimental studies, supraphysiological doses of ghrelin have been shown to increase GH secretion (Kojima *et al.*, 1999; Takaya *et al.*, 2000), whereas at physiological levels of ghrelin, such an effect has not been demonstrated convincingly. In pregnancy, no clear-cut associations between maternal ghrelin and PGH levels have been observed (Bouhours-Nouet *et al.*, 2006; Fuglsang *et al.*, 2005, 2006a). Apparently, PGH levels vary between the fasting and nonfasting states (Fuglsang *et al.*, 2006a). We therefore used nonfasting PGH as a marker of basal PGH secretion, and found a significant association between ghrelin and basal PGH secretion, but not fasting PGH (Fuglsang *et al.*, 2006a). It could be speculated, whether such an association merely reflects the impact of both hormones on the maternal energy metabolism rather than implying any direct effects of ghrelin on PGH secretion. GH-releasing hormone (GHRH) has no effect on PGH secretion (de Zegher *et al.*, 1990; Evain-Brion *et al.*, 1990). This is in clear contrast to GH, which is secreted after GHRH administration and, furthermore, endogenous GHRH seems essential for the GH response to ghrelin (Tannenbaum *et al.*, 2003). During pregnancy, GHRH levels appear fairly constant (Mazlan *et al.*, 1990).

In intrauterine growth restriction (IUGR), PGH levels are low and a compensatory reactivation of GH secretion may occur (Caufriez *et al.*, 1993).

The role of ghrelin in IUGR pregnancies and GH secretion in such pregnancies is unknown. Cases of IUGR may have widely different aetiology and the role for ghrelin, if any, may be confined to certain subgroups of IUGR cases.

Acromegalic subjects have high GH and IGF-1 levels. These patients also have high insulin levels and decreased levels of ghrelin (Cappiello et al., 2002). During an oral glucose tolerance test, the ghrelin response was blunted among acromegalic subjects in contrast to the declining ghrelin levels among controls and obese subjects. Similarly, the rise in insulin levels were augmented in the acromegalic group, a finding also observed after ghrelin injection (Arosio et al., 2004; Cappiello et al., 2002). The lack of an inverse correlation between insulin and ghrelin levels in acromegalic subjects (Cappiello et al., 2002) is similar to the observation in pregnant women at term (Fuglsang et al., 2005, 2006a). In pregnant women with a mean gestational age of 29 weeks, an inverse association was found between acylated ghrelin and C-peptide. In the same report, insulin requirements in third trimester gestational diabetic subjects were inversely correlated to ghrelin levels (Palik et al., 2007).

In GH deficiency, GH replacement therapy induces weight loss, a diminished amount of body fat, and increased insulin resistance. Concomitantly, ghrelin levels decline (Eden et al., 2003). Thus, several lines of evidence indicate that in nonpregnant as well as pregnant individuals, ghrelin levels decline with increasing levels of GH. With regard to ghrelin physiology, it could be speculated that pregnancy represents a state of GH hypersecretion and therefore mimics acromegaly more closely than it mimics obesity.

D. Hypertensive disorders: Preeclampsia

Ghrelin has effects on the cardiovascular system with the lower ghrelin levels associated with the higher blood pressure levels (Fagerberg et al., 2003; Poykko et al., 2003), and ghrelin infusion leads to decrease in blood pressure in healthy volunteers (Nagaya et al., 2001). GHSR1a and 1b is found in the myocardium, (Gnanapavan et al., 2002) and labeled GHS binds to the myocardium and vascular endothelium (Papotti et al., 2000).

Hypertension is a frequent complication in pregnancy, and with concomitant proteinuria eventually defining preeclampsia. One study reported higher ghrelin levels in women with hypertensive complications, giving birth to newborns with birth weights in the low-normal range (Makino et al., 2002). The mechanism for this observation is speculative. The increase in circulating ghrelin levels may be compensatory to the increasing maternal blood pressure. Alternatively, ghrelin levels may increase to act on pituitary function and GH secretion in response to a low level of placental hormones, which expectedly occurs in preeclamptic pregnancies complicated by IUGR.

Of notice, preeclampsia occurs more often in overweight subjects (Sebire et al., 2001), which in turn are expected to have the lower levels of ghrelin. Future studies will be needed to scrutinize the relationship between ghrelin and hypertensive disorders in pregnancy.

E. Direct effects of maternal ghrelin levels on the fetus

From animal studies, evidence suggests that maternal ghrelin levels affect the pregnancy product. In rodents, ghrelin treatment of pregnant females affects litter size and fetal growth. The development of preimplantation embryos in mice was inhibited *in vitro* by addition of ghrelin to culture media (Kawamura et al., 2003), a finding that was corroborated by the observed decrease in litter size in chronically ghrelin-treated female rats (Fernandez-Fernandez et al., 2005). It may be speculated that elevated ghrelin levels as expectedly seen under conditions of malnutrition or famine would reserve available energy for the mother or to a reduced number of fetuses.

In contrast, ghrelin treatment in pregnancy results in an increase in birth weights of the pups that do develop (Fernandez-Fernandez et al., 2005; Hayashida et al., 2002; Nakahara et al., 2006). This also occurs when food intake is restricted, demonstrating a more direct effect of ghrelin. Accordingly, animals passively immunized against ghrelin gave birth to pups of lower birth weight (Nakahara et al., 2006). Finally it was demonstrated that when ghrelin was injected in the mother, circulating fetal ghrelin increased within minutes, implying a rapid placental transfer of ghrelin. These findings clearly suggest an effect of maternal ghrelin on the growth of the fetus; however, evidence only derives from animal studies using pharmacological doses of ghrelin. As described below, ghrelin may not pass the placenta in humans.

F. Birth

As serum ghrelin levels in the last period of pregnancy are comparable to, or slightly lower than, the nonpregnant levels, it could be inferred that circulating ghrelin does not play a major systemic role in the last weeks of pregnancy or immediately after birth. Placental functions at birth do not appear to be dependent on ghrelin, as suggested by the nearly absent ghrelin immunostaining and GHS binding to term placentas (Gualillo et al., 2001; Papotti et al., 2000).

G. Other effects

In the maternal organism not only the pregnancy product displays rapid cell proliferation, but also many maternal tissues such as haemopoietical tissues. Ghrelin may be of importance in cell proliferation or differentiation, for example in the bone marrow or in the immune system (Hattori et al., 2001; Thompson et al., 2004).

With advancing gestational age, the uterus takes up increasing amounts of intra-abdominal space. This minimizes gastric volume capacity, and gastric reflux is a common complaint in late pregnancy. In animal studies, gastric motility and acid secretion is enhanced by ghrelin (Date et al., 2001; Masuda et al., 2000), why third trimester low ghrelin levels might not unanimously work to provide relief for reflux symptoms.

VII. Lactation

In the lactational period, the mother ships substantial amounts of energy into the breast milk for the baby. This in turn is followed by a high maternal energy intake and usually also a weight loss. Maternal appetite regulation is therefore important in the puerperium. Interestingly, ghrelin levels do not appear to imitate a state of hunger as judged from the very limited number of studies assessing ghrelin in lactation. Instead, ghrelin levels after birth are slightly lower than in nonpregnant subjects, are stable for the first days postpartum (Fuglsang et al., 2006a), and then an increase in serum levels are observed during the (initial) lactational period (Aydin et al., 2006; Makino et al., 2002). This increase may reflect the need for a sufficient caloric intake, especially when the milk volume increases, but it may also be related to the simultaneous weight loss which in the first period after birth is due to the loss of excess fluid and probably redistribution of fat mass (Dewey et al., 1993).

Additionally, ghrelin administration stimulates release of prolactin (PRL) and adrenocorticotropic hormone (ACTH) (Arosio et al., 2004; Broglio et al., 2004; Takaya et al., 2000). PRL secretion is important for lactation. The increase in ghrelin levels after birth in the breast-feeding period may therefore facilitate PRL secretion.

In rats, circulating ghrelin decreases after birth and is lower in lactating dams compared to nonlactating control rats (Shibata et al., 2004). Bromocriptine, haloperidol, and oxytocin receptor antagonists have been injected to attenuate the central regulation of lactation; however, circulating ghrelin levels were unchanged (Shibata et al., 2004). Thus, neither PRL nor oxytocin seems to be responsible for the decreased ghrelin levels during lactation. Nevertheless, ghrelin may be important in the lactational period. When ghrelin was administered repeatedly to lactating rats, milk secretion and litter weight gain increased (Nakahara et al., 2003). Interchanging pups between ghrelin treated and control dams induced an increased body weight gain in pups suckling from a ghrelin treated mother and vice versa. Dams increased their food intake and body weight and the milk yield was higher than in the control group (Nakahara et al., 2003). The presence of breast tissue GHSR mRNA in these dams indicates a direct impact of ghrelin in the breast, and GHSR mRNA has been observed in human breast tissue as well (Gnanapavan et al., 2002).

Ghrelin mRNA has been demonstrated in human breast tissue (Gnanapavan et al., 2002) and mammary epithelial cells (Kierson et al., 2006), and ghrelin is found in colostrum and milk (Aydin et al., 2006; Kierson et al., 2006). The role of ghrelin in breast milk is unknown. To exert systemic effects in the neonate, ghrelin must pass either the epithelium of the sublingual mouth or the stomach and gut, then enter the blood stream or even pass into brain tissues in the newborn—a rather complicated route, especially in light of the numerous reports suggesting ghrelin secretion in the fetus.

VIII. Fetal Ghrelin

Fetal ghrelin will be briefly discussed here: In the fetus, circulating ghrelin is detectable from week 20 to 23 (Cortelazzi et al., 2003; Ng et al., 2005). Circulating fetal ghrelin may be derived either from the fetus itself, from the placenta, or from the maternal circulation. Rodents have a placental capacity for ghrelin transport, enabling transfer of ghrelin from the mother to fetuses (Nakahara et al., 2006). In humans, the presence of placental ghrelin and ghrelin mRNA (Gualillo et al., 2001) argues against the need for transport of ghrelin from mother to fetus. Cord blood sampling have revealed discrepant results, despite most studies having samples from eight or less subjects (Cortelazzi et al., 2003; Kitamura et al., 2003; Yokota et al., 2005). We reported on the umbilical arteriovenous difference in ghrelin levels in 27 newborns, demonstrating higher levels in the umbilical arteries (Fuglsang et al., 2006a) and thereby making a fetal source of ghrelin most likely. Also, ghrelin levels in (arterial) cord blood is higher than in maternal blood (Bellone et al., 2004, 2006; Fuglsang et al., 2006a)—eventually, placental ghrelin transfer could even be in the direction fetus to mother. A positive correlation between ghrelin levels in the maternal and the fetal circulation has been found by some, but not all investigators (Bellone et al., 2004; Cortelazzi et al., 2003; Fuglsang et al., 2006a).

The fetal pancreas may be a candidate as a source for circulating fetal ghrelin (Wierup and Sundler, 2004). This is consistent with findings in rat fetuses, which demonstrated higher ghrelin contents in pancreas as compared to the stomach (Chanoine and Wong, 2004). Furthermore, an increase in fetal active ghrelin was observed after maternal fasting (Chanoine and Wong, 2004), indicating that the fetoplacental unit rapidly senses the nutritional status of the mother (Jansson and Powell, 2006). However, maternal fasting affected only the tissue ghrelin contents not the plasma levels of ghrelin (Chanoine and Wong, 2004). In rats, maternally injected ghrelin reached the fetal circulation and increased fetal cell proliferation but, however, had no impact on fetal pituitary GH mRNA expression or on circulating IGF-1 levels (Nakahara et al., 2006). Amniotic fluid has been suggested to serve as a reservoir for fetal ghrelin, at least in rodents (Nakahara et al., 2006), and this could represent a buffer capacity for fluctuating ghrelin levels in the fetus. The occurrence of ghrelin or ghrelin isoforms in amniotic fluid in human pregnancies has not been investigated.

In humans, some maternal conditions such as diabetes affect neonatal plasma ghrelin levels (Ng et al., 2004). In contrast, mode of delivery, and thereby the labour process, does not appear to affect cord blood ghrelin levels (Bellone et al., 2004; Ng et al., 2005; Onal et al., 2004). After birth, circulating ghrelin levels increase with advancing age (Bellone et al., 2006; Savino et al., 2005, 2006).

Finally, the active form of ghrelin in fetuses is unclarified. In rats, the relative levels of desacyl ghrelin were higher in the fetuses, suggesting differentiated effects of ghrelin in the mother and the fetus, using different signaling pathways (Nakahara et al., 2006).

A. Fetal ghrelin and fetal and neonatal growth

In singleton pregnancies, several studies have described an inverse association between birth weight and cord or newborn levels of ghrelin (Cortelazzi et al., 2003; Farquhar et al., 2003; Kitamura et al., 2003; Ng et al., 2004, 2005; Onal et al., 2004; Yokota et al., 2005). Some of these studies, though, suffer from insufficient information on gestational age at delivery and do not compensate for the impact of gestational age at birth. Lower birth weights may therefore be caused by lower gestational ages at birth. However, ghrelin levels in cord blood appear to increase with advancing gestational age and lower levels are found in fetuses born preterm (Bellone et al., 2004; Yokota et al., 2005), but controversy exists (Ng et al., 2005). Still, preterm birth may be caused by conditions that also have an influence on ghrelin levels, at least theoretically. Using gestational age, adjusted size at birth would be an appropriate method of comparing fetal size and circulating ghrelin levels, but this approach has mostly been used rather roughly by categorizing newborns, for example appropriate for gestational age or small for gestational age.

In rats, food-restricted pregnant dams delivered IUGR pups with increased ghrelin levels (Desai et al., 2005). Assuming an impact on appetite, it appears intuitively reasonable that fetuses subjected to intrauterine caloric deprivation attempt to compensate with a postpartum readiness for increased caloric intake. A study among human twins exhibiting discordant growth at birth demonstrated that the difference in ghrelin levels at birth was significantly associated with the difference in body weights at 1 year of age. Thus, the small-for-gestational-age twins with high ghrelin levels displayed improved catch-up growth in the first year after birth (Gohlke et al., 2005).

It has been suggested that the physiological actions of circulating ghrelin in the fetus only reache maturity in the very last weeks of pregnancy (Ng et al., 2005), and an immature ghrelin response to meals has been reported in premature newborns (Bellone et al., 2006). Breast-fed infants appear to have slightly lower serum levels of ghrelin in the first months of life compared to formula-fed infants (Savino et al., 2005) and curiously, infant weight gain among breast-fed newborns was inversely correlated to serum ghrelin levels in this study (Savino et al., 2005). Whether this finding is influenced by a different composition of breast milk and formula milk, different volumes ingested, or differences in gestational age at birth in the two groups remains unrevealed. In general, newborns with limited resources such as growth retarded and premature babies are more often exposed to formula feeding, either supplementary or as the only energy source. Formula feeding appears to be associated with higher energy intake compared to breast feeding (Cripps et al., 2005; Heinig et al., 1993).

Thus, the relationship between cord blood ghrelin levels and markers of fetal growth has not been resolved.

IX. Future Directions for Research

The regulation of ghrelin secretion appears to be complex, and the discovery of obestatin, which derives from the same gene but exerts opposite effects on food intake as compared to ghrelin (Zhang et al., 2005), makes the role for ghrelin and the ghrelin gene even more intriguing. The various isoforms of ghrelin that occur physiologically also need clarification in pregnancy as well.

Pregnancy induces an occurrence of several hormones that are not found during nonpregnant circumstances. Such hormones may interfere with the regulation of ghrelin, adipokines, GHs, and many other hormones relevant to maternal energy intake and redistribution. Through the maternal energy metabolism, effects on fetal growth may be induced, and even during the breast-feeding period, maternal energy intake may have consequences for neonatal growth (Ozanne and Hales, 2004). The Barker hypothesis renders the importance of the intrauterine milieu for later health (Cripps et al., 2005; Hales and Barker, 2001) highly probable, and future studies of ghrelin in pregnancy and lactation may therefore contribute to a greater understanding of the mechanisms underlying fetal and neonatal growth.

X. Summary

During pregnancy, ghrelin levels change dynamically with peek serum levels observed at mid-gestation. Paradoxically, an increase in ghrelin levels is seen during the first half of pregnancy. Afterward, ghrelin levels decline and in late pregnancy low levels are observed concomitantly with increasing maternal weight, increasing levels of PGH, increasing basal insulin levels, and decreasing insulin sensitivity. Thus, changes in circulating ghrelin in late pregnancy are similar to many observations in nonpregnant individuals.

After birth, low ghrelin levels are seen initially. Low ghrelin levels are seen even after normalization of insulin levels. Gradually, ghrelin levels are restored and this normalization occurs when maternal body weight normalizes. The impact of breast feeding on ghrelin secretion is unrevealed.

Ghrelin may well be involved in appetite regulation during pregnancy and breast feeding. Autocrine and paracrine effects of ghrelin in pregnancy are likely but hitherto hypothetical.

ACKNOWLEDGMENTS

The valuable criticism from Dr. Jan Frystyk, D.M.Sc., Ph.D., of the present work was highly appreciated.

REFERENCES

Arosio, M., Ronchi, C. L., Gebbia, C., Pizzinelli, S., Conte, D., Cappiello, V., Epaminonda, P., Cesana, B. M., Beck-Peccoz, P., and Peracchi, M. (2004). Ghrelin administration affects circulating pituitary and gastro-entero-pancreatic hormones in acromegaly. *Eur. J. Endocrinol.* **150,** 27–32.

Aydin, S., Aydin, S., Ozkan, Y., and Kumru, S. (2006). Ghrelin is present in human colostrum, transitional and mature milk. *Peptides* **27,** 878–882.

Baumann, G., Davila, N., Shaw, M. A., Ray, J., Liebhaber, S. A., and Cooke, N. E. (1991). Binding of human growth hormone (GH)-variant (placental GH) to GH-binding proteins in human plasma. *J. Clin. Endocrinol. Metab.* **73,** 1175–1179.

Bellone, S., Rapa, A., Vivenza, D., Vercellotti, A., Petri, A., Radetti, G., Bellone, J., Broglio, F., Ghigo, E., and Bona, G. (2004). Circulating ghrelin levels in the newborn are positively associated with gestational age. *Clin. Endocrinol. (Oxf.)* **60,** 613–617.

Bellone, S., Baldelli, R., Radetti, G., Rapa, A., Vivenza, D., Petri, A., Savastio, S., Zaffaroni, M., Broglio, F., Ghigo, E., and Bona, G. (2006). Ghrelin secretion in preterm neonates progressively increases and is refractory to the inhibitory effect of food intake. *J. Clin. Endocrinol. Metab.* **91,** 1929–1933.

Bjorklund, A. O., Adamson, U. K., Carlstrom, K. A., Hennen, G., Igout, A., Lins, P. E., and Westgren, L. M. (1998). Placental hormones during induced hypoglycaemia in pregnant women with insulin-dependent diabetes mellitus: Evidence of an active role for placenta in hormonal counter-regulation. *Br. J. Obstet. Gynaecol.* **105,** 649–655.

Bodart, V., Febbraio, M., Demers, A., McNicoll, N., Pohankova, P., Perreault, A., Sejlitz, T., Escher, E., Silverstein, R. L., Lamontagne, D., and Ong, H. (2002). CD36 mediates the cardiovascular action of growth hormone-releasing peptides in the heart. *Circ. Res.* **90,** 844–849.

Bouros, D., Tzouvelekis, A., Anevlavis, S., Doris, M., Tryfon, S., Froudarakis, M., Zournatzi, V., and Kukuvitis, A. (2006). Smoking acutely increases plasma ghrelin concentrations. *Clin. Chem.* **52,** 777–778.

Bouhours-Nouet, N., Boux, D. C., Rouleau, S., Douay, O., Mathieu, E., Bouderlique, C., Gillard, P., Limal, J. M., Descamps, P., and Coutant, R. (2006). Maternal and cord blood

ghrelin in the pregnancies of smoking mothers: Possible markers of nutrient availability for the fetus. *Horm. Res.* **66,** 6–12.
Broglio, F., Benso, A., Castiglioni, C., Gottero, C., Prodam, F., Destefanis, S., Gauna, C., van der Lely, A. J., Deghenghi, R., Bo, M., Arvat, E., and Ghigo, E. (2003a). The endocrine response to ghrelin as a function of gender in humans in young and elderly subjects. *J. Clin. Endocrinol. Metab.* **88,** 1537–1542.
Broglio, F., Gottero, C., Benso, A., Prodam, F., Destefanis, S., Gauna, C., Maccario, M., Deghenghi, R., van der Lely, A. J., and Ghigo, E. (2003b). Effects of ghrelin on the insulin and glycemic responses to glucose, arginine, or free fatty acids load in humans. *J. Clin. Endocrinol. Metab.* **88,** 4268–4272.
Broglio, F., Gottero, C., Prodam, F., Gauna, C., Muccioli, G., Papotti, M., Abribat, T., van der Lely, A. J., and Ghigo, E. (2004). Non-acylated ghrelin counteracts the metabolic but not the neuroendocrine response to acylated ghrelin in humans. *J. Clin. Endocrinol. Metab.* **89,** 3062–3065.
Broglio, F., Prodam, F., Riganti, F., Muccioli, G., and Ghigo, E. (2006). Ghrelin: From somatotrope secretion to new perspectives in the regulation of peripheral metabolic functions. *Front Horm. Res.* **35,** 102–114.
Caixas, A., Bashore, C., Nash, W., Pi-Sunyer, F., and Laferrere, B. (2002). Insulin, unlike food intake, does not suppress ghrelin in human subjects. *J. Clin. Endocrinol. Metab.* **87,** 1902.
Cappiello, V., Ronchi, C., Morpurgo, P. S., Epaminonda, P., Arosio, M., Beck-Peccoz, P., and Spada, A. (2002). Circulating ghrelin levels in basal conditions and during glucose tolerance test in acromegalic patients. *Eur. J. Endocrinol.* **147,** 189–194.
Cassoni, P., Papotti, M., Ghe, C., Catapano, F., Sapino, A., Graziani, A., Deghenghi, R., Reissmann, T., Ghigo, E., and Muccioli, G. (2001). Identification, characterization, and biological activity of specific receptors for natural (ghrelin) and synthetic growth hormone secretagogues and analogs in human breast carcinomas and cell lines. *J. Clin. Endocrinol. Metab.* **86,** 1738–1745.
Caufriez, A., Frankenne, F., Hennen, G., and Copinschi, G. (1993). Regulation of maternal IGF-I by placental GH in normal and abnormal human pregnancies. *Am. J. Physiol.* **265,** E572–E577.
Chan, J. L., Bullen, J., Lee, J. H., Yiannakouris, N., and Mantzoros, C. S. (2004). Ghrelin levels are not regulated by recombinant leptin administration and/or three days of fasting in healthy subjects. *J. Clin. Endocrinol. Metab.* **89,** 335–343.
Chanoine, J. P., and Wong, A. C. (2004). Ghrelin gene expression is markedly higher in fetal pancreas compared with fetal stomach: Effect of maternal fasting. *Endocrinology* **145,** 3813–3820.
Chellakooty, M., Skibsted, L., Skouby, S. O., Andersson, A. M., Petersen, J. H., Main, K. M., Skakkebaek, N. E., and Juul, A. (2002). Longitudinal study of serum placental GH in 455 normal pregnancies: Correlation to gestational age, fetal gender, and weight. *J. Clin. Endocrinol. Metab.* **87,** 2734–2739.
Chellakooty, M., Vangsgaard, K., Larsen, T., Scheike, T., Falck-Larsen, J., Legarth, J., Andersson, A. M., Main, K. M., Skakkebaek, N. E., and Juul, A. (2004). A longitudinal study of intrauterine growth and the placental growth hormone (GH)-insulin-like growth factor I axis in maternal circulation: Association between placental GH and fetal growth. *J. Clin. Endocrinol. Metab.* **89,** 384–391.
Chen, E. Y., Liao, Y. C., Smith, D. H., Barrera-Saldana, H. A., Gelinas, R. E., and Seeburg, P. H. (1989). The human growth hormone locus: Nucleotide sequence, biology, and evolution. *Genomics.* **4,** 479–497.
Cortelazzi, D., Cappiello, V., Morpurgo, P. S., Ronzoni, S., Nobile De Santis, M. S., Cetin, I., Beck-Peccoz, P., and Spada, A. (2003). Circulating levels of ghrelin in human fetuses. *Eur. J. Endocrinol.* **149,** 111–116.

Coutant, R., Boux, D. C., Douay, O., Mathieu, E., Rouleau, S., Beringue, F., Gillard, P., Limal, J. M., and Descamps, P. (2001). Relationships between placental GH concentration and maternal smoking, newborn gender, and maternal leptin: Possible implications for birth weight. *J. Clin. Endocrinol. Metab.* **86,** 4854–4859.

Cripps, R. L., Martin-Gronert, M. S., and Ozanne, S. E. (2005). Fetal and perinatal programming of appetite. *Clin. Sci. (Lond.)* **109,** 1–11.

Cummings, D. E., Purnell, J. Q., Frayo, R. S., Schmidova, K., Wisse, B. E., and Weigle, D. S. (2001). A preprandial rise in plasma ghrelin levels suggests a role in meal initiation in humans. *Diabetes* **50,** 1714–1719.

Date, Y., Nakazato, M., Murakami, N., Kojima, M., Kangawa, K., and Matsukura, S. (2001). Ghrelin acts in the central nervous system to stimulate gastric acid secretion. *Biochem. Biophys. Res. Commun.* **280,** 904–907.

de Zegher, F., Vanderschueren-Lodeweyckx, M., Spitz, B., Faijerson, Y., Blomberg, F., Beckers, A., Hennen, G., and Frankenne, F. (1990). Perinatal growth hormone (GH) physiology: Effect of GH-releasing factor on maternal and fetal secretion of pituitary and placental GH. *J. Clin. Endocrinol. Metab.* **71,** 520–522.

Desai, M., Gayle, D., Babu, J., and Ross, M. G. (2005). Programmed obesity in intrauterine growth-restricted newborns: Modulation by newborn nutrition. *Am. J. Physiol. Regul. Integr. Comp. Physiol.* **288,** R91–R96.

Dewey, K. G., Heinig, M. J., and Nommsen, L. A. (1993). Maternal weight-loss patterns during prolonged lactation. *Am. J. Clin. Nutr.* **58,** 162–166.

Dimaraki, E. V., Jaffe, C. A., Bowers, C. Y., Marbach, P., and Barkan, A. L. (2003). Pulsatile and nocturnal growth hormone secretions in men do not require periodic declines of somatostatin. *Am. J. Physiol. Endocrinol. Metab.* **285,** E163–E170.

Dye, J. F., Jablenska, R., Donnelly, J. L., Lawrence, L., Leach, L., Clark, P., and Firth, J. A. (2001). Phenotype of the endothelium in the human term placenta. *Placenta* **22,** 32–43.

Eden, E. B., Burman, P., Holdstock, C., and Karlsson, F. A. (2003). Effects of growth hormone (GH) on ghrelin, leptin, and adiponectin in GH-deficient patients. *J. Clin. Endocrinol. Metab.* **88,** 5193–5198.

Eriksson, L., Eden, S., Frohlander, N., Bengtsson, B. A., and Von Schoultz, B. (1988). Continuous 24-hour secretion of growth hormone during late pregnancy. A regulator of maternal metabolic adjustment? *Acta Obstet. Gynecol. Scand.* **67,** 543–547.

Eriksson, L., Frankenne, F., Eden, S., Hennen, G., and Von Schoultz, B. (1989). Growth hormone 24-h serum profiles during pregnancy—lack of pulsatility for the secretion of the placental variant. *Br. J. Obstet. Gynaecol.* **96,** 949–953.

Evain-Brion, D., Alsat, E., Mirlesse, V., Dodeur, M., Scippo, M. L., Hennen, G., and Frankenne, F. (1990). Regulation of growth hormone secretion in human trophoblastic cells in culture. *Horm. Res.* **33,** 256–259.

Fagerberg, B., Hulten, L. M., and Hulthe, J. (2003). Plasma ghrelin, body fat, insulin resistance, and smoking in clinically healthy men: The atherosclerosis and insulin resistance study. *Metabolism* **52,** 1460–1463.

Farquhar, J., Heiman, M., Wong, A. C., Wach, R., Chessex, P., and Chanoine, J. P. (2003). Elevated umbilical cord ghrelin concentrations in small for gestational age neonates. *J. Clin. Endocrinol. Metab.* **88,** 4324–4327.

Fernandez-Fernandez, R., Navarro, V. M., Barreiro, M. L., Vigo, E. M., Tovar, S., Sirotkin, A. V., Casanueva, F. F., Aguilar, E., Dieguez, C., Pinilla, L., and Tena-Sempere, M. (2005). Effects of chronic hyperghrelinemia on puberty onset and pregnancy outcome in the rat. *Endocrinology* **146,** 3018–3025.

Flanagan, D. E., Evans, M. L., Monsod, T. P., Rife, F., Heptulla, R. A., Tamborlane, W. V., and Sherwin, R. S. (2003). The influence of insulin on circulating ghrelin. *Am. J. Physiol. Endocrinol. Metab.* **284,** E313–E316.

Fuglsang, J., Lauszus, F., Flyvbjerg, A., and Ovesen, P. (2003). Human placental growth hormone, insulin-like growth factor I and -II, and insulin requirements during pregnancy in type 1 diabetes. *J. Clin. Endocrinol. Metab.* **88**, 4355–4361.

Fuglsang, J., Skjaerbaek, C., Espelund, U., Frystyk, J., Fisker, S., Flyvbjerg, A., and Ovesen, P. (2005). Ghrelin and its relationship to growth hormones during normal pregnancy. *Clin. Endocrinol. (Oxf.)* **62**, 554–559.

Fuglsang, J., Sandager, P., Moller, N., Fisker, S., Frystyk, J., and Ovesen, P. (2006a). Peripartum maternal and fetal ghrelin, growth hormones, IGFs and insulin interrelations. *Clin. Endocrinol. (Oxf.)* **64**, 502–509.

Fuglsang, J., Sandager, P., Moller, N., Fisker, S., Orskov, H., and Ovesen, P. (2006b). Kinetics and secretion of placental growth hormone around parturition. *Eur. J. Endocrinol.* **154**, 449–457.

Gaytan, F., Morales, C., Barreiro, M. L., Jeffery, P., Chopin, L. K., Herington, A. C., Casanueva, F. F., Aguilar, E., Dieguez, C., and Tena-Sempere, M. (2005). Expression of growth hormone secretagogue receptor type 1a, the functional ghrelin receptor, in human ovarian surface epithelium, mullerian duct derivatives, and ovarian tumors. *J. Clin. Endocrinol. Metab.* **90**, 1798–1804.

Ghigo, E., Broglio, F., Arvat, E., Maccario, M., Papotti, M., and Muccioli, G. (2005). Ghrelin: More than a natural GH secretagogue and/or an orexigenic factor. *Clin. Endocrinol. (Oxf.)* **62**, 1–17.

Gnanapavan, S., Kola, B., Bustin, S. A., Morris, D. G., McGee, P., Fairclough, P., Bhattacharya, S., Carpenter, R., Grossman, A. B., and Korbonits, M. (2002). The tissue distribution of the mRNA of ghrelin and subtypes of its receptor, GHS-R, in humans. *J. Clin. Endocrinol. Metab.* **87**, 2988.

Gohlke, B. C., Huber, A., Hecher, K., Fimmers, R., Bartmann, P., and Roth, C. L. (2005). Fetal insulin-like growth factor (IGF)-I, IGF-II, and ghrelin in association with birth weight and postnatal growth in monozygotic twins with discordant growth. *J. Clin. Endocrinol. Metab.* **90**, 2270–2274.

Goodman, H. M., Tai, L. R., Ray, J., Cooke, N. E., and Liebhaber, S. A. (1991). Human growth hormone variant produces insulin-like and lipolytic responses in rat adipose tissue. *Endocrinology* **129**, 1779–1783.

Gualillo, O., Caminos, J., Blanco, M., Garcia-Caballero, T., Kojima, M., Kangawa, K., Dieguez, C., and Casanueva, F. (2001). Ghrelin, a novel placental-derived hormone. *Endocrinology* **142**, 788–794.

Gualillo, O., Caminos, J. E., Nogueiras, R., Seoane, L. M., Arvat, E., Ghigo, E., Casanueva, F. F., and Dieguez, C. (2002). Effect of food restriction on ghrelin in normal-cycling female rats and in pregnancy. *Obes. Res.* **10**, 682–687.

Hales, C. N., and Barker, D. J. (2001). The thrifty phenotype hypothesis. *Br. Med. Bull.* **60**, 5–20.

Hattori, N., Saito, T., Yagyu, T., Jiang, B. H., Kitagawa, K., and Inagaki, C. (2001). GH, GH receptor, GH secretagogue receptor, and ghrelin expression in human T cells, B cells, and neutrophils. *J. Clin. Endocrinol. Metab.* **86**, 4284–4291.

Hayashida, T., Nakahara, K., Mondal, M. S., Date, Y., Nakazato, M., Kojima, M., Kangawa, K., and Murakami, N. (2002). Ghrelin in neonatal rats: Distribution in stomach and its possible role. *J. Endocrinol.* **173**, 239–245.

Heinig, M. J., Nommsen, L. A., Peerson, J. M., Lonnerdal, B., and Dewey, K. G. (1993). Energy and protein intakes of breast-fed and formula-fed infants during the first year of life and their association with growth velocity: The DARLING Study. *Am. J. Clin. Nutr.* **58**, 152–161.

Henson, M. C., and Castracane, V. D. (2000). Leptin in pregnancy. *Biol. Reprod.* **63**, 1219–1228.

Herrera, E. (2002). Lipid metabolism in pregnancy and its consequences in the fetus and newborn. *Endocrine* **19**, 43–55.

Highman, T. J., Friedman, J. E., Huston, L. P., Wong, W. W., and Catalano, P. M. (1998). Longitudinal changes in maternal serum leptin concentrations, body composition, and resting metabolic rate in pregnancy. *Am. J. Obstet. Gynecol.* **178,** 1010–1015.

Ho, K. Y., Veldhuis, J. D., Johnson, M. L., Furlanetto, R., Evans, W. S., Alberti, K. G., and Thorner, M. O. (1988). Fasting enhances growth hormone secretion and amplifies the complex rhythms of growth hormone secretion in man. *J. Clin. Invest.* **81,** 968–975.

Jansson, T., and Powell, T. L. (2006). IFPA 2005 Award in Placentology Lecture. Human placental transport in altered fetal growth: Does the placenta function as a nutrient sensor?—a review. *Placenta* **27**(Suppl. A), S91–S97.

Johnstone, L. E., and Higuchi, T. (2001). Food intake and leptin during pregnancy and lactation. *Prog. Brain Res.* **133,** 215–227.

Kawamura, K., Sato, N., Fukuda, J., Kodama, H., Kumagai, J., Tanikawa, H., Nakamura, A., Honda, Y., Sato, T., and Tanaka, T. (2003). Ghrelin inhibits the development of mouse preimplantation embryos in vitro. *Endocrinology* **144,** 2623–2633.

Kellokoski, E., Poykko, S. M., Karjalainen, A. H., Ukkola, O., Heikkinen, J., Kesaniemi, Y. A., and Horkko, S. (2005). Estrogen replacement therapy increases plasma ghrelin levels. *J. Clin. Endocrinol. Metab.* **90,** 2954–2963.

Kierson, J. A., Dimatteo, D. M., Locke, R. G., Mackley, A. B., and Spear, M. L. (2006). Ghrelin and cholecystokinin in term and preterm human breast milk. *Acta Paediatr.* **95,** 991–995.

Kitamura, S., Yokota, I., Hosoda, H., Kotani, Y., Matsuda, J., Naito, E., Ito, M., Kangawa, K., and Kuroda, Y. (2003). Ghrelin concentration in cord and neonatal blood: Relation to fetal growth and energy balance. *J. Clin. Endocrinol. Metab.* **88,** 5473–5477.

Kojima, M., Hosoda, H., Date, Y., Nakazato, M., Matsuo, H., and Kangawa, K. (1999). Ghrelin is a growth-hormone-releasing acylated peptide from stomach. *Nature* **402,** 656–660.

Lawrence, C. B., Snape, A. C., Baudoin, F. M., and Luckman, S. M. (2002). Acute central ghrelin and GH secretagogues induce feeding and activate brain appetite centers. *Endocrinology* **143,** 155–162.

Lewandowski, K., Horn, R., O'Callaghan, C. J., Dunlop, D., Medley, G. F., O'Hare, P., and Brabant, G. (1999). Free leptin, bound leptin, and soluble leptin receptor in normal and diabetic pregnancies. *J. Clin. Endocrinol. Metab.* **84,** 300–306.

Liebhaber, S. A., Urbanek, M., Ray, J., Tuan, R. S., and Cooke, N. E. (1989). Characterization and histologic localization of human growth hormone-variant gene expression in the placenta. *J. Clin. Invest.* **83,** 1985–1991.

Linnemann, K., Malek, A., Sager, R., Blum, W. F., Schneider, H., and Fusch, C. (2000). Leptin production and release in the dually *in vitro* perfused human placenta. *J. Clin. Endocrinol. Metab.* **85,** 4298–4301.

Makino, Y., Hosoda, H., Shibata, K., Makino, I., Kojima, M., Kangawa, K., and Kawarabayashi, T. (2002). Alteration of plasma ghrelin levels associated with the blood pressure in pregnancy. *Hypertension* **39,** 781–784.

Masuda, Y., Tanaka, T., Inomata, N., Ohnuma, N., Tanaka, S., Itoh, Z., Hosoda, H., Kojima, M., and Kangawa, K. (2000). Ghrelin stimulates gastric acid secretion and motility in rats. *Biochem. Biophys. Res. Commun.* **276,** 905–908.

Mazlan, M., Spence-Jones, C., Chard, T., Landon, J., and McLean, C. (1990). Circulating levels of GH-releasing hormone and GH during human pregnancy. *J. Endocrinol.* **125,** 161–167.

McIntyre, H. D., Serek, R., Crane, D. I., Veveris-Lowe, T., Parry, A., Johnson, S., Leung, K. C., Ho, K. K., Bougoussa, M., Hennen, G., Igout, A., Chan, F. Y., *et al.* (2000). Placental growth hormone (GH), GH-binding protein, and insulin-like growth factor axis in normal, growth-retarded, and diabetic pregnancies: Correlations with fetal growth. *J. Clin. Endocrinol. Metab.* **85,** 1143–1150.

Mirlesse, V., Frankenne, F., Alsat, E., Poncelet, M., Hennen, G., and Evain-Brion, D. (1993). Placental growth hormone levels in normal pregnancy and in pregnancies with intrauterine growth retardation. *Pediatr. Res.* **34,** 439–442.

Moller, N., Nygren, J., Hansen, T. K., Orskov, H., Frystyk, J., and Nair, K. S. (2003). Splanchnic release of ghrelin in humans. *J. Clin. Endocrinol. Metab.* **88,** 850–852.

Monti, V., Carlson, J. J., Hunt, S. C., and Adams, T. D. (2006). Relationship of ghrelin and leptin hormones with body mass index and waist circumference in a random sample of adults. *J. Am. Diet. Assoc.* **106,** 822–828.

Nagaya, N., Kojima, M., Uematsu, M., Yamagishi, M., Hosoda, H., Oya, H., Hayashi, Y., and Kangawa, K. (2001). Hemodynamic and hormonal effects of human ghrelin in healthy volunteers. *Am. J. Physiol. Regul. Integr. Comp. Physiol.* **280,** R1483–R1487.

Nakahara, K., Hayashida, T., Nakazato, M., Kojima, M., Hosoda, H., Kangawa, K., and Murakami, N. (2003). Effect of chronic treatments with ghrelin on milk secretion in lactating rats. *Biochem. Biophys. Res. Commun.* **303,** 751–755.

Nakahara, K., Nakagawa, M., Baba, Y., Sato, M., Toshinai, K., Date, Y., Nakazato, M., Kojima, M., Miyazato, M., Kaiya, H., Hosoda, H., Kangawa, K., *et al.* (2006). Maternal ghrelin plays an important role in rat fetal development during pregnancy. *Endocrinology* **147,** 1333–1342.

Nakazato, M., Murakami, N., Date, Y., Kojima, M., Matsuo, H., Kangawa, K., and Matsukura, S. (2001). A role for ghrelin in the central regulation of feeding. *Nature* **409,** 194–198.

Ng, P. C., Lee, C. H., Lam, C. W., Wong, E., Chan, I. H., and Fok, T. F. (2004). Plasma ghrelin and resistin concentrations are suppressed in infants of insulin-dependent diabetic mothers. *J. Clin. Endocrinol. Metab.* **89,** 5563–5568.

Ng, P. C., Lee, C. H., Lam, C. W., Chan, I. H., Wong, E., and Fok, T. F. (2005). Ghrelin in preterm and term newborns: Relation to anthropometry, leptin and insulin. *Clin. Endocrinol. (Oxf.)* **63,** 217–222.

Onal, E. E., Cinaz, P., Atalay, Y., Turkyilmaz, C., Bideci, A., Akturk, A., Okumus, N., Unal, S., Koc, E., and Ergenekon, E. (2004). Umbilical cord ghrelin concentrations in small- and appropriate-for-gestational age newborn infants: Relationship to anthropometric markers. *J. Endocrinol.* **180,** 267–271.

Ozanne, S. E., and Hales, C. N. (2004). Lifespan: Catch-up growth and obesity in male mice. *Nature* **427,** 411–412.

Palik, E., Baranyi, E., Melczer, Z., Audikovszky, M., Szocs, A., Winkler, G., and Cseh, K. (2007). Elevated serum acylated (biologically active) ghrelin and resistin levels associate with pregnancy-induced weight gain and insulin resistance. *Diabetes Res. Clin. Pract.* **76,** 351–357.

Papotti, M., Ghe, C., Cassoni, P., Catapano, F., Deghenghi, R., Ghigo, E., and Muccioli, G. (2000). Growth hormone secretagogue binding sites in peripheral human tissues. *J. Clin. Endocrinol. Metab.* **85,** 3803–3807.

Parretti, E., Mecacci, F., Papini, M., Cioni, R., Carignani, L., Mignosa, M., La Torre, P., and Mello, G. (2001). Third-trimester maternal glucose levels from diurnal profiles in nondiabetic pregnancies: Correlation with sonographic parameters of fetal growth. *Diabetes Care* **24,** 1319–1323.

Phelps, R. L., Metzger, B. E., and Freinkel, N. (1981). Carbohydrate metabolism in pregnancy. XVII. Diurnal profiles of plasma glucose, insulin, free fatty acids, triglycerides, cholesterol, and individual amino acids in late normal pregnancy. *Am. J. Obstet. Gynecol.* **140,** 730–736.

Poykko, S. M., Kellokoski, E., Horkko, S., Kauma, H., Kesaniemi, Y. A., and Ukkola, O. (2003). Low plasma ghrelin is associated with insulin resistance, hypertension, and the prevalence of type 2 diabetes. *Diabetes* **52,** 2546–2553.

Saad, M. F., Bernaba, B., Hwu, C. M., Jinagouda, S., Fahmi, S., Kogosov, E., and Boyadjian, R. (2002). Insulin regulates plasma ghrelin concentration. *J. Clin. Endocrinol. Metab.* **87,** 3997–4000.
Savino, F., Liguori, S. A., Fissore, M. F., Oggero, R., Silvestro, L., and Miniero, R. (2005). Serum ghrelin concentration and weight gain in healthy term infants in the first year of life. *J. Pediatr. Gastroenterol. Nutr.* **41,** 653–659.
Savino, F., Grassino, E. C., Fissore, M. F., Guidi, C., Liguori, S. A., Silvestro, L., Oggero, R., and Miniero, R. (2006). Ghrelin, motilin, insulin concentration in healthy infants in the first months of life: Relation to fasting time and anthropometry. *Clin. Endocrinol. (Oxf.)* **65,** 158–162.
Scippo, M. L., Frankenne, F., Hooghe-Peters, E. L., Igout, A., Velkeniers, B., and Hennen, G. (1993). Syncytiotrophoblastic localization of the human growth hormone variant mRNA in the placenta. *Mol. Cell. Endocrinol.* **92,** R7–R13.
Sebire, N. J., Jolly, M., Harris, J. P., Wadsworth, J., Joffe, M., Beard, R. W., Regan, L., and Robinson, S. (2001). Maternal obesity and pregnancy outcome: A study of 287,213 pregnancies in London. *Int. J. Obes. Relat. Metab. Disord.* **25,** 1175–1182.
Shibata, K., Hosoda, H., Kojima, M., Kangawa, K., Makino, Y., Makino, I., Kawarabayashi, T., Futagami, K., and Gomita, Y. (2004). Regulation of ghrelin secretion during pregnancy and lactation in the rat: Possible involvement of hypothalamus. *Peptides* **25,** 279–287.
Sivan, E., Homko, C. J., Chen, X., Reece, E. A., and Boden, G. (1999). Effect of insulin on fat metabolism during and after normal pregnancy. *Diabetes* **48,** 834–838.
Spellacy, W. N., Buhi, W. C., and Birk, S. A. (1978). Human placental lactogen levels in multiple pregnancies. *Obstet. Gynecol.* **52,** 210–212.
Sun, Y., Ahmed, S., and Smith, R. G. (2003). Deletion of ghrelin impairs neither growth nor appetite. *Mol. Cell. Biol.* **23,** 7973–7981.
Takaya, K., Ariyasu, H., Kanamoto, N., Iwakura, H., Yoshimoto, A., Harada, M., Mori, K., Komatsu, Y., Usui, T., Shimatsu, A., Ogawa, Y., Hosoda, K., *et al.* (2000). Ghrelin strongly stimulates growth hormone release in humans. *J. Clin. Endocrinol. Metab.* **85,** 4908–4911.
Tanaka, K., Minoura, H., Isobe, T., Yonaha, H., Kawato, H., Wang, D. F., Yoshida, T., Kojima, M., Kangawa, K., and Toyoda, N. (2003). Ghrelin is involved in the decidualization of human endometrial stromal cells. *J. Clin. Endocrinol. Metab.* **88,** 2335–2340.
Tannenbaum, G. S., Epelbaum, J., and Bowers, C. Y. (2003). Interrelationship between the novel peptide ghrelin and somatostatin/growth hormone-releasing hormone in regulation of pulsatile growth hormone secretion. *Endocrinology* **144,** 967–974.
Thompson, N. M., Gill, D. A., Davies, R., Loveridge, N., Houston, P. A., Robinson, I. C., and Wells, T. (2004). Ghrelin and des-octanoyl ghrelin promote adipogenesis directly *in vivo* by a mechanism independent of the type 1a growth hormone secretagogue receptor. *Endocrinology* **145,** 234–242.
Tschop, M., Smiley, D. L., and Heiman, M. L. (2000). Ghrelin induces adiposity in rodents. *Nature* **407,** 908–913.
Veldhuis, J. D., Iranmanesh, A., Ho, K. K., Waters, M. J., Johnson, M. L., and Lizarralde, G. (1991). Dual defects in pulsatile growth hormone secretion and clearance subserve the hyposomatotropism of obesity in man. *J. Clin. Endocrinol. Metab.* **72,** 51–59.
Villar, J., Cogswell, M., Kestler, E., Castillo, P., Menendez, R., and Repke, J. T. (1992). Effect of fat and fat-free mass deposition during pregnancy on birth weight. *Am. J. Obstet. Gynecol.* **167,** 1344–1352.
Volante, M., Allia, E., Gugliotta, P., Funaro, A., Broglio, F., Deghenghi, R., Muccioli, G., Ghigo, E., and Papotti, M. (2002). Expression of ghrelin and of the GH secretagogue receptor by pancreatic islet cells and related endocrine tumors. *J. Clin. Endocrinol. Metab.* **87,** 1300–1308.

Wierup, N., and Sundler, F. (2004). Circulating levels of ghrelin in human fetuses. *Eur. J. Endocrinol.* **150,** 405.

Wren, A. M., Seal, L. J., Cohen, M. A., Brynes, A. E., Frost, G. S., Murphy, K. G., Dhillo, W. S., Ghatei, M. A., and Bloom, S. R. (2001). Ghrelin enhances appetite and increases food intake in humans. *J. Clin. Endocrinol. Metab.* **86,** 5992.

Yildiz, B. O., Suchard, M. A., Wong, M. L., McCann, S. M., and Licinio, J. (2004). Alterations in the dynamics of circulating ghrelin, adiponectin, and leptin in human obesity. *Proc. Natl. Acad. Sci. USA* **101,** 10434–10439.

Yogev, Y., Ben Haroush, A., Chen, R., Rosenn, B., Hod, M., and Langer, O. (2004). Diurnal glycemic profile in obese and normal weight nondiabetic pregnant women. *Am. J. Obstet. Gynecol.* **191,** 949–953.

Yokota, I., Kitamura, S., Hosoda, H., Kotani, Y., and Kangawa, K. (2005). Concentration of the n-octanoylated active form of ghrelin in fetal and neonatal circulation. *Endocr. J.* **52,** 271–276.

Zhang, J. V., Ren, P. G., Avsian-Kretchmer, O., Luo, C. W., Rauch, R., Klein, C., and Hsueh, A. J. (2005). Obestatin, a peptide encoded by the ghrelin gene, opposes ghrelin's effects on food intake. *Science* **310,** 996–999.

CHAPTER TWELVE

Ghrelin and Reproduction: Ghrelin as Novel Regulator of the Gonadotropic Axis

Manuel Tena-Sempere*

Contents

I.	Introduction: Ghrelin Is a Multifunctional Regulator with Key Roles in Energy Balance	286
II.	Neuroendocrine Control of Reproduction: The Gonadotropic Axis	287
III.	Reproduction and the Energy Status Are Functionally Linked	288
IV.	Ghrelin as Putative Regulator of the Gonadotropic Axis	289
V.	Role of Ghrelin in the Control of Gonadotropin Secretion	289
VI.	Putative Roles of Ghrelin in Puberty Onset	291
VII.	Molecular Diversity of Ghrelin: Reproductive Effects of UAG and Obestatin	292
VIII.	Expression and Direct Actions of Ghrelin in the Gonads	293
IX.	Futures Perspectives and Conclusions	295
	Acknowledgments	297
	References	298

Abstract

Identification of ghrelin in late 1999, as the endogenous ligand of the growth hormone secretagogue receptor (GHSR), opened up a new era in our understanding of the regulatory mechanisms of several neuroendocrine systems, including growth and energy homeostasis. Based on similarities with other endocrine integrators and its proposed role as signal for energy insufficiency, it appeared tempting to hypothesize that ghrelin might also operate as regulator of reproductive function. Yet, contrary to other of its biological actions the reproductive "dimension" of ghrelin has remained largely unexplored. Nonetheless, experimental evidence, coming mostly from animal studies, have been gathered during the last years suggesting that ghrelin may actually function as a metabolic modulator of the gonadotropic axis, with predominant inhibitory

---

* Physiology Section, Department of Cell Biology, Physiology and Immunology, University of Córdoba, 14004 Córdoba, Spain

Vitamins and Hormones, Volume 77   © 2008 Elsevier Inc.
ISSN 0083-6729, DOI: 10.1016/S0083-6729(06)77012-1   All rights reserved.

effects in line with its role as signal of energy deficit. These effects likely include inhibition of luteinizing hormone (LH) secretion (which has been reported in different species and developmental stages), as well as partial suppression of normal puberty onset. In addition, expression and/or direct gonadal actions of ghrelin have been reported in the human, rat, and chicken. Altogether, those findings document a novel reproductive facet of ghrelin, which may cooperate with other neuroendocrine integrators, as leptin, in the joint control of energy balance and reproduction. © 2008 Elsevier Inc.

Abbreviations

FSH, follicle-stimulating hormone; GHSR, growth hormone secretagogue receptor; GnRH, gonadotropin-releasing hormone; LH, luteinizing hormone.

I. Introduction: Ghrelin Is a Multifunctional Regulator with Key Roles in Energy Balance

Identification of ghrelin by reverse pharmacology in late 1999 was the endpoint of the long search for the endogenous natural ligand of the growth hormone secretagogue (GHS) receptor (GHSR) (Kojima *et al.*, 1999), and a major breakthrough in contemporary neuroendocrinology. As direct proof of the extraordinary interest drawn by this molecule, more than 1850 research articles have been published on this topic over the last 7 years. The major structural and functional features of ghrelin have been extensively reviewed in other chapters of this volume of Vitamins and Hormones. For the purpose of this chapter, it is important to stress as unique structural attribute the addition of an *n*-octanoyl group at Ser3 of mature ghrelin (acylation); a posttranslational modification that was the first of this type described in a secreted molecule and essential for ghrelin to bind to its receptor (the type 1a GHSR) and to stimulate GH secretion (Gualillo *et al.*, 2003; Kojima *et al.*, 1999; van der Lely *et al.*, 2004). Interestingly, due to its inability to bind to the canonical ghrelin receptor, the unacylated form of ghrelin (UAG) was originally considered biologically inactive. Yet, evidence suggest that UAG, whose levels largely exceed those of acylated ghrelin, is provided with a wide spectrum of biological actions, which are either similar or distinct to those of the acylated molecule (van der Lely *et al.*, 2004). As additional level of complexity, generation of a novel peptide, distinct from mature ghrelin, has been reported by alternative processing of ghrelin precursor (Zhang *et al.*, 2005). This has been termed obestatin and has been proposed to operate as antagonist for some of ghrelin functions (Section VII).

From a functional standpoint, a fascinating characteristic of ghrelin is its ubiquitous expression and functional diversity. Indeed, while circulating ghrelin is mostly secreted by the stomach, expression of ghrelin has been reported in a large array of tissues and cell types including small intestine, pancreas, lymphocytes, placenta, kidney, lung, pituitary, brain, and the gonads (Barreiro and Tena-Sempere, 2004; Gualillo *et al.*, 2003; van der Lely *et al.*, 2004). Moreover, although ghrelin was initially regarded solely as the endogenous counterpart of GHS, with ability to stimulate GH secretion (Kojima *et al.*, 1999), it soon became evident that the biological actions of ghrelin are much more diverse than those originally described, including both endocrine and nonendocrine effects. These range from the regulation of diverse neuroendocrine axes to the local control of cell proliferation. By far, one of the functional features of ghrelin that has attracted more attention is its ability to stimulate food intake and to promote body weight gain. In fact, as orexigenic signal, expression of ghrelin has been shown to increase after food deprivation, and its plasma levels are (in most cases) negatively correlated with the body mass index. On this basis, ghrelin has been proposed as a unique circulating signal for energy insufficiency (the only known circulating orexigen), which may play a major role in the short- and long-term control of body weight (Cummings, 2006; Zigman and Elmquist, 2003).

II. NEUROENDOCRINE CONTROL OF REPRODUCTION: THE GONADOTROPIC AXIS

Reproductive capacity, defined by the ability to generate fertilizable gametes and to sustain pregnancy and lactation, critically depends on a concerted series of developmental events and the coordinated function of different endocrine elements which compose the so-called hypothalamic-pituitary-gonadal (HPG) or gonadotropic axis. Three major groups of hormonal factors can be identified in this system: the hypothalamic decapeptide gonadotropin-releasing hormone (GnRH), the pituitary gonadotropins, luteinizing hormone (LH) and follicle-stimulating hormone (FSH), and sex steroids and peptide hormones from the gonads (Tena-Sempere and Huhtaniemi, 2003). The function of the HPG axis, and hence of the gonads, primarily depends on the interaction of these three major groups of signals, through positive and negative feedforward and feedback regulatory loops. In addition, a plethora of central and peripheral signals, of stimulatory and inhibitory nature, are involved in the modulation of the function of the gonadotropic axis by changes in internal conditions and external cues. Considering the central position of GnRH in the hierarchical control of the HPG axis, many of those regulators are known to target hypothalamic GnRH neurons to regulate fertility. However, despite significant progress in the field, the effectors and mechanisms for the precise control of the hypothalamic GnRH pulse generator are yet to be fully characterized.

III. Reproduction and the Energy Status Are Functionally Linked

The capacities to maintain the state of energy reserves and to reproduce are pivotal factors for the survival of individuals and species. Accordingly, these functions are under the precise control of different regulatory networks, which are partially overlapping. Indeed, on the basis of intuitive knowledge, the contention that reproduction is metabolically gated, especially in the female, had been well settled since old ages. Thus, fecundity was often symbolized by states of energy abundance (such as mild obesity), whereas situations of persistent energy deficit, such as starvation or extreme physical exercise, were known to be coupled to impaired reproductive capacity. Extensive experimental and epidemiological work conducted in 1960s–1970s led to the formulation of the so-called critical (fat) mass hypothesis, which defined the need of a certain degree of adiposity (i.e., energy stores) in order to proceed into puberty and to maintain proper reproductive function in adulthood (Frisch and McArthur, 1974; Frisch and Revelle, 1970). Despite such a scientific formulation for the link between fuel reserves and reproductive function, the neuroendocrine and molecular basis for such a phenomenon remained largely unknown until mid-1990s, when cloning of leptin paved the way for the deciphering of the signals and mechanisms responsible for the endocrine control of food intake and energy balance (Ahima et al., 2000; Casanueva and Dieguez, 1999; Zhang et al., 1994).

The biological properties of leptin in terms of structure, regulation, effects, and mechanisms of action have been extensively revised elsewhere (Ahima et al., 2000; Casanueva and Dieguez, 1999; Wauters et al., 2000). In terms of energy homeostasis, the most salient features are that of leptin, which is primarily produced by the white adipose tissue, is secreted in proportion to the amount of body energy (fat) stores, and functions as satiety factor in the regulation of body weight (Ahima et al., 2000; Casanueva and Dieguez, 1999). From a more general perspective, it is also remarkable that leptin operates as modulator of a wide array of biological functions (other than energy balance), which include the regulation of different neuroendocrine axes. Thus, leptin was considered as genuine neuroendocrine integrator, linking the control of different essential biological functions and hormonal systems to the state of energy reserves of the organism (Wauters et al., 2000). In this context, it soon became evident that leptin plays an important role in the control of the central networks governing reproduction; leptin being an essential permissive factor for puberty onset and normal functioning of the HPG axis (Ahima et al., 2000; Casanueva and Dieguez, 1999; Rosenbaum and Leibel, 1998). These reproductive actions of leptin provided a mechanistic explanation for the state of hypogonadism observed in conditions of energy insufficiency (when leptin levels are suppressed). Nevertheless, considering the complexity and redundancy of the signaling systems controlling body weight and reproduction, it was predictable that other neuroendocrine integrators might cooperate with leptin in the joint regulation of energy balance and fertility (Fernandez-Fernandez *et al.*, 2006).

IV. Ghrelin as Putative Regulator of the Gonadotropic Axis

Following the prototypic case of leptin, different research groups, including ours (Fernandez-Fernandez *et al.*, 2006), have searched for potential neuroendocrine integrators involved in the joint control of energy homeostasis and reproduction. Initial characterization of the biological profile of ghrelin made it a suitable candidate for such a function. Thus, ghrelin appeared to be involved not only in the regulation of growth but also energy homeostasis; ghrelin being an orexigenic signal promoting body weight gain (Gualillo *et al.*, 2003; van der Lely *et al.*, 2004). Moreover, as described in Section I, ghrelin levels were found to inversely correlate (in most cases) with body mass index. These observations led to the proposal that ghrelin operates as peripheral signal for energy insufficiency, whose levels rise in conditions of negative energy balance in order to activate homeostatic responses in terms of increased food intake and decreased spontaneous locomotor activity (Tang-Christensen *et al.*, 2004; Zigman and Elmquist, 2003). Interestingly based on its biological effects and mechanisms of action, ghrelin has been proposed as functional antagonist of the effects of leptin on energy balance; ghrelin and leptin playing a *yin and yang* role in the control of food intake and body weight. We hypothesized that such a dynamic interaction might also involve the regulation of reproductive function, where ghrelin could contribute, in conjunction with other metabolic signals, to the physiological coupling of reproduction and the state of energy reserves. Experimental data from different groups, which are reviewed in detail in the following sections, strongly suggest that, indeed, ghrelin may function as modulator of different reproductive parameters such as gonadotropin secretion, puberty onset, and gonadal function.

V. Role of Ghrelin in the Control of Gonadotropin Secretion

Among its potential actions on the reproductive axis, the effects of ghrelin on gonadotropin secretion have been now explored in different species. Overall, the studies so far conducted on the rat, rhesus monkey, and sheep have demonstrated that central administration of ghrelin is able to suppress different aspects of pulsatile LH secretion (Fernandez-Fernandez *et al.*, 2004; Furuta *et al.*, 2001; Iqbal *et al.*, 2006; Vulliémoz *et al.*, 2004). Thus, central injection of ghrelin reduced LH pulse frequency, without major changes in pulse amplitude, in ovariectomized rats and rhesus monkeys (Furuta *et al.*, 2001; Vulliémoz *et al.*, 2004), whereas ghrelin administration to intact rats and sheep resulted also in reduction of basal LH levels (Fernandez-Fernandez *et al.*, 2004; Iqbal *et al.*, 2006). In good agreement, ghrelin administration has been reported to suppress LH pulsatility (both in terms of pulse height and frequency) as well as LH baseline in healthy humans (Lanfranco *et al.*, 2006).

The predominant inhibitory effects of ghrelin on LH secretion *in vivo* have been characterized in detail by our group in the rat, with studies in both males and females, at different phases of development and functional states of the gonadotropic axis. Central administration of ghrelin inhibited LH secretion in intact prepubertal males, as well as in gonadectomized male and female rats (Fernandez-Fernandez *et al.*, 2004). Likewise, intracerebral injection of ghrelin decreased pulsatile LH secretion in freely moving adult male rats (Martini *et al.*, 2006), and partially suppressed LH release in adult cyclic females (across the estrous cycle) and ovariectomized rats (Fernandez-Fernandez *et al.*, 2005a). Moreover, chronic intravenous infusion of ghrelin for 7 days induced a significant decrease in circulating LH levels (Martini *et al.*, 2006). Altogether, these observations substantiate a major inhibitory role of ghrelin in the control of gonadotropin secretion, which is likely to manifest in conditions of hyperghrelinemia, such as those of energy deficit (Fernandez-Fernandez *et al.*, 2006). Of note, ghrelin was also able to decrease GnRH release *ex vivo* by hypothalamic explants from ovariectomized female rats (Fernandez-Fernandez *et al.*, 2005a). This observation, together with its reported effects on LH pulsatilty, strongly suggests a major central (hypothalamic) site of action for the inhibitory effects of ghrelin on the gonadotropic axis.

Noteworthy, the actions of ghrelin on FSH secretion *in vivo* remain less well characterized. In general, the effects of ghrelin on FSH release have been either not reported (in some of the studies; see Furuta *et al.*, 2001; Iqbal *et al.*, 2006; Vulliémoz *et al.*, 2004) or were mostly undetectable (Fernandez-Fernandez *et al.*, 2004, 2005a; Lanfranco *et al.*, 2006). In our studies on the rat, central administration of ghrelin to cyclic female rats evoked marginal inhibitory responses only at estrus, whereas in the adult males significant inhibition of FSH levels was only detected after infusion of high doses (1 nmol/h for 7 days) of ghrelin (Martini *et al.*, 2006). The above dissociation in LH and FSH responses to ghrelin might reflect the differences in the profiles of secretion and regulation between gonadotropins; FSH being more constitutively secreted than LH.

In addition to central (hypothalamic) effects, another potential site for the actions of ghrelin on the gonadotropic axis is the pituitary, where abundant expression of its functional receptor has been reported (Gualillo *et al.*, 2003; van der Lely *et al.*, 2004). Interestingly, in our studies using rat pituitary explants, ghrelin seems to conduct dual, apparently opposing, actions in the direct control of gonadotropin secretion. Thus, in line with its reported inhibitory actions after *in vivo* administration (see above), ghrelin significantly decreased GnRH-induced LH release *ex vivo* by pituitaries from prepubertal animals and adult cyclic female rats, at different stages of the estrous cycle (Fernandez-Fernandez *et al.*, 2004, 2005a). However in our settings, ghrelin was also able to evoke clear-cut stimulatory responses in terms of LH and FSH secretion, both in prepubertal and adult male and female rats (Fernandez-Fernandez *et al.*, 2004, 2005a). The physiological significance and mechanisms for this antithetical mode of action remain to be defined. Nonetheless, it is tempting to speculate that such combination of inhibitory and stimulatory effects might derive from the differential roles of systemically derived and locally produced ghrelin in the regulation of gonadotropin secretion. In this sense, in the rat, the stimulatory effects of ghrelin on pituitary LH release are evident only at high concentrations (micromolar range). Yet, ghrelin expression has been reported at the pituitary (Caminos *et al.*, 2003a), thus allowing achievement of high levels locally. In any event, when results from *in vivo* (central and systemic administration) and *ex vivo* (direct pituitary effects) experiments are compared, it becomes evident that the predominant role of ghrelin in the control of the gonadotropic axis is an inhibitory action that is likely conducted at central levels and could be reinforced by the ability of ghrelin to suppress GnRH-induced LH release at the pituitary.

VI. Putative Roles of Ghrelin in Puberty Onset

Other facet of ghrelin actions on the gonadotropic system that has been experimentally explored is its ability to modulate the activation of the HPG axis at puberty. In this sense, it is well known that puberty is exquisitely sensitive to the state of energy reserves, situations of energy insufficiency being associated to the delay or absence of puberty (Fernandez-Fernandez *et al.*, 2006). Based on the proposed role of ghrelin as signal of energy insufficiency, involved the long-term control of energy balance, and the nature of the parameter under analysis (i.e., pubertal maturation, which takes place as continuum from immature stages up to adulthood), protocols of repeated injection of ghrelin were implemented in male and female rats along puberty, and different indices of activation of the reproductive axis (such as gonadotropin and sex steroid levels, as well as external markers—vaginal opening in the female and balanopreputial separation in the male) were monitored (Fernandez-Fernandez *et al.*, 2005b).

Repeated administration of ghrelin, at a dose of 0.5 nmol/12 h for 7 days during pubertal transition, significantly decreased serum LH and testosterone levels, and partially delayed balanopreputial separation (as external index of puberty onset) in male rats (Fernandez-Fernandez *et al.*, 2005b); observations that have been confirmed by another study from our group using higher doses of ghrelin (Martini *et al.*, 2006). In contrast, a similar protocol of repeated injections of ghrelin to peripubertal females did not induce major changes in serum levels of gonadotropins or estradiol, nor did it significantly alter the timing of puberty, as estimated by the ages of vaginal opening and first estrus (Fernandez-Fernandez *et al.*, 2005b). Two major conclusions can be drawn from these studies. First, persistently elevated ghrelin levels, as putative signal for energy insufficiency, are not only able to inhibit LH secretion but also to impair the normal timing of puberty, as evidenced by different hormonal and phenotypic biomarkers. Second, the female rat is apparently less sensitive than the male to the effects of ghrelin at puberty; a phenomenon that is opposite to that described for leptin, which appears more relevant for the permissive control of female puberty (Tena-Sempere and Barreiro, 2002). Nonetheless, the female rat seems to be also sensitive to the inhibitory effects of high doses of ghrelin at puberty, as evidenced by our recent data showing that twice daily administration of 1-nmol ghrelin for 7 days was able to delay vaginal opening in pubertal female rats (M.T.-S., manuscript in preparation).

VII. Molecular Diversity of Ghrelin: Reproductive Effects of UAG and Obestatin

One of the most intriguing facets of ghrelin physiology is related with the molecular diversity of the elements of this signaling system, which applies both to the ligand and its receptor. Concerning the ligands, processing of ghrelin precursor results in the generation of acylated and unacylated (UAG) forms of ghrelin (Gualillo *et al.*, 2006; see also Section I); UAG being far more abundant in the circulation than the acylated molecule. In addition, the differential posttranslational processing of proghrelin was reported to generate a distinct 23-amino acid peptide, termed obestatin (Gualillo *et al.*, 2006; Zhang *et al.*, 2005). In terms of function, ghrelin was cloned on the basis of its ability to activate the canonical GHSR type 1a, leading to the hypothesis that the biological actions of ghrelin are solely mediated by this receptor. In contrast, UAG was shown to be unable to bind and activate GHSR1a; thus, it was initially regarded as biologically inert (van der Lely *et al.*, 2004). Both contentions have been refuted by pharmacological evidence, showing that some of the effects of ghrelin are likely mediated via receptors other than the classical GHSR1a (van der Lely *et al.*, 2004), and that UAG is provided with a wide array of biological actions, which are not always similar to those of the acylated molecule (Gauna *et al.*, 2006; Martini *et al.*, 2006; Toshinai *et al.*, 2006). Concerning obestatin, initial reports described its ability to bind GPR39 and to oppose to ghrelin effects on food intake; obestatin being a potent inhibitor of appetite and body weight gain (Zhang *et al.*, 2005). However, the capacity of obestatin to decrease food intake, and even to bind GPR39, has been questioned (Holst *et al.*, 2006; Nogueiras *et al.*, 2006). In any event, the fact that a common prohormone might generate different ligands, acting through different receptors, provides an additional level of sophistication to this system that merits further investigation.

In the context of our ongoing studies on the characterization of the effects of acylated ghrelin (Sections V and VI), we found it relevant to comparatively evaluate the roles, if any, of other proghrelin-derived peptides in the control of the gonadotropic axis. Concerning UAG, our studies involving acute administration to adult male rats in freely moving conditions, as well as chronic infusions, demonstrated that UAG is able to mimic the inhibitory effects of acylated ghrelin on LH secretion (Martini *et al.*, 2006). Likewise, repeated administration of UAG was as effective as acylated ghrelin in inducing a partial suppression of the activation of the gonadotropic axis at puberty, as evidenced by similar decreases in serum LH levels and percentages of balanopreputial separation in ghrelin-treated groups (Martini *et al.*, 2006). To our knowledge, these data are the first to demonstrate the ability of UAG to mimic the effects of ghrelin on a "classical" neuroendocrine axis, and strongly suggest the contribution of mechanisms independent of the classical type 1a GHS receptor in the observed effects of ghrelin on LH secretion and puberty onset.

As has been described for UAG, we have recently initiated also the analysis of potential effects of obestatin on the reproductive axis, by means of acute or repeated administration of the peptide in male and female rats, at different stages of development. Although some of these studies are still in progress, the experimental data obtained so far cast doubts on any significant function of obestatin in the modulation of key aspects of the gonadotropic axis in the rat such as LH and FSH secretion and timing of puberty (M.T.-S., manuscript in preparation). This would imply that the reproductive roles of the ghrelin system, at least in terms of central gonadotropic regulation, are apparently conducted via ghrelin peptides, in both acylated and unacylated forms. The relative contribution of those to the metabolic control of reproduction awaits to be elucidated.

VIII. Expression and Direct Actions of Ghrelin in the Gonads

Besides central actions on the reproductive axis, several lines of evidence indicate that ghrelin is also expressed and conducts specific biological effects directly at the gonadal level; a phenomenon which illustrates the multifaceted mode of action of ghrelin in the control of the gonadotropic axis. Detailed description of those facets of ghrelin physiology can be found elsewhere (Barreiro and Tena-Sempere, 2004; Tena-Sempere, 2005). For the purpose of the present chapter, it is important to stress that expression of ghrelin has been documented in the rat and human testis (Barreiro et al., 2002; Gaytan et al., 2004; Gnanapavan et al., 2002; Tena-Sempere et al., 2002). Likewise, ovarian expression of ghrelin has been also reported in the rat, chicken, and human (Caminos et al., 2003b; Gaytan et al., 2003; Sirotkin et al., 2006). In addition, expression of the canonical ghrelin receptor, GHSR type 1a, has been described (at the mRNA and/or protein level) in the rat and human testis (Barreiro et al., 2003; Gaytan et al., 2004; Tena-Sempere et al., 2002), as well as in the human and chick ovary (Gaytan et al., 2003, 2005; Sirotkin et al., 2006). These observations strongly suggest that the gonads are primary sources of ghrelin expression and that ghrelin, either systemically derived or locally produced, might conduct specific, direct actions at the gonadal level. The latter was reinforced by the observation that both ghrelin and GHSR1a mRNAs in rat testis are precisely regulated by hormonal signals, which include pituitary LH (major regulator of ghrelin), FSH, and ghrelin itself (major regulators of GHSR expression) (for a review see Tena-Sempere, 2005).

To shed further light on the potential direct gonadal actions of ghrelin, we used the rat testis as model, and measurements of different indices of testicular function (as testosterone secretion, seminiferous tubule gene expression, and cell proliferation) after ghrelin stimulation in different *in vivo* and *ex vivo* settings. Results from those experiments evidenced the ability of ghrelin to conduct modulatory actions on both the interstitial and tubular compartments of the testis. On the former, ghrelin was able to inhibit, in a dose-dependent manner, stimulated testosterone secretion by Leydig cells *ex vivo*, as well as to decrease the expression levels of several key factors of the steroidogenic route such as StAR, P450scc, 3β-HSD, and testis-specific 17β-HSD type III (Tena-Sempere et al., 2002). In addition, intratesticular injection of ghrelin *in vivo* induced a significant decrease in the proliferative activity of immature Leydig cells, both during pubertal development and after selective ablation of preexisting mature Leydig cells by administration of the cytotoxic compound ethylene dimethane sulfonate (Barreiro et al., 2004). Overall, it is tempting to speculate that ghrelin, whose testicular expression is selectively restricted to Leydig cells (Barreiro et al., 2002; Tena-Sempere et al., 2002), operates as local modulator of essential functions of this cell type, as steroidogenesis and proliferation.

Besides its effects on Leydig cells, expression of GHSR1a in the tubular compartment of the testis (Barreiro et al., 2003; Gaytan et al., 2004) strongly suggested the ability of ghrelin to modulate seminiferous tubule functions. In this sense, our analyses using staged fragments of seminiferous tubules demonstrated a finely regulated pattern of expression of GHSR1a mRNA along the seminiferous epithelial cycle, which is suggestive of a precise control of ghrelin signaling in the tubular compartment of the testis (Barreiro et al., 2003). Moreover by a combination of *in vivo* and *in vitro* approaches, we demonstrated that ghrelin is able to inhibit the expression of the Sertoli cell product stem cell factor (SCF), which is the major paracrine stimulator of germ cell development (Barreiro et al., 2004). Of note, the ability of ghrelin to inhibit SCF mRNA might also have mechanistic implications for its reported effects on Leydig cell proliferation, as SCF has been involved in the control of development and survival of this cell population (Yan et al., 2000). Overall, our functional data strongly suggest that, in addition to central neuroendocrine actions on the centers governing gonadotropin secretion, ghrelin conducts also direct regulatory effects at the testicular level, involving key aspects of testis physiology such as steroidogenesis, Leydig cell proliferation, and tubular functions.

IX. Futures Perspectives and Conclusions

Since its cloning 7 years ago, ghrelin has evolved from the endogenous counterpart of synthetic GH secretagogues into a fundamental pleiotropic regulator of different biological functions, including food intake and energy homeostasis. Based on its biological profile, as signal of energy insufficiency whose circulating levels negative correlate with body mass index, and following the prototypic example of leptin, it was hypothesized that ghrelin could operate also as modulator of some facets of reproductive function. The experimental evidence gathered during the last years supports this hypothesis, as ghrelin expression has been identified at all levels of the HPG axis, and effects of ghrelin on gonadotropin secretion, puberty onset, and gonadal function have been described. Concerning the central control of the gonadotropic axis, the reported effects of ghrelin are mostly inhibitory, in opposition to those described previously for leptin that function as permissive/positive signal at central levels of the reproductive axis (Casanueva and Dieguez, 1999; Tena-Sempere and Barreiro, 2002). Considering the proposed mutual antagonism between leptin and ghrelin in the long-term control of energy balance (Zigman and Elmquist, 2003), it is tempting to speculate that these two factors operate also, in a tightly coupled, reciprocal manner, in the regulation of reproductive function and its modulation by metabolic cues.

Notwithstanding, the analysis of the reproductive dimension of ghrelin is still quite incomplete. Thus, while inhibitory effects of ghrelin on LH secretion have been reported in a number of species, including the human (as preliminarily reported recently; see Lanfranco et al., 2006), the neuroendocrine circuitries for such a phenomenon remains mostly unexplored. Thus, initial evidence suggested the ability of ghrelin to inhibit GnRH

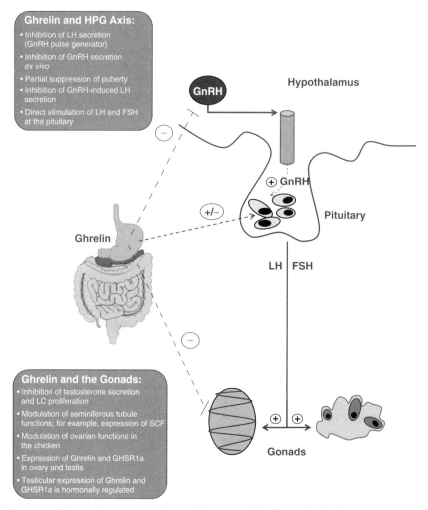

Figure 1 Schematic representation of the potential effects of ghrelin in the control of the reproductive axis. Ghrelin, a hormone secreted by the stomach as signal of energy insufficiency, conducts specific regulatory effects at different levels of the hypothalamic-pituitary-gonadal (HPG) axis, which include the inhibition of pulsatile secretion of luteinizing hormone (LH), the suppression of gonadotropin-releasing hormone (GnRH) secretion, and the reduction of LH responsiveness to GnRH at the pituitary level. In addition, ghrelin has been reported to partially blunt the activation of the reproductive axis at (male) puberty. However, stimulatory effects of ghrelin on basal LH and FSH release directly at the pituitary level have been also described. Besides those central actions, systemically derived ghrelin might directly operate at the gonadal level, as expression of canonical ghrelin receptors has been described in the ovary and the testis. Moreover, ghrelin is also locally produced in the gonads, where it may conduct additional regulatory actions, such as inhibition of Leydig cell (LC) proliferation and testosterone secretion, as well as regulation of tubular functions, such as expression of stem cell factor (SCF) gene. In addition, specific regulatory actions of ghrelin on relevant ovarian functions have been recently described in the chicken. Altogether, these observations allow us to propose that ghrelin is a pleiotropic modulator of the gonadotropic system, which might contribute to the physiological coupling of reproductive function to the state of energy reserves of the organism. (See Color Insert.)

secretion by hypothalamic explants from ovariectomized rats *ex vivo*; yet, whether this effect is directly conducted on GnRH neurons or is mediated via regulatory networks controlling GnRH secretion has not been elucidated. In this sense, it has been proposed that hypothalamic KiSS-1 neurons are essential gatekeepers of GnRH secretion (Tena-Sempere, 2006), which makes it worthy to explore the potential interaction between ghrelin and KiSS-1 at central levels. As preliminary finding, we have observed that ghrelin might moderately suppress the potent LH-releasing activity of kisspeptin in male rats (Martini *et al.*, 2006). In addition, the effects of ghrelin on gonadotropic function at certain relevant states of the reproductive axis, such as the preovulatory surge of gonadotropins, are yet to be fully evaluated.

Another reproductive facet of ghrelin that merits further investigation is its role in the direct control of gonadal function, especially at the ovary. Thus, while several studies have investigated ghrelin effects in the testis, its function, if any, in the direct control of mammalian ovary remains totally unexplored. Studies from our group, however, have demonstrated that functional ghrelin receptors are present in the chicken ovary, where ghrelin gene is also expressed (Sirotkin *et al.*, 2006). Moreover, ghrelin has been shown to modulate essential ovarian functions in this species such as cell proliferation, apoptosis, and hormone (steroids and peptides) secretion. Similar functional analyses in the rat and/or human are eagerly needed in order to define the potential role of ghrelin in the control of female gonadal function in mammals.

In summary, the data reviewed herein on the effects of ghrelin at different levels of the gonadotropic axis, in different species and at different developmental and functional states of the system, support the contention that ghrelin is a putative modulator of relevant aspects of reproductive physiology such as puberty onset, gonadotropin secretion, and gonadal function. Some of these "novel" facets of ghrelin are schematically summarized in Fig. 1. Overall, although the physiological relevance of ghrelin in the control of reproductive function and its regulation by the energy status is yet to be fully determined, the available evidence strongly suggests that ghrelin does participate in the complex network of neuroendocrine integrators responsible for the joint control of energy balance and reproduction.

ACKNOWLEDGMENTS

The author is indebted with E. Aguilar and L. Pinilla (University of Cordoba, Spain), and with C. Dieguez and F. F. Casanueva (University of Santiago de Compostela, Spain) for continuous support and helpful discussions during preparation of this chapter. The experimental work conducted in the author's laboratory was supported by grants BFI 2002-00176 and BFI 2005-07446 from Ministerio de Educacion y Ciencia (Spain), funds from Instituto de Salud Carlos III (Project PI042082 and CIBER-03 *Physiopathology of Obesity and Nutrition*), and EU research contract EDEN QLK4-CT-2002-00603.

REFERENCES

Ahima, R. S., Saper, C. B., Flier, J. S., and Elmquist, J. K. (2000). Leptin regulation of neuroendocrine systems. *Front. Neuroendocrinol.* **21,** 263–307.

Barreiro, M. L., and Tena-Sempere, M. (2004). Ghrelin and reproduction: A novel signal linking energy status and fertility? *Mol. Cell. Endocrinol.* **226,** 1–9.

Barreiro, M. L., Gaytan, F., Caminos, J. E., Pinilla, L., Casanueva, F. F., Aguilar, E., Dieguez, C., and Tena-Sempere, M. (2002). Cellular location and hormonal regulation of ghrelin expression in rat testis. *Biol. Reprod.* **67,** 1768–1776.

Barreiro, M. L., Suominen, J. S., Gaytan, F., Pinilla, L., Chopin, L. K., Casanueva, F. F., Diéguez, C., Aguilar, E., Toppari, J., and Tena-Sempere, M. (2003). Developmental, stage-specific and hormonally regulated expression of growth hormone secretagogue receptor messenger RNA in rat testis. *Biol. Reprod.* **68,** 1631–1640.

Barreiro, M. L., Gaytan, F., Castellano, J. M., Suominen, J. S., Roa, J., Gaytan, M., Aguilar, E., Dieguez, C., Toppari, J., and Tena-Sempere, M. (2004). Ghrelin inhibits the proliferative activity of immature Leydig cells *in vivo* and regulates stem cell factor messenger ribonucleic acid expression in rat testis. *Endocrinology* **145,** 4825–4834.

Caminos, J. E., Nogueiras, R., Blanco, M., Seoane, L. M., Bravo, S., Alvarez, C. V., Garcia-Caballero, T., Casanueva, F. F., and Dieguez, C. (2003a). Cellular distribution and regulation of ghrelin messenger ribonucleic acid in the rat pituitary gland. *Endocrinology* **144,** 5089–5097.

Caminos, J. E., Tena-Sempere, M., Gaytan, F., Sanchez-Criado, J. E., Barreiro, M. L., Nogueiras, R., Casanueva, F. F., Aguilar, E., and Dieguez, C. (2003b). Expression of ghrelin in the cyclic and pregnant rat ovary. *Endocrinology* **144,** 1594–1602.

Casanueva, F. F., and Dieguez, C. (1999). Neuroendocrine regulation and actions of leptin. *Front. Neuroendocrinol.* **20,** 317–363.

Cummings, D. E. (2006). Ghrelin and the short- and long-term regulation of appetite and body weight. *Physiol. Behav.* **89,** 71–84.

Fernandez-Fernandez, R., Tena-Sempere, M., Aguilar, E., and Pinilla, L. (2004). Ghrelin effects on gonadotropin secretion in male and female rats. *Neurosci. Lett.* **362,** 103–107.

Fernandez-Fernandez, R., Tena-Sempere, M., Navarro, V. M., Barreiro, M. L., Castellano, J. M., Aguilar, E., and Pinilla, L. (2005a). Effects of ghrelin upon gonadotropin-releasing hormone and gonadotropin secretion in adult female rats: *In vivo* and *in vitro* studies. *Neuroendocrinology* **82,** 245–255.

Fernandez-Fernandez, R., Navarro, V. M., Barreiro, M. L., Vigo, E. M., Tovar, S., Sirotkin, A. V., Casanueva, F. F., Aguilar, E., Dieguez, C., Pinilla, L., and Tena-Sempere, M. (2005b). Effects of chronic hyperghrelinemia on puberty onset and pregnancy outcome in the rat. *Endocrinology* **146,** 3018–3025.

Fernandez-Fernandez, R., Martini, A. C., Navarro, V. M., Castellano, J. M., Dieguez, C., Aguilar, E., Pinilla, L., and Tena-Sempere, M. (2006). Novel signals for the integration of energy balance and reproduction. *Mol. Cell. Endocrinol.* **254–255,** 127–132.

Frisch, R., and McArthur, J. (1974). Menstrual cycles: Fatness as a determinant of minimum weight for height necessary for their maintenance or onset. *Science* **185,** 949–951.

Frisch, R., and Revelle, R. (1970). Height and weight at menarche: A hypothesis of critical body weights and adolescent events. *Science* **169,** 397–399.

Furuta, M., Funabashi, T., and Kimura, F. (2001). Intracerebroventricular administration of ghrelin rapidly suppresses pulsatile luteinizing hormone secretion in ovariectomized rats. *Biochem. Biophys. Res. Commun.* **288,** 780–785.

Gauna, C., Delhanty, P. J., van Aken, M. O., Janssen, J. A., Themmen, A. P., Hofland, L. J., Culler, M., Broglio, F., Ghigo, E., and van der Lely, A. J. (2006). Unacylated ghrelin is active on the INS-1E rat insulinoma cell line independently of the growth hormone secretagogue receptor type 1a and the corticotropin releasing factor 2 receptor. *Mol. Cell. Endocrinol.* **251,** 103–111.

Gaytan, F., Barreiro, M. L., Chopin, L. K., Herington, A. C., Morales, C., Pinilla, L., Casanueva, F. F., Aguilar, E., Dieguez, C., and Tena-Sempere, M. (2003). Immunolocalization of ghrelin and its functional receptor, the type 1a growth hormone secretagogue receptor, in the cyclic human ovary. *J. Clin. Endocrinol. Metab.* **88,** 879–887.

Gaytan, F., Barreiro, M. L., Caminos, J. E., Chopin, L. K., Herington, A. C., Morales, C., Pinilla, L., Paniagua, R., Nistal, M., Casanueva, F. F., Aguilar, E., Dieguez, E., et al. (2004). Expression of ghrelin and its functional receptor, the type 1a growth hormone secretagogue receptor, in normal human testis and testicular tumors. *J. Clin. Endocrinol. Metab.* **89,** 400–409.

Gaytan, F., Morales, C., Barreiro, M. L., Jeffery, P., Chopin, L. K., Herington, A. C., Casanueva, F. F., Aguilar, E., Dieguez, C., and Tena-Sempere, M. (2005). Expression of growth hormone secretagogue receptor type 1a, the functional ghrelin receptor, in human ovarian surface epithelium, mullerian duct derivatives, and ovarian tumors. *J. Clin. Endocrinol. Metab.* **90,** 1798–1804.

Gnanapavan, S., Kola, B., Bustin, S. A., Morris, D. G., McGee, P., Fairclough, P., Bhattacharya, S., Carpenter, R., Grossman, A. B., and Korbonits, M. (2002). The tissue distribution of the mRNA of ghrelin and subtypes of its receptor, GHS-R, in humans. *J. Clin. Endocrinol. Metab.* **87,** 2988–2991.

Gualillo, O., Lago, F., Gomez-Reino, J., Casanueva, F. F., and Dieguez, C. (2003). Ghrelin, a wide-spread hormone: Insights into molecular and cellular regulation of its expression and mechanism of action. *FEBS Lett.* **552,** 105–109.

Gualillo, O., Lago, F., Casanueva, F. F., and Dieguez, C. (2006). One ancestor, several peptides post-translational modifications of preproghrelin generate several peptides with antithetical effects. *Mol. Cell. Endocrinol.* **256,** 1–8.

Holst, B., Egerod, K. L., Schild, E., Vickers, S. P., Cheetham, S., Gerlach, L. O., Storjohann, L., Stidsen, C. E., Jones, R., Beck-Sickinger, A. G., and Schwartz, T. W. (2006). GPR39 signaling is stimulated by zinc ions but not by obestatin. *Endocrinology* **148**(1), 13–20.

Iqbal, J., Kurose, Y., Canny, B., and Clarke, I. J. (2006). Effects of central infusion of ghrelin on food intake and plasma levels of growth hormone, luteinizing hormone, prolactin, and cortisol secretion in sheep. *Endocrinology* **147,** 510–519.

Kojima, M., Hosada, H., Date, Y., Nakazato, M., Matsuo, H., and Kangawa, K. (1999). Ghrelin is a growth-hormone-releasing acylated peptide from stomach. *Nature* **402,** 656–660.

Lanfranco, F., Bonelli, L., Broglio, F., Me, E., Baldi, M., di Bisceglie, C., Tagliabue, M., Manieri, C., and Ghigo, E. (2006). Ghrelin inhibits LH pulsatility in humans. 88th Endocrine Society Meeting (ENDO2006), Boston, P3–P807.

Martini, A. C., Fernandez-Fernandez, R., Tovar, S., Navarro, V. M., Vigo, E., Vazquez, M. J., Davies, J. S., Thompson, N. M., Aguilar, E., Pinilla, L., Wells, T., Dieguez, T., et al. (2006). Comparative analysis of the effects of ghrelin and un-acylated ghrelin upon luteinizing hormone secretion in male rats. *Endocrinology* **147,** 2374–2382.

Nogueiras, R., Pfluger, P., Tovar, S., Myrtha, A., Mitchell, S., Morris, A., Perez-Tilve, D., Vazquez, M. J., Wiedmer, P., Castaneda, T. R., Dimarchi, R., Tschop, R., et al. (2006).

Effects of obestatin on energy balance and growth hormone secretion in rodents. *Endocrinology* **148**(1), 21–26.

Rosenbaum, M., and Leibel, R. L. (1998). Leptin: A molecule integrating somatic energy stores, energy expenditure and fertility. *Trends Endocrinol. Metab.* **9**, 117–124.

Sirotkin, A. V., Grossmann, R., María-Peon, M. T., Roa, J., Tena-Sempere, M., and Klein, S. (2006). Novel expression and functional role of ghrelin in chicken ovary. *Mol. Cell. Endocrinol.* **257–258**, 15–25.

Tang-Christensen, M., Vrang, N., Ortmann, S., Bidlingmaier, M., Horvath, T. L., and Tschop, M. (2004). Central administration of ghrelin and agouti-related protein (83–132) increases food intake and decreases spontaneous locomotor activity in rats. *Endocrinology* **145**, 4645–4652.

Tena-Sempere, M. (2005). Exploring the role of ghrelin as novel regulator of gonadal function. *Growth Horm. IGF Res.* **15**, 83–88.

Tena-Sempere, M. (2006). GPR54 and kisspeptin in reproduction. *Hum. Reprod. Update* **12**, 631–639.

Tena-Sempere, M., and Barreiro, M. L. (2002). Leptin in male reproduction: The testis paradigm. *Mol. Cell. Endocrinol.* **188**, 9–13.

Tena-Sempere, M., and Huhtaniemi, I. (2003). Gonadotropins and gonadotropin receptors. *In* "Reproductive Medicine—Molecular, Cellular and Genetic Fundamentals" (B. C. J. M. Fauser, Ed.), pp. 225–244. Parthenon Publishing.

Tena-Sempere, M., Barreiro, M. L., Gonzalez, L. C., Gaytan, F., Zhang, F. P., Caminos, J. E., Casanueva, F. F., Dieguez, C., and Aguilar, E. (2002). Novel expression and functional role of ghrelin in rat testis. *Endocrinology* **143**, 717–725.

Toshinai, K., Yamaguchi, H., Sun, Y., Smith, R. G., Yamanaka, A., Sakurai, T., Date, Y., Mondal, M. S., Shimbara, T., Kawagoe, T., Murakami, N., Miyazato, N., *et al.* (2006). Des-acyl ghrelin induces food intake by a mechanism independent of the growth hormone secretagogue receptor. *Endocrinology* **147**, 2306–2314.

van der Lely, A. J., Tschop, M., Heiman, M. L., and Ghigo, E. (2004). Biological, physiological, pathophysiological, and pharmacological aspects of ghrelin. *Endocr. Rev.* **25**, 426–457.

Vulliémoz, N. R., Xiao, E., Xia-Zhang, L., Germond, M., Rivier, J., and Ferin, M. (2004). Decrease in luteinizing hormone pulse frequency during a five-hour peripheral ghrelin infusion in the ovariectomized rhesus monkey. *J. Clin. Endocrinol. Metab.* **89**, 5718–5723.

Wauters, M., Considine, R. V., and van Gaal, L. F. (2000). Human leptin: From an adipocyte hormone to an endocrine mediator. *Eur. J. Endocrinol.* **143**, 293–311.

Yan, W., Kero, J., Huhtaniemi, I., and Toppari, J. (2000). Stem cell factor functions as a survival factor for mature Leydig cells and a growth factor for precursor Leydig cells after ethylene dimethane sulfonate treatment: Implication of a role of the stem cell factor/c-kit system in Leydig cell development. *Dev. Biol.* **227**, 169–182.

Zhang, J. V., Ren, P. G., Avsian-Kretchmer, O., Luo, C. W., Rauch, R., Klein, C., and Hsueh, A. J. (2005). Obestatin, a peptide encoded by the ghrelin gene, opposes ghrelin's effects on food intake. *Science* **310**, 996–999.

Zhang, Y., Proenca, R., Maffei, M., Barone, M., Leopold, L., and Friedman, J. M. (1994). Positional cloning of the mouse obese gene and its human homologue. *Nature* **372**, 425–432.

Zigman, J. M., and Elmquist, J. K. (2003). From anorexia to obesity—The Yin and Yang of body weight control. *Endocrinology* **144**, 3749–3756.

CHAPTER THIRTEEN

Ghrelin and Prostate Cancer

Fabio Lanfranco,* Matteo Baldi,* Paola Cassoni,[†] Martino Bosco,[†] Corrado Ghé,[‡] and Giampiero Muccioli[‡]

Contents

I. Prostate Cancer	302
II. Influences of Hormones on Prostate Cancer Progression	303
III. The Peptide Hormone Ghrelin and Prostate Cancer	307
A. The many faces of the ghrelin biology	307
B. Effects on cell growth	311
C. Ghrelin and its receptors in human prostate tumors and related cell lines	312
D. Functional role of ghrelin in prostate cancer	314
IV. Conclusions	315
Acknowledgments	316
References	316

Abstract

Ghrelin, a 28-amino acid octanoylated peptide predominantly produced by the stomach, has been discovered to be a natural ligand of the type 1a growth hormone secretagogue receptor (GHSR1a). Ghrelin has recently attracted the interest as a new GH-releasing and orexigenic factor. However, ghrelin exerts several other activities, including regulation of tissue growth and development and control of neoplastic cell proliferation. Several endocrine and nonendocrine cancer cells (pituitary adenomas; gastroenteropancreatic and pulmonary carcinoids; colorectal neoplasms, thyroid tumors; lung, breast, and pancreatic carcinomas) as well as their related cell lines have been shown able to express ghrelin both at mRNA and at protein level. Many of the above-listed tumors express GHSR1a and/or alternative GHS receptor subtypes such as the type 1b GHSR, a truncated isoform of GHSR1a, and binding sites able to recognize ghrelin independently of its acylation. Evidence that ghrelin and multiple

---

* Department of Internal Medicine, Division of Endocrinology and Metabolism, University of Turin, 10126 Turin, Italy
[†] Department of Biomedical Science and Oncology, Division of Pathology, University of Turin, 10126 Turin, Italy
[‡] Department of Anatomy, Pharmacology and Forensic Medicine, Division of Pharmacology, University of Turin, 10125 Turin, Italy

ghrelin/GHS receptors are coexpressed in cancer cell lines and tumoral tissues from organs, such as the breast, that do not express these receptors in physiological conditions suggests that the ghrelin system is likely to play an important autocrine/paracrine role in some cancers. This chapter highlights the evidence for the expression of ghrelin and its receptors in one of the most frequent human malignancies, the prostate cancer, and information regarding their potential functional role in related cell lines. © 2008 Elsevier Inc.

I. Prostate Cancer

Prostate cancer is the third most frequent cancer in men worldwide with the highest incidence in the United States, Canada, and northwestern Europe, but less common in Asian countries and South America (Quinn and Babb, 2002). The incidence of prostate cancer rose in the late 1980s and early 1990s partly due to increased life expectancy, earlier and more accurate diagnosis, and increased public awareness of the disease. Prostate cancer constitutes about 11% of all male cancers in Europe (Bray, 2002) and accounts for 9% of all cancer deaths among men within the European Union (EU) (Black et al., 1997). The relatively slow rate of growth of prostate cancer and the improvements in treatment have resulted in a 5-year cancer-specific survival rate of 97% after diagnosis (Jemal et al., 2002). About 190,000 new cases occur each year (15% of all cancers in men) (Ferlay et al., 2001). In Europe, the annual incidence rates (ASW) in 2000 ranged between 19 (Eastern Europe) and 55 (Western Europe) per 100,000. Despite earlier diagnosis in younger men, prostate cancer remains primarily a disease of the aging population. Clinical prostate cancer is most common among men aged 65 and above, with about 80% of all diagnosed cases found in this range of age (Underwood et al., 2005).

The ultimate cause of prostate cancer has not been identified yet. However, family history, race, diet, and environmental factors have been associated with the disease (Gann et al., 1994; Steinberg et al., 1990). Risk of prostate cancer doubles among men having a first-degree relative with the disease and rises up to more than eightfold greater if both a first- and second-degree relative had prostate cancer. Wide variation in incidence has been reported between different races and ethnic groups, with incidence rates higher in Scandinavian and African–American men. A high dietary intake of animal and polyunsaturated fats has been associated with prostate cancer, though how dietary fat is related to higher risk remains unclear. Exposure to cadmium, a trace mineral found in cigarette smoke and alkaline batteries, has been postulated to weakly increase the risk of prostate cancer. The molecular pathogenesis of prostatic cancer is complex and involves many aspects of the cell biology of prostatic epithelium and stroma, including dysregulation of inflammatory response, defense toward genotoxic damage, cell proliferation, differentiation, apoptosis, and hormonal dependence. Postulated precursor lesions for prostatic cancer include prostatic intraepithelial neoplasia and proliferative inflammatory atrophy.

Prostatic intraepithelial neoplasia (PIN) is defined as a neoplastic transformation of the secretory epithelium confined within the epithelium. Based on architectural and, more importantly, cytological criteria, two grades are recognized: low-grade PIN (LGPIN) and high-grade PIN (HGPIN). There are many morphological, epidemiological, and molecular lines of evidences suggesting the presence of a pathogenetic link between HGPIN and prostatic cancer (Montironi et al., 2002). The vast majority of prostatic cancers are of secretory cell origin and are therefore known as acinar adenocarcinomas. Prostatic adenocarcinomas range from well-differentiated tumors, which simulate normal prostatic glands, to poorly differentiated lesions, which have completely lost the glandular architecture and cannot be easily recognized as of prostatic origin. Prostatic carcinoma can metastasize through lymphatic or hematogenous dissemination and can spread locally into the urethra, the bladder neck, the seminal vesicles, or the bladder trigone. The most common sites of lymphatic metastasis are the obturator, hypogastric, iliac, presacral, and periaortic lymph nodes. Bone metastases constitute the most common consequence of hematogenous spread. The most frequent sites of bony metastatic involvement are the pelvis, lumbar spine, femora, thoracic spine, and ribs. Although uncommon, visceral metastases can also be found in the lung and the liver (Carter and Partin, 1998). Transrectal ultrasound-guided methods are the usual and preferred means of obtaining prostate tissue for histological diagnosis. Because prostate cancer rarely causes symptoms until advanced disease is evident, a suspicion of prostate cancer resulting in a recommendation for prostatic biopsy is often raised through abnormalities found on digital rectal examination (DRE) or elevations in serum prostatic-specific antigen (PSA) (>4.0 ng/ml). Although controversial, it has been demonstrated that an early diagnosis of prostate cancer is best achieved through a combination of DRE and PSA measurements.

II. INFLUENCES OF HORMONES ON PROSTATE CANCER PROGRESSION

The development of the prostate is known to be regulated by testicular steroid hormones, which might be involved in the development of benign prostate hyperplasia and the pathogenesis of prostate cancer as well. The principal circulating androgen is testosterone. In several androgen target tissues, among which the prostate, testosterone is converted to

5α-dihydrotestosterone (DHT), which is the most potent natural androgen. Testosterone and DHT bind to an identical receptor, but they play distinct physiological functions: testosterone regulates sexual differentiation and maintains libido and sexual functions, whereas DHT plays a major role in embryonic and pubertal and external virilization (Grumbach et al., 2003). Androgens are necessary for the initiation of prostate cancer and the balance between androgen-induced cell proliferation, and apoptosis is thought to regulate the growth of the normal and cancerous prostate (Soronen et al., 2004). In normal conditions, a steady state exists between synthesis and inactivation of active androgens (Denmeade et al., 1996), and this balance is regulated by coordinated network of steroidogenic and steroid metabolizing enzymes (Labrie et al., 2000; Russel and Wilson, 1994). Changes in the balance may be important in cancer development since hormones may act as carcinogens by increasing cellular proliferation and thereby the chance of random DNA copy errors (Soronen et al., 2004). The functional integrity of androgen receptor (AR) and the pathways of transduction of its signal are clearly important in the prostate cancer progression, which is likely to be specifically dependent on structurally modified AR. The AR, in fact, contains CAG repeats in exon 1, repeats which are associated with many genetic diseases. In addition, the length of the CAG-rich regions appears to be a determinant of tumorigenesis, as short CAG repeats are associated with a higher risk of developing prostate cancer (Giovannucci et al., 1997). However, uncertainty persists about the possible relationship of CAG repeats lengths to tumor grade or progression, thus requiring further work in this area. Prostate cancer progression is also accompanied by the transition from an androgen-dependent to an androgen-independent state, which may be correlated to either functional alterations of AR accruing deriving from mutation in the protein sequence (Thompson et al., 2003) or, alternatively, to the activation of specific pathways by AR, activated by factors other than androgens.

The results of epidemiological and experimental studies have suggested that estrogens could be involved in the induction of prostate cancer. Estrogen receptor β has been suggested to play a role in the differentiation and proliferation of prostatic cells as well as possibly to modulate the initial phases of prostate carcinogenesis and androgen-independent prostatic growth (Signoretti and Loda, 2001). The progression of prostatic carcinogenesis is believed to involve epithelial-mesenchymal/stromal interaction which takes place with the participation of both testosterone and estradiol (Cunha et al., 2004), but the precise roles of these two classes of steroid hormones are not entirely clear. However, estrogens alone or in conjunction with androgens do induce aberrant prostate growth and neoplastic transformation (Ho, 2004).

In recent years, there has been increasing recognition that some cytokine-like polypeptide factors widely secreted by adipose tissue, including leptin, heparin-binding epidermal growth factor-like growth factor (HB-EGF), interleukin-6 (IL-6), vascular endothelial growth factor (VEGF), and adiponectin, exert multiple effects on the biological behavior of tumor cells. Leptin is one of the major cytokines that is produced by adipocytes and plays a major role in controlling body weight homeostasis, regulating satiety and energy expenditure (Friedman and Halaas, 1998). High expression of the leptin receptor has been observed in the prostate (Cioffi et al., 1996), and circulating leptin levels increase in parallel with prostate growth at puberty in the rat (Nazian and Cameron, 1999). In cell culture experiments, exogenous leptin stimulated the growth of the androgen-independent DU-145 and PC-3 human prostate cancer cell lines (Onuma et al., 2003; Somasundar et al., 2003, 2004), but the androgen-dependent LNCaP prostate cancer cell line, although it possesses the two leptin receptor isoforms, showed no mitogenic response to the ligand (Onuma et al., 2003). The proliferative response of DU-145 and PC-3 prostate cancer cells to leptin was shown to involve phosphatidylinositol 3-kinase (PI-3K) and mitogen-activated protein kinase (MAPK) activation (Somasundar et al., 2004), and the participation of activated c-Jun N-terminal kinase (JNK); the pharmacological inhibition of JNK blocked the proliferative response to leptin (Onuma et al., 2003). It is of particular interest that JNK has been shown to be elevated in animal models of obesity and to induce insulin resistance (Hirosumi et al., 2002; Ozcan et al., 2004). Lin et al. (1999) demonstrated that the PI-3K pathway is also critical for the survival of androgen-dependent prostate cancer cells by showing that apoptosis of LNCaP cells occurred rapidly in the presence of a PI-3K inhibitor, an antiproliferative effect which was attenuated by another adipocytokine, HB-EGF. Interestingly, leptin stimulates angiogenesis (Sierra-Honigmann et al., 1998). Stattin et al. (2001) showed a significantly increased risk for prostate cancer in the intermediate range of circulating leptin concentrations. These authors also found immunoreactivity for leptin receptors in prostatic cancer specimens, with a strong expression in HGPIN lesions, a precursor lesion for manifest cancer (Sakr et al., 1996). Unlike a previous investigation (Lagiou et al., 1998), this study showed for the first time a possibile association between leptin and cancer risk.

Results from a population-based, case-control study conducted in China (Hsing et al., 2000) suggest that higher serum levels of insulin are associated with an increased risk of prostate cancer. The observed association of insulin with prostate cancer risk is independent of overall and abdominal adiposity. Even among men in the lowest tertile of WHR (<0.873), those in the highest tertile of insulin levels had a fourfold risk of prostate cancer, suggesting that insulin may modulate the risk of prostate cancer through mechanisms other than obesity. One such mechanism may involve an interplay between insulin and the insulin-like growth factor I (IGF-I) system. IGF-I is secreted by adipocytes as well as by androgen-independent prostate cancer cell lines. IGF-I has been implicated in the regulation of prostate epithelial cell proliferation and in the etiology of prostate cancer (Chan et al., 1998; Chokkalingam et al., 2001; Giovannucci, 1999; Pollak et al., 1998–1999; Wolk et al., 1998). Plasma IGF-I levels have been associated with prostate cancer risk and may be associated with the risk of developing advanced-stage prostate cancer (Chan et al., 1998, 2002).

Insulin also decreases IGF-binding protein-1 (IGFBP-1) production and secretion, thereby increasing the bioavailability of IGF-I (McCarty, 1997). Moreover, insulin may affect prostate cancer risk through the obesity–sex hormone pathway by regulating the production and metabolism of testosterone and sex hormone-binding globulin (SHBG) (Haffner, 2000). Decreased levels of SHBG result in increased levels of the bioactive free fraction of testosterone, which, in turn, may increase the risk of prostate cancer. The reported lower prostate cancer risk among diabetic patients in some studies (Giovannucci et al., 1998; Will et al., 1999), but not in all, (Hsieh et al., 1999) further supports a potential role for insulin in prostate cancer etiology. The basis for the reduced prostate cancer risk among diabetics is unclear, but potential explanations include lower levels of testosterone and IGF-I, as well as insulin insensitivity, in people with non-insulin-dependent diabetes mellitus (type 2 diabetes) (Barrett-Connor et al., 1990; Clauson et al., 1998; Haffner et al., 1996).

IL-6 demonstrated additive effects to those of leptin stimulation while IGF-I had a synergistic effect on cell growth. IL-6 is secreted by adipocytes and other cell types in the body, the level correlating with body mass index (BMI). Serum IL-6 levels have been reported to be elevated in hormone-refractory prostate cancer patients and have been noted to be secreted by androgen-independent prostate cancer cell lines (Drachenberg et al., 1999).

Adiponectin is a 30,000 Da collagen-like plasma protein which is synthesized exclusively in adipocytes (Scherer et al., 1995) and circulates in the blood at a concentration which amounts to 0.05% of the total serum proteins (Scherer et al., 1995). The plasma adiponectin levels, like those of leptin, are higher in women than in men (Cnop et al., 2003; Nishizawa et al., 2002). Adiponectin levels are higher in hypogonadal than in eugonadal men and are reduced by testosterone replacement therapy (Lanfranco et al., 2004). There are a number of situations where adiponectin and leptin have antagonistic effects, including their opposing actions on insulin signaling. The opposing effects of adiponectin and leptin on various aspects of metabolism and cell biology suggest that, in contrast to leptin, adiponectin may possess anticancer properties which are lost in clinical situations associated with hypoadiponectinemia. In support of this proposition are reports that subnormal plasma concentrations of adiponectin are related to increased risks of carcinoma of the breast (Mantzoros et al., 2004) and endometrium (Dal Maso et al., 2004). A small case-control study performed in Turkey by Goktas *et al.* (2005) has now extended the relationship between hypoadiponectinemia and cancer to include carcinoma of the prostate. Plasma adiponectin levels were significantly lower in a group of 30 prostate cancer patients than in 41 patients with benign prostatic hyperplasia, and 36 healthy controls. Further, the adiponectin concentrations were inversely correlated with the tumor grade and disease stage. Interestingly, although obese men were excluded from this study, and there were no significant differences in the BMI values between the three groups, the plasma insulin levels were higher, and insulin resistance as indicated by the homeostasis model assessment method was greater in the cancer patients. There have been no published studies to determine if adiponectin has a direct effect, inhibitory or otherwise, on prostate, breast, endometrial, or colon cancer cell proliferation, but Yokota *et al.* (2000) did report that it suppressed the growth of myelocyte lines and induced apoptosis in myelomonocytic leukemia cell lines.

III. THE PEPTIDE HORMONE GHRELIN AND PROSTATE CANCER

A. The many faces of the ghrelin biology

Ghrelin is a 28-amino acid peptide predominantly produced by the stomach, but also expressed by several other tissues such as bowel, pancreas, kidney, immune system, placenta, testis, lung, and hypothalamus (Kojima and Kangawa, 2005). Ghrelin is the first natural hormone to be identified in which the hydroxyl group of one of its Ser3 residue is acylated by *n*-octanoic acid (Kojima *et al.*, 1999). This acylation is essential for hormone's binding to the type 1a growth hormone secretagogue receptor (GHSR1a), for the GH-releasing capacity of ghrelin, and most likely for its other endocrine actions. Unacylated ghrelin (UAG), which is present in circulation in far greater amount than acylated ghrelin, does not bind GHSR1a and therefore is devoid of any endocrine action. However, UAG is able to exert some nonendocrine activities, including cardiovascular, antilipolytic, and antiproliferative effects (Ghigo *et al.*, 2005; van der Lely *et al.*, 2004), probably by binding a GHSR subtype common for both acylated and unacylated ghrelin forms. This receptor has been identified and characterized in cardiomyocytes, adipocytes, and some breast cancer cell lines (Muccioli *et al.*, 2004). Another type of ghrelin peptide has been purified and identified as des-Gln14-ghrelin that, except for the deletion of Gln14, is identical to ghrelin. Des-Gln14-ghrelin is the result of alternative splicing of the ghrelin gene and it seems to possess the same potency of activities as that of ghrelin (Kojima and Kangawa, 2005). Additionally, during the course of purification, other minor forms of ghrelin peptides were isolated and classified by the type of acylation observed at Ser3: decanoylated and decenoylated (Hosoda et al., 2003).

Ghrelin acts through the type 1a GHSR, a specific G-protein–coupled receptor (Howard et al., 1996), which is expressed by a single gene found at human chromosomal location 3q26.2 (McKee et al., 1997). Two types of GHSR cDNAs, which are presumably the result of alternate processing of a pre-mRNA, have been identified and designated as receptors 1a and 1b. The GHSR1a is a polypeptide of 366 amino acids with a molecular mass of ~41,000 Da and it is the functional form of the ghrelin receptor. Synthetic peptidyl (hexarelin) and non-peptidyl (MK0677) GHS as well as all the endogenous acylated ghrelin derivatives bind with high affinity to GHSR1a (Bednarek et al., 2000). GHSR1a was shown to be expressed in a wide variety of tissues such as hypothalamus, anterior pituitary gland, stomach, intestine, pancreas, kidney, heart, and aorta, as well as in different human pituitary adenomas and various endocrine neoplasms of lung, stomach, and pancreas (Papotti et al., 2004). The GHSR1b is a truncated form of GHSR1a and consists of a 298-amino acid protein containing only the first five-transmembrane domains plus a unique 24-amino acid "tail" encoded by alternative spliced intronic sequence. GHSR1b expression was also widespread in several normal (Gnanapavan et al., 2002) and neoplastic human tissues (Papotti et al., 2004). Surprisingly, no binding of synthetic GHS ligands or ghrelin molecules to this GHSR isoform has been demonstrated and thus, GHSR1b has been considered to be a nonfunctional GHSR subtype. However, a recent study on non-small lung cancer cells (NSCLC) showed a novel binding site for neuromedin U (NMU), a neuropeptide whose known receptors (NMU1R and NMU2R) show a high degree of homology to GHSRs (Tan et al., 1998), and neurotensin receptor 1 (NTSR1): such receptor, which mediated its proliferative effect on the cells, happen to be composed of GHSR1b and NTSR1. In this study, the majority of cancer cell lines and clinical NSCLCs that expressed NMU, but not its receptors, also expressed GHSR1b and NTSR1, and in response to NMU, GHSR1b and NTSR1 cointernalized and heterodimerized (Takahashi et al., 2006). It has also been reported that coexpression of GHSR1a and GHSR1b results in an attenuation of signaling capability of GHSR1a, again suggesting an interaction between the two receptors through heterodimerization (Chan and Cheng, 2004). Thus, the splice variant GHSR1b, despite unable to mediate per se the biological activities of ghrelin and GHS, is capable of altering the pharmacological and functional properties of closely related receptors through a direct protein–protein interaction, that is, creating a new receptor.

The GH-releasing property was the first-recognized effect of ghrelin (Kojima et al., 1999). Ghrelin displays strong GH-releasing activity (Broglio et al., 2003) that takes place both directly on pituitary cells (Hashizume et al., 2003) and through modulation of growth hormone release hormone (GHRH) from the hypothalamus (Tannenbaum et al., 2003). Ghrelin has been demonstrated to be much more than a natural GHS. Ghrelin stimulates prolactin secretion in humans, independent of both gender and age and probably involving both a direct action on somatomammotroph cells and indirect hypothalamic actions (Arvat et al., 2001; Ghigo et al., 2005). Moreover, ghrelin and synthetic GHS possess an acute stimulatory effect on the activity of the hypothalamic-pituitary-adrenal (HPA) axis in humans, which is similar to that of the opioid antagonist naloxone, arginine-vasopressin (AVP), and corticotrophin-releasing hormone (CRH) (Ghigo et al., 2005). In the context of the interaction of ghrelin with the neuroendocrine axes, it has been suggested that ghrelin may participate in the modulation of the hypothalamic-pituitary-gonadal function, with a predominantly inhibitory effect on the reproductive system in rats, sheep, and primates (Barreiro and Tena-Sempere, 2004; Fernandez-Fernandez et al., 2004, 2005; Furuta et al., 2001; Iqbal et al., 2006; Kawamura et al., 2003; Tena-Sempere et al., 2002; Vulliémoz et al., 2004). Animal studies have shown that ghrelin is able to suppress luteinizing-hormone (LH) secretion *in vivo* and to decrease LH responsiveness to gonadotropin-releasing hormone (GnRH) *in vitro* (Fernandez-Fernandez et al., 2004; Furuta et al., 2001; Iqbal et al., 2006; Vulliémoz et al., 2004). Moreover, repeated administration of ghrelin induced a partial delay in the timing of puberty in male rats (Fernández-Fernández et al., 2005). Finally, transcripts for ghrelin and its cognate receptor have also been identified in rat and human gonads (Tena-Sempere et al., 2002; van der Lely et al., 2004), and ghrelin has been reported to inhibit stimulated testicular testosterone secretion (Iqbal et al., 2006). Vulliémoz et al. (2004) have reported that ghrelin infusion significantly decreases LH pulse frequency in adult ovariectomized rhesus monkeys whereas LH pulse amplitude is not affected, thus indicating that ghrelin can inhibit GnRH pulse activity.

Ghrelin is one of the most powerful orexigenic and adipogenic agents known in mammalian physiology (Kojima and Kangawa, 2005; Muccioli et al., 2002). First, Locke et al. (1995) showed an increase in food intake after intracerebroventricular administration of a synthetic peptidyl GHS, such as GHRP-6, in rats without affecting plasma GH response. Similar appetite-stimulating effects have been reported by several other groups (Lall et al., 2001; Okada et al., 1996; Torsello et al., 1998, 2000) after central and/or peripheral administration of other ghrelin mimetic substances. The orexigenic activity of ghrelin is GH independent and it cannot be prevented by blocking the GHRH pathway (van der Lely et al., 2004). Ghrelin stimulated food intake in rodents in a dose-dependent way and more powerfully after central than after peripheral administration: This finding indicates a predominat central mechanism of action (Horvath et al., 2001). It has been demonstrated that two major hypothalamic pathways are the predominant mediators of ghrelin's influence on energy balance and appetite (Tschöp et al., 2002). One involves the neuropeptide Y (NPY) neurons, and the other involves the melanocortin receptors and their agonistic and antagonistic ligands, the anorexigenic proopiomelanocortin (POMC)-derived α-melanocyte stimulating hormone (α-MSH), and the orexigenic agouti-related protein (AGRP), which is expressed in NPY neurons. Ghrelin increases AGRP and NPY after acute and chronic administration, and hypothalamic AGRP-mRNA expression levels are found to be upregulated after chronic activation of the GHSR for several weeks (Kamegai et al., 2000).

Apart from its role in the regulation of food intake and energy homeostasis, several studies have shown that ghrelin is also involved in the regulation of sleep, memory, and anxiety-behavioral responses (Ghigo et al., 2005). In rats, there is evidence that ghrelin affects sleep–wake patterns. Furthermore, ghrelin itself has been reported to be a sleep-promoting factor in humans (Weikel et al., 2003). Besides regulating sleep pattern, it has also been shown that ghrelin has an anxiogenic action in rodents, via mechanisms involving the HPA axis. Finally, injections of ghrelin in the hippocampus, amygdala, and dorsal raphe nucleus clearly increase memory retention in rats (Cummings, 2006).

In agreement with the existence of specific ghrelin receptors in peripheral tissues, metabolic and nonmetabolic activities of ghrelin molecules have widely been demonstrated. Ghrelin acts at the gastroenteropancreatic level, where GHSR1a and GHSR1b expressions have been found. Ghrelin stimulates gastric acid secretion and gut motility in rats (Masuda et al., 2000; Murakami et al., 2002; Trudel et al., 2002), and circulating ghrelin levels are positively correlated with gastric emptying time in humans (Gondo et al., 2001). Ghrelin and GHSR1a mRNA are present both in esocrine and endocrine pancreas (van der Lely et al., 2004). Regarding the exocrine pancreas, it has been shown that ghrelin is a potent inhibitor of pancreatic cholecystokin-induced exocrine secretion in rats and in pancreatic lobes in vitro (Zhang et al., 2001). Regarding the endocrine pancreas, ghrelin exerts a tonic inhibitory effect of insulin secretion and induces a significant increase in human plasma glucose levels (Arosio et al., 2003; Ghigo et al., 2005). In addition, ghrelin has been shown to induce a rapid increase of glucose output by primary human hepatocytes (Gauna et al., 2005).

In the cardiovascular system, there is already evidence that ghrelin and/or peptidyl GHS mediate GH-independent cardiovascular activities, both in animals and in humans. In fact, ghrelin or GHS administration in normal human subjects and even in patients with chronic heart failure significantly decreases systemic vascular resistance and increases cardiac index and stroke volume index (Benso et al., 2004; Enomoto et al., 2003). These events are accompanied by a concomitant reduction in mean arterial pressure, but not by any change in heart rate, mean pulmonary arterial pressure, or pulmonary capillary wedge pressure. Ghrelin, as well as UAG, prevents cell death of cultured cardiomyocytes and endothelial cells induced by either doxorubicin, serum withdrawal, or activation of FAS, stimulating survival intracellular signaling pathways such as tyrosine phosphorylation of intracellular proteins and activation of extracellular-signal-regulated kinase-1 and -2 and kinase B/Akt (Benso et al., 2004). Moreover, data indicate that ghrelin either in acylated or unacylated form (UAG) shows similar negative inotropic effect on isolated guinea pig papillary muscle (Bedendi et al., 2003). These evidences suggest the possible existence of another GHSR subtype that recognizes both acylated and unacylated ghrelin in the cardiovascular system (Muccioli et al., 2004).

B. Effects on cell growth

Ghrelin has been demonstrated to exert a proliferative effect on cardiac, pancreatic, adipose, adrenal, pituitary, and immune system cells (Nogueiras et al., 2006). Furthermore, reports showed that ghrelin also stimulates neurogenesis in the dorsal motor nucleus of the vagus (Zhang et al., 2004) as well as in the rat fetal spinal cord (Sato et al., 2006). Conversely, an antiproliferative activity of the hormone has been observed in differentiating immature Leydig cells (Lago et al., 2005). In the last few years, it was rapidly shown that ghrelin and its receptors occur also in different endocrine and nonendocrine neoplasms (Papotti et al., 2004). Specific binding sites for peptidyl and non-peptidyl GHS are present in normal and neoplastic human thyroid tissue. Binding sites for GHS have been demonstrated in all follicular and most parafollicular thyroid carcinomas, as well as in different human thyroid tumor cell lines. Moreover, medullary, but not follicular, thyroid carcinomas and carcinoma cell lines remarkably express ghrelin. The significance of ghrelin production by thyroid tumor cells is poorly understood, but the reported presence of ghrelin/GHS binding sites in thyroid tumor suggests that the hormone may act directly on the tumor cells. In fact, ghrelin or synthetic GHS inhibit [^{3}H]-thymidine incorporation and cell growth in human thyroid tumor cell lines, already within 24 h (Cassoni et al., 2000, 2002; Kanamoto et al., 2001). Ghrelin has also been found in pituitary adenomas, with a lower expression in corticotroph adenomas and relatively high in somatotroph tumors, as compared with that of normal pituitary tissue. Other studies found the highest ghrelin expression in nonfunctioning adenomas and the lowest in prolactinomas (Papotti et al., 2004). Pancreatic well-differentiated endocrine tumors express ghrelin in variable percentages. The presence of ghrelin in pancreatic endocrine tumors is not surprising since several groups documented the presence of ghrelin in fetal and adult islets of Langerhans, and the existence of a separate cell type, the "ghrelin cell," has been proposed in the pancreatic islets (Prado et al., 2004). Ghrelin-producing cells are mainly represented in the fetal pancreas, but may occasionally persist in the adult (Papotti et al., 2004). Korbonits et al. (1998) showed that endocrine pancreatic tumors contain GHSR1b and only some of them also express GHSR1a transcript. All

GHSR1a positive tumors were insulinomas, while those positive for type 1b included either insulinomas or other functioning or nonfunctioning neoplasms. Information on ghrelin production by nonendocrine tumors is largely confined to the analysis of human tumor cell lines. The occurrence of ghrelin mRNA and/or protein was demostrated in leukemic cell lines (De Vriese *et al.*, 2004), as well as in pulmonary squamous cell or adenocarcinomas, pancreatic ductal adenocarcinomas, gastric and colorectal adenocarcinomas, and hepatocellular carcinomas, where ghrelin has been demonstrated to have mitogenic effects (Papotti *et al.*, 2004). GHSRs have also been found in tumoral tissues from organs that do not express these receptors in physiological conditions such as mammary gland tissue (Cassoni *et al.*, 2001). The presence of specific GHSR was shown in breast cancer, but not in fibroadenomas or normal mammary parenchyma. In breast tumors, the highest binding activity is present in well-differentiated invasive breast carcinomas and is progressively reduced in moderately to poorly differentiated tumors. GHSR is also present in both estrogen-dependent (MCF7, T47D) and estrogen-independent (MDA-MB231) breast cancer cell lines in which ghrelin and synthetic GHS cause inhibition of cell proliferation. Other data indicate that even adenocarcinomas of the lung contain specific GHS binding sites (Ghè *et al.*, 2002). These sites are also present in a human lung cancer cell line CALU-1 of which the proliferation is inhibited by different synthetic peptidyl GHS (Ghè *et al.*, 2002).

C. Ghrelin and its receptors in human prostate tumors and related cell lines

The involvement of the ghrelin system in prostate cancer has been demonstrated only in recent years. Cassoni *et al.* (2004) have shown that ghrelin and GHS receptors are expressed in human prostatic carcinomas and benign hyperplasias and that ghrelin and some of its natural (UAG) and synthetic (hexarelin) derivatives modulate cell proliferation in different prostatic carcinoma cell lines, both androgen dependent (LNCaP) and androgen independent (DU-145 and PC-3). Interestingly, ghrelin mRNA was found in all the prostatic benign (hyperplastic) and malignant (carcinomatous) tissues. In contrast, ghrelin mRNA was not detected in a normal prostate cDNA library, suggesting very low mRNA levels in normal prostatic tissue (Jeffery *et al.*, 2002). In human surgical specimens of both prostate carcinomas and benign prostatic hyperplasias, ghrelin was demonstrated by RT-PCR and its localization within epithelial cells, either neoplastic or hyperplastic, was confirmed by *in situ* hybridization (Cassoni *et al.*, 2004). Despite containing the specific mRNA, neither prostate carcinomas nor hyperplastic lesions displayed any ghrelin immunoreactivity, suggesting the inability of these cells to effectively synthesize the peptide or to retain it in the cytoplasm once ghrelin has been synthesized. It cannot be excluded that such difference between mRNA and peptide expression could rely on the presence of different proghrelin-deleted isoforms from which the mature ghrelin peptide origins and that, in particular the exon 3-deleted proghrelin, has been found upregulated in prostate as well as in breast cancer tissues when compared to normal histological tissue (Jeffery et al., 2003, 2005). This proghrelin variant could be recognized by molecular biology techniques only since deletion of exon 3 determines the creation of an alternative C-terminal peptide which could be unreactive with the commonly used anti-ghrelin antibodies which are directed toward the C-terminal peptide portion. Whatever the reason for such discrepancy between ghrelin mRNA and peptide expression reported in literature, it can be concluded that in prostate the modified epithelial cells (both neoplastic or hyperplastic) contain higher amount of mRNA for ghrelin or propeptide isoforms than normal prostate, suggesting the possible existence of an autocrine/paracrine role of the ghrelin system in participating to the neoplastic processes. When ghrelin production was investigated in various prostate cancer cell lines, such as ALVA-41, LNCaP, DU-145, or PC-3 cells, conflicting results were obtained. Using a Western analysis of cell lysates, Jeffery et al. (2002) demonstrated the presence of a single band of 3000 Da corresponding to the acylated peptide ghrelin in all cell lines examined. In contrast, immunoreactive ghrelin was found by immunocytochemistry in PC-3 cells exclusively by Cassoni et al. (2004). More specifically, in PC-3 cells, the presence of ghrelin was detected only within the neoplastic cell but not in the culture medium, suggesting that the neoplastic PC-3 cells synthesize the peptide but are unable to secrete it in the culture medium (Cassoni et al., 2004). The reason of these divergent results remains unclear. Possible explanations could include different methodological approaches and/or variations in cell culture conditions, such as time of cell cultures and passages, and low sensitivity of the immunoenzymatic assay for ghrelin detection.

Despite the possible functional autocrine role of ghrelin, if any, could a role for circulating ghrelin be envisaged in prostate cancer? The first premise to make this hypothesis concrete is the existence of specific ghrelin binding sites in prostate neoplastic cells. As previously described for ghrelin production in prostate cancer, divergent observations have been reported in literature on the GHSR expression in prostate carcinoma cell lines and tissues. Jeffery et al. (2002) reported that mRNA for type 1a and/or type 1b GHSR was expressed in four prostate carcinoma cell lines (ALVA41, DU-145, LNCaP, and PC-3) and that the GHSR1a protein was also detectable by immunocytochemistry. In contrast to these data, neither GHSR1a mRNA nor GHSR1a protein has been detected in the same cell lines (except DU-145) by Cassoni et al. (2004). Similarly, primary human prostate carcinomas were lacking GHSR1a and GHSR1b mRNA. However, despite the absence of GHSR mRNA, binding studies performed by the same group demonstrated the existence of specific high-affinity [^{125}]Tyr4-ghrelin binding sites in prostate tumor, PC-3, and DU-145 cell membranes. Moreover, in tissue or cell membranes radiolabeled ghrelin binding was displaced in a dose-dependent manner by unlabeled ghrelin, synthetic GHS, or unacylated ghrelin, an endogenous ghrelin variant devoid of GHSR1a binding affinity (Cassoni et al., 2004). This finding opens the challenging possibility of an alternative still unidentified ghrelin receptor subtype in prostate cancer, commonly shared by ghrelin and its unacylated form. A similar binding pattern has been observed in small cell lung carcinoma cell lines, further sustaining the hypothesis of a ghrelin receptor different from GHSR1a in neoplastic cells (Cassoni et al., 2006). Whatever the receptor, GHSR1a, GHSR1b, and/or other GHSR subtypes, such as binding sites common for acylated and unacylated ghrelin, the presence of ghrelin receptors in some human prostate carcinomas and related cell lines strongly supports the concept that ghrelin and its derivatives could exert direct regulatory effects on tumor cell growth.

D. Functional role of ghrelin in prostate cancer

As previously reported for ghrelin and GHSR expression, the data on the ghrelin effects on prostate carcinoma cell growth are contradictory. Jeffery et al. (2002, 2003) and Yeh et al. (2005) tested the effect of exogenous ghrelin on cell growth of PC-3 and LNCaP prostate cancer cell lines and reported that 5 and 10 nM ghrelin concentrations added to the culture medium over a 3-day period stimulated cell proliferation of both cell lines, without any evidence of a dose-dependent effect. Differently, Cassoni et al. (2004) demonstrated that ghrelin had a biphasic effect on the growth of PC-3 cells, with a stimulating activity at concentrations (10 and 100 pM) quite close to those present in circulation and an inhibiting effect at pharmacological doses (1 μM). However, in other prostate cancer cell lines (DU-145 and LNCaP) studied, ghrelin inhibited cell proliferation of DU-145 cells at concentrations ranging from 1 nM to 1 μM and it was ineffective on LNCaP cells at all concentration tested (1 nM–1 μM). The lack of ghrelin effect on LNCaP cell proliferation is in agreement with the absence of ghrelin binding sites in these cells, whereas in PC-3 and DU-145 cell lines the modulatory effect of ghrelin on cell growth fits well with the presence of nonclassical type 1a GHSR recognized by both acylated and unacylated ghrelin. Since ghrelin and its unacylated form determined identical effects on the proliferation of PC-3 and DU-145 cells, the hypothesis of a novel, as yet unidentified, receptor common for acylated and unacylated ghrelin that could mediate the activity of both peptides is further strengthened (Cassoni et al., 2004). It still remains to be clarified why different ghrelin concentrations in the culture medium exert opposite effect on cell proliferation, as observed in the PC-3 cells, and why the same peptide produces stimulating or inhibiting effects on cell proliferation in different prostate cancer cell lines. The first question can be partially answered suggesting for ghrelin a biological behavior similar to that already described for other molecules, such as somatostatin and its analogues, that also determine biphasic effects on cell proliferation as a function of their concentrations (Hoelting et al., 1996). Concerning the second point, the discrepant response to ghrelin (and UAG) among the three cell lines could depend on many factors, such as a different androgen-dependent status (Gasparian et al., 2002) and/or different expression of signal transducers and transcription activators (Mora et al., 2002) and/or different responsiveness to Fas-mediated apoptosis (Gewies et al., 2000).

The mechanism underlying the mitogenic effect of ghrelin on prostatic cancer cells is still unclear and could involve activation of the cell-survival ERK1/2 MAPK pathway. Activation of this signal transduction pathway has been correlated with disease progression in several cancer models (Reddy et al., 2003). Histopathological studies have detected high levels of ERK1/2 activation in more advanced, metastatic, and androgen-independent prostate cancers, suggesting a potential contribution of this cascade to prostate cancer progression (Gioeli et al., 1999). Elevated levels of MAPK activity have also been shown in androgen-independent prostate cancer cells (Onishi et al., 2001). Ghrelin-induced activation of MAPK pathways also accounts for the antiapoptotic effect of ghrelin treatment reported in cardiomyocytes and endothelial cells (Benso et al., 2004) or in β pancreatic cells (Granata et al., 2006). However, no immediate effect of ghrelin treatment on apoptosis has been reported in LNCaP and PC-3 prostate cancer cell lines, as ghrelin acted by promoting cell growth via ERK1/2 phosphorylation, without directly inhibiting apoptosis (Yeh et al., 2005).

IV. Conclusions

Current data indicate that prostate cancer cells express mRNA for GHSR1a and/or its truncated isoform GHSR1b and contain also a novel, as yet unidentified, high-affinity ghrelin binding site recognized by both acylated and unacylated ghrelin. In the prostate, modified epithelial cells (both neoplastic or hyperplastic) have higher amount of mRNA for ghrelin or propeptide isoforms in comparison with nontumoral prostatic cells, suggesting the possible existence of an autocrine/paracrine role of the peptide in the neoplastic processes. A stimulatory role on prostatic cancer cell growth is exerted *in vitro* by ghrelin at concentrations quite close to those found in circulation (10 and 100 pM) or in ghrelin-producing tissues (5 and 10 nM), whereas higher doses of peptide have an inhibitory effect on cell proliferation. In addition, UAG, which is also present in circulation in far greater amount than ghrelin, shares with the acylated peptide the same effects on prostate cancer cell growth through binding to a receptor common for both ghrelin forms. Interestingly, a number of observations also report a differential expression of components of the ghrelin system, particularly exon 3-deleted proghrelin and GHSR1b, between normal and malignant prostatic tissue. Altogether, these findings suggest that ghrelin may be involved in the control of prostate cancer growth and provide a basis for additional studies that are critical to evaluate the potential of the ghrelin system as diagnostic marker and/or as target for new antineoplastic agents.

ACKNOWLEDGMENTS

This work was supported by grants to G. Muccioli (FIN-60% 2005 University of Turin, Turin, Italy; CIPE 2005 Regione Piemonte, Italy; MIUR 2005 Italian Ministry of the University and Research, Rome, Italy) and the SMEM Foundation, Turin, Italy.

REFERENCES

Arosio, M., Ronchi, C. L., Gebbia, C., Cappiello, V., Beck-Peccoz, P., and Peracchi, M. (2003). Stimulatory effects of ghrelin on circulating somatostatin and pancreatic polypeptide levels. *J. Clin. Endocrinol. Metab.* **88,** 701–704.

Arvat, E., Maccario, M., Di Vito, L., Broglio, F., Benso, A., Gottero, C., Papotti, M., Muccioli, G., Dieguez, C., Casanueva, F. F., Deghenghi, R., Camanni, F., et al. (2001). Endocrine activities of ghrelin, a natural growth hormone secretagogue (GHS), in humans: Comparison and interactions with hexarelin, a nonnatural peptidyl GHS, and GH-releasing hormone. *J. Clin. Endocrinol. Metab.* **86,** 1169–1174.

Barreiro, M. L., and Tena-Sempere, M. (2004). Ghrelin and reproduction: A novel signal linking energy status and fertility? *Mol. Cell. Endocrinol.* **226,** 1–9.

Barrett-Connor, E., Khaw, K. T., and Yen, S. S. (1990). Endogenous sex hormone levels in older adult men with diabetes mellitus. *Am. J. Epidemiol.* **132,** 895–901.

Bedendi, I., Alloatti, G., Marcantoni, A., Malan, D., Catapano, F., Ghè, C., Deghenghi, R., Ghigo, E., and Muccioli, G. (2003). Cardiac effects of ghrelin and its endogenous derivatives des-octanoyl ghrelin and des-Gln^{14}-ghrelin. *Eur. J. Pharmacol.* **476,** 87–95.

Bednarek, M. A., Feighner, S. D., Pong, S. S., McKee, K. K., Hreniuk, D. L., Silva, M. V., Warren, V. A., Howard, A. D., Van der Ploeg, L. H., and Heck, J. V. (2000). Structure-function studies on the new growth hormonereleasing peptide, ghrelin: Minimal sequence of ghrelin necessary for activation of growth hormone secretagogue receptor 1a. *J. Med. Chem.* **43,** 4370–4376.

Benso, A., Broglio, F., Marafetti, L., Lucatello, B., Scardo, M. A., Granata, R., Martina, V., Papotti, M., Muccioli, G., and Ghigo, E. (2004). Ghrelin and synthetic growth hormone secretagogues are cardioactive molecules with identities and differences. *Semin. Vasc. Med.* **4,** 107–114.

Black, R. J., Bray, F., Ferlay, J., and Parkin, D. M. (1997). Cancer incidence and mortality in the European Union: Cancer registry data and estimates of national incidence for 1990. *Eur. J. Cancer* **33,** 1075–1107.

Bray, G. A. (2002). The underlying basis for obesity: Relationship to cancer. *J. Nutr.* **132,** 3451S–3455S.

Broglio, F., Benso, A., Castiglioni, C., Gottero, C., Prodam, F., Destefanis, S., Gauna, C., Van der lely, A. J., Deghenghi, R., Bo, M., Arvat, E., and Ghigo, E. (2003). The endocrine response to ghrelin as a function of gender in humans in young and elderly subjects. *J. Clin. Endocrinol. Metab.* **88**, 1537–1542.

Carter, H. B., and Partin, A. W. (1998). Diagnosis and staging of prostate cancer. In "Campbell's Urology" (P. C. Walsh, A. B. Retik, E. D. Vaughan, and A. J. Wein, eds.), 7th Ed., pp. 2519–2537. Philadelphia, W.B. Saunders.

Cassoni, P., Papotti, M., Catapano, F., Ghè, C., Deghenghi, R., Ghigo, E., and Muccioli, G. (2000). Specific binding sites for synthetic growth hormone secretagogues in non-tumoral and neoplastic human thyroid tissue. *J. Endocrinol.* **165**, 139–146.

Cassoni, P., Papotti, M., Ghè, C., Catapano, F., Sapino, A., Graziani, A., Deghenghi, R., Reissmann, T., Ghigo, E., and Muccioli, G. (2001). Identification, characterization, and biological activity of specific receptors for natural (ghrelin) and synthetic growth hormone secretagogues and analogs in human breast carcinomas and cell lines. *J. Clin. Endocrinol. Metab.* **86**, 1738–1745.

Cassoni, P., Muccioli, G., Marrocco, T., Volante, M., Allia, E., Ghigo, E., Deghenghi, R., and Papotti, M. (2002). Cortistatin-14 inhibits cell proliferation of human thyroid carcinoma cell lines of both follicular and parafollicular origin. *J. Endocrinol. Invest.* **25**, 362–368.

Cassoni, P., Ghè, C., Marrocco, T., Tarabra, E., Allia, E., Catapano, F., Deghenghi, R., Ghigo, E., Papotti, M., and Muccioli, G. (2004). Expression of ghrelin and biological activity of specific receptors for ghrelin and des-acyl ghrelin in human prostate neoplasms and related cell lines. *Eur. J. Endocrinol.* **150**, 173–184.

Cassoni, P., Allia, E., Marrocco, T., Ghè, C., Ghigo, E., Muccioli, G., and Papotti, M. (2006). Ghrelin in lung cancer: Expression in human primari tumors and *in vitro* effect on H345 smal cell carcinoma cell line. *J. Endocrinol. Invest.* **29**, 781–790.

Chan, C. B., and Cheng, C. H. K. (2004). Identification and functional characterization of two alternatively spliced growth hormone secretagogue receptor transcripts from the pituitary of black seabream *Achantopagrus schlegeli*. *Mol. Cell Endocrinol.* **214**, 81–95.

Chan, J. M., Stampfer, M. J., Giovannucci, E., Gann, P. H., Ma, J., Wilkinson, P., Hennekens, C. H., and Pollak, M. (1998). Plasma insulin-like growth factor-I and prostate cancer risk: A prospective study. *Science* **279**, 563–566.

Chan, J. M., Stampfer, M. J., Ma, J., Gann, P., Gaziano, J. M., Pollak, M., and Giovannucci, E. (2002). Insulin-like growth factor-I (IGF-I) and IGF binding protein-3 as predictors of advanced-stage prostate cancer. *J. Natl. Cancer Inst.* **94**, 1099–1106.

Chokkalingam, A. P., Pollak, M., Fillmore, C. M., Gao, Y. T., Stanczyk, F. Z., Deng, J., Sesterhenn, I. A., Mostofi, F. K., Fears, T. R., Madigan, M. P., Ziegler, R. G., Fraumeni, J. F., Jr., *et al.* (2001). Insulin-like growth factors and prostate cancer: A population-based case-control study in China. *Cancer Epidemiol. Biomarkers Prev.* **10**, 421–427.

Cioffi, J. A., Shafer, A. W., Zupancic, T. J., Smith-Gbur, J., Mikhail, A., Platika, D., and Snodgrass, H. R. (1996). Novel B219/OB receptor isoforms: Possible role of leptin in hematopoiesis and reproduction. *Nat. Med.* **2**, 585–589.

Clauson, P. G., Brismar, K., Hall, K., Linnarsson, R., and Grill, V. (1998). Insulin-like growth factor-I and insulin-like growth factor binding protein-1 in a representative population of type 2 diabetic patients in Sweden. *Scand. J. Clin. Lab. Invest.* **58**, 353–360.

Cnop, M., Havel, P. J., Utzschneider, K. M., Sinha, M. K., Boyko, E. J., Retzlaff, B. M., Knopp, R. H., Brunzell, J. D., and Kahn, S. E. (2003). Relationship of adiponectin to body fat distribution, insulin sensitivity and plasma lipoproteins: Evidence for independent roles of age and sex. *Diabetologia* **46**, 459–469.

Cummings, D. E. (2006). Ghrelin and the short- and long-term regulation of appetite and body weight. *Physiol. Behav.* **89**, 71–84.

Cunha, G. R., Ricke, W., Thomson, A., Marker, P. C., Risbridger, G., Hayward, S. W., Wang, Y. Z., Donjacour, A. A., and Kurita, T. (2004). Hormonal, cellular, and molecular regulation of normal and neoplastic prostatic development. *J. Steroid Biochem. Mol. Biol.* **92**, 221–236.

Dal Maso, L., Augustin, L. S., Karalis, A., Talamini, R., Franceschi, S., Trichopoulos, D., Mantzoros, C. S., and La Vecchia, C. (2004). Circulating adiponectin and endometrial cancer risk. *J. Clin. Endocrinol. Metab.* **89**, 1160–1163.

Denmeade, S. R., Lin, X. S., and Isaacs, J. T. (1996). Role of programmed (apoptotic) cell death during the progression and therapy for prostate cancer. *Prostate* **28**, 251–265.

De Vriese, C., Gregoire, F., De Neef, P., Robberecht, P., and Delporte, C. (2004). Ghrelin is produced by the human erythroleukemic HEL cell line and involved in an autocrine pathway leading to cell proliferation. *Endocrinology* **146**, 1514–1522.

Drachenberg, D. E., Elgamal, A. A., Rowbotham, R., Peterson, M., and Murphy, G. P. (1999). Circulating levels of interleukin-6 in patients with hormone refractory prostate cancer. *Prostate* **41**, 127–133.

Enomoto, M., Nagaya, N., Uematsu, M., Okumura, H., Nakagawa, E., Ono, F., Hosoda, H., Oya, H., Kojima, M., Kanmatsuse, K., and Kangawa, K. (2003). Cardiovascular and hormonal effects of subcutaneous administration of ghrelin, a novel growth hormone-releasing peptide, in healthy humans. *Clin. Sci. (Lond.)* **105**, 431–435.

Ferlay, J., Bray, F., Pisani, P., and Parkin, D. M. (2001). GLOBOCAN 2000: Cancer incidence, mortality and prevalence worldwide, version 1.0. *In* "IARC Cancer Base No. 5 book." Lyon, IARC Press.

Fernandez-Fernandez, R., Tena-Sempere, M., Aguilar, E., and Pinilla, L. (2004). Ghrelin effects on gonadotropin secretion in male and female rats. *Neurosci. Lett.* **362**, 103–107.

Fernandez-Fernandez, R., Navarro, V. M., Barreiro, M. L., Vigo, E. M., Tovar, S., Sirotkin, A. V., Casanueva, F. F., Aguilar, E., Dieguez, C., Pinilla, L., and Tena-Sempere, M. (2005). Effects of chronic hyperghrelinemia on puberty onset and pregnancy outcome in the rat. *Endocrinology* **146**, 3018–3025.

Friedman, J. M., and Halaas, J. L. (1998). Leptin and the regulation of body weight in mammals. *Nature* **395**, 763–770.

Furuta, M., Funabashi, T., and Rimura, F. (2001). Intracerebroventricular administration of ghrelin rapidly suppresses pulsatile luteinizing hormone secretion in ovariectomized rats. *Biochem. Biophys. Res. Commun.* **288**, 780–785.

Gann, P. H., Hennekens, C. H., Sacks, F. M., Grodstein, F., Giovannucci, E. L., and Stampfer, M. J. (1994). Prospective study of plasma fatty acids and risk of prostate cancer. *J. Natl. Cancer Inst.* **86**, 281–286.

Gasparian, A. B., Yao, Y. J., Kowalczyk, D., Lyakh, L. A., Karseladze, A., Slaga, T. J., and Budunova, I. V. (2002). The role of IKK in constitutive activation of NFkB transcription factor in prostate carcinoma cells. *J. Cell. Sci.* **115**, 141–151.

Gauna, C., Delhanty, P. J., Hofland, L. J., Janssen, J. A., Broglio, F., Ross, R. J., Ghigo, E., and van der Lely, A. J. (2005). Ghrelin stimulates, whereas des-octanoyl ghrelin inhibits, glucose output by primary hepatocytes. *J. Clin. Endocrinol. Metab.* **90**, 1055–1060.

Gewies, A., Rokhlin, O. W., and Cohen, M. B. (2000). Cytochrome c is involved in Fas-mediated apoptosis of prostatic carcinoma cell lines. *Cancer Res.* **60**, 2163–2168.

Ghè, C., Cassoni, P., Catapano, F., Marrocco, T., Deghenghi, R., Ghigo, E., Muccioli, G., and Papotti, M. (2002). The antiproliferative effect of synthetic peptidyl GH secretagogues in human CALU-1 lung carcinoma cells. *Endocrinology* **143**, 484–491.

Ghigo, E., Broglio, F., Arvat, E., Maccario, M., Papotti, M., and Muccioli, G. (2005). Ghrelin: More than a natural GH secretagogue and/or an orexigenic factor. *Clin. Endocrinol.* **62**, 1–17.

Gioeli, D., Mandell, J. W., Petroni, G. R., Frierson, H. F., Jr., and Weber, M. J. (1999). Activation of mitogen-activated protein kinase associated with prostate cancer progression. *Cancer Res.* **59**, 279–284.

Giovannucci, E. (1999). Insulin-like growth factor-I and binding protein-3 and risk of cancer. *Horm. Res.* **51**(Suppl. 3), 34–41.
Giovannucci, E., Stampfer, M. K., Krithivas, K., Brown, M., Dahl, D., Brufsky, A., Talcott, J., Hennekens, C. H., and Kantoff, P. W. (1997). The CAG repeat within the androgen receptor gene and its relationship to prostate cancer. *Proc. Natl. Acad. Sci. USA* **94,** 3320–3323.
Giovannucci, E., Rimm, E. B., Stampfer, M. J., Colditz, G. A., and Willett, W. C. (1998). Diabetes mellitus and risk of prostate cancer (United States). *Cancer Causes Control* **9,** 3–9.
Gnanapavan, S., Kola, B., Bustin, S. A., Morris, D. G., McGee, P., Fairclough, P., Bhattacharya, S., Carpenter, R., Grossmann, A. B., and Korbonits, M. (2002). The tissue distribution of the mRNA of ghrelin and subtypes of its receptors, GHS-R, in humans. *J. Clin. Endocrinol. Metab.* **87,** 2988–2991.
Goktas, S., Yilmaz, M. I., Caglar, K., Sonmez, A., Kilic, S., and Bedir, S. (2005). Prostate cancer and adiponectin. *Urology* **65,** 1168–1172.
Gondo, R. G., Aguiar-Oliveira, M. H., Hayashida, C. Y., Toledo, S. P., Abelin, N., Levine, M. A., Bowers, C. Y., Souza, A. H., Pereira, R. M., Santos, N. L., and Salvatori, R. (2001). Growth hormone-releasing peptide-2 stimulates GH secretion in GH-deficient patients with mutated GH-releasing hormone receptor. *J. Clin. Endocrinol. Metab.* **86,** 3279–3283.
Granata, R., Settanni, F., Biancone, L., Trovato, L., Nano, R., Bertuzzi, F., Destefanis, S., Annunziata, M., Martinetti, M., Catapano, F., Ghè, C., Isgaard, J., *et al.* (2007). Acylated and unacylated ghrelin promote proliferation and inhibit apoptosis of pancreatic β cells and human islets. Involvement of cAMP/PKA, ERK1/2 and PI3K/AKT signalling. *Endocrinology* **148,** 512–529.
Grumbach, M. M., Hughes, I. A., and Conte, F. A. (2003). Disorders of sex differentiation. *In* "Williams Textbook of Endocrinology" (P. R. Larsen, H. M. Kronenberg, S. Melmed, and K. S. Polonsky, eds.), 10th Ed., pp. 842–1002. W.B. Saunders, Philadelphia.
Haffner, S. M. (2000). Sex hormones, obesity, fat distribution, type 2 diabetes and insulin resistance: Epidemiological and clinical correlation. *Int. J. Obes. Relat. Metab. Disord.* **24**(Suppl. 2), S56–S58.
Haffner, S. M., Shaten, J., Stern, M. P., Smith, G. D., and Kuller, L. (1996). Low levels of sex hormone-binding globulin and testosterone predict the development of non-insulin-dependent diabetes mellitus in men. MRFIT Research Group. Multiple Risk Factor Intervention Trial. *Am. J. Epidemiol.* **143,** 889–897.
Hashizume, T., Horiuchi, M., Tate, N., Nonaka, S., Mikami, U., and Kojima, M. (2003). Effects of Ghrelin on growth hormone secretion from cultured adenohypophysial cells in pigs. *Domest. Anim. Endocrinol.* **24,** 209–218.
Hirosumi, J., Tuncman, G., Chang, L., Gorgun, C. Z., Uysal, K. T., Maeda, K., Karin, M., and Hotamisligil, G. S. (2002). A central role for JNK in obesity and insulin resistance. *Nature* **420,** 333–336.
Ho, S. M. (2004). Estrogens and antiestrogens: Key mediators of prostate carcinogenesis and new therapeutic candidates. *J. Cell. Biochem.* **91,** 491–503.
Hoelting, T., Duh, Q. Y., Clark, O. H., and Herfarth, C. (1996). Somatostatin analog octreotide inhibits the growth of differentiated thyroid cancer cells *in vitro* but not *in vivo*. *J. Clin. Endocrinol. Metab.* **81,** 2638–2641.
Horvath, T. L., Diano, S., Sotonyi, P., Heiman, M., and Tschop, M. (2001). Minireview: Ghrelin and the regulation of energy balance—a hypothalamic perspective. *Endocrinology* **142,** 4163–4169.
Hosoda, H., Kojima, M., Mizushima, T., Shimizu, S., and Kangawa, K. (2003). Structural divergence of human ghrelin. Identification of multiple ghrelin-derived molecules produced by post-translational processing. *J. Biol. Chem.* **278,** 64–70.

Howard, A. D., Feighner, S. D., Cully, D. F., Arena, J. P., Liberator, P. A., Rosenblum, C. I., Hamelin, M., Hreniuk, D. L., Palyha, O. C., Anderson, J., Paress, P. S., Diaz, C., *et al.* (1996). A receptor in pituitary and hypothalamus that functions in growth hormone release. *Science* **273,** 974–977.

Hsieh, C. C., Thanos, A., Mitropoulos, D., Deliveliotis, C., Nantzoros, C. S., and Trichopoulos, D. (1999). Risk factors for prostate cancer: A case–control study in Greece. *Int. J. Cancer* **80,** 699–703.

Hsing, A. W., Deng, J., Sesterhenn, I. A., Mostofi, F. K., Stanczyk, F. Z., Benichou, J., Xie, T., and Gao, Y. T. (2000). Body size and prostate cancer: A population-based case-control study in China. *Cancer Epidemiol. Biomarkers Prev.* **9,** 1335–1341.

Iqbal, J., Kurose, Y., Canny, B., and Clarke, I. J. (2006). Effects of central infusion of ghrelin on food intake and plasma levels of growth hormone, luteinizing hormone, prolactin, and cortisol secretion in sheep. *Endocrinology* **147,** 510–519.

Jeffery, P. L., Herington, A. C., and Chopin, L. K. (2002). Expression and action of the growth hormone releasing peptide ghrelin and its receptor in prostate cancer cell lines. *J. Endocrinol.* **172,** R7–R11.

Jeffery, P. L., Herington, A. C., and Chopin, L. K. (2003). The potential autocrine/paracrine roles of ghrelin and its receptor in hormone-dependent cancer. *Cytokine Growth Factor Rev.* **14,** 113–122.

Jeffery, P. L., Murray, R. E., Yeh, A. H., McNamara, J. F., Duncan, R. P., Francis, G. D., Herington, A. C., and Chopin, L. K. (2005). Expression and function of the ghrelin axis, including a novel preproghrelin isoform, in human breast cancer tissues and cell lines. *Endocr. Relat. Cancer* **12,** 839–850.

Jemal, A., Thomas, A., Murray, T., and Thun, M. (2002). Cancer statistics, 2002. *CA Cancer J. Clin.* **52,** 23–47.

Kamegai, J., Tamura, H., Shimizu, T., Ishii, S., Sugihara, H., and Wakabayashi, I. (2000). Central effect of ghrelin, an endogenous growth hormone secretagogue, on hypothalamic peptide gene expression. *Endocrinology* **141,** 4797–4800.

Kanamoto, N., Akamizu, T., Hosoda, H., Hataya, Y., Ariyasu, H., Takaya, K., Hosoda, K., Saijo, M., Moriyama, K., Shimatsu, A., Kojima, M., Kangawa, K., *et al.* (2001). Substantial production of ghrelin by a human medullary thyroid carcinoma cell line. *J. Clin. Endocrinol. Metab.* **86,** 4984–4990.

Kawamura, K., Sato, N., Fukuda, J., Kodama, H., Kumagai, J., Tanikawa, H., Nakamura, A., Honda, Y., Sato, T., and Tanaka, T. (2003). Ghrelin inhibits the development of mouse preimplantation embryos *in vitro*. *Endocrinology* **144,** 2623–2633.

Kojima, M., and Kangawa, K. (2005). Ghrelin: Structure and function. *Physiol. Rev.* **85,** 495–522.

Kojima, M., Hosoda, H., Date, Y., Nakazato, M., Matsuo, H., and Kangawa, K. (1999). Ghrelin is a growth-hormone-releasing acylated peptide from stomach. *Nature* **402,** 656–660.

Korbonits, M., Jacobs, R. A., Aylwin, S. J., Burrin, J. M., Dahia, P. L., Monson, J. P., Honegger, J., Fahlbush, R., Trainer, P. J., Chew, S. L., Besser, G. M., and Grossman, A. B. (1998). Expression of the growth hormone secretagogue receptor in pituitary adenomas and other neuroendocrine tumors. *J. Clin. Endocrinol. Metab.* **83,** 3624–3630.

Labrie, F., Luu, V., Lin, S. X., Simard, J., and Labrie, C. (2000). The role of 17β-hydroxysteroid dehydrogenases in sex steroid formation in peripheral intracrine tissues. *Trends Endocrinol. Metab.* **11,** 421–427.

Lagiou, P., Signorello, L. B., Trichopoulos, D., Tzonou, A., Trichopoulou, A., and Mantzoros, C. S. (1998). Leptin in relation to prostate cancer and benign prostatic hyperplasia. *Int. J. Cancer* **76,** 25–28.

Lago, F., Gonzalez-Juanatey, J. R., Casanueva, F. F., Gomez-Reino, J., Dieguez, C., and Gualillo, O. (2005). Ghrelin, the same peptide for different functions: Player or bystander? *Vitam. Horm.* **71,** 405–432.

Lall, S., Tung, L. Y., Ohlsson, C., Jansson, J. O., and Dickson, S. L. (2001). Growth hormone (GH)-independent stimulation of adiposity by GH secretagogues. *Biochem. Biophys. Res. Commun.* **280,** 132–138.

Lanfranco, F., Zitzmann, M., Simoni, M., and Nieschlag, E. (2004). Serum adiponectin levels in hypogonadal males: Influence of testosterone replacement therapy. *Clin. Endocrinol.* **60,** 500–507.

Lin, J., Adam, R. M., Santiestevan, E., and Freeman, M. R. (1999). The phosphatidylinositol 3′-kinase pathway is a dominant growth factor-activated cell survival pathway in LNCaP human prostate carcinoma cells. *Cancer Res.* **59,** 2891–2897.

Locke, W., Kirgis, H. D., Bowers, C. Y., and Abdoh, A. A. (1995). Intracerebroventricular growth-hormone-releasing peptide-6 stimulates eating without affecting plasma growth hormone responses in rats. *Life Sci.* **56,** 1347–1352.

Mantzoros, C. S., Petridou, E., Dessypris, N., Chavelas, C., Dalamaga, M., Alexe, D. M., Papadiamantis, Y., Markopoulos, C., Spanos, E., Chrousos, G., and Trichopoulos, D. (2004). Adiponectin and breast cancer risk. *J. Clin. Endocrinol. Metab.* **89,** 1102–1107.

Masuda, Y., Tanaka, T., Inomata, N., Ohnuma, N., Tanaka, S., Itoh, Z., Hosoda, H., Kojima, M., and Kangawa, K. (2000). Ghrelin stimulates gastric acid secretion and motility in rats. *Biochem. Biophys. Res. Commun.* **276,** 905–908.

McCarty, M. F. (1997). Up-regulation of IGF binding protein-1 as an anticarcinogenic strategy: Relevance to caloric restriction, exercise, and insulin sensitivity. *Med. Hypotheses* **48,** 297–308.

McKee, K. K., Tan, C. P., Palyha, O. C., Liu, J., Feighner, S. D., Hreniuk, D. L., Smith, R. G., Howard, A. D., and Van der Ploeg, L. H. (1997). Cloning and characterization of two human G protein-coupled receptor genes (GPR38 and GPR39) related to the growth hormone secretagogue and neurotensin receptors. *Genomics* **46,** 426–434.

Montironi, R., Mazzucchelli, R., and Scarpelli, M. (2002). Precancerous lesions and conditions of the prostate: From morphological and biological characterization to chemoprevention. *Ann. N. Y. Acad. Sci.* **963,** 169–184.

Mora, L. B., Buettner, R., Seigne, J., Diaz, J., Ahmad, N., Garcia, R., Bowman, T., Falcone, R., Fairclough, R., Cantor, A., Muro-Cacho, C., Livingston, S., *et al.* (2002). Constitutive activation of Stat3 in human prostate tumors and cell lines: Direct inhibition of Stat3 signaling induces apoptosis of prostate cancer cells. *Cancer Res.* **62,** 6659–6666.

Muccioli, G., Tschop, M., Papotti, M., Deghenghi, R., Heiman, M., and Ghigo, E. (2002). Neuroendocrine and peripheral activities of ghrelin: Implications in metabolism and obesity. *Eur. J. Pharmacol.* **440,** 235–254.

Muccioli, G., Broglio, F., Tarabra, E., and Ghigo, E. (2004). Known and unknown growth hormone secretagogue receptors and their ligands. *In* "Ghrelin" (E. Ghigo, ed.), pp. 27–45. Dordrecht, Kluwer Academic Publishers.

Murakami, N., Hayashida, T., Kuroiwa, T., Nakahara, K., Ida, T., Mondal, M. S., Nakazato, M., Kojima, M., and Kangawa, K. (2002). Role for central ghrelin in food intake and secretion profile of stomach ghrelin in rats. *J. Endocrinol.* **174,** 283–288.

Nazian, S. J., and Cameron, D. F. (1999). Temporal relation between leptin and various indices of sexual maturation in the male rat. *J. Androl.* **20,** 487–491.

Nishizawa, H., Shimomura, I., Kishida, K., Maeda, N., Kuriyama, H., Nagaretani, H., Matsuda, M., Kondo, H., Furuyama, N., Kihara, S., Nakamura, T., Tochino, Y., *et al.* (2002). Androgens decrease plasma adiponectin, an insulin-sensitizing adipocyte-derived protein. *Diabetes* **51,** 2734–2741.

Nogueiras, R., Perez-Tilve, D., Wortley, K. E., and Tschöp, M. (2006). Growth hormone secretagogue (ghrelin-) receptors—A complex drug target for the regulation of body weight. *CNS Neurol. Disord. Drug Targets* **5**, 335–343.

Okada, K., Ishii, S., Minami, S., Sugihara, H., Shibasaki, T., and Wakabayashi, I. (1996). Intracerebroventricular administration of the growth hormone-releasing peptide KP-102 increases food intake in freefeeding rats. *Endocrinology* **137**, 5155–5158.

Onishi, T., Yamakawa, K., Franco, O. E., Kawamura, J., Watanabe, M., Shiraishi, T., and Kitazawa, S. (2001). Mitogen-activated protein kinase pathway is involved in α6 integrin gene expression in androgen-independent prostate cancer cells: Role of proximal Sp1 consensus sequence. *Biochim. Biophys. Acta* **1538**, 218–227.

Onuma, M., Bub, J. D., Rummel, T. L., and Iwamoto, Y. (2003). Prostate cancer cell-adipocyte interaction: Leptin mediates androgen-independent prostate cancer cell proliferation through c-Jun NH2-terminal kinase. *J. Biol. Chem.* **278**, 42660–42667.

Ozcan, U., Cao, Q., Yilmaz, E., Lee, A. H., Iwakoshi, N. N., Ozdelen, E., Tuncman, G., Gorgun, C., Glimcher, L. H., and Hotamisligil, G. S. (2004). Endoplasmic reticulum stress links obesity, insulin action, and type 2 diabetes. *Science* **306**, 457–461.

Papotti, M., Ghè, C., Volante, M., and Muccioli, G. (2004). Ghrelin and tumors. *In* "Ghrelin" (E. Ghigo, ed.), pp. 143–164. Dordrecht, Kluwer Academic Publishers.

Pollak, M., Beamer, W., and Zhang, J. C.(1998–1999). Insulin-like growth factors and prostate cancer. *Cancer Metastasis Rev.* **17**, 383–390.

Prado, C. L., Pugh-Bernard, A. E., Elghazi, L., Sosa-Pineda, B., and Sussel, L. (2004). Ghrelin cells replace insulin-producing beta cells in two mouse models of pancreas development. *Proc. Natl. Acad. Sci. USA* **101**, 2924–2929.

Quinn, M., and Babb, P. (2002). Patterns and trends in prostate cancer incidence, survival, prevalence and mortality. Part 1: International comparisons. *BJU Int.* **90**, 162–173.

Reddy, K. B., Nabha, S. M., and Atanaskova, N. (2003). Role of MAP kinase in tumor progression and invasion. *Cancer Metastasis Rev.* **22**, 395–403.

Russel, D. W., and Wilson, J. D. (1994). Steroid 5-α-reductase: Two genes/two isoenzymes. *Annu. Rev. Biochem.* **63**, 25–61.

Sakr, W. A., Grignon, D. J., Haas, G. P., Heilbrun, L. K., Pontes, J. E., and Crissman, J. D. (1996). Age and racial distribution of prostatic intraepithelial neoplasia. *Eur. Urol.* **30**, 138–144.

Sato, M., Nakahara, K., Goto, S., Kaiya, H., Mivazato, M., Date, Y., Nakazato, M., Kangawa, K., and Murakami, N. (2006). Effects of ghrelin and des-acyl ghrelin on neurogenesis of the rat fetal spinal cord. *Biochem. Biophys. Res. Commun.* **350**, 598–603.

Scherer, P. E., Williams, S., Fogliano, M., Baldini, G., and Lodish, H. F. (1995). A novel serum protein similar to C1q, produced exclusively in adipocytes. *J. Biol. Chem.* **270**, 26746–26749.

Sierra-Honigmann, M. R., Nath, A. K., Murakami, C., Garcia-Cardena, G., Papapetropoulos, A., Sessa, W. C., Madge, L. A., Schechner, J. S., Schwabb, M. B., Polverini, P. J., and Flores-Riveros, J. R. (1998). Biological action of leptin as an angiogenic factor. *Science* **281**, 1683–1686.

Signoretti, S., and Loda, M. (2001). Estrogen receptor β in prostate cancer: Brake pedal or accelerator? *Am. J. Pathol.* **159**, 13–16.

Somasundar, P., Yu, A. K., Vona-Davis, L., and McFadden, D. W. (2003). Differential effects of leptin on cancer *in vitro*. *J. Surg. Res.* **113**, 50–55.

Somasundar, P., Frankenberry, K. A., Skinner, H., Vedula, G., McFadden, D. W., Riggs, D., Jackson, B., Vangilder, R., Hileman, S. M., and Vona-Davis, L. C. (2004). Prostate cancer cell proliferation is influenced by leptin. *J. Surg. Res.* **118**, 71–82.

Soronen, P., Laiti, M., Törn, S., Härkönen, P., Patrikainen, L., Li, Y., Pulkka, A., Kurkela, R., Herrala, A., Kaija, H., Isomaa, V., and Vihko, P. (2004). Sex steroid hormone metabolism and prostate cancer. *J. Steroid Biochem. Mol. Biol.* **92**, 281–286.

Stattin, P., Soderberg, S., Hallmans, G., Bylund, A., Kaaks, R., Stenman, U. H., Bergh, A., and Olsson, T. (2001). Leptin is associated with increased prostate cancer risk: A nested case-referent study. *J. Clin. Endocrinol. Metab.* **86,** 1341–1345.

Steinberg, G. D., Carter, B. S., Beaty, T. H., Childs, B., and Walsh, P. C. (1990). Family history and the risk of prostate cancer. *Prostate* **17,** 337–347.

Takahashi, K., Furukawa, C., Takano, A., Ishikawa, N., Kato, T., Hayama, S., Suzuki, C., Yasui, W., Inai, K., Sone, S., Ito, T., Nishimura, H., *et al.* (2006). The neuromedin u-growth hormone secretagogue receptor 1b/neurotensin receptor 1 oncogenic signaling pathway as a therapeutic target for lung cancer. *Cancer Res.* **66,** 9408–9419.

Tan, C. P., McKee, K. K., Liu, Q., Palyha, O. C., Feighner, S. D., Smith, R. G., and Howard, A. D. (1998). Cloning and characterization of a human and murine T-cell orphan G-protein-coupled receptor similar to the growth hormone secretagogue and neurotensin receptors. *Genomics* **52,** 223–229.

Tannenbaum, G. S., Epelbaum, J., and Bowers, C. Y. (2003). Interrelationship between the novel peptide ghrelin and somatostatin/growth hormone-releasing hormone in regulation of pulsatile growth hormone secretion. *Endocrinology* **144,** 967–974.

Tena-Sempere, M., Barreiro, M. L., Gonzalez, L. C., Gaytan, F., Zhang, F. P., Caminos, J. E., Pinilla, L., Casanueva, F. F., Dieguez, C., and Aguilar, E. (2002). Novel expression and functional role of ghrelin in rat testis. *Endocrinology* **143,** 717–725.

Thompson, J., Hyytinen, E. R., Haapala, K., Rantala, I., Helin, H. J., Janne, O. A., Palvimo, J. J., and Koivisto, P. A. (2003). Androgen receptor mutations in high-grade prostate cancer before hormonal therapy. *Lab. Invest.* **83,** 1709–1713.

Torsello, A., Luoni, M., Schweiger, F., Grilli, R., Guidi, M., Bresciani, E., Deghenghi, R., Muller, E. E., and Locatelli, V. (1998). Novel hexarelin analogs stimulate feeding in the rat through a mechanism not involving growth hormone release. *Eur. J. Pharmacol.* **360,** 123–129.

Torsello, A., Locatelli, V., Melis, M. R., Succu, S., Spano, M. S., Deghenghi, R., Muller, E. E., and Argiolas, A. (2000). Differential orexigenic effects of hexarelin and its analogs in the rat hypothalamus: Indication for multiple growth hormone secretagogue receptor subtypes. *Neuroendocrinology* **72,** 327–332.

Trudel, L., Tomasetto, C., Rio, M. C., Bouin, M., Plourde, V., Eberling, P., and Poitras, P. (2002). Ghrelin/motilin-related peptide is a potent prokinetic to reverse gastric postoperative ileus in rat. *Am. J. Physiol. Gastrointest. Liver Physiol.* **282,** G948–G952.

Tschöp, M., Statnick, M. A., Suter, T. M., and Heiman, M. L. (2002). GH-releasing peptide-2 increases fat mass in mice lacking NPY: Indication for a crucial mediating role of hypothalamic agouti-related protein. *Endocrinology* **143,** 558–568.

Underwood, W., III, Jackson, J., Wei, J. T., Dunn, R., Baker, E., Demonner, S., and Wood, D. P. (2005). Racial treatment trends in localized/regional prostate carcinoma: 1992–1999. *Cancer* **103,** 538–545.

van der Lely, A. J., Tschop, M., Heiman, M. L., and Ghigo, E. (2004). Biological, physiological, pathophysiological, and pharmacological aspects of ghrelin. *Endocr. Rev.* **25,** 426–457.

Vulliémoz, N. R., Xiao, E., Xia-Zhang, L., Germond, M., Rivier, J., and Ferin, M. (2004). Decrease in luteinizing hormone pulse frequency during a five-hour peripheral ghrelin infusion in the ovariectomized rhesus monkey. *J. Clin. Endocrinol. Metab.* **89,** 5718–5723.

Weikel, J. C., Wichniak, A., Ising, M., Brunner, H., Friess, E., Held, K., Mathias, S., Schmid, D. A., Uhr, M., and Steiger, A. (2003). Ghrelin promotes slow-wave sleep in humans. *Am. J. Physiol. Endocrinol. Metab.* **284,** E407–E415.

Will, J. C., Vinicor, F., and Calle, E. E. (1999). Is diabetes mellitus associated with prostate cancer incidence and survival? *Epidemiology* **10,** 313–318.

Wolk, A., Mantzoros, C. S., Andersson, S. O., Bergstrom, R., Signorello, L. B., Lagiou, P., Adami, H. O., and Trichopoulos, D. (1998). Insulin-like growth factor 1 and prostate cancer risk: A population-based, case–control study. *J. Natl. Cancer Inst.* **90,** 911–915.

Yeh, A. H., Jeffery, P. L., Duncan, R. P., Herington, A. C., and Chopin, L. K. (2005). Ghrelin and a novel preproghrelin isoform are highly expressed in prostate cancer and ghrelin activates mitogen-activated protein kinase in prostate cancer. *Clin. Cancer Res.* **11,** 8295–8303.

Yokota, T., Oritani, K., Takahashi, I., Ishikawa, J., Matsuyama, A., Ouchi, N., Kihara, S., Funahashi, T., Tenner, A. J., Tomiyama, Y., and Matsuzawa, Y. (2000). Adiponectin, a new member of the family of soluble defense collagens, negatively regulates the growth of myelomonocytic progenitors and the functions of macrophages. *Blood* **96,** 1723–1732.

Zhang, W., Chen, M., Chen, X., Segura, B. J., and Mulholland, M. W. (2001). Inhibition of pancreatic protein secretion by ghrelin in the rat. *J. Physiol.* **537,** 231–236.

Zhang, W., Lin, T. R., Hu, Y., Fan, Y., Zhao, L., Stuenkel, E. L., and Mulholland, M. W. (2004). Ghrelin stimulates neurogenesis in the dorsal motor nucleus of the vagus. *J. Physiol.* **559,** 729–737.

CHAPTER FOURTEEN

NOVEL CONNECTIONS BETWEEN THE NEUROENDOCRINE AND IMMUNE SYSTEMS: THE GHRELIN IMMUNOREGULATORY NETWORK

Dennis D. Taub[*]

Contents

I. Neuroendocrine–Immune Interactions	326
II. The Growth Hormone Secretagogue Receptor	327
A. Growth hormone secretagogues	327
B. GHSR properties	328
C. GHSR expression on immune cells	328
III. Ghrelin: Hormone or Cytokine?	329
A. Properties of ghrelin	329
B. Sources of ghrelin production	330
C. Biological and functional activities of ghrelin	332
IV. Conclusions	341
Acknowledgments	342
References	342

Abstract

There appears to be bidirectional communication between the neuroendocrine and immune systems. This communication is mediated by way of an array of cytokines, hormones, and neuropeptides. Inflammatory cytokines released by immune cells have been shown to act on the central nervous system to control food intake and energy homeostasis. Decrease in food intake or anorexia is one of the most common symptoms of illness, injury, or inflammation. The adipocyte-derived hormone, leptin, is considered a critical sensory anorexigenic mediator that signals to the brain changes in stored energy, determined by an altered balance between food intake and energy expenditure and has been shown to exert certain proinflammatory effects on immune cells. In contrast, ghrelin, the endogenous ligand for growth hormone secretagogue receptors (GHSRs), is produced primarily from stomach serving as a potent circulating

---

[*] Laboratory of Immunology, National Institute on Aging (NIH), Baltimore, Maryland 21224

Vitamins and Hormones, Volume 77
ISSN 0083-6729, DOI: 10.1016/S0083-6729(06)77014-5

orexigen controlling energy expenditure, adiposity, and GH secretion. However, the functional role of ghrelin and GHS in immune cell function remains unclear. Here, we review the current literature supporting a role for ghrelin in controlling inflammation and immunity and the potential therapeutic use of ghrelin and GHSR agonists in the management of inflammation and in restoration of thymic function in immunocompromised individuals. © 2008 Elsevier Inc.

I. Neuroendocrine–Immune Interactions

The cross talk between the neuroendocrine and immune systems is now well established. These systems utilize similar ligands and receptors to establish intra- and intersystem communication that plays an important role in physiological homeostasis (Steinman, 2004). Increasing evidence indicates that many hormones and neuropeptides are potent immunoregulatory molecules that participate in various aspects of immune function, in healthy and diseased individuals. Immune cells express receptors for many of these ligands and, on ligand exposure, their functions are modulated by a variety of hormones, neurotransmitters, and neuropeptides including corticosteroids, insulin, prolactin, growth hormone (GH), androgens, opioids, substance P, and somatostatins. Similarly, receptors for many immune-derived cytokines, chemokines, and growth factors have been identified on neuronal cells and endocrine glands under normal physiological conditions and in response to stress and disease. Further, immune and neuronal cells express or are induced to express many similar cytokines, hormones, and growth factors, thus mediating communications between these systems.

The pituitary hormone, GH, has been shown to exhibit a number of effects on the cells and organs of the immune function. GH is classically defined as a peptide hormone that is synthesized and secreted primarily by somatotrophic cells in the anterior pituitary. The production of GH is pulsatile, primarily nocturnal, and is controlled by hypothalamic hormones such as GH-releasing hormone (GHRH), hypothalamic GH release-inhibiting factor, and somatostatin. Circulating levels of GH are highest in the immediate neonatal period, decreasing during childhood but peaking again during puberty. GH secretion falls precipitously during aging. It has been demonstrated that GH is produced by normal lymphocytes (Welniak et al., 2002). In humans, unstimulated peripheral blood lymphocytes (PBL) express GH mRNA, and up to 10% of the cells secrete biologically active protein. One of the most important effects of GH is its ability to stimulate growth of skeleton and soft tissues. Moreover, GH has been shown to mediate the proliferation of a number of cell types, including chondrocytes, fibroblasts, adipocytes, myoblasts, and T lymphocytes. Many of these GH effects appear to be mediated indirectly via the production of insulin-like growth factor-1 (IGF-1). Several *in vitro* or *in vivo* studies have suggested that GH could play a role in immune function, inducing the survival and/or the proliferation of lymphoid cells. Earlier observations documented that thymic atrophy was often associated with ablation of the pituitary gland and mice with the dwarf mutation had reduced immune functions (Welniak *et al.*, 2002; Woody *et al.*, 1999). Hypophysectomized rats or Snell dwarf mice (dw/dw), which are both defective in the production of GH (and also of prolactin and thyroid hormones), displayed deficiencies in lymphocyte development and function, which were corrected on administration of exogenous GH to these animals. Similarly, administration of GH was found to enhance development of the thymus in aged mice and promote the engraftment of murine or human T cells in severe, combined immunodeficiency (SCID) mice (Murphy *et al.*, 1992a,b; Taub *et al.*, 1994). Some of the GH effects are indirectly mediated by IGF-1, synthesized in the liver, and induced by GH. Furthermore, GH and IGF-1 could synergize with other cytokines such as GM-CSF in hematopoiesis (Welniak *et al.*, 2002), possibly leading to the various effects attributed to GH. Thus, based on these data, it was proposed that GH administration would be valuable in therapeutic treatment of transplantation and states of immunosuppression (Murphy and Longo, 2000). Among the most promising immune effects of GH relevant to aging is its potential to improve thymic function, promote thymic and bone marrow engraftment, and stimulate hematopoiesis in immunosuppressed and aged animals (French *et al.*, 2002; Murphy *et al.*, 1992b, 1993). While these data seem convincing, a role for GH in immune function remains controversial especially with the lack of immunologic defects in GH receptor knockout mice (Zhou *et al.*, 1997). However, despite this knockout mouse data, there is ample evidence for a role of GH and its receptor in immune modulation including data supporting an immunostimulatory role for GH under conditions of stress and immunosuppression (e.g., with age and posttransplantation).

II. THE GROWTH HORMONE SECRETAGOGUE RECEPTOR

A. Growth hormone secretagogues

Despite the beneficial effects of GH, the many side effects of hormone administration including fluid retention, joint pain, and nerve compression as well as concerns regarding an increased risk of developing diabetes or cancer prompted a search for alternative strategies to obtain the beneficial effects of GH without the toxicities.

Cyril Bowers and coworkers made the startling discovery that several synthetic opioid analogues elicited significant GH release from pituitary cells on binding to unknown receptor(s) distinct from known endogenous opioid and GHRH-binding sites (Bowers, 2001). This class of synthetic nonpeptidyl compounds is known as growth hormone secretagogues (GHSs) (Smith et al., 2001). GHS induces calcium flux in rat pituitary cells and causes the release of GH. It is believed that GHS may be utilized therapeutically to mimic the *in vivo* effects of GH. Preclinical and clinical studies have demonstrated the efficacy of GHS in facilitating muscle development and the strengthening bones in the elderly (Dixit and Taub, 2005). *In vitro* experiments have also demonstrated that GHS promote the release of GH by acting on a specific G-protein–coupled receptor, called the "growth hormone secretagogue receptor" (GHSR) (Howard et al., 1996). However, the significance of this receptor in mammalian physiology was unknown until the natural or endogenous ligand for this receptor, ghrelin, in 1999 (Kojima et al., 1999).

B. GHSR properties

GHSR gene is located within the quantitative trait locus (QTL) on chromosome 3q26-q29, a locus that is associated with specific traits of metabolic syndrome and body mass index (BMI) (Kissebah et al., 2000). This receptor is alternatively spliced and exists in two forms, namely GHSR1a and GHSR1b. The GHSR1a gene encodes a 366-amino acid functional seven-transmembrane receptor with ligand-binding pocket in the third transmembrane loop (Smith et al., 1999). The GHSR1b form of the receptor is predicted to be a 289-amino acid five-transmembrane receptor, although its expression remains unclear. Furthermore, the lack of calcium release by GHSR1b-transfected COS-7 cells in response to GHS treatment suggests that this receptor is nonfunctional (Petersenn et al., 2001). Experiments in mice lacking the GHSR1a gene proved definitively that the orexigenic and GH-stimulatory effects of GHS are mediated via this receptor (Sun et al., 2003). The canonical signaling pathway for GHSR1a is through G_q/phospholipase by the observation that mice lacking $G_{q\alpha}$-subunits in neurons and glia have impaired hypothalamic responses to ghrelin (Cummings and Overduin, 2007; Gil-Campos et al., 2006). In addition, GHSR has also been shown to be associated with the G_s/protein kinase A pathways in several tissues (Gil-Campos et al., 2006). However, in a recent study by Holst et al. (2004), the overexpression of GHSR in cell lines revealed that this receptor appears to be constitutively active and may operate in a ligand-independent manner. The exact physiological significance of this activity remains to be determined and to date, no subsequent studies supporting these findings in nontransfected or primary cells have been reported.

C. GHSR expression on immune cells

The GHSR1a was originally thought to be exclusively expressed only in the pituitary and hypothalamus; however, more recent studies have demonstrated that this receptor is widely expressed in the brain and peripheral tissues (especially the pituitary, stomach, intestine, pancreas, thymus, gonads, thyroid, and heart) and in various forms of cancers (Gnanapavan et al., 2002; Jeffery et al., 2003). Moreover, GHSR mRNA is expressed in several lymphoid organs (Gnanapavan et al., 2002) and in various leukocyte subsets including T and B cells, monocytes (Dixit et al., 2004; Hattori et al., 2001), and dendritic cells (personal observation), suggesting that GHSR might play some role in the generation and/or control of immune interactions. In our work with resting human T cells, we have characterized the expression GHSR protein on human T cells and monocytes (Dixit et al., 2004). These studies revealed the following:

1. T lymphocytes express high affinity, functional GHSR receptors on their cell surface and these receptors colocalize with lipid rafts on cellular activation. These receptors are expressed on the cell surface in distinct heterogeneous patterns ranging from "crescent moon" shaped to a more punctate and diffuse. In resting T cells, the majority of these receptors appear to be segregated from the GM1-rich cholesterol and sphingolipid-rich lipid raft microdomains. However, on T cell activation, the GHSRs also redistribute into lipid rafts displaying a polarized capped phenotype similar to chemokine receptors. Such raft localization and receptor redistribution suggests that GHSR may play an important role in cellular signaling and motility.
2. Multiple T cell subsets including CD4+ and CD8+ and memory and naive T cells constitutively express GHSR mRNA and protein, which is significantly upregulated on T-cell receptor (TCR) activation.
3. Ligation of these receptors with GHSR ligands results in increased receptor-specific calcium mobilization and actin polymerization in a fashion similar to many chemokine receptors.

The relevance of increased numbers of functional GHSRs during T cells activation remains to be defined. To date, no definitive data exists linking GHSR or its ligands to any immunologic or inflammatory disorders. However, we believe that immune-associated GHSRs play a novel immunoregulatory role in controlling cytokine expression, cellular activation and trafficking, and apoptosis. These possibilities shall be discussed below.

III. Ghrelin: Hormone or Cytokine?

A. Properties of ghrelin

While GHSR was initially identified based on its ability to bind and signal in response to GHS ligands, no endogenous ligand was known for this receptor until 1999 when Kojima et al. (1999) identified and cloned the GHSR ligand, ghrelin. The ghrelin gene encodes a 117-amino acid peptide, pre-proghrelin. This pre-pro form is subsequently cleaved into the mature 28-amino acid form after which it is secreted. Ghrelin function was primarily described by its ability to strongly stimulate the GH release from the pituitary. However, more recent studies have revealed that ghrelin is a potent inducer of food intake and also increases adiposity (Cummings *et al.*, 2002; Dixit and Taub, 2005; Tschop *et al.*, 2000). Ghrelin is predominantly expressed by the X/A-like cells of the stomach (Kojima *et al.*, 1999; Smith *et al.*, 2005); and during states of fasting, ghrelin is released from the stomach into the circulation where it impinges on feeding centers in hypothalamus to induce a hunger response (Cummings and Overduin, 2007; Dixit and Taub, 2005; Gil-Campos *et al.*, 2006). Ghrelin mediates its orexigenic effects via inhibition of leptin expression and the signaling of the NPY, AGRP, and orexin pathways. Ghrelin administration typically stimulates food intake within an hour of administration, and continuous ghrelin treatment results in sustained feeding eventually leading to increased body weight and adiposity. Ghrelin is believed to be the major regulator in humans as a preprandial rise in serum ghrelin levels induces feeding behavior (Cummings *et al.*, 2001).

Ghrelin is currently the only known hormone that is acylated on its third serine residue. This octanoylation is critical for the binding of ghrelin to the GHSR and for inducing GH secretion and food intake (Kojima *et al.*, 1999). Endogenous esterases such as lysophospholipase-1 in blood and various cells are believed to cleave the acyl linkage rendering the hormone inactive (Dixit and Taub, 2005). The desacyl form of ghrelin is currently believed to be biologically inactive and is present at three times higher than acylated ghrelin in stomach. A report has suggested that desghrelin can be acylated *in vivo* by ingested long-chain fatty acids, giving rise to the speculation that an exquisite balance between the circulating levels of desacyl and acylated ghrelin is maintained in the body (Nishi *et al.*, 2005). While desacyl ghrelin does not bind GHSR1a and appears to be devoid of any known endocrine activities, both acylated and desacyl ghrelin bind to common sites on these cells and inhibit apoptosis of cardiomyocytes and endothelial cells *in vitro* (Baldanzi *et al.*, 2002). These findings suggest that both forms of ghrelin, at least on cardiomyocytes, may bind and signal through a novel receptor other than GHSR1a. Moreover, Ariyasu *et al.* (2005) have reported that transgenic mice with up to 44-fold higher levels of plasma desghrelin compared to wild-type mice exhibit a dwarf phenotype with a suppressed GH-IGF-1 axis. This phenotype was actually quite surprising in that it is exactly opposite phenotype of what one would observe with acylated ghrelin transgenic mice. These opposite phenotypes suggest a possible role for desghrelin as a counterbalance to the effects of acylated ghrelin.

B. Sources of ghrelin production

While the stomach is considered the major source of peripheral ghrelin, recent studies have demonstrated ghrelin to be widely distributed in body throughout several major organ systems. Ghrelin mRNA is expressed in hypothalamus, jejunum, duodenum, colon, lungs, liver, fat, and placenta as well as the spleen, lymph nodes (Gnanapavan et al., 2002), thymus, and peripheral T cells, monocytes, and dendritic cells (Dixit et al., 2004; personal observations). Reports have also hypothesized that the ghrelin-GHSR axis may operate through similar autocrine/paracrine role in cancer biology (Dixit and Taub, 2005; Dixit et al., 2006). Ghrelin and GHSR have now been reported to be expressed in a variety of cancers, including endocrine tumors, breast carcinoma, prostate cancer cells, lung carcinomas, hepatoma, thyroid carcinomas, ovarian cancer, and gastrointestinal cancer (Dixit et al., 2006). We have demonstrated a direct role for ghrelin and GHSR in astrocytoma cell growth, motility, and function (Dixit et al., 2006). These findings provide the first evidence for the existence of ghrelin–GHSR-signaling pathways in astrocytomas and demonstrate that an endogenous hormonal loop may play a critical role in astrocytoma motility and invasiveness. It is unclear if other forms of cancer similarly utilize endogenous and exogenous ghrelin to promote cellular motility and growth.

Lymphocytes are known to produce many well-characterized hormones, including GH, PRL, ACTH, and GnRH, which exert a number of autocrine and paracrine effects on the immune system. Given the potent effect of ghrelin on cytokine expression by immune cells, the possible presence of endogenously produced ghrelin by immune cells was hypothesized. Dixit et al. (2004) have reported that human T cells express the pre-pro form of ghrelin in their Golgi bodies where it is presumably processed and cleaved to its mature form. However, on cellular activation, both acylated and desghrelin are produced and secreted by T cells as well as B cells, monocytes, and dendritic cells. Activated human T cells and monocytes can produce nanogram quantities of total ghrelin in the culture supernatants. Given that only activated T cells express and secrete the ghrelin protein, it is believed that the pre-pro ghrelin peptide must be actively cleaved in T cells to yield the active ghrelin peptide. Similar to several cytokines (e.g., TGF-β) and hormones (e.g., TSH), these precursor proteins are synthesized and subsequently stored for immediate cleavage and use when needed. Furthermore, immune-derived ghrelin is expressed in a polarized fashion in association with $GM1^+$ lipid rafts, in a fashion similar to GHSR expression in activated T cells. These results suggest that ghrelin may be selectively targeted to plasma membrane to facilitate interaction with its own transmembrane receptor to optimally mediate receptor–ligand interactions. These results also demonstrate that human T cells are capable of expressing and secreting bioactive ghrelin, which may play a role in controlling their own activities or may operate to influence other cell types within a tissue microenvironment. Thus, localized production of ghrelin may play a critical role in the immediate control of ongoing and leptin-mediated responses within the local microenvironment. As lipid rafts are known to preferentially accumulate acylated proteins in their cholesterol-rich microdomains, it seems feasible that the unique acylation of ghrelin peptide may specifically target ghrelin to the immune synapse and thus may play an important role in T cell signaling. Based on the capacity of multiple immune subpopulations to produce this orexigen, ghrelin concentrations within the local microenvironment may reach significantly high levels without undergoing the classic dilution effects typically observed with its release into the peripheral circulation from stomach. Thus, "normal" circulating ghrelin levels may not reflect the levels necessary to mediate specific effects within an immune microenvironment.

C. Biological and functional activities of ghrelin

1. Ghrelin functions

Since its discovery, ghrelin has been shown to exert diverse biological actions, including effects on hormone secretion, glucose homeostasis, pancreatic function, gastrointestinal motility, cardiovascular function, neurogenesis, immunity, inflammation, cell proliferation and survival, bone metabolism, reproductive organ functions, memory, sleep, gastric emptying and gastric acid secretion, and more (Fig. 1; Dixit and Taub, 2005). The physiological relevance of many of these actions remains to be defined.

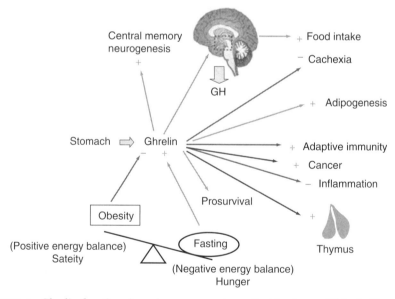

Figure 1 Ghrelin functions in various organ systems. Besides the established effects of ghrelin on food intake and GH production, ghrelin has also been shown to mediate a number of functions on a variety of other organ systems, including promoting adipogenesis, cardiomyocytes survival, enhancement of memory, and neurogenesis. In addition, ghrelin has been shown to inhibit inflammatory cytokine production and promote thymic function. (See Color Insert.)

However, ghrelin's roles in energy homeostasis are currently considered its most important functions. Moreover, while ghrelin is considered to be a potent GHS both *in vitro* and *in vivo*, it is unclear whether this property is physiologically important. Current hypotheses suggest that ghrelin may act as an adjuvant enhancing the magnitude of GH pulse in response to GHRH (Cummings and Overduin, 2007; Gil-Campos *et al.*, 2006). Ghrelin administration has also been shown to attenuate cachexia associated with chronic heart failure in rats (Nagaya *et al.*, 2001a,b), while the GHSR analogue, GHRP-2, counteracts protein hypercatabolism, skeletal muscle proteolysis, and osteoporosis in critically ill patients with wasting condition (Van den Berghe *et al.*, 1999). Despite ghrelin's participation in a variety of biological functions, no metabolic or physiological abnormalities were found in the ghrelin or GHSR knockout mice (Sun *et al.*, 2003, 2004; Wortley *et al.*, 2004). Perhaps more detailed studies in these mice using various disease and stress models or with progressive aging may reveal more specific defects in various organ compartments.

2. Ghrelin effects on leptin- and activation-induced inflammation

Cytokines such as interleukin-1 (IL-1), IL-6, TNF-α, and IFN-γ have been implicated in wasting associated with inflammation, chronic low-grade inflammation during aging, atherosclerosis, endotoxin shock and sepsis, and cancer (Dixit and Taub, 2005; Kelley *et al.*, 2003). Inflammatory cytokines released by immune cells have been shown to act on the central nervous system (CNS) to control food intake and energy homeostasis. Decrease in food intake or anorexia is one of the most common symptoms of illness, injury, or inflammation. Studies of the cytokine network involved in cachexia continue to be a major area of research. With recent advances in neurobiology research, peptidergic neural circuits involving leptin (an anorexigenic hormone secreted from adipose tissue) and hypothalamic neurotransmitters (e.g., the orexigen, neuropeptide Y, and ghrelin) have been delineated. This homeostatic circuit appears to regulate food intake and metabolism of energy derived from nutrients. Leptin is released from adipocytes as a function of the amount of fat and reduces food intake by stimulating the anorexigenic hypothalamic pathway and inhibiting the orexigenic pathway, both of them originate in the arcuate nucleus of the hypothalamus and project to the paraventricular nucleus and the lateral hypothalamic area. Dysfunction of this circuit has been implicated in both anorexia and cachexia. Regulation of these hormones and inflammatory cytokine production by endogenous factors holds promise in the amelioration of a wide variety of ailments and disease conditions.

Leptin is considered a critical sensory hormone that signals to the brain changes in stored energy, determined by an altered balance between food intake and energy expenditure. This molecule encoded by the obese (ob) gene is a 16-kDa nonglycosylated protein (Bernotiene *et al.*, 2006;

Dixit and Taub, 2005; Tilg and Moschen, 2006). Structurally, it belongs to the type I cytokine family and is characterized by a long-chain four-helical bundle structure, such as GH, PRL, erythropoietin, IL-3, IL-11, and LIF. Leptin is synthesized primarily by white adipose tissue in proportion with body fat mass and at low levels by tissues as placenta, gastric fundic mucosa, mammary epithelium, skeletal muscles, and neurons. Leptin synthesis increases in response to acute infection, sepsis, and secretion of inflammatory mediators such as IL-1, TNF-α, and LIF. Leptin has been shown to influence a wide spectrum of biological functions, as lipid and glucose metabolism, synthesis of glucocorticoids, insulin, $CD4^+$ T lymphocyte proliferation, cytokine secretion, phagocytosis, and synaptic transmission (Bernotiene et al., 2006; Tilg and Moschen, 2006). Previous reports have demonstrated that immune cells express leptin receptors (Dixit et al., 2003, 2004; Lord et al., 1998) and that these receptors play an important role in mediating the effects of leptin on immune function. In addition, this hormone regulates the hypothalamic-pituitary-adrenal axis, maturation of the reproductive system, hematopoiesis, angiogenesis, and fetal development. Naturally leptin-deficient obese *ob/ob* mice display many abnormalities similar to those observed in starved animals and malnourished humans, including impaired $CD4^+$ T lymphocyte functions and thymic atrophy. Indeed, these animals exhibit impaired cell-mediated immunity and thymic atrophy. More specifically, chronic leptin deficiency results in reduced secretion of Th1 cytokines (IL-2, IFN-γ, IL-18) and the increased production of the Th2 cytokine, IL-4, in mice. Furthermore, serum leptin levels are increased during inflammation and cell-mediated diseases in humans, such as endometriosis, type 1 diabetes (IDDM), chronic pulmonary inflammation, and in animal models, as intestinal experimental inflammation, carrageenan-induced arthritis, and experimental allergic encephalomyelitis (EAE) (Sanna et al., 2003). Thus, leptin appears to be one of the major players in the immunoendocrine scenario, regulating nutritional status, basal metabolism, and immune function. We have recently reported that leptin directly stimulates the expression and release of IL-1α, IL-1β, IL-6, and TNF-α from both monocytes and T cells (Dixit et al., 2004). Leptin also promotes T-helper immune responses by promoting the release of IL-2 and IFN-γ and inhibiting IL-4 secretion (Lord et al., 1998).

The initial focus of ghrelin research has been associated with its ability to regulate food intake and GH release. However, given the wide distribution of functional GHSR on various immune cell subsets, it was hypothesized that this peptide may exert immunoregulatory effects on immune cell subpopulations. Immune cells are highly sensitive to changes in the energy balance in a tissue microenvironment and thus require proper receptors' and ligands' sense and signal to alert or adjust immune responses. Conversely, inflammatory cytokines from immune cells have been shown to act on the CNS to control food intake and energy homeostasis. Considering the critical role played by proinflammatory cytokines in controlling metabolic activity, we initially examined the ability of ghrelin to regulate the expression of IL-1β, IL-6, and TNF-α by activated monocytes, T cells, and human peripheral blood mononuclear cells. Ghrelin was found to exert potent inhibitory effects on the mRNA and protein expression of the proinflammatory cytokines, IL-1β, IL-6, and TNF-α, via a GHSR-specific mechanism (Dixit et al., 2004). This regulatory effect appears to be both dose- and time dependent. While diminished cytokine mRNA and protein levels are found in T cells and accessory cells postghrelin treatment, it remains unclear if this is due to a direct inhibition of transcription or possible effects on RNA stability. Posttranscriptional mechanisms of gene regulation, particularly those affecting mRNA stability, are emerging as critical effectors of gene expression changes during immune cell activation and cellular proliferation. Although the mechanisms determining mRNA turnover are poorly understood, they are generally believed to involve RNA-binding proteins, recognizing specific RNA sequences. Best characterized among the RNA sequences that influence mRNA stability are AU-rich elements (AREs), usually found in the 3$'$-untranslated regions (UTR) of short-lived mRNAs, such as those encoding cytokines (IFN, ILs, TNF-α, CSFs) (Wang et al., 2003, 2004). Several studies have provided increasing support for the notion that mRNA stability is regulated through mechanisms akin to those controlling gene transcription (i.e., signal transduction pathways involving phosphorylation events). Early reports described the altered turnover of ARE-containing mRNAs in response to extracellular as well as internally generated signals, such as PMA, TCR ligation, and TNF-α. AMP-activated protein kinase (AMPK), an enzyme involved in responding to metabolic stresses, has been shown to be a potent regulator of RNA stability as inhibition of AMPK increases the levels of cytoplasmic RNA-binding proteins and enhances their binding to transcripts elevating their expression and half-lives (Wang et al., 2003, 2004). Conversely, AMPK activation decreases these levels and transcript half-lives. We have found that ghrelin is capable of activating AMPK in T cells and monocytes, which we believe may actually inhibit the cytoplasmic levels of certain RNA-binding proteins (personal observations). This in turn may influence the mRNA stability and the expression of HuR target transcripts.

Interestingly, ghrelin was also found to significantly attenuate leptin-induced proinflammatory and Th1 responses in human mononuclear and T cells (Dixit et al., 2004; personal observations). These data have established a novel role for ghrelin in immune cell function as a negative regulator of inflammatory cytokine expression induced by cell activation by antigen, mitogens, or leptin (Fig. 2). The ability of ghrelin to control cytokine expression and the reciprocal regulatory effects of this hormone on leptin-induced activity on immune cells suggests the existence of a novel immunoregulatory network, controlling cytokine expression, cellular activation, trafficking, and apoptosis. This may have implications on the

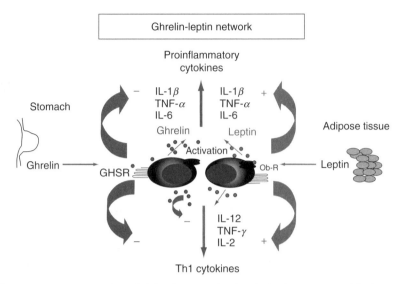

Figure 2 Hypothetical model for functional role of ghrelin as a signal linking the immune and endocrine systems in control of food intake. Immune-derived ghrelin and leptin may play a role in balancing the inflammatory status of a host. When a metabolic imbalance occurs, these systems may come into play assisting in the regulation of food intake by controlling inflammatory cytokine expression. Conversely, circulating hormones may influence ongoing immune responses by enhancing or dampening the level of activation and cytokine secretion. Ghrelin appears to mediate anti-inflammatory effects on IL-1, TNF, and IL-6 cytokine expression by T cells and mononuclear cells, while leptin appears to promote inflammatory cytokine expression. Moreover, leptin has been shown to promote Th1 cytokine response, while data within our laboratory has demonstrated that ghrelin inhibits leptin-induced Th1 development. Perhaps these hormones and immune cell responses to these hormones are a form of "checks-and-balances" by which the neuroendocrine, metabolic, and immune systems can control each other's activities, preventing one system from interfering with another system. (See Color Insert.)

control of the cellular and inflammatory responses. These findings also have implications on the degree of "cross talk" between the neuroendocrine, metabolic, and immune systems. Such a reciprocal immunoregulatory effects may be critical in maintaining immune cell homeostasis, preventing aberrant cytokine production that may potentially result in illness and pathology. Regulation of these hormones and inflammatory cytokine production by endogenous factors also holds promise in the amelioration of a wide variety of ailments and disease conditions.

In addition to our published work with T cells and monocytes (Dixit *et al.*, 2004), we have also established a role for the ghrelin–leptin network in dendritic cells. Both monocytes and dendritic cells constitutively express GHSR and Ob-R on their cell surface, and these receptors are upregulated on cellular activation (personal observations). Moreover, ghrelin treatment of these populations also results in diminished proinflammatory cytokine and chemokine expression in these populations in response to cellular activation and leptin treatment. Similar to T cells, both monocytes and dendritic cells also express ghrelin and leptin mRNA and protein at rest and greater levels after activation. Again, the relevance of GHSR and ghrelin expression by dendritic and other accessory cells remains to be defined.

3. Ghrelin effects on systemic inflammation

a. Obesity Obesity is a state of chronic positive energy balance and is associated with increase in leptin and a decrease in circulating ghrelin levels (Badman and Flier, 2005). A report has demonstrated a linkage, and association of SNPs and haplotypes within GHSR gene region plays a role in human obesity (Baessler *et al.*, 2005). Moreover, a mutation in pre-proghrelin gene sequence (Arg51Gln) has also been found to be prevalent in obese subjects (Ukkola *et al.*, 2001). There is increasing evidence linking obesity and inflammation and impaired immune responses (Clement *et al.*, 2004; Wellen and Hotamisligil, 2003; Xu *et al.*, 2003). Given that ghrelin inhibits and leptin increases proinflammatory cytokines, it is quite feasible that the chronic inflammation observed in obese individuals (such as increased IL-6 levels) may be associated with reduced ghrelin levels and increased leptin levels in the circulation. This increase in systemic inflammation can be even higher with progressive aging and frailty. While this hypothesis remains to be proven, it seems feasible that therapeutic administration of ghrelin or ghrelin mimetics may actually reduce BMI- and age-associated inflammation and anorexia.

b. Endotoxemia and inflammatory bowel disease Lipopolysaccharide (LPS) is derived from cell wall of gram-negative bacteria and is a major component in the initiation and systemic effects observed in endotoxic shock. LPS acts on many cells within the body, including endothelial cells, fibroblasts, keratinocytes, monocytes, dendritic cells, and B cells to induce an acute phase response *in vivo* resulting in excessive production of proinflammatory cytokines such as IL-1, IL-6, and TNF-α. The amplification of these cytokines results in an array of proinflammatory and anorexigenic effects contributing to pathogenesis of sepsis and multiple organ failure. In a rat model of LPS toxemia, ghrelin administration was found to exert significant protective effects on the survival and also attenuated the hypotensive effects of sepsis (Chang *et al.*, 2003). Moreover, ghrelin administration exerts prokinetic effects in septic gastric ileus (De Winter *et al.*, 2004). Similarly, Hataya *et al.* (2003) reported that ghrelin treatment attenuated wasting in sepsis. Studies from our laboratory have revealed that ghrelin exerts potent anti-inflammatory activity in a murine model of endotoxemia by inhibiting circulating and tissue-associated levels of IL-1α, IL-1β, IL-6, and TNF-α post-LPS challenge (Dixit *et al.*, 2004). Ghrelin also inhibits inflammatory cytokines produced by endothelial cells in response to LPS, and these effects are associated with a decrease in NF-κB activity (Li et al., 2004). Interestingly, circulating ghrelin levels also appear to decline during states of sepsis and endotoxin challenge (Dixit and Taub, 2005). During uncomplicated postoperative responses and intra-abdominal sepsis, both ghrelin and leptin plasma levels have been shown to be elevated and positively correlate with both inflammatory cytokines and C-reactive protein (Maruna et al., 2005). Similarly, plasma ghrelin levels were significantly higher in patients with peptic ulcer than in those with gastritis without ulcer, suggesting a possible relationship between mucosal injury susceptibility and elevated fasting plasma ghrelin (Suzuki et al., 2006). Moreover, studies examining model of pancreaticobiliary obstruction revealed that ghrelin administration was able to organ injury and systemic inflammation. In contrast to saline-treated group with severe pancreatic damage, these investigators found that ghrelin-treated rats demonstrated a moderate pancreatic and hepatic destruction accompanied with reduced pulmonary and renal damages. These results further illustrate that ghrelin protects the hepatic and pancreatic tissues, as well as remote organs against oxidative injury, typically mediated by neutrophils (Kasimay et al., 2006). Ghrelin administration was also found to ameliorate the clinical and histopathologic severity of the trinitrobenzene sulfonic acid-induced colitis and also abrogated body weight loss, diarrhea, and inflammation as well as increasing survival (Gonzalez-Rey et al., 2006). This effect was associated with a downregulation of inflammatory and Th1-driven autoimmune response through the regulation of inflammatory mediators. These studies suggest that ghrelin administration represents a novel therapeutic approach for the treatment of Crohn's disease and other Th1-mediated inflammatory diseases such as rheumatoid arthritis and multiple sclerosis.

c. Arthritis and bone loss Further support for anti-inflammatory role for ghrelin comes from a study where ghrelin and the synthetic GHSR agonist, GHRP-2, significantly inhibited the release of IL-6 and nitric oxide (NO) from activated macrophages and attenuated arthritis in a rat model (Granado et al., 2005), while leptin-stimulated NO release from mononuclear cells (Dixit et al., 2003) and leptin-deficient mice are protected from collagen-induced arthritis (Busso et al., 2002). Osteoblasts were found to express GHSR, and on treatment with ghrelin, a significant increase in osteoblast proliferation and differentiation was observed through GHSR-specific signaling (Fukushima et al., 2005). Furthermore, ghrelin infusion led to an increase in bone formation via GH-independent pathways. Inflammatory cytokines play a vital role in bone remodeling, and persistent high levels of proinflammatory mediators are known to cause bone loss associated with aging and other inflammatory joint diseases (Rodan and Martin, 2000). Given the reciprocal-regulatory role for ghrelin and leptin on inflammation, it is conceivable that these hormones also regulate bone formation in a similar manner. Additional research in this area holds promise in development of therapeutic strategies to manage joint inflammation as well as bone loss.

4. Effects of ghrelin and GHS on thymopoiesis and thymic involution

The thymus is critical for the development, selection, and maintenance of the peripheral T cell pool possessing a broad spectrum of TCR specificities. The mammalian thymus is capable of generating T cells throughout the life span; however, postpuberty and with advancing age, the thymic space becomes progressively filled with adipocytes coupled with a dramatic loss of thymocytes in the cortical and medullary areas, leading to a reduction in output of naive T cells. This process is called as thymic involution (Taub and Longo, 2005). The lack of a thymus in humans (DiGeorge's syndrome) and neonatal mice after thymectomy leads to severe immunodeficiency due to paucity of mature T cells. During physiological aging, the total peripheral T cell pool is maintained by homeostatic expansion of preexisting T cells rather than replenishment by thymic export. Consequently, the long-lived naive T cell repertoire is reduced with an expansion of memory phenotype T cells, and thereby limiting a host's ability to mount responses against new antigenic challenges. It has been demonstrated that defects in naive $CD4^+$ T cells are also due to the chronological age of the naive cells itself rather than the chronological age of the host. Involution of the thymus with age and lack or paucity of newly formed naive $CD4^+$ T cells is therefore believed to be responsible for much of the deterioration in adaptive immunity and the resultant immune dysfunction in the elderly. However, several recent studies have demonstrated that the old thymus still retains the capacity for T cell lymphopoiesis with ability to mount functional immune responses, albeit to a limited extent (Taub and Longo, 2005). Thus, restoration of thymic function and T cell export in the aged host remains a promising therapeutic goal. The thymus lacks a pool of self renewing progenitors and needs continuous seeding of lymphoid progenitors from the bone marrow. Accordingly, it has been elegantly demonstrated that early thymocyte progenitors (ETPs) decline markedly with age, and these progenitors have increased rate of apoptosis with reduced proliferative capacity. The precise mechanisms responsible for thymic involution with age remain to be identified and presumably involve complex multisystem interactions, wherein various cytokines, neuropeptides, and hormones influence immune compartment.

A variety of factors have been demonstrated to promote pro-T cell survival and differentiation, notably GH, IGF-1, IL-7, and stem cell factor (SCF) (Taub and Longo, 2005). As noted earlier in this chapter, several studies using neuroendocrine GH-deficient dwarf mice or rat receiving hypophysectomy demonstrated impaired immune responses, thymic involution, and, in some cases, premature death (Murphy *et al.*, 1992b). This impaired thymic function could be restored by the administration of GH. Neuroendocrine hormones have long been associated with effects on immune cell function, including growth-promoting effects on multiple immune cell lineages. Similarly, leptin administration has also been shown to protect ob/ob mice from the lymphoid atrophy associated with starvation and reverses the inhibitory effect of starvation on the development of DTH reactions (Bernotiene *et al.*, 2006; Tilg and Moschen, 2006). Thus, some or all of these factors may play a role in mediating the thymopoietic or hematopoietic effects observed on hormone administration. The use of these hormones to promote immune reconstitution is attractive considering their pleiotropic effects and low toxicity after systemic administration. However, more information is required to adequately evaluate their appropriateness as agents to promote T cell recovery in immunosuppressed subjects.

Based on the ability of GHS to stimulate the GH-IGF-1 axis and a study by Poppi *et al.* (2002) demonstrating that synthetic GHSR analogues stimulate GH secretion directly by peripheral immune cells, Koo *et al.* (2001) treated 14-month-old mice with a synthetic GHSR agonist for 3 weeks and examined the effects on the thymus and bone marrow. Their results revealed that oral GHS administration resulted in a significant increase in thymic cellularity in old mice along with improved bone marrow engraftment using an SCID model. These investigators proposed that the observed effects were being mediated through GH induction in response to the GHS administration. However, the precise role for GH in mediating these effects remains to be defined. More recently, we have found that administration of acylated but not desacylated ghrelin into aged (6-, 12-, 14-, 18-, 22-, and 24-month old) but not young (2- and 4-month old) mice resulted in a partial reversal of thymic involution (Dixit *et al.*, 2007). Infusion of ghrelin also significantly improved the age-associated changes in thymic architecture and thymocyte counts along with an increase in recent thymic emigrants (RTE), an improvement of TCR diversity of peripheral $CD4^+$ and $CD8^+$ T cells and enhanced numbers of ETPs and bone marrow-derived pluripotential stem cells. Furthermore, the ghrelin and GHSR-deficient mice displayed enhanced age-associated thymic involution with reduced thymopoiesis, contraction of stem cells, and major perturbations in the TCR repertoire of peripheral T lymphocytes. Our findings demonstrate a novel role for ghrelin and its receptor in thymic biology and T cell development. Overall, the interplay between these hormones and their receptors in the thymic and BM compartments appears to play an important role in rejuvenation of thymic output in old animals. These data also support the existence of a functional immunoregulatory network involving orexigenic and anorexigenic hormones that appear to play a significant role in cytokine regulation, cellular activation, and survival. Furthermore, it should be noted that the expression of ghrelin, leptin, and their receptors as well as GH appears to be significantly diminished with age within specific immune subsets and lymphoid organs, including the thymus. While the relevance of immune-derived ghrelin deficiency with age remains to be defined, we believe that this loss of hormone secretion may be associated with a loss of control of cytokine expression, resulting in increased levels of circulating and tissue-associated IL-6 and other proinflammatory cytokines commonly observed in older subjects, possibly mediating thymic loss. We have hypothesized that the loss of lymphoid organ-associated ghrelin may also result in significant changes in the thymic anatomy and microenvironment as well as the hormonal and cytokine mediators that play a role in thymic maintenance. Thus, ghrelin or GHSR (and thymic GH for that matter) loss with age may actually influence age-associated thymic involution and thymopoiesis. Moreover, these data also support the potential therapeutic use of ghrelin and GHSR agonists in the management of acute and chronic inflammation and cancer and in restoration of thymic function in aged and immunocompromised individuals. Ghrelin and ghrelin mimetics may actually be quite valuable in the patients undergoing bone marrow transplantation where it is necessary to engraft and replenish the host after radiation treatment with donor cells. Graft-versus-host disease (GVHD) is a major problem in these subjects and is mediated by many inflammatory cytokines. Administration of ghrelin in such patents may inhibit GVHD development while also promoting thymic and bone marrow engraftment.

IV. Conclusions

Signals originating from the immune system may directly influence the neuronal and metabolic axes, while alterations in the energy balance may potentially exert potent effects on immune function and the systemic control of inflammation. Interactions between these systems are especially sensitive to states of stress and disease and work together in an effort to maintain physiological homeostasis. We believe that ghrelin and other orexigenic hormones are potent anti-inflammatory mediators, while anorexigenic hormones such as leptin promote and enhance inflammatory responses. Detailed studies on the role of orexigens and anorexigens on regulating immune function and controlling inflammation are still in their early stages. Also, a greater understanding of how these various ligands and receptors signal on immune cells, how their receptors are regulated, how these ligands regulate cytokine and growth factor expression and the specific role(s) of immune-expressed hormones in the genesis of immune responses should provide great insight into the true physiological roles of these hormones in the immune system. In addition, the pharmaceutical industry is working hard to develop new and more potent ghrelin agonists for possible therapeutic use in the treatment of anorexia and cachexia. Ghrelin and other GHSR agonists may prove valuable in the treatment of a number of acute and chronic autoimmune and inflammatory disorders, including sepsis, multiple sclerosis, inflammatory bowel disease, Crohn's disease, rheumatoid arthritis, and perhaps even age-associated inflammation and thymic involution. Moreover, these ligands may prove useful in the therapeutic treatment of GVHD that occurs during bone marrow transplantation. Continued bench-to-bedside efforts should provide valuable data into the therapeutic potential of ghrelin and GHS agonists.

ACKNOWLEDGMENTS

I would like to thank Dr. Vishwa Deep Dixit for his efforts on these many studies with ghrelin and leptin performed during his time in my laboratory. This research was supported (in part) by the Intramural Research Program of the NIH, NIA.

REFERENCES

Ariyasu, H., Takaya, K., Iwakura, H., Hosoda, H., Akamizu, T., Arai, Y., Kangawa, K., and Nakao, K. (2005). Transgenic mice overexpressing des-acyl ghrelin show small phenotype. *Endocrinology* **146,** 355–364.

Badman, M. K., and Flier, J. S. (2005). The gut and energy balance: Visceral allies in the obesity wars. *Science* **307,** 1909–1914.

Baessler, A., Hasinoff, M. J., Fischer, M., Reinhard, W., Sonnenberg, G. E., Olivier, M., Erdmann, J., Schunkert, H., Doering, A., Jacob, H. J., Comuzzie, A. G., Kissebah, A. H., *et al.* (2005). Genetic linkage and association of the growth hormone secretagogue receptor (ghrelin receptor) gene in human obesity. *Diabetes* **54,** 259–267.

Baldanzi, G., Filigheddu, N., Cutrupi, S., Catapano, F., Bonissoni, S., Fubini, A., Malan, D., Baj, G., Granata, R., Broglio, F., Papotti, M., Surico, N., *et al.* (2002). Ghrelin and desacyl ghrelin inhibit cell death in cardiomyocytes and endothelial cells through ERK1/2 and PI 3-kinase/AKT. *J. Cell Biol.* **159,** 1029–1037.

Bernotiene, E., Palmer, G., and Gabay, C. (2006). The role of leptin in innate and adaptive immune responses. *Arthritis Res. Ther.* **8**(5), 217–227.

Bowers, C. Y. (2001). Unnatural growth hormone–releasing peptide begets natural ghrelin. *J. Clin. Endocrinol. Metab.* **86,** 1464–1469.

Brzozowski, T., Konturek, P. C., Konturek, S. J., Kwiecien, S., Drozdowicz, D., Bielanski, W., Pajdo, R., Ptak, A., Nikiforuk, A., Pawlik, W. W., and Hahn, E. G. (2004). Exogenous and endogenous ghrelin in gastroprotection against stress-induced gastric damage. *Regul. Pept.* **120,** 39–51.

Busso, N., So, A., Chobaz-Peclat, V., Morard, C., Martinez-Soria, E., Talabot-Ayer, D., and Gabay, C. (2002). Leptin signaling deficiency impairs humoral and cellular immune responses and attenuates experimental arthritis. *J. Immunol.* **168,** 875–882.

Chang, L., Zhao, J., Yang, J., Zhang, Z., Du, J., and Tang, C. (2003). Therapeutic effects of ghrelin on endotoxic shock in rats. *Eur. J. Pharmacol.* **473,** 171–176.

Clement, K., Viguerie, N., Poitou, C., Carette, C., Pelloux, V., Curat, C. A., Sicard, A., Rome, S., Benis, A., Zucker, J. D., Vidal, H., Laville, M., *et al.* (2004). Weight loss regulates inflammation-related genes in white adipose tissue of obese subjects. *FASEB J.* **18,** 1657–1669.

Cummings, D. E., and Overduin, J. (2007). Gastrointestinal regulation of food intake. *J. Clin. Invest.* **117**(1), 13–23.

Cummings, D. E., Purnell, J. Q., Frayo, R. S., Schmidova, K., Wisse, B. E., and Weigle, D. S. (2001). A preprandial rise in plasma ghrelin levels suggests a role in meal initiation in humans. *Diabetes* **50**, 1714–1719.

Cummings, D. E., Weigle, D. S., Frayo, R. S., Breen, P. A., Ma, M. K., Dellinger, E. P., and Purnell, J. Q. (2002). Plasma ghrelin levels after diet-induced weight loss or gastric bypass surgery. *N. Engl. J. Med.* **346**, 1623–1630.

De Winter, B. Y., De Man, J. G., Seerden, T. C., Depoortere, I., Herman, A. G., Peeters, T. L., and Pelckmans, P. A. (2004). Effect of ghrelin and growth hormone-releasing peptide 6 on septic ileus in mice. *Neurogastroenterol. Motil.* **16**, 439–446.

Dixit, V. D., and Taub, D. D. (2005). Ghrelin and immunity: A young player in an old field. *Exp. Gerontol.* **40**(11), 900–910.

Dixit, V. D., Mielenz, M., Taub, D. D., and Parvizi, N. (2003). Leptin induces growth hormone secretion from peripheral blood mononuclear cells via a protein kinase C- and nitric oxide-dependent mechanism. *Endocrinology* **144**, 5595–5603.

Dixit, V. D., Schaffer, E. M., Pyle, R. S., Collins, G. D., Sakthivel, S. K., Palaniappan, R., Lillard, J. W., Jr., and Taub, D. D. (2004). Ghrelin inhibits leptin- and activation-induced proinflammatory cytokine expression by human monocytes and T cells. *J. Clin. Invest.* **114**, 57–66.

Dixit, V. D., Weeraratna, A. T., Yang, H., Bertak, D., Cooper-Jenkins, A., Riggins, G. J., Eberhart, C. G., and Taub, D. D. (2006). Ghrelin and the growth hormone secretagogue receptor constitute a novel autocrine pathway in astrocytoma motility. *J. Biol. Chem.* **281**(24), 16681–16690.

Dixit, V. D., Yang, H., Sun, Y., Weeraratna, A. T., Youm, Y. H., Smith, R. G., and Taub, D. D. (2007). Ghrelin promotes thymopoiesis during aging. *J. Clin. Invest.* Sept. 6, [Epub ahead of print].

French, R. A., Broussard, S. R., Meier, W. A., Minshall, C., Arkins, S., Zachary, J. F., Dantzer, R., and Kelley, K. W. (2002). Age-associated loss of bone marrow hematopoietic cells is reversed by GH and accompanies thymic reconstitution. *Endocrinology* **143**, 690–699.

Fukushima, N., Hanada, R., Teranishi, H., Fukue, Y., Tachibana, T., Ishikawa, H., Takeda, S., Takeuchi, Y., Fukumoto, S., Kangawa, K., Nagata, K., and Kojima, M. (2005). Ghrelin directly regulates bone formation. *J. Bone Miner. Res.* **20**, 790–798.

Gil-Campos, M., Aguilera, C. M., Canete, R., and Gil, A. (2006). Ghrelin: A hormone regulating food intake and energy homeostasis. *Br. J. Nutr.* **96**(2), 201–226.

Gnanapavan, S., Kola, B., Bustin, S. A., Morris, D. G., McGee, P., Fairclough, P., Bhattacharya, S., Carpenter, R., Grossman, A. B., and Korbonits, M. (2002). The tissue distribution of the mRNA of ghrelin and subtypes of its receptor, GHS-R, in humans. *J. Clin. Endocrinol. Metab.* **87**, 2988–2995.

Gonzalez-Rey, E., Chorny, A., and Delgado, M. (2006). Therapeutic action of ghrelin in a mouse model of colitis. *Gastroenterology* **130**(6), 1707–1720.

Granado, M., Priego, T., Martin, A. I., Villanua, M. A., and Lopez-Calderon, A. (2005). Anti-inflammatory effect of the ghrelin agonist growth hormone-releasing peptide-2 (GHRP-2) in arthritic rats. *Am. J. Physiol. Endocrinol. Metab.* **288**, E486–E492.

Hataya, Y., Akamizu, T., Hosoda, H., Kanamoto, N., Moriyama, K., Kangawa, K., Takaya, K., and Nakao, K. (2003). Alterations of plasma ghrelin levels in rats with lipopolysaccharide-induced wasting syndrome and effects of ghrelin treatment on the syndrome. *Endocrinology* **144**, 5365–5371.

Hattori, N., Saito, T., Yagyu, T., Jiang, B. H., Kitagawa, K., and Inagaki, C. (2001). GH, GH receptor, GH secretagogue receptor, and ghrelin expression in human T cells, B cells, and neutrophils. *J. Clin. Endocrinol. Metab.* **86**, 4284–4291.

Holst, B., Holliday, N. D., Bach, A., Elling, C. E., Cox, H. M., and Schwartz, T. W. (2004). Common structural basis for constitutive activity of the ghrelin receptor family. *J. Biol. Chem.* **279**, 53806–53817.

Howard, A. D., Feighner, S. D., Cully, D. F., Arena, J. P., Liberator, P. A., Rosenblum, C. I., Hamelin, M., Hreniuk, D. L., Palyha, O. C., Anderson, J., Paress, P. S., Diaz, C., et al. (1996). A receptor in pituitary and hypothalamus that functions in growth hormone release. *Science* **273,** 974–977.

Jeffery, P. L., Herington, A. C., and Chopin, L. K. (2003). The potential autocrine/paracrine roles of ghrelin and its receptor in hormone-dependent cancer. *Cytokine Growth Factor Rev.* **14,** 113–122.

Kasimay, O., Iseri, S. O., Barlas, A., Bangir, D., Yegen, C., Arbak, S., and Yegen, B. C. (2006). Ghrelin ameliorates pancreaticobiliary inflammation and associated remote organ injury in rats. *Hepatol. Res.* **36**(1), 11–19.

Kelley, K. W., Bluthe, R. M., Dantzer, R., Zhou, J. H., Shen, W. H., Johnson, R. W., and Broussard, S. R. (2003). Cytokine-induced sickness behavior. *Brain Behav. Immun.* **17** (Suppl. 1), S112–S118.

Kissebah, A. H., Sonnenberg, G. E., Myklebust, J., Goldstein, M., Broman, K., James, R. G., Marks, J. A., Krakower, G. R., Jacob, H. J., Weber, J., Martin, L., Blangero, J., et al. (2000). Quantitative trait loci on chromosomes 3 and 17 influence phenotypes of the metabolic syndrome. *Proc. Natl. Acad. Sci. USA* **97,** 14478–14483.

Kojima, M., Hosoda, H., Date, Y., Nakazato, M., Matsuo, H., and Kangawa, K. (1999). Ghrelin is a growth-hormone-releasing acylated peptide from stomach. *Nature* **402,** 656–660.

Koo, G. C., Huang, C., Camacho, R., Trainor, C., Blake, J. T., Sirotina-Meisher, A., Schleim, K. D., Wu, T. J., Cheng, K., Nargund, R., and McKissick, G. (2001). Immune enhancing effect of a growth hormone secretagogue. *J. Immunol.* **166,** 4195–4201.

Li, W. G., Gavrila, D., Liu, X., Wang, L., Gunnlaugsson, S., Stoll, L. L., McCormick, M. L., Sigmund, C. D., Tang, C., and Weintraub, N. L. (2004). Ghrelin inhibits proinflammatory responses and nuclear factor-kappaB activation in human endothelial cells. *Circulation* **109**(181), 2221–2226.

Lord, G. M., Matarese, G., Howard, J. K., Baker, R. J., Bloom, S. R., and Lechler, R. I. (1998). Leptin modulates the T-cell immune response and reverses starvation-induced immunosuppression. *Nature* **394,** 897–901.

Maruna, P., Gurlich, R., Frasko, R., and Rosicka, M. (2005). Ghrelin and leptin elevation in postoperative intra-abdominal sepsis. *Eur. Surg. Res.* **37**(6), 354–359.

Murphy, W. J., and Longo, D. L. (2000). Growth hormone as an immunomodulating therapeutic agent. *Immunol. Today* **21,** 211–213.

Murphy, W. J., Durum, S. K., and Longo, D. L. (1992a). Human growth hormone promotes engraftment of murine or human T cells in severe combined immunodeficient mice. *Proc. Natl. Acad. Sci. USA* **89,** 4481–4485.

Murphy, W. J., Durum, S. K., and Longo, D. L. (1992b). Role of neuroendocrine hormones in murine T cell development. Growth hormone exerts thymopoietic effects in vivo. *J. Immunol* **149,** 3851–3857.

Murphy, W. J., Durum, S. K., and Longo, D. L. (1993). Differential effects of growth hormone and prolactin on murine T cell development and function. *J. Exp. Med.* **178,** 231–236.

Nagaya, N., Miyatake, K., Uematsu, M., Oya, H., Shimizu, W., Hosoda, H., Kojima, M., Nakanishi, N., Mori, H., and Kangawa, K. (2001a). Hemodynamic, renal, and hormonal effects of ghrelin infusion in patients with chronic heart failure. *J. Clin. Endocrinol. Metab.* **86,** 5854–5859.

Nagaya, N., Uematsu, M., Kojima, M., Ikeda, Y., Yoshihara, F., Shimizu, W., Hosoda, H., Hirota, Y., Ishida, H., Mori, H., and Kangawa, K. (2001b). Chronic administration of ghrelin improves left ventricular dysfunction and attenuates development of cardiac cachexia in rats with heart failure. *Circulation* **104**, 1430–1435.

Nishi, Y., Hiejima, H., Hosoda, H., Kaiya, H., Mori, K., Fukue, Y., Yanase, T., Nawata, H., Kangawa, K., and Kojima, M. (2005). Ingested medium-chain Fatty acids are directly utilized for the acyl modification of ghrelin. *Endocrinology* **146**, 2255–2264.

Petersenn, S., Rasch, A. C., Penshorn, M., Beil, F. U., and Schulte, H. M. (2001). Genomic structure and transcriptional regulation of the human growth hormone secretagogue receptor. *Endocrinology* **142**, 2649–2659.

Poppi, L., Dixit, V. D., Baratta, M., Giustina, A., Tamanini, C., and Parvizi, N. (2002). Growth hormone secretagogue (GHS) analogue, hexarelin stimulates GH from peripheral lymphocytes. *Exp. Clin. Endocrinol. Diabetes* **110**, 343–347.

Rodan, G. J., and Martin, T. J. (2000). Therapeutic approaches to bone diseases. *Science* **289**, 1508–1514.

Sanna, V., Di Giacomo, A., La Cava, A., Lechler, R. I., Fontana, S., Zappacosta, S., and Matarese, G. (2003). Leptin surge precedes onset of autoimmune encephalomyelitis and correlates with development of pathogenic T cell responses. *J. Clin. Invest.* **111**, 241–250.

Smith, R. G., Feighner, S., Prendergast, K., Guan, X., and Howard, A. (1999). A new orphan receptor involved in pulsatile growth hormone release. *Trends Endocrinol. Metab.* **10**, 128–135.

Smith, R. G., Leonard, R., Bailey, A. R., Palyha, O., Feighner, S., Tan, C., Mckee, K. K., Pong, S. S., Griffin, P., and Howard, A. (2001). Growth hormone secretagogue receptor family members and ligands. *Endocrine* **14**(1), 9–14.

Smith, R. G., Betancourt, L., and Sun, Y. (2005). Molecular endocrinology and physiology of the aging central nervous system. *Endocr. Rev.* **26**, 203–250.

Steinman, L. (2004). Elaborate interactions between the immune and nervous systems. *Nat. Immunol.* **5**(6), 575–581.

Sun, Y., Ahmed, S., and Smith, R. G. (2003). Deletion of ghrelin impairs neither growth nor appetite. *Mol. Cell. Biol.* **23**, 7973–7981.

Sun, Y., Wang, P., Zheng, H., and Smith, R. G. (2004). Ghrelin stimulation of growth hormone release and appetite is mediated through the growth hormone secretagogue receptor. *Proc. Natl. Acad. Sci. USA* **101**, 4679–4684.

Suzuki, H., Masaoka, T., Nomoto, Y., Hosoda, H., Mori, M., Nishizawa, T., Minegishi, Y., Kangawa, K., and Hibi, T. (2006). Increased levels of plasma ghrelin in peptic ulcer disease. *Aliment Pharmacol. Ther.* **4**, 120–126.

Taub, D. D., and Longo, D. L. (2005). Insights into thymic aging and regeneration. *Immunol. Rev.* **205**, 72–93.

Taub, D. D., Tsarfaty, G., Lloyd, A. R., Durum, S. K., Longo, D. L., and Murphy, W. J. (1994). Growth hormone promotes human T cell adhesion and migration to both human and murine matrix proteins *in vitro* and directly promotes xenogeneic engraftment. *J. Clin. Invest.* **94**(1), 293–300.

Tilg, H., and Moschen, A. R. (2006). Adipocytokines: Mediators linking adipose tissue, inflammation and immunity. *Nat. Rev. Immunol.* **6**(10), 772–783.

Tschop, M., Smiley, D. L., and Heiman, M. L. (2000). Ghrelin induces adiposity in rodents. *Nature* **407**, 908–913.

Ukkola, O., Ravussin, E., Jacobson, P., Snyder, E. E., Chagnon, M., Sjostrom, L., and Bouchard, C. (2001). Mutations in the preproghrelin/ghrelin gene associated with obesity in humans. *J. Clin. Endocrinol. Metab.* **86**, 3996–3999.

Van den Berghe, G., Wouters, P., Weekers, F., Mohan, S., Baxter, R. C., Veldhuis, J. D., Bowers, C. Y., and Bouillon, R. (1999). Reactivation of pituitary hormone release and metabolic improvement by infusion of growth hormone-releasing peptide and thyrotropin-releasing hormone in patients with protracted critical illness. *J. Clin. Endocrinol. Metab.* **84,** 1311–1323.

Wang, W., Yang, X., Lopez de Silanes, I., Carling, D., and Gorospe, M. (2003). Increased AMP:ATP ratio and AMP-activated protein kinase activity during cellular senescence linked to reduced HuR function. *J. Biol. Chem.* **278**(29), 27016–27023.

Wang, W., Yang, X., Kawai, T., Lopez de Silanes, I., Mazan-Mamczarz, K., Chen, P., Chook, Y. M., Quensel, C., Kohler, M., and Gorospe, M. (2004). AMP-activated protein kinase-regulated phosphorylation and acetylation of importin alpha1: Involvement in the nuclear import of RNA-binding protein HuR. *J. Biol. Chem.* **279**(46), 48376–48388.

Wellen, K. E., and Hotamisligil, G. S. (2003). Obesity-induced inflammatory changes in adipose tissue. *J. Clin. Invest.* **112,** 1785–1788.

Welniak, L. A., Sun, R., and Murphy, W. J. (2002). The role of growth hormone in T-cell development and reconstitution. *J. Leukoc. Biol.* **71**(3), 381–387.

Woody, M. A., Welniak, L. A., Richards, S., Taub, D. D., Tian, Z., Sun, R., Longo, D. L., and Murphy, W. J. (1999). Use of neuroendocrine hormones to promote reconstitution after bone marrow transplantation. *Neuroimmunomodulation* **6**(1–2), 69–80.

Wortley, K. E., Anderson, K. D., Garcia, K., Murray, J. D., Malinova, L., Liu, R., Moncrieffe, M., Thabet, K., Cox, H. J., Yancopoulos, G. D., Wiegand, S. J., and Sleeman, M. W. (2004). Genetic deletion of ghrelin does not decrease food intake but influences metabolic fuel preference. *Proc. Natl. Acad. Sci. USA* **101,** 8227–8232.

Xu, H., Barnes, G. T., Yang, Q., Tan, G., Yang, D., Chou, C. J., Sole, J., Nichols, A., Ross, J. S., Tartaglia, L. A., and Chen, H. (2003). Chronic inflammation in fat plays a crucial role in the development of obesity-related insulin resistance. *J. Clin. Invest.* **112,** 1821–1830.

Zhou, Y., Xu, B. C., Maheshwari, H. G., He, L., Reed, M., Lozykowski, M., Okada, S., Cataldo, L., Coschigamo, K., Wagnar, T. E., Baumann, G., and Kopchick, J. J. (1997). A mammalian model for Laron syndrome produced by targeted disruption of the mouse growth hormone receptor/binding protien gene (the Laron mouse). *Proc. Natl. Acad. Sci. USA.* **94,** 13215–13220.

Index

Page numbers followed by f and t indicate figures and tables, respectively.

A

AC. *See* Adenylate cyclase
Acetyl CoA carboxylase (ACC), 130
Acinar adenocarcinomas, 303
Acromegaly, 266, 270–271
ACTH. *See* Adrenocorticotropin hormone
β-actin mRNAs expression, RT-PCR detection of, 190f
Actinomycin D, 185f, 191, 194f, 196
Acylated ghrelin, 126. *See also* Ghrelin
  biological actions of, 17–18
  D-acyl ghrelin, leptin-releasing activity of, 191f
Adenosine, 97–98
  as GHSR agonist, 75–76
  on intracellular calcium levels, 76
Adenosine monophosphate-activated protein kinase, 103–104, 122
  activation of, 132
  appetite, role on, 133
  cannabinoids effects on, 135
  cardiovascular function, role on, 134–135
  effects on energy metabolism, 132f
  ghrelin effects on, 135, 136f
  on lipid metabolism, 133
  peripheral metabolism, role on, 133–134
  structure of, 131–133
Adenylate cyclase, 97
Adipocyte *ob* mRNA expression, 192t
Adipocyte(s), 172
  culture system, 177
  RNA isolation and analyses, 177–178
  as source of adipokines and factors, 173f
Adipogenesis, 175, 195
  desacyl ghrelin effect on, 187
Adiponectin, 305, 306
Adiponectin levels, 130
Adipose tissue cells. *See* Adipocyte(s)
Adiposity, 155
  desacyl ghrelin as potential physiological modulator of, 183
Adrenocorticotropin hormone, 62
Adrenomedullin, 223

AG. *See* Acylated ghrelin
Aging, 126
  anorexia of, 71
Agoutirelated protein (AGRP), 175, 310
Ala204Glu, 95
Aldosterone, 62
Alkaline phosphatase (ALP) activity, 245–246
Amphibian ghrelins and reptile ghrelin, amino acid sequences of, 40f
AMPK. *See* Adenosine monophosphate-activated protein kinase
AMP, molecular structure of, 131, 132f
Anabolic therapy, 240
Androgen-induced cell proliferation, 304
Androgen receptor, 304
Angiogenesis, 305
Anguilla japonica, 38
Anorexia nervosa, 244
ANP. *See* Atrial natriuretic peptide
Apoptosis, 304–305, 307, 315
Appetite regulation, mechanism of, 123f
AR. *See* Androgen receptor
Arcuate nucleus (ARC), 64, 124
AREs. *See* AU-rich elements
Argininevasopressin, 309
β-arrestin, 99
Arthritis, 338
ATP, 131
Atrial natriuretic peptide, 17
AU-rich elements, 335
AVP. *See* Argininevasopressin

B

BBB. *See* Blood-brain barrier
B cells, 331
Biguanides, 133
Bioluminescence resonance energy transfer assay, 100
Biphasic model, 243
Bird ghrelin, 36–37
  amino acid sequence comparison of, 37f
Blood–brain barrier, 20, 66, 158

BMC. *See* Bone mineral content
BMD. *See* Bone mineral density
Body mass index (BMI), 153, 263, 289
Bone
  gastrectomy/fundectomy effect on, 240–242
  and gastrointestinal system, 240
Bone balance, resorption and formation of, 239–240
Bone loss, 241, 338
Bone metabolism, GH and GHS effects on, 241–244
Bone mineral content, 240
Bone mineral density, 240
Brachmann-de Lange syndrome, 51
Brain neurotransmitters, 157
BRET assay. *See* Bioluminescence resonance energy transfer assay

C

Cachexia, 70
Calcium mobilization *in vitro*, 210
Calcium release, 58–59
Calmodulin kinase kinase (CaMKK), 132
Calvarial osteoblasts, 248
cAMP production, 105
cAMP-responsive element, 94
Cannabinoids
  AMPK, effects on, 135
  appetite effects of, 128–129
  cardiovascular system, effects on, 130
  function of, 127–128
  metabolic effects of, 129–130
Cannabis sativa, 127
Capsaicin, 158
Cardiovascular disorders, 130
Cardiovascular system, 51
CART neurons. *See* Cocaine-and amphetamine-regulated transcript neurons
CB1 antagonist SR141716. *See* Rimonabant
CB1 receptor, 127, 128f
  in brain, 129
  in hepatocytes, 130
  knockout animals, 135
CB2 receptor, 127
CCK. *See* Cholecystokinin
CD36, 108, 218
  to mediate effects of GHS, 261
C-ghrelin, 20–22
Channel catfish ghrelin, 39
Chemical shifts, 6
Chicken ghrelin, 36
Cholecystokinin, 156
Cholera toxin, 60
Cocaine-and amphetamine-regulated transcript neurons, 129
Colitis, 69
Collagen I, 246

Corin, 17
Corticotroph adenomas, 311
Corticotrophin-releasing hormone, 75, 129, 309
Corticotropin-releasing factor receptor type 2, 107
Cortisol, 62
Cortistatin, 77, 98
COS7 cells, 57
CRE. *See* cAMP-responsive element
CRF-R2. *See* Corticotropin-releasing factor receptor type 2
CRH. *See* Corticotrophin-releasing hormone
Critical (fat) mass hypothesis, 288
CTX. *See* Cholera toxin
Cycloheximide, 185f, 192, 194f
Cyclooxygenase, 216
Cytokines, 331, 333, 337
Cytotrophoblast, 265

D

DAG. *See* Diacylglycerol
Decidualization, in ghrelin induction, 262
Desacyl ghrelin, 34, 159
  on adipogenesis, 187
  on adiposity, 196
  bioactivity of, 17–20
  biological effects, 210
  in blood, 210
  leptin-releasing activity (LRA) of, 197
  orexigenic properties of, 74
  as potential physiological modulator of adiposity, 183
  receptors for, 107
  in vitro acylation of, 197
  in vitro differentiated, RP adipocytes, 188–191
Desacyl ghrelin-induced *ob* production, effects of RNA, protein synthesis inhibitors, and GHSR1a antagonist on, 191
Des-Gln14-ghrelin, 15, 33, 74, 174, 307
Dexamethasone, 179, 181
DHT. *See* 5α dihydrostestosterone
Diacylglycerol, 58, 101
DiGeorge's syndrome, 339
Digital rectal examination, 303
5α dihydrostestosterone, 303
Dimyristoyl phosphatidylcholine (DMPC) molecules, 7
D-Lys3-GHRP-6, 227
DNA copy errors, 304
Dopamine D1 receptor (D1R), 100
Dopamine signaling, 72, 73f
DRE. *See* Digital rectal examination

E

Early thymocyte progenitors, 339
Eel ghrelin, 38
EFS. *See* Electric field stimulation

Index

Electric field stimulation, 227
Endocannabinoids, 122
  blood levels of, 130
  in obesity, 130
Endocrine cells, types of, 150
Endocrine control, of food intake and energy balance, 288
Endothelial dysfunction, 224
Endothelin-1 (ET-1), 104, 223
Endotoxemia, 337
Energy balance, 64
Energy homeostasis, 96, 172
Energy metabolism, 64, 172
Energy sensor, 103
Enterochroman-like (ECL)-acid, 241
EP-40737, 245
ERK1/2 activity, 102
Estradiol, 304
Ethylene dimethane sulfonate, 294
ETPs. *See* Early thymocyte progenitors
Euglycemic hyperinsulinemic clamp, 153
Extracellular signal-regulated protein kinases (ERK1/2), 216

F

Fasting, 152f, 195
Fat-free mass (FFM), 126
Fatty acid amide hydrolase (FAAH) expression, 130
Fatty acid oxidation, 135
Fatty acid synthase (FAS), 129, 130
Feeding behavior, regulation of, 64
Fenfluramine, 157
Fetal ghrelin, 274–276
Fetal rat osteoblasts, 246
Fibroblast growth factor-2 (FGF-2)-induced cell proliferation, 105
Fish ghrelins, 37
  amino acid sequence comparison of, 38f
Follicle-stimulating hormone (FSH), 287, 293–294
Food intake, 91, 124, 175
Frog ghrelin, 39

G

Gastric ghrelin, 33, 155
Gastrin, 241
Gastrocalcin, 241
Gastrointestinal smooth muscles
  effects on ghrelin receptor, 226
  prokinetic action of ghrelin on, 226–227
Gastrointestinal system, and bone, 240
Gastrointestinal tract, 69
GH axis, 268
GH-induced ghrelin regulation, 154f
GH-releasing activity, ghrelin, 308
GH releasing hormone (GHRH), 48

GH-releasing peptide 6, 156
Ghrelin. *See also* Ghrelin structure
  AMPK, effects on, 135, 136f
  anti-inflammatory properties of, 67
  appetite effects of, 123–125
  in appetite regulation and body weight, 267, 289
  autocrine and paracrine effect on, 265–266
  biology of, 307–311
  and bone parameters in clinical studies, 244
  cardiovascular system, effects on, 126–127, 137
  in cat stomach, 36
  cell growth, effects on, 311–312
  contractile effects of, 209f, 217, 221t–222t
  desoctanoylation of, 18
  discovery of, 14
  distribution of, 210–211
  diurnal profile of, 264
  3D structure of, 4
  effects on fetus, 272, 274–276
  effects on osteoblastic cells *in vitro*, 245
  endothelial actions of, 220f
  energy minimization (EM) of, 4
  in fallopian tubes and uterus, 262
  fat mass in pregnancy, association with, 267
  fatty acid chain length affecting, 41
  forms of, 159
  functional role in prostate cancer, 314–315
  function of, 122
  functions in various organ systems, 332f, 333
  gonadotropin secretion and, 289–291
  hemodynamic effects of, 209
  in human stomach, 34
  and hypertensive disorders in pregnancy., 272
  immune system modulation of, 67
  as immunomodulator, 70f
  in inhibitory effects on LH secretion, 290
  insulin sensitivity, association to, 268
  intracerebroventricular administration of, 64
  intravenous injection of, 32, 218
  levels in last period of pregnancy, 272
  lipid-binding properties of, 7
  in mammalian brain, 23
  MD simulation of, 4, 6–7, 8f, 9
  metabolic effects of, 125–126
  as model for signal linking in immune and endocrine systems, 336f
  molecular diversity of, 292–293
  molecular forms
    in rat stomach, 33
    in tissues, 35
  and motilin family, 40, 41f
  nonacylated form of (*See* Desacyl ghrelin)
  nuclear magnetic resonance (NMR) spectroscopy of, 2, 5
  orexigenic effect of, 135, 266
  orexigenic signal expression of, 286–287
  peptide backbone structure of, 7f

Ghrelin. *See also* Ghrelin structure (*continued.*)
  PGH and GH regulation, role in, 268, 269*f*, 270–271
  physiological actions in pregnancy, 266–268, 269*f*, 270–273
  physiological functions of, 32
  placenta as source of, 261–262, 265–266
  PRL and ACTH, effect on, 273–274
  proliferative effect of, 311–312
  properties of, 329–330
  in rat stomach, 14
  and receptors in human prostate tumors, 312–314
  receptors of (*See* Ghrelin receptors)
  28-residue amino acid sequence of, 3*f*
  residues of, 10
  role in decidualization of endometrial stromal cells, 261
  SAMD simulation of, 5
  secretion in pregnancy, 262–264, 263*f*
  Ser3-side chain of, 42*t*
  sources of production, 330–332
  structure of, 51*f*, 210
  in vivo administration of, 104
  X-ray crystallography of, 2
Ghrelin-A, 197
Ghrelin afferent pathways, from stomach to hypothalamus, 158
Ghrelin feedback systems, 150
Ghrelin gene-derived peptides, structural chemical formula of, 174*f*
Ghrelin gene precursors
  exons of, 15, 16*f*
  organization of, 15
  promoter activity of, 15–16
Ghrelin/GHSR1a system, role of, 96–97
Ghrelin/GHSR system, in hypothalamic circuits, 66*f*
Ghrelin modulation, of *ob* gene function, GHSR1a antagonist on, 182
Ghrelin mRNA, in pituitary cells, 66
Ghrelin receptors, 107, 214
  in human tissues, 212*t*–213*t*
  muscle localization of, 215*t*
  myocardial actions on, 217, 219*f*
  skeletal muscle effects of, 227, 229*f*
  vascular smooth muscle effects on, 223, 225*f*
Ghrelin-related substances, leptin-releasing activity of, 177
Ghrelin secretion
  brain 5-HT systems and, 154*f*
  regulation of
    body weight, 152
    brain neurotransmitters, 157
    efferent vagus nerve, 152
    fasting, 152*f*
    gastric bypass, 157
    hormones and peptides, 153–156
    starvation and feeding, 151, 152*f*
    regulators of, 150*t*
Ghrelin structure, 1
  in aqueous solution, 6
  computational methods of, 4–5
  experimental methods of, 2
GH replacement therapy, 243
GHRH-expressing neurons, 63
GHRH-induced cAMP accumulation, 105
GH secretion, pathways leading to, 56, 58
GHSR1a antagonist, 185*f*, 194*f*
GHSR1a–GHSR1b heterodimer interactions, 77
GHSR1b, 51, 92, 214
  cellular localization of, 53*f*
GHS receptors (GHSRs), 260, 312
  expressing neurons, 56
  expression on immune cells, 328–329
  GHSR1b, 308, 310–311
  GHSR type 1a, 286, 292, 294, 307–308, 310–311
  homology model, ghrelin model on, 10*f*
  properties, 328
GHSR gene, 49
  genetic locus of, 51
  hormonal regulation of, 56
  ligand-independent activity of, 57
  promoter sequences of, 55
  in rat pituitary cells, 56
GHSR mRNA, 91
GHSR null mice, 63
GHSRs in pregnancy, 260
  localization in relation to pregnancy and lactation, 261*t*
Glitazones, 134
Glucagon, 154
Glucocorticoid, circulating levels of, 197–198
Gluconeogenesis, 135
Glucose concentration, 185
Glucose-dependent insulinotropic polypeptide (GIP), 134
Glucose metabolism, 125, 133
Glucose-sensing neurons, 125
Glycoprotein IV, 261
Gly-Ser-Ser (n-octanoyl)-Phe, 54
Gly-Ser-Ser (n-octanoyl)-Phe-Leu, 96
Goldfish ghrelin, 39
Gonadotropic axis, ghrelin as regulator of, 287
Gonadotropin-releasing hormone (GnRH), 287
Gonadotropin secretion, role of Ghrelin in control of, 289–291
Gonads, expression and direct actions of ghrelin on, 293–295
G-protein-coupled receptor (GPCR), 2, 11, 49, 91, 99
GPR39 receptor, 21
Graft-*versus*-host disease (GVHD), 341
Green fluorescent protein (GFP), 92

Growth factor receptor-bound protein 2 (GRB2), 61, 102
Growth hormone (GH), 260
Growth hormone/insulin-like growth factor-I (GH/IGF-1), 217
Growth hormone secretagogue (GHS), 260
Growth hormone secretagogue receptor 1a (GHSR1a), 19, 48, 51, 89
  in arcuate nucleus, 93
  binding cavity of, 94
  cellular localization of, 53f
  constitutive activity of, 77, 95
  desensitization and internalization, 214
  endocytosis of, 98–99
  features of, 93
  G-protein-signaling pathways, 101–107
  in HEK293 cells, 98
  homodimers, 100
  homo-or heteromeric complexes of, 99–101
  in hypothalamic-pituitary axis, 62
  in hypothalamus, 92
  internalization of, 98–99
  internalization pathway of, 216f
  ligands for, 97–98
  model, tube structure of, 54f
  in osteoblastic cells, 103
  in prostate cancer cell line, 103
  protein levels of, 69
  structure of, 52–55, 91–96
  tissue distribution and functions, in organ systems, 62
Growth hormone secretagogue receptor (GHSR), 2, 7, 48, 328
  amino acid sequence of, 50f
  in cardiovascular system, 69
  in cells of immune system, 67
  endogenous ligand, 74, 208
  on immune cells, 68
  on monocytes, 68
  in peripheral tissues, 74
  in plasma membrane, 52f
  processing of, 53f
  regulatory elements of, 55
  signal transduction mechanism of, 58, 59f
  signal transduction pathways of, 214
  in T cells, 67
Growth hormone secretagogues (GHSs), 90, 327–328

H

Heart failure (HF), 218, 220
Heart glucose metabolism, 134
HEK293 cells, 52, 76–77
  GHSR1a in, 98
Heparin-binding epidermal growth factor-like growth factor (HB-EGF), 304–305
Heterodimerization, 98, 105

Hexarelin, 77, 217, 244, 262
High-carbohydrate diet, 151
High-density lipoprotein (HDL), 18, 210
High-fat (HF) feeding, 65
High-grade PIN (HGPIN), 303, 305
Histamine, vasodilatating effect of, 130
Homeostasis, 288, 307, 333
Hormonerefractory prostate cancer patients, 306
HPG axis, 287
Human ghrelin
  localization in relation to pregnancy and lactation, 261t
  structure of, 33f
Human placental lactogen (hPL), 265
Human umbilical vein endothelial cells (HUVECs), 68
Hydrophobic octanoyl group, 9
5-Hydroxytryptamine (5-HT), 157
Hyperinsulinemia, 155, 268
Hyperphagia, 159
Hyperthyroidism, 155
Hypoadiponectinemia, 306
Hypogonadism, 288
Hypothalamic endocannabinoid levels, 129
Hypothalamic gene expression, and plasma desacyl ghrelin, 161
Hypothalamic 5-HT2C/1B receptor, and SGK-1gene expression, 160f
Hypothalamic pathways, ghrelin's influence on energy balance and appetite, 309
Hypothalamic-pituitary-adrenal (HPA) axis, 309
Hypothalamicpituitary- gonadal (HPG), 287

I

Ictalurus punctatus, 39
IGF-1 and IGF-2, in linear bone growth, 242–243
IGF-1 mRNA expression, 103
Immune dysfunction, 71
Inflammatory bowel disease, 337
Inositol phosphate, 57, 94
Inositol 1,4,5-triphosphate (IP_3)-receptors, 58, 97, 101
Insulin, 94, 153, 179, 181
  resistance, 268
  risk on prostate cancer, 305–306
Insulin-like growth factor-1 (IGF-1) levels, 51
Insulin-like growth factors (IGFs), 266
Insulin receptor substrate-1 (IRS-1), 61
Interleukin-1, 157
Interleukin (IL)–8, 106
Interleukins, 305–306, 333, 337, 341
International Union of Pharmacology, 55
Intestinal anandamide, 129
Intracellular calcium mobilization, 97, 101
Intragastric glucose, 151

Intraocular smooth muscles, effects on ghrelin receptors, 225–226
Intrauterine growth restriction (IUGR), 270
Ipamorelin, 244
Islets of Langerhans, 311
Isoproterenol-induced lipolysis, 126
Isoproterenol-induced myocardial injury, 219

J

c-Jun N-terminal Au6 kinase (JNK), 305

K

Krebs–Ringer–MOPS medium, 176

L

Lactation, 273–274
Left ventricular mass (LVM), 223
Leptin, 65, 70, 94, 123f, 129, 266, 288, 305–306
  and ghrelin regulation, 155f
  measurement of, 178
Leptin and activation-induced inflammation, ghrelin effects on, 333–335, 336f, 337
Leptin-releasing activity (LRA), 179
  adipocyte protein synthesis and GHSR1a blockage on, 181
  of D-acyl ghrelin, 191f
  of desacyl ghrelin, 197
  on RP adipocytes, 183t, 184f, 195
Leydig cells (LC), 294, 311
Lipid bilayer, 7
Lipid metabolism, AMPK's effect on, 133
Lipid oxidation, 135
Lipogenesis, 130, 134
Lipolysis, 134
Lipopolysaccharide (LPS), 69, 157, 337
Lipoproteinlipase (LPL)
  mRNA expression, 187
  RT-PCR detection of, 190f
Low-grade PIN (LGPIN), 303
Luteinizing hormone (LH), 287, 293–294, 309
Lysophospholipase I, 18

M

Mammalian ghrelin. *See also* Ghrelin
  amino acid sequence of, 35, 36f
  site of maximum gene expression of, 22
  structure of, 33
Mammalian proghrelin precursors, amino acid sequences of, 17f
Matrigel tube formation, 105
Mature ghrelin, 17, 24f. *See also* Ghrelin
m-Chlorophenylpiperazine (mCPP), 157
MC3T3-E1 cells, 246
MD simulation
  of ghrelin, 4, 6–7, 8f, 9

limitations of, 5
Medium chain triacylglycerols (MCTs), 42
Melanocortin receptors, 309
α Melanocyte stimulating hormone (α-MSH), 309
Melatonin, 156
Metformin, 134
Mitogen-activated protein kinase (MAPK), 19, 61, 102, 305
MK0677, 60, 68, 73, 243
Monocyte chemoattractant protein-1 (MCP-1), 106
Motilin, 214
Motilin-related peptide (MTLRP), 40
Myelomonocytic leukemia, 307
Myocardial ischemia reperfusion (I/R) injury, 131
Myocardium
  chronic effects of ghrelin, 218–219
  clinical implications, 220
  effects of ghrelin receptors, 217, 219f

N

Neuroendocrine axes, 288
Neuroendocrine control of reproduction, 287
Neuroendocrine hormones, 48, 67
Neuroendocrine–immune interactions, 326–327
Neurogenesis, 311
Neuromedin U (NMU), 214, 308
Neuromicrovascular endothelial cells, 105
Neuropeptides, 93
Neuropeptide Y (NPY), 61, 104, 175
  neurons, 309
Neurotensin receptor 1 (NTSR1), 308
Neurotensin receptors, 214
Nifedipine, 60
Nitric oxide (NO), 338
Nitric oxide synthase (NOS), 104
Noladin ether, 128
Non-insulin-dependent diabetes mellitus, 306
Nonmammalian ghrelin, structure of, 36–40
Non-small lung cancer cells (NSCLC), 308
Nontransformed colonic epithelial cells (NCM460), 61
Nuclear factor-kappa B (NF-κB), 106
Nuclear magnetic resonance spectroscopy, of ghrelin, 5–6
Nuclear overhauser effect (NOE), 4

O

ob and β-actin mRNAs
  Northern blot detections of, 186f
  RT-PCR analysis of, 186f
Obesity, 124, 159, 337
  endocannabinoid system in, 130
  treatment of, 137
Obestatin, 20–22, 286, 292–293

Index

n-Octanoyl ghrelin, 33, 95
Octanoyl-Ser3, 6
Omeprazole, 241
Oreochromis mossambicus, 39
Orexigenic neuropeptides, 122
Osteoarthritic bone biopsies, 247
Osteoblastic cells *in vitro*, ghrelin effects on, 245
Osteoblasts differentiation, 245, 247
Osteocalcin (OC) production, 245
Osteomalacia, 240
Osteopenia, 241
Osteoporosis, 240

P

Pancreas, 211
Pancreatic cholecystokin, 310
Pancreatic hormones, 153–155
Paraventricular nucleus (PVN), 129
PD98059, 247
Peptide hormone ghrelin, 307
Peroxisome proliferator-activated receptor γ (PPARγ), 106, 216
Phe4, 6
Phe279Leu, 95
Phenylmethylsulfonyl fluoride (PMSF), 18
Phosphatidylinositol 4,5-biphosphate (PIP$_2$), 101
Phosphoenolpyruvate carboxykinase (PEPCK) mRNA expression, 102
Phosphoinositide-3 kinase (PI-3K), 19
PI-3K/Akt pathway, 102–103
PIN. *See* Prostatic intraepithelial neoplasia
Pituitary adenoma, 242, 311
Pituitary gland, 92
Pituitary gonadotropins, 287
Pituitary-specific transcription factor Pit-1, 62
PKA/cAMP pathway, 60
Placental growth hormone (PGH), 264
Plasma ghrelin levels, 149, 151
PLC signaling pathway, 58
Porcine ghrelin cDNA, 34
Potassium currents, 60
PPARγ2, 187
  RT-PCR detection of, 190*f*
Prader–Willi syndrome, 152
Preeclampsia, 271–272, 272
Proghrelin
  C-terminal region of (*See* C-ghrelin)
  posttranslational products of, 17–22
Proghrelin peptides
  distribution of, 22–23
  in lower vertebrates, 24–25
Prolactin, 62
Proopiomelanocortin (POMC), 309
  neurons, 65
Prostaglandins, 157, 226
Prostate cancer, 302, 307
Prostatic adenocarcinomas, 303
Prostatic intraepithelial neoplasia, 302
Prostatic-specific antigen (PSA), 303
Protein kinase A (PKA), 97, 131
Protein kinase C (PKC), 49
Puberty, roles of ghrelin in, 291–292
Pulmonary hypertension, monocrotaline model of, 219
Putative ghrelin acyl-modifying enzyme, 41

Q

Quantum mechanical calculations, 4

R

Rainbow trout ghrelin, 38
Rat retroperitoneal adipocyte endocrine functions, 176
Reproductive axis, effects of ghrelin in control of, 296*f*
Reptile ghrelin, 40
Retroperitoneal (RP) adipocyte isolation, 173
Reverse-phase high performance liquid chromatography (RP-HPLC), 14
Rimonabant, 137*f*, 138
Rimonabant in Obesity (RIO) program, 138
Root mean square deviation (RMSD), of peptide backbone, 7, 8*f*
RP adipocyte(s)
  endocrine function, 179–183
  glucose concentration in, 185
  glucose consumption by, 193*f*
  insulin-stimulated leptin output by, 179, 180*f*
  leptin concentration after 12 h incubation of, 192*t*
  leptin secretagogues by, 179
  leptin secretion by, 179, 181*t*, 182*f*
  LRA activity on, 183*t*, 184*f*
  in vitro differentiated, desacyl ghrelin effect on, 188–191
RP preadipocytes
  isolation and differentiation of, 184
  lipid accumulation in, 188*f*

S

Sepsis, 69
Serotonin, 157
Serum-and glucocorticoid-induced protein kinase (SGK), 161
Serum-responsive element (SRE), 94
Severe, combined immunodeficiency (SCID) mice, 68
Sex hormone-binding globulin (SHBG), 306
Sex steroids, 93
Siberian hamster, 124
Simulated annealing (SA)MD simulation
  of ghrelin, 5, 7
  limitations of, 5

Single nucleotide polymorphisms, 51
Skeletal integrity, 240
Skeletal muscle
  acute and chronic effects of ghrelin, 228, 230
  contractile function of, 227
  effects of ghrelin receptors, 227, 229f
Skeletal muscle sodium channels, 228
Smooth muscle. *See* Gastrointestinal smooth muscles; Intraocular smooth muscles; Vascular smooth muscle
Sodium dodecyl sulfate (SDS), 178
Somatomammotroph cells, 308
Somatomedin hypothesis, 242
Somatostatin, 154
Somatotroph tumors, 311
Stem cell factor (SCF), 295, 339
Sterol regulatory element binding protein-1 (SREBP1), 130
SV-HFO cells, 246, 248
Synthetic peptidyl GHS, 309
Systemic inflammation, ghrelin effects on, 337–341

T

T cells, 329, 331
Testosterone, 291, 294, 304, 309
Testosterone replacement therapy, 306
Δ^9-tetrahydrocannabinoid (Δ^9-THC), 127
Thiazolidinediones, 133
Thymic involution, effects of ghrelin and GHS on, 339–341
Thymopoiesis, effects of ghrelin and GHS on, 339–341
Thyroid hormone, 93, 155–156
Thyroid stimulating hormone-releasing hormone (TRH), 159

Tilapia ghrelin, 39
T lymphocytes, 329, 340
Trachemys scripta elegans, 40
Transmembrane (TM) regions, 91
Trifluoroacetic acid, 6
Triiodothyronine (T3), 56
Tumor necrosis factor-α (TNF-α), 68, 106, 335, 337
Type 2 diabetes, 133

U

"Ultimate anabolic hormone", 123
UMR-106, 247
Unacylated form of ghrelin (UAG), 286, 292–293, 307, 315
Urocortin-1, 20, 156

V

Vagotomy, 152, 158
Vascular endothelial growth factor (VEGF), 305
Vascular smooth muscle
  acute and chronic effects of ghrelin on, 223–224
  effects of ghrelin receptor, 223, 225f
Vasoconstrictor effect, 218
Vasodilatation, 126, 209

W

Wasting syndromes, 96

Z

Zebrafish ghrelin, 39

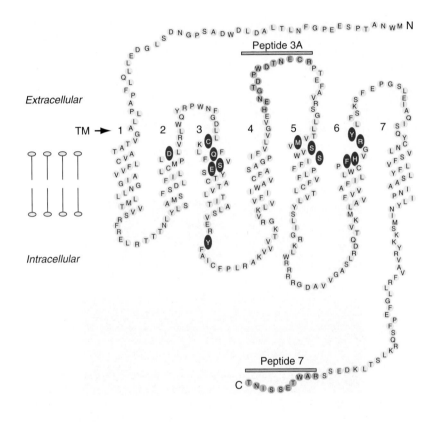

CRUZ AND SMITH, FIGURE 3. Schematic representation of the ghrelin receptor illustrating peptide epitopes (green) used to generate antibodies to extracellular loop 2 (peptide 3A) and the C-terminus (peptide 7) for determining orientation of the GHSR in the plasma membrane. Also highlighted are residues mutated to investigate the binding pocket occupied by MK0677. Reprinted from *Molecular Endocrinology*, Copyright 1998, *The Endocrine Society* (Feighner et al., 1998).

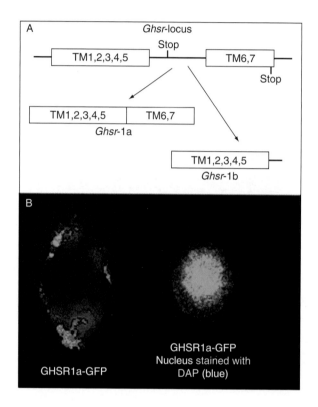

CRUZ AND SMITH, FIGURE 4. (A) Processing of GHSR primary transcript illustrating formation of GHSR subtypes. (B) Cellular localization of GHSR1a and GHSR1b following expression of GHSR1a and GHSR1b cDNAs tagged with green fluorescent protein (GFP) in HEK293 cells. GHSR1a is shown localized on the plasma membrane and GHSR1b is localized to the nucleus. Reprinted from *Trends in Endocrinology and Metabolism*, Copyright 2005, with permission from *Elsevier* (Smith *et al.*, 2005).

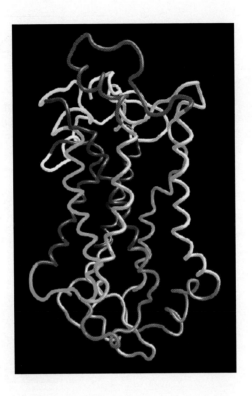

CRUZ AND SMITH, FIGURE 5. Tube structure of the GHSR1a model colored by segment. Color legend (NT, N-terminal domain; TM1–7, transmembrane domains 1–7; CL1–3, cytoplasmic loops 1–3; EL1–3, extracellular loops 1–3; CT, C-terminal domain): NT = white, TM1 = red, CL1 = green, TM2 = azure, EL1 = yellow, TM3 = dark red, CL2 = violet, TM4 = pink, EL2 = indigo, TM5 = gray, CL3 = orange, TM6 = dark green, EL3 = dark yellow, TM7 = brown, CT = blue. Reprinted with permission, Copyright 2006, *American Chemical Society* (Pedretti *et al.*, 2006).

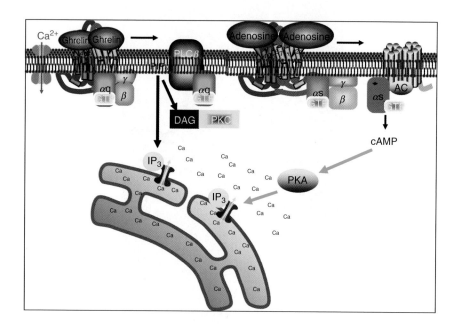

CRUZ AND SMITH, FIGURE 6. Signal transduction mechanisms employed by the GHSR illustrating the dominant PLC-PKC-inositol triphosphate (PLC/PKC/IP$_3$) pathway activated by ghrelin and ghrelin mimetics such as MK-0677 and GHRP-6, and the adenylate cyclase/PKA/cAMP (AC/PKA/cAMP) pathway activated by the partial agonist adenosine. The GHSR (ghrelin receptor) is represented as homodimers.

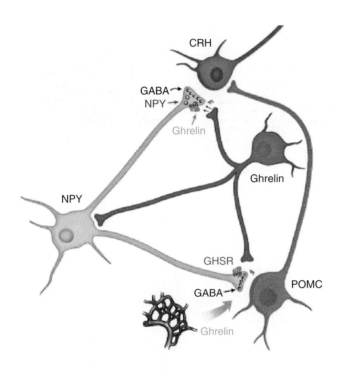

CRUZ AND SMITH, FIGURE 7. Hypothalamic circuits employed by the ghrelin/GHSR system during energy homeostasis illustrating ghrelin activation of NPY-containing neurons and inhibition of pro-opiomelanocortin (POMC) neurons. Corticotrophin-releasing hormone, CRH; γ-aminobutyric acid, GABA. Reprinted from Neuron, Copyright 2003, with permission from *Elsevier* (Cowley *et al.*, 2003).

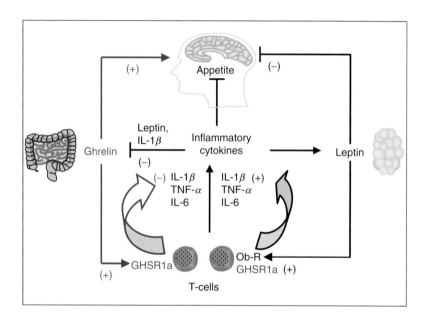

CRUZ AND SMITH, FIGURE 8. Ghrelin as an immunomodulator. Ghrelin inhibits the production of proinflammatory cytokines such as IL-1β, TNF-α, IL-6, and leptin; ghrelin stimulates appetite, whereas leptin and the proinflammatory cytokines inhibit food intake.

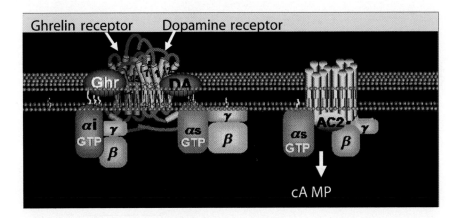

CRUZ AND SMITH, FIGURE 9. Model of ghrelin amplification of dopamine signaling involving GHSR/D1R heterodimerization causing a switch in GHSR G-protein coupling from $G_{\alpha q11}$ to $G_{\alpha i}$, resulting in synergistic activation of adenylyl cyclase-2 via $G_{\alpha s}$ (D1R coupling) and $\beta\gamma$-subunits derived from $G_{\alpha i}$ (GHSR coupling). Modified from *Molecular Endocrinology*, Copyright 2006, *The Endocrine Society* (Jiang *et al.*, 2006).

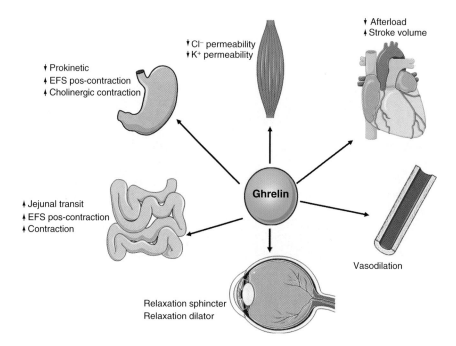

LEITE-MOREIRA ET AL., FIGURE 1. Global view of contractile effects of ghrelin. Ghrelin has described action on the cardiac muscle, vascular smooth muscle, ocular smooth muscle, gastrointestinal smooth muscle, and skeletal muscle.

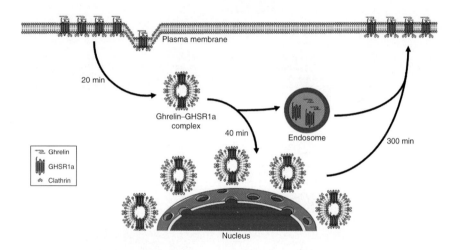

LEITE-MOREIRA ET AL., FIGURE 2. Internalization pathway of GHSR1a. Under unstimulated conditions, GHSR1a is mainly expressed at the plasma membrane. After 20 min exposure to ghrelin, the ghrelin–GHSR1a complex progressively disappears by endocytosis via clathrin-coated pits. The complex accumulates in the perinuclear region after 60 min. GHSR1a shows a slow recycling and its membrane levels return to normal after 360 min (Camina et al., 2004).

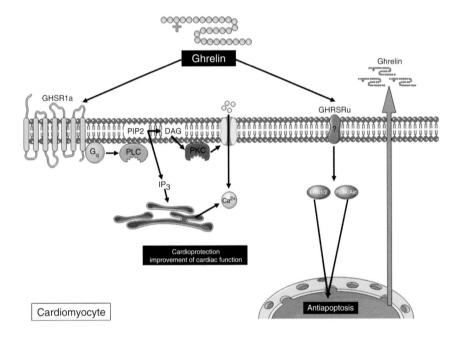

LEITE-MOREIRA ET AL., FIGURE 3. Myocardial actions of ghrelin and its pathways. Activation of GHSR1a by ghrelin stimulates a G-protein that activates PLC-signaling pathway producing IP_3 and DAG. Both IP_3 and DAG lead to an elevation of myocardial Ca^{2+} levels via stimulation of Ca^{2+} influx through the voltage-gated Ca^{2+} channel and Ca^{2+} release from sarcoplasmic reticulum. A putative receptor, GHSRu, with unknown structure mediates the effects of ghrelin on the survival of myocytes through extracellular signal-regulated kinase 1/2 (ERK1/2) and PI-3K activation (Baldanzi et al., 2002). Ghrelin is also synthesized and secreted by cardiomyocytes, probably mediating paracrine/autocrine effects (Iglesias et al., 2004).

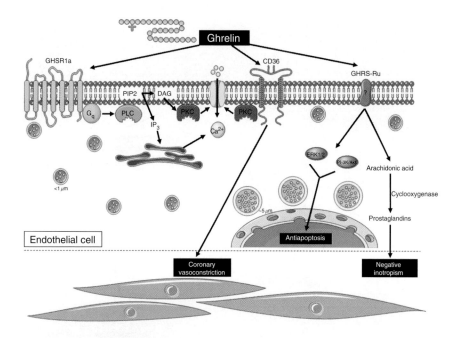

LEITE-MOREIRA ET AL., FIGURE 4. Endothelial actions of ghrelin and its pathways. Ghrelin can be detected in the cytoplasm of endothelial cells in two subcellular compartments: Vesicle-like structures (diameter <1 μm) within the cytoplasm and structures of ∼5 μm diameter localized closed to the cell nucleus (Kleinz et al., 2006). Activation of GHSR1a induces an increase in cytosolic Ca^{2+} levels. Ghrelin binding to GHSRu inhibits cell death in endothelial cells through extracellular-signal-regulated kinase 1/2 (ERK1/2) and PI-3K/Akt (Baldanzi et al., 2002). Ghrelins have a negative inotropic effect mediated by cyclooxygenase metabolites produced by endothelial cells by interaction with GHSRu (Bedendi et al., 2003). CD36 is a new ghrelin receptor that mediates coronary vasoconstriction that seems to involve PKC and a calcium channel (Bodart et al., 2002).

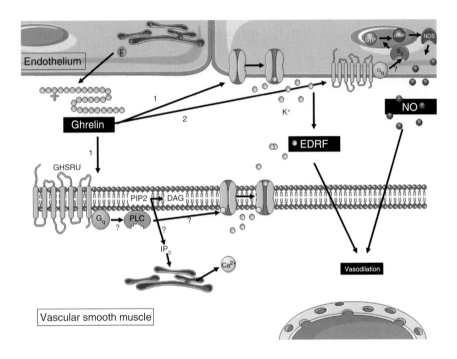

LEITE-MOREIRA ET AL., FIGURE 5. Smooth muscle effects of ghrelin and its pathways. In vascular smooth muscle, ghrelin provides vasodilation through two GHSR1a-independent pathways. Ghrelin stimulates an unknown receptor (GHSRu) which through a G-protein–coupled pathway will close the Ca^{2+}-dependent K^+ channels of the endothelium or vascular smooth muscle and release the endothelial-derived relaxing factor (EDRF) which will relax the vascular smooth muscle (pathway 1) (Shinde et al., 2005). Another pathway for the relaxation is the stimulation of a G-protein–coupled system and activation of the guanidyl cyclase (Gc). This will increase the production of cGMP with consequent induction of the NO synthase promoting the NO-mediated vasodilation (pathway 2) (Shimizu et al., 2003).

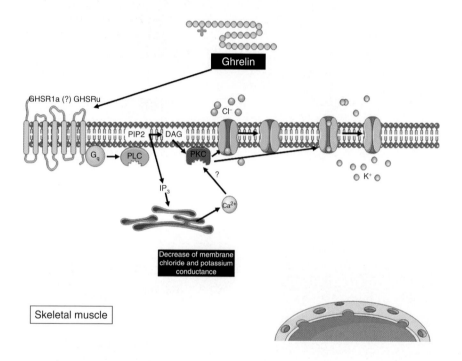

LEITE-MOREIRA ET AL., FIGURE 6. Ghrelin's skeletal muscle effects and its subcellular pathways. Ghrelin stimulates a membrane receptor (possibly other than GHSR1a), coupled to a G-protein–coupled system. That system activates a PLC-signaling pathway producing IP$_3$ and DAG. Both IP$_3$ and DAG produce persistent increase of the Ca^{2+} levels which will stimulate the PKC. PKC produces a phosphorylation of the Cl$^-$ and K$^+$ channels with decrease in chloride and potassium membrane conductivity (Pierno et al., 2003).

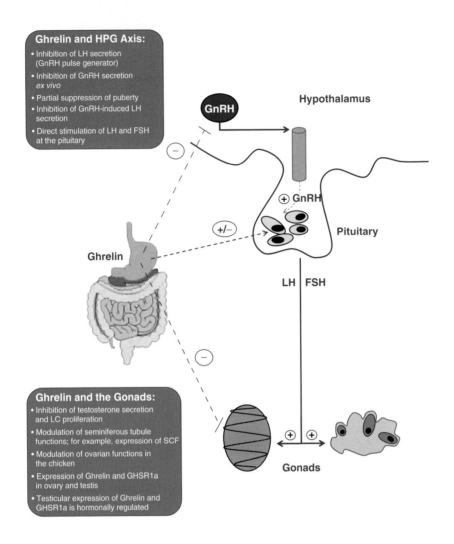

TENA-SEMPERE, FIGURE 1. Schematic representation of the potential effects of ghrelin in the control of the reproductive axis. Ghrelin, a hormone secreted by the stomach as signal of energy insufficiency, conducts specific regulatory effects at different levels of the hypothalamic-pituitary-gonadal (HPG) axis, which include the inhibition of pulsatile secretion of luteinizing hormone (LH), the suppression of gonadotropin-releasing hormone (GnRH) secretion, and the reduction of LH responsiveness to GnRH at the pituitary level. In addition, ghrelin has been reported to partially blunt the activation of the reproductive axis at (male) puberty. However, stimulatory effects of ghrelin on basal LH and FSH release directly at the pituitary level have been also described. Besides those central actions, systemically derived ghrelin might directly operate at the gonadal level, as expression of canonical ghrelin receptors has been described in the ovary and the testis. Moreover, ghrelin is also locally produced in the gonads, where it may conduct additional regulatory actions, such as inhibition of Leydig cell (LC) proliferation and testosterone secretion, as well as regulation of tubular functions, such as expression of stem cell factor (SCF) gene. In addition, specific regulatory actions of ghrelin on relevant ovarian functions have been recently described in the chicken. Altogether, these observations allow us to propose that ghrelin is a pleiotropic modulator of the gonadotropic system, which might contribute to the physiological coupling of reproductive function to the state of energy reserves of the organism.

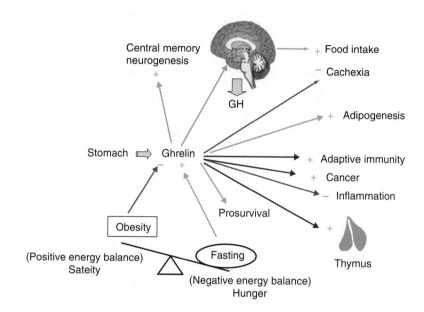

TAUB, FIGURE 1. Ghrelin functions in various organ systems. Besides the established effects of ghrelin on food intake and GH production, ghrelin has also been shown to mediate a number of functions on a variety of other organ systems, including promoting adipogenesis, cardiomyocytes survival, enhancement of memory, and neurogenesis. In addition, ghrelin has been shown to inhibit inflammatory cytokine production and promote thymic function.

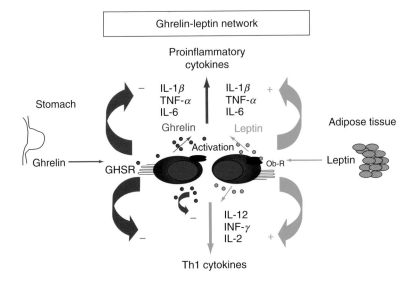

TAUB, FIGURE 2. Hypothetical model for functional role of ghrelin as a signal linking the immune and endocrine systems in control of food intake. Immune-derived ghrelin and leptin may play a role in balancing the inflammatory status of a host. When a metabolic imbalance occurs, these systems may come into play assisting in the regulation of food intake by controlling inflammatory cytokine expression. Conversely, circulating hormones may influence ongoing immune responses by enhancing or dampening the level of activation and cytokine secretion. Ghrelin appears to mediate anti-inflammatory effects on IL-1, TNF, and IL-6 cytokine expression by T cells and mononuclear cells, while leptin appears to promote inflammatory cytokine expression. Moreover, leptin has been shown to promote Th1 cytokine response, while data within our laboratory has demonstrated that ghrelin inhibits leptin-induced Th1 development. Perhaps these hormones and immune cell responses to these hormones are a form of "checks-and-balances" by which the neuroendocrine, metabolic, and immune systems can control each other's activities, preventing one system from interfering with another system.